PLANT MITOCHONDRIA

Structural, Functional, and Physiological Aspects

PLANT MITOCHONDRIA

Structural, Functional, and Physiological Aspects

Edited by

A. L. Moore
University of Sussex
Brighton, England

and

R. B. Beechey
University College of Wales
Aberystwyth, Wales

PLENUM PRESS • NEW YORK AND LONDON

Library of Congress Cataloging in Publication Data

International Meeting on Plant Mitochondria (2nd: 1986: Aberystwyth, Dyfed)
Plant mitochondria.

"Proceedings of the Second International Meeting on Plant Mitochondria... held July 20–24, 1986, in Aberystwyth, Wales"—T.p. verso.
Bibliography: p.
Includes index.
1. Plant mitochondria—Congresses. I. Moore, A. L. (Anthony L.), date. II. Beechey, R. B. (R. Brian) III. Title.
QK725.I4946 1986 581.87′342 87-15303
ISBN 0-306-42572-6

Proceedings of the Second International Meeting on Plant Mitochondria: Structural, Functional, and Physiological Aspects, held July 20–24, 1986, in Aberystwyth, Wales

Printed in the United States of America

PREFACE

Eight years have elapsed since the first International Meeting on Plant Mitochondria was held in Marseilles. Since this date numerous important developments have occurred within the field and hence a further conference on this fundamental area of research was considered well overdue. This volume summarises the lecture and poster sessions of the second International Meeting on Plant Mitochondria held in Aberystwyth, July 20-24th, 1986. The meeting was held not only to bring together plant scientists interested in the bioenergetics of plant mitochondria but also those who are interested in the regulatory role of mitochondria in plant growth and respiration. A further important aspect of this conference was to introduce plant physiologists and biochemists to the plant molecular biologists in an attempt to not only discuss problems of mutual interest but to also learn much more about the real questions which the biochemists and physiologists wish to answer. Hopefully the volume reflects much of the current excitement and advances being made in the field. Although many of the participants of the first meeting were present the expertise of Walter Bonner, Jack Hanson and Gaston Ducet, to name but a few, was sorely missed.

The conference consisted of forty-five minute review lectures followed by thirty minute research lectures, the summaries of which are found in the longer articles. The meeting was divided into four sessions, namely, organisation of the electron transport chain; mitochondrial interactions; mitochondrial biogenesis and plant growth and development. The poster abstracts have been arranged in such a manner that they are associated with each of the four sessions.

We gratefully acknowledge the financial support recieved from Bio-Rad Laboratories Ltd., Ciba-Geigy Agrichemicals, Du Pont (UK) Ltd., Flow Laboratories Ltd., Gelman Sciences Ltd., GIBCO-BRL., Hansatech Ltd., Millipore (UK) Ltd., Severn Science Ltd., Shell Research Ltd., Unilever Research & Engineering Division and Wellcome Research Laboratories. We are also extremely grateful to Mike Proudlove, Alyson Tobin, Naheed Kaderbhai, Mike Brailsford, Mike Walsh, Simon Dobbs, Andy Thompson and Mike Jones for their help with the day-to-day conference running. The help and enthusiasm of the Advisory Panel, consisting of Tadashi Asahi, David Day, Ingemar Ericson, Hans Heldt, Hans Lambers, Claude Lance, Chris Leaver, Don Moreland, John Palmer, Anna Rychter, Jim Siedow and last but not least the instigator of this conference Joe Wiskich, is also gratefully acknowledged.

We thoroughly enjoyed organising the meeting and look forward with eager anticipation to the next conference, to be organised by Hans Lambers, in the not too distant future.

Tony Moore
Brian Beechey
1986

CONTENTS

ORGANISATION OF THE ELECTRON TRANSPORT CHAIN

MITOCHONDRIAL INTERACTIONS

MITOCHONDRIAL BIOGENESIS

PLANT GROWTH AND RESPIRATION

GENERAL ORGANIZATION OF THE RESPIRATORY CHAIN AND MATRIX-ASSOCIATED SPECIFIC DEHYDROGENASES IN HIGHER PLANT MITOCHONDRIA

Roland Douce and Michel Neuburger

Laboratoire de Physiologie Cellulaire Végétale
Unité associée au CNRS n° 576, CEN-G and USTM-G, 85 X
F-38041 Grenoble-cédex, France

INTRODUCTION

One of the major functions of mitochondria from all organisms is to provide ATP as the principal energy source for the cell. This is true also of plant mitochondria and it is therefore no surprise that many basic features of mitochondrial membranes have been conserved between animals and plants despite a billion years of divergent evolution. Thus the morphology of plant mitochondria closely resembles that of their animal counterparts, as do their cyt chain, ATPase complex, energy conservation (H^+ ejection) mechanisms, and membrane phospholipid composition (Douce, 1985). Presumably these basic features of mitochondrial membranes are essential for their functioning in energy transduction.

Nonetheless it is becoming increasingly apparent that mitochondria isolated from higher plants have a considerably more complex respiratory chain and a more sophisticated matrix dehydrogenases equipment.

This chapter shall consider the general organization of the respiratory chain in plant mitochondria and set out the unique features of matrix dehydrogenases which one assumes reflect their functioning in autotrophic metabolism.

GENERAL ORGANIZATION OF THE RESPIRATORY CHAIN

For many years the organization of the respiratory chain of plant mitochondria was thought to be very similar to mitochondria from more extensively studied animal sources such as rat liver or beef heart. In fact, the plant mitochondrial cyanide-sensitive electron pathway (i.e. the sequence of electron carriers wich mediate the flow of electrons from respiratory substrates to O_2 via cytochrome oxidase) appears similar to that found in mitochondria from animal tissues (Douce, 1985). However, it is now recognized that there are a number of distinct differences between plant and animal mitochondria. These include the cyanide-resistant electron pathway, which is also encountered in the mitochondria of micro-organisms (Lloyd, 1974) and the rotenone-resistant electron pathway (Palmer, 1976). At one time these differences were not felt to be real but rather to be artifacts due to difficulties associated with isolating mitochondria from plant tissues. Fortunately this view is no longer widely held.

Cytochrome oxidase pathway

The immense complexity of the respiratory chain becomes apparent when it is realized that a single respiratory chain unit (Mr $\simeq 1.5 \times 10^6$) contains as many as 40 redox centers and 50 polypeptides together with significant amounts of phospholipids. The mitochondrial respiratory chain which is remarkably compact consists of only four protein complexes and cytochrome *c* (cyt *c*) : complex I (responsible for electron transfer from NADH formed in the matrix to ubiquinone), complex II (responsible for electron transfer from succinate to ubiquinone), complex III (usually called the cytochrome b-c_1 complex and responsible for electron transfer from ubiquinol to cytochrome *c*) and complex IV or cytochrome *c* oxidase (responsible for electron transfer from ferrocytochrome *c* to O_2) (Tzagoloff, 1982). Except for cyt *c*, these complexes are very hydrophobic (the fractionation procedure involves the use of detergent) and are soluble in the "fluid" lipid bilayer medium of the mitochondrial cristal membrane. Electron transport is coupled to the lateral diffusion of all redox components and does not require ordered chains assemblies, or aggregates of redox components (Gupte et al, 1984). In support of this interesting suggestion Bligny and Douce (1977) have shown that copper deficiency in suspension-cultured sycamore cells inhibits cytochrome aa_3 appearance. Under these conditions the low amount of cytochrome *c* oxidase present in copper-deficient mitochondria does not limit electron flow at the level of the inner mitochondrial membrane. An ordered chain assemblies would imply a marked decrease in the rate of electron flow. Several lines of evidence have pointed to the free quinone itself (localized in the hydrophobic core of the lipid bilayer) as being the mobile agent, which electronically connects the multiprotein donors (complex I and II) and acceptor (complex III) (Moore and Rich, 1985). On the other hand, the ionic interaction between the cyt *c* molecule and the membrane outer surface is weak giving the cyt *c* two-dimensional mobility such that it may reversibly and alternately interacts with its physiological oxidoreductase (i.e. complex III and complex IV). The transfer of electrons from substrate to O_2 via the various complexes is coupled to an electrogenic translocation of protons acrosss the inner mitochondrial membrane. According to the chemiosmotic hypothesis the respiration-linked proton pump (at the level of complex I, III and IV) provides a link between oxidoreduction and ATP synthesis (Mitchell, 1980 ; Moore and Rich, 1985).

Mitochondrial complex I. Complex I is the entry point for the redox equivalents of NADH produced in the matrix space during the course of substrate (malate, α-ketoglutarate, pyruvate, citrate and glycine) oxidation catalyzed by the various NAD^+-linked dehydrogenases of the tricarboxylic acid cycle, NAD^+-linked malic enzyme and the glycine decarboxylase complex. Biochemical characterization of complex I in plant mitochondria has not been undertaken so far. Cammack and Palmer (1973) and Brouquisse et al (1986) however have demonstrated unambiguously that complex I in Jerusalem artichoke and potato tubers mitochondria contains at least four distinguishable iron-sulfur centers corresponding closely to center N-1b (apparent $g_{x,y,z}$ = 1.93, 1.93, 2.017), N-2 (apparent $g_{x,y,z}$ = 1.92, 1.92, 2.05), N 3 (apparent $g_{x,z}$ = 1.87, 2.031), and N-4 (apparent $g_{x,z}$ = 1.88, 2.10), found in the NADH-ubiquinone reductase segment of the animal mitochondrial respiratory chain. Potentiometric titrations suggest that the sequence of iron-sulfur centers in complex I of plant mitochondria is N-1→N-3→N-2 (Cammack and Palmer, 1977).

Mitochondrial complex II. Complex II is the segment of the respiratory chain responsible for electron transfer from succinate to ubiquinone. Succinate dehydrogenase in mitochondria isolated from plant tissues can exist in at least two forms. The inactive form is stabilized by oxaloacetate. The enzyme can be converted into the active form by incubation with an activating ligand such as ATP (for a review see Douce, 1985). Recently, Douce et al (1986) have demonstrated that NAD^+ supplied to purified *Solanum tuberosum* mitochondria caused progressive inhibition of succinate oxidation in state III. They demonstrated that NAD^+-induced inhibition of succinate oxidation was due to an inhibition of succinate dehydrogenase resulting from increased accumulation of oxaloacetate generated from malate oxidation via malate dehydrogenase.

Complex II isolated from sweet potato root (Hattori and Asahi, 1982 ; Hattori et al, 1983) and mung bean hypocotyl and soybean cotyledon (Burke et al, 1982) mitochondria can be dissociated by means of chaotropic agents into two sub-complexes. The first sub-complex containing FAD and several non-heme iron centers is water soluble and consists of two large polypeptides (Fp subunit, Mr 70,000 ; Ip subunit, Mr 27,000). The fluorescence of the flavin moiety permits its ready localization in the Fp subunit (Hattori and Asahi, 1982). The other sub-complex is very hydrophobic and consists of two small polypeptides (Mr 7,000 and 13,000). It has been suggested that one of these hydrophobic subunits carries semiquinone-binding site(s) (Singer and Johnson, 1985). The two small peptides are needed for binding of enzyme to the membrane and for the reduction of ubiquinone but are not required for the enzyme to oxidize succinate in the presence of artificial electron acceptors (Hatefi and Galante, 1980). In addition, Cammack and Palmer (1977), Moore et al (1976), Rich and Bonner (1978), Burke et al (1982) and Brouquisse et al (1986) have also identified three iron-sulfur centers in plant mitochondria which correspond closely to the iron-sulfur centers S-1 $[(2\ Fe\text{-}2\ S)^{2+,1+}]$, S-2 $(4\ Fe\text{-}4\ S)^{2+,1+}]$, and S-3 associated with complex II in mammalian. In the spectra of oxidized mitochondria and submitochondrial particles an intense and nearly isotropic, EPR signal was seen at $g = 2.014$ with a low field maximum of $g = 2.02$ (Cammack and Palmer, 1977 ; Rich and Bonner, 1978 ; Brouquisse et al, 1986). This type of signal is of the type expected for 3 Fe cluster in the oxidized state and assigned to center S-3 in succinate dehydrogenase. Brouquisse et al (1986) provided evidence for the existence of an iron-sulfur center in plant mitochondria yielding and EPR signal typical of oxidized 3 Fe clusters with a peak at a g value of approximately 2.03 and distinct from center S-3. This signal, present in the matrix space and attributed to aconitase, caused an apparent broadening of center S-3 peak in intact mitochondria (Fig. 1).

Mitochondrial complex III. Complex III is the segment of the respiratory chain responsible for electron transfer from ubiquinol (QH_2) to cyt *c* in an antimycin-sensitive fashion. The electron transfer between the dehydrogenases and complex III was affected by two bimolecular collision processes, firstly a reaction between oxidized Q and the respective dehydrogenases, and secondly a reaction between reduced QH_2 and complex III (Kröger and Klingenberg, 1973a,b). Q acts kinetically homogeneously as a pool and controls the electron transfer rates (Cottingham and Moore, 1983). Complex III has been isolated from various plant tissues (Ducet and Diano, 1978 ; Nakajima et al, 1984 ; Degli Espoti et al, 1985). Complex III isolated from sweet potato mitochondria (Nakajima et al, 1984) and Jerusalem artichoke mitochondria (Zannoni, unpublished data) can be resolved into several polypeptides. Three polypeptides of intermediate molecular weight have been characterized as cyt *b* (Mr 32,000-36,000) cyt c_1 (Mr 30,000-33,000) and Rieske iron-sulfur protein (Mr 27,000) first characterized in plant mitochondria by Bonner and Prince (1984).

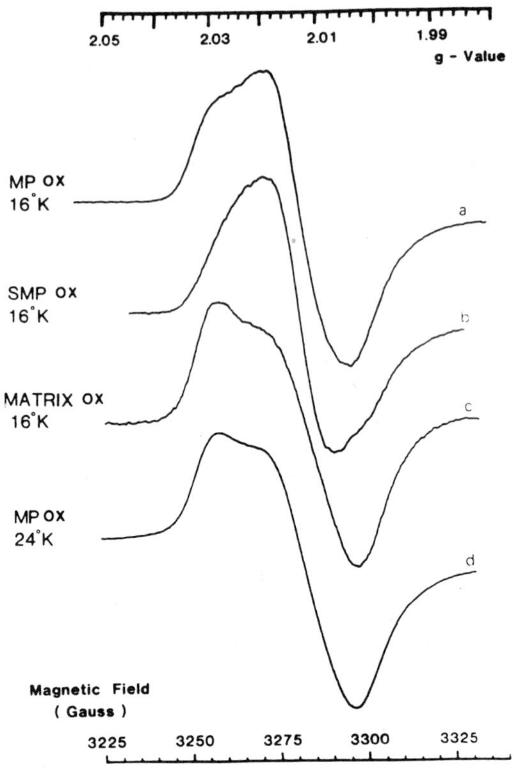

Fig. 1. First derivative EPR spectra of mitochondria (MP) and mitochondrial fractions (submitochondrial particules, SMP ; matrix) isolated from potato tubers. Mitochondria and mitochondrial particules were resuspended to an appropriate protein concentration (MP, 55 mg/ml ; SMP, 18 mg/ml ; Matrix, 18 mg/ml) in suspending medium. Oxygen was blown over the suspension for 30 s before it was transferred to a quartz EPR tube and frozen in liquid nitrogen. Conditions of measurement were : modulation amplitude, 8 Gauss ; modulation frequency, 100 KHz ; microwave frequency, 9252 MHz ; microwave power, a b and c : 0.5 mW, d : 5 mW ; temperature as indicated.

It is clear, therefore, that complex III isolated from plant mitochondria resembles the complexes from other sources with respect to electron-carrier bearing subunits. However the two largest polypeptides (Mr 46,000 and 43,000) or "core proteins" present in large amount in mammalian complex III are barely detectable in complex III from plants (Zannoni this volume). Interestingly Ducet and Diano (1978) reported that plant cyt b solubilized in presence of guanidine exhibits one α-band at low temperature whereas three cyt b α-bands are seen in the low temperature absorption spectrum of intact mitochondria (Lance and Bonner, 1968). They concluded, therefore, that spectral and potentiometic differences attributed to the three cyt b can best be rationalized in terms of different environments conferred on the hemes by the same protein. According to the Q-cycle scheme the cytochrome b_{566}-cytochrome b_{562} sequence serves only to conserve the second electron released during the oxidation of QH_2 by returning it across the inner membrane to the Q pool at center i. This role of cyt b appears oversimplified however considering the multiplicity of spectral and kinetic perturbations displayed by the two cyt b heme groups present in the QH_2 reductase complex (see Rieske, 1986). In fact considerable evidence has been reported to support a proton-pumping function of the cyt b chain to supplement or replace vectorial proton translocation by QH_2 as proposed in the Q cycle hypothesis (Wikström et al, 1981). Interestingly the results of Degli Esposti et al (1985) demonstrated that deoxycholate fractionation of plant mitochondria led to the isolation of a complex III-enriched fraction bearing many of the functional characteristics retained by mammalian and yeast b/c complexes such as the oxidant-induced extra-reduction of b-type cytochrome. Taken together these observations demonstrate the close analogy between the plant and animal complex III.

The data obtained by Cottingham and Moore (1983) are also good evidence in favor of the idea that plant mitochondria, like their mammalian counterparts, use a mobile quinone pool as a carrier of reducing equivalents in the respiratory chain. Finally, it seems that complex III in plant mitochondria has an absolute preference for electrons generated from complex I (for a review see Douce, 1985). This last result suggests that diffusion distance between complex I and III is shorter than that between complex II and III.

Complex IV. Complex IV (cytochrome c oxidase) is the terminal complex of the electron transport chain. Sweet potato cytochrome c oxidase containing two cytochromes (Lance and Bonner, 1968) and two atoms of copper (CU_A and CU_B) (Richaud and Denis, 1984) was purified with a high yield by solubilization of the enzyme from submitochondrial particles with deoxycholate (Maeshima and Asahi, 1978, 1981). Five polypeptides were present with the following molecular weights : 39,000 ; 33,500 ; 26,000 ; 20,000 and 5,700. In addition according to Maeshima and Asahi (private communication) subunit V band of sweet potato cytochrome oxidase comprises at least 3 kinds of polypeptides which differ from one another in the mobility after prolonged electrophoresis of the purified enzyme on an SDS-urea polyacrylamide gel. These data suggest therefore that plant cytochrome c oxidase consists of 7 subunits. In addition the purified enzyme contained approximately 12 nmoles of heme a/mg of protein. Interestingly cytochrome oxidase is proposed to contain 7 different subunits in 1:1 ratios in both mammalian and *Neurospora* mitochondria (Wikström et al, 1981). However subunit composition of complex IV has been the subject to some controversy perhaps attributable to heterogeneity both with respect to more or less extensive lost of phosthetic groups during purification and to contaminating polypeptides.

Cytochrome oxidase in plant mitochondria shows some optical differences compared with the mammalian enzyme. For example, the combined α-band of cytochrome c oxidase has an absorption maximum at 602-603 nm rather than at 605 nm (Lance and Bonner, 1968). In addition kinetic studies after flash photolysis at low temperature of CO rebinding to the mammalian reduced heme a_3 revealed an essentially monophasic process whereas four steps have been resolved in the case of plant reduced heme a_3 (Denis and Richaud, 1982). It is clear from the optical features of cytochrome c oxidase in potato mitochondria in the near-infrared region that Cu_B (known as the invisible copper), which is characterized by a positive band centered at 812 nm, must play an important role in this unexpected mechanism (Richaud and Denis, 1984).

Cyanide resistant electron pathway

Practically, all the plant mitochondria isolated sofar show a residual respiration in the presence of CO,N_3^- or CN^- (Henry and Nyns, 1975). Bendal and Bonner (1971) established that CN^- and antimycin A-insensitive respiration was mediated by an additional electron transport pathway consisting of the same set of dehydrogenase than the respiratory chain, but entirely by-passing the cytochromes probably via a second oxidase (alternative oxidase). In fact, the cyanide-resistant electron transport system consists of a branch point from the conventional electron transport system beginning with ubiquinone and terminating with an "alternative oxidase" (Storey, 1976). According to Bahr and Bonner (1973) the cyanide-insensitive pathway seems to be regulated by the activity of the normal pathway. Electrons from endogenous NADH are diverted to the alternative pathway only when the cytochrome pathway approaches saturation. In contrast, De Troomstenbergh and Nyns (1978) suggested that electrons are partitioned between cytochrome oxidase and alternative pathway according to the rate constants associated with the reactions between ubiquinol and either the alternative oxidase or complex III respectively. Such a result would be expected to display kinetic associated with "Q pool" behavior as originally demonstrated by Kröger and Klingenberg (1973a,b). In fact, Cottingham and Moore (1983) and Stegink and Siedow (1986a) have provided good evidence for Q-pool behavior with isolated plant mitochondria during electron flow through the cytochrome and alternative pathways. However this model does not explain the well known fact that the oxidation of external NADH, which is inhibited by antimycin A, does not seem to be connected with the alternative oxidase (except in the case of mitochondria from *Arum* lilies). A specific protein called "engaging factor" can be envisioned as being essential for coupling electron flow between the ubiquinone pool and the alternative pathway (Stegink and Siedow, 1986b).

According to Huq and Palmer (1978) and Rich (1978) the alternative oxidase might be a quinol oxidase. On the other hand, according to Rustin et al (1983) ubiquinol -or a pool of it- can be oxidized in the presence of free unsaturated fatty acid peroxyl radicals (ROO°) acting as electron acceptors. In this scheme (non-enzymatic process) the most probable reduction product of ROO° is ROH, and its reoxidation is cyanide resistant and yields H_2O not H_2O_2, as a terminal product. Furthermore Vanderleyden et al (1980) produced data which suggested that the solubilized quinol oxidase activity might be a rather non-specific effect caused merely by a general catalysis of quinol autooxidation by proteins in the solution. In fact according to Bonner et al (1986) the oxidase activity is most likely of a protein nature, probably with Fe as an internal component, exhibiting an extraordinary thermal lability.

The physiological role of the alternative oxidase is still uncertain (Douce, 1985). Several workers have concluded that energy conservation is not associated with the alternative oxidase and therefore ascribe a thermogenic role to this independent "non cytochrome" oxidase. In a few cases such as in *Arum* lilies the inhibitor resistant electron transport is apparently directly related to thermogenic metabolism (Meeuse, 1975). However, in many tissues there is no such correlation and the existence of a cyanide-insensitive pathway appears to be a wasteful energy process. Obviously much yet remains to be done in order to understand how the control of the alternative pathway operates under physiological situations. The mechanism whereby the alternative path is engaged and the extent to which it operates are of overiding importance in the physiological role of the alternative path. Finally the biochemical characterization of the enzyme (ouinol oxidase ?) involved is absolutely required.

Rotenone-resistant electron pathway

Lehninger (1964) found that NADH added to liver mitochondria was not oxidized. However, if the mitochondria were gently disrupted by hypotonic swelling, oxidation of NADH was considerably enhanced via the universally distributed rotenone-sensitive respiratory chain-linked NADH dehydrogenase (Complex I). Indeed isotopic studies clearly showed that NADH readily penetrated the outer mitochondrial membrane but exchanged at an insignificant rate with NADH in the matrix compartment (Von Jagow and Klingenberg, 1970). It is now generally accepted that in animal mitochondria the inner membrane is totally impermeable to NADH and NAD^+.

Oxidation of exogenous NAD(P)H by plant mitochondria. The mitochondria of higher plants catalyze a rapid oxidation of exogenous NADH in the absence of added cytochrome c (Douce, 1985). Exogenous NADH is oxidized by an external NADH dehydrogenase located on the outer surface of the inner membrane. This dehydrogenase is specific for the β-hydrogen of NADH and feeds electrons directly to complex III (Douce, 1985) by-passing complex I and the first site of H^+ translocation. This pathway, which is inhibited by antimycin A does not seem to be connected with the alternative oxidase (except in the case of mitochondria from Arum lilies). Consequently NADH oxidation by this external NADH dehydrogenase, which does not require NADH translocase, is insensitive to rotenone and has an ADP/O ratio similar to that of succinate. Platanetin, a 3,5,7,8-tetrahydroxy,6-isoprenyl flavone isolated from the bud scales of plane tree is a potent inhibitor of the external NADH dehydrogenase of the inner membrane (Ravanel et al, 1986). This activity is probably due to the flavonoid structure and the high lipophilicity of platanetin associated with the presence of the isoprenyl chain. The activity of this external NADH dehydrogenase is regulated by micromolar concentrations of Ca^{2+}, inhibited by calcium chelators such as EGTA and does not appear to involve calmodulin (Møller and Lin, 1986). Parenthetically plant mitochondria also oxidize exogenous NADPH, apparently via a Ca^{2+}-dependent dehydrogenase located on the outer surface of the inner membrane (Møller and Lin, 1986). The metabolic significance of these external dehydrogenases capable of oxidizing cytosolic NAD(P)H very rapidly and present in all the plant mitochondria isolated so far, is unknown.

Oxidation of endogenous NADH by plant mitochondria. The oxidation of endogenous NADH in plant mitochondria appears to be more complex than its counterpart in mammalian mitochondria (Møller and Lin, 1986). The most obvious indication of this is that inhibitors such as piericidin A or rotenone, which inhibit the oxidation of endogenous NADH in animal mitochondria by interacting with the iron-sulphur centres associated with complex I, only cause a partial and sometimes an imperceptible inhibition in the plant mitochondrial system. By measuring the ATP formation, it is clear that the rotenone-resistant pathway is not coupled to the first site of ATP synthesis, whereas the rotenone-sensitive pathway is (Palmer, 1976). It seems that plant mitochondria, in contrast with animal mitochondria, possess two internal NADH dehydrogenases on the inner surface of the inner membrane. One of these internal dehydrogenases oxidizes endogenous NADH readily in a rotenone-sensitive manner. This dehydrogenase is therefore coupled to the synthesis of three molecules of ATP and is probably similar to complex I characterized in mammalian mitochondria. Current evidence suggests that this dehydrogenase has an apparent Km for NADH of 8 μM (Møller and Lin, 1986). We also believe that complex I which operates in close relationship with all the NAD^+-linked TCA cycle dehydrogenases utilizes a common pool of NAD^+ present in the matrix space (Neuburger and Douce, 1983). The second dehydrogenase connected to the respiratory chain via the ubiquinone pool is insensitive to inhibition by rotenone and is coupled to the synthesis of only two moles of ATP (Palmer, 1976). This dehydrogenase, in contrast with complex I, exhibits a low affinity for internal NADH (Møller and Lin, 1986) and differs from the rotenone-resistant NADH dehydrogenase associated with the outer face of the inner membrane inasmuch as it is not sensitive to EGTA or Ca^{++} (Møller and Lin, 1986). The physiological significance of the rotenone-resistant internal NADH dehydrogenase is not understood. It has been suggested that complex I may be associated with the cyanide-sensitive oxidase, whereas the internal non-phosphorylating dehydrogenase is associated with the cyanide-resistant oxidase providing a totally non-phosphorylating pathway for the oxidation of endogenous NADH when the energy charge is high. Thus the rotenone-resistant dehydrogenase may play a role when the phosphate potential restricts electron flow through the normal respiratory chain. Such an idea is very attractive especially in the case of mitochondria from thermogenic tissues. Again it is clear that the mechanism whereby the rotenone-insensitive pathway is engaged and the extent to which it operates are of the utmost importance in the physiological role of this pathway. The concentration of NADH in the matrix space seem to play an important role in the regulation of the pathways responsible for endogenous NADH oxidation because the affinity of the rotenone-sensitive NADH dehydrogenase for NADH is greater than the affinity of the internal rotenone-resistant NADH dehydrogenase.

Transport of NAD^+ in plant mitochondria. Plant mitochondria isolated from a number of tissues are capable of actively accumulating NAD^+ from the external medium (Tobin et al, 1980). The slow rate of NAD^+ accumulation in the matrix space is concentration-dependent, exhibits Michaelis-Menten kinetics and is strongly inhibited by the analogue N-4-azido-2-nitrophenyl-4-aminobutyryl-NAD^+ (NAP_4 - NAD^+) (Neuburger and Douce, 1983). When suspended in a medium that avoided rupture of the outer membrane, intact purified mitochondria progressively lost their NAD^+ content by passive diffusion (Neuburger and Douce, 1983). This led to a marked decline in the state-3 rate of NAD^+-linked substrates oxidation by isolated mitochondria. The rate of NAD^+ efflux from the matrix space is strongly temperature dependent and is inhibited by the analog inhibitor of NAD^+ transport indicating that a carrier is required for net flux in either direction (Neuburger et al, 1985). It is possible that this carrier has an important regulatory function *in vivo* by allowing manipulation of matrix NAD^+ concentration and thus regulating the activity of all NAD-linked enzymes (Neuburger and Douce, 1983 ; Neuburger et al, 1985). Since the intramitochondrial concentration of NAD^+ has such a profound influence on matrix enzyme activity and O_2-uptake via the rotenone-insensitive pathway by isolated mitochondria, it is potentially a very powerful regulator of plant respiration *in vivo* and could play and important role in the coarse control of metabolism, particularly during transition from a dormant stage to a stage of active growth (and *vice versa*). In this context, the mitochondria from young growing tissues, including the shoots of sprouting potato tubers, have higher matrix NAD^+ contents than those from storage tissues (Tobin et al, 1980). It is also interesting to note that preliminary experiments carried out in our laboratory indicate that over the time of potato tuber storage, the endogenous NAD^+ content of the mitochondria first declined from an original value of 3 nmol/mg protein to 0.3 nmol after 90 d of storage, and thereafter rose again during sprouting. It is possible, therefore, that the rates of respiration in these tissues would be affected by the concentration of NAD^+ in the extramitochondrial cytoplasm and this concentration might differ significantly from one tissue to the other, or even between different physiological situations of the same tissue.

MATRIX-ASSOCIATED SPECIFIC DEHYDROGENASES

The mitochondrial matrix is the site of tricarboxylic cycle enzymes that provide reducing equivalents to the electron transport chain and also, via ancillary reactions, provide numerous substrates for biosynthetic reactions in the cytoplasm (for review see Douce, 1985). The fundamental structure of the tricarboxylic acid cycle dehydrogenases and their regulation are remarkably similar to those found in mammalian system. Nonetheless it is becoming increasingly apparent that mitochondria isolated from higher plants have several specific NAD-linked dehydrogenases such as NAD^+-linked malic enzyme and glycine decarboxylase which one assumes play an important role in autotrophic metabolism.

NAD^+- linked malic enzyme

Plant mitochondria in contrast with mammalian mitochondria, readily oxidize malate without the necessity of removing oxaloacetate because they possess a specific NAD^+-linked malic enzyme and a specific oxaloacetate carrier. In the absence of thiamin pyrophosphate, O_2 uptake with malate as substrate is attributed solely to malate dehydrogenase and/or NAD^+-linked malic enzyme. Pyruvate and/or oxaloacetate are therefore the major products formed during the course of malate oxidation (Palmer, 1976). NAD^+-linked malic enzyme was discovered in plant mitochondria by Macrae and Moorhouse (1970) and is not found in most animal mitochondria. This enzyme is specific for L-malate, has an absolute requirement for Mn^{2+} and is characterized by low substrate affinity. It is inhibited by bicarbonate and this inhibition is relieved by CoASH (Neuburger and Douce, 1980). The ratio of the products, oxaloacetate and pyruvate, during the course of malate oxidation reflects the balance of the two malate-oxidizing enzymes (Neuburger and Douce, 1980). When the activity of the NAD^+-linked malic enzyme is weakened [high bicarbonate concentration (alkaline pH) low CoA concentration], oxaloacetate is preferentially excreted and there is a decrease in the rate of malate oxidation as the reaction proceeds owing to the accumulation of oxaloacetate and the unfavorable equilibrium of the reaction catalysed by malate dehydrogenase. One of the reasons why potato mitochondria are able to consume O_2 at significant rates, at high pH with malate as substrate in the absence of a system to remove oxaloacetate is because they excrete oxaloacetate in the external medium (Neuburger and Douce, 1980). Phthalonate is a potent inhibitor of oxaloacetate uptake and efflux in plant mitochondria : consequently adding phthalonate to plant mitochondria respiring malate at alkaline pH induces a marked inhibition of O_2 uptake (Day et al, 1984). On the other hand, when the activity of the NAD^+-linked malic enzyme is powerful [low bicarbonate concentration (acidic pH), high CoA concentration] oxaloacetate concentration is maintained at a low level and pyruvate is rapidly excreted. In other words, with all the plant mitochondria isolated so far, whenever the NAD^+-linked malic enzyme activity is weakened, the rate of oxaloacetate production is higher than that of pyruvate. Consequently it is clear that malate dehydrogenase and NAD^+-linked malic enzyme are competing at the level of pyridine nucleotide pool and that the regulation in vivo of malate dehydrogenase can be readily accounted for by equilibrium effects alone (Neuburger and Douce, 1980).

O_2 uptake by potato and cauliflower bud mitochondria oxidizing malate was progressively inhibited as the pH of the external medium was increased in response to accumulation of oxaloacetate (Macrae, 1971 ; Neuburger and Douce, 1980). Adding 0.5 mM coenzyme A to the medium reversed this trend by stimulating intramitochondrial NAD-linked malic enzyme at alkaline pH (Day et al, 1984). In potato mitochondria, coenzyme A stimulation of O_2 uptake at alkaline pH was only observed when NAD^+ was also provided exogenously. These results show that pH, NAD^+ and coenzyme A levels in the matrix act together to regulate malate oxidation and that coenzyme A can be taken up by intact mitochondria. In fact, exogenous coenzyme A was accumulated by plant mitochondria in a manner sensitive to uncouplers (at low external coenzyme A concentration) and low temperature. This coenzyme A uptake followed saturation kinetics with an apparent K_m of 0.2 mM and a V of 4-6.5 nmol min^{-1} (Neuburger et al, 1984). This transport was insensitive to an inhibitor of NAD^+ transport. It is suggested that plant mitochondria possess a specific carrier for the net accumulation of coenzyme A.

Crover et al (1981) have isolated NAD-malic enzyme to near homogeneity from potato tuber mitochondria. Sodium dodecylsulfate gel electrophoresis of the enzyme reveals two main bands with molecular weights of 61,000 and 58,000. We have confirmed this result in our laboratory (Dumas, Neuburger and Douce, unpublished data). Interestingly, NAD malic enzyme can exist in dimer ($\alpha\ \beta$, the least active form), tetramer ($\alpha_2\ \beta_2$ the most active form) and octamer ($\alpha_4\ \beta_4$). Malate, ionic strength, coenzyme A, and perhaps bicarbonate may regulate NAD malic enzyme by controlling its state of oligomerization.

Glycine decarboxylase

Glycine and serine are intermediates in the photorespiratory metabolism of glycolate to 3-phosphoglycerate. During photorespiration in the leaves of higher plants having the C_3 pathway of photosynthesis, glycine is rapidly oxidatively decarboxylated in the mitochondria to yield CO_2, serine and NADH (Tolbert, 1980). No glycine oxidation activity is present in mitochondria from non-green or etiolated tissues (Gardeström et al, 1980 ; Day et al, 1985b). Glycine is oxidized in the matrix space by the glycine cleavage system (aminomethyl transferase EC 2.1.2.10) to CO_2, NH_3 and 5,10 methylene-tetrahydrofolate (Kikuchi and Hiraga, 1982 ; Walker and Oliver, 1986). The latter compound produced reacts with a second mole of glycine to form serine in a reaction catalyzed by serine hydroxymethyltransferase (SHMT, EC 2.1.2.1) (Schirch and Peterson, 1980).

The polypeptide pattern of the membranes from green leaf mitochondria were virtually identical to that of mitochondria from etiolated leaves, but marked differences were observed between the matrix proteins from both tissues. In particular, the matrix of Pea leaf mitochondria is distinguished by the presence of five pronounced bands at 94, 63, 51, 41 and 15,5 kDa which are barely detectable in the mitochondria from etiolated leaves (Day et al, 1985b). Similar differences have been reported by Ericson et al (1984) for the mitochondria from leaves and roots of spinach. These four polypeptides belong to the glycine decarboxylase/serine hydroxymethyl transferase complex and represent a large proportion of the matrix protein in green leaf mitochondria (approx 50 % of the total amount of matrix proteins). This observation explains the relatively lower lipid-to-protein ratio in the leaf mitochondria compared to those, from non-photosynthetic tissues (Gardestrom et al, 1983 ; Day et al, 1985b) and the lower cytochrome-to-protein ratio in spinach leaf mitochondria (Gardestrom et al, 1983). It also explains the higher density of green leaf mitochondria compared to that of mitochondria from other tissues (Douce and Neuburger, unpublished data).

Neuburger et al (1986) demonstrated for the first time that high molecular weight proteins from the green leaf mitochondrial matrix retained on an XM-300 diaflo membrane produced an enzymatic extract exhibiting high rates of glycine oxidation in the presence of tetrahydrofolate (THF) and pyridoxal-P. During the course of glycine oxidation NAD^+ and serine were formed equally indicating that glycine cleavage system and SHMT are present in the enzymatic extract and operate in a concerted manner. It is very likely that *in vivo* glycine decarboxylase and SHMT are "linked together" via a soluble pool of THF, which does not react with O_2. These results also demonstrated that high ionic strength and pH disrupt in a reversible way, the fairly stable complex involved in glycine oxidation. Interestingly the rate of glycine oxidation, exhibited saturation with increasing NAD^+ concentration with an apparent K_m of 75 µM (Fig. 2). The product of the glycine oxidation, NADH, competitively inhibited the reaction when NAD^+ was the varied substrate at saturating concentrations of glycine. Furthermore the apparent K_i value for NADH, was 5 times lower than the K_m for NAD^+ (Fig. 2). These results demonstrate that _in vivo_ glycine decarboxylase activity in green leaf mitochondria

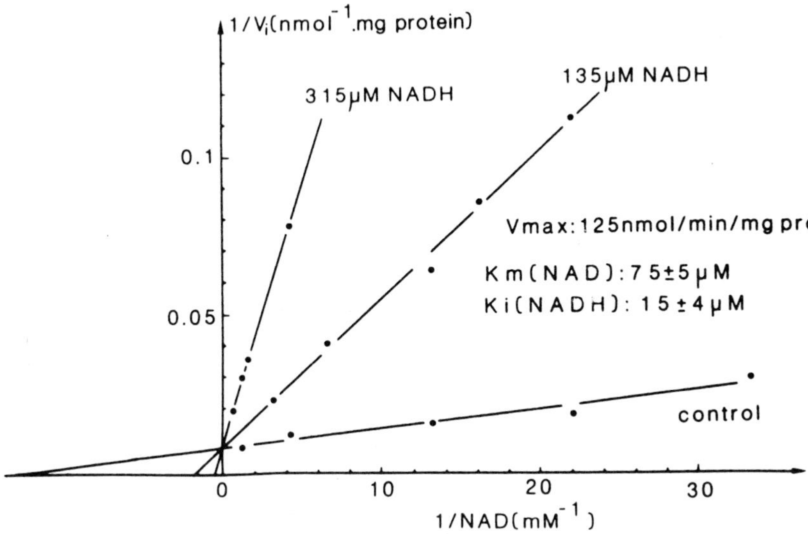

Fig. 2. Initial velocity patterns of glycine oxidation by an enzymatic extract isolated from pea leaf mitochondria. The reaction medium and the preparation of the enzymatic extract are described in the text. NAD^+ was the variable substrate and NADH fixed at 0 µM (control) ; 135 µM and 315 µM. Glycine oxidation was assayed at 25°C by measuring the formation of NADH at 340 nm using a Kontron (Uvikon-810) spectrophotometer.

is regulated by NADH-NAD^+ molar ratio. It is clear therefore that the preferential oxidation of glycine observed by several groups is achieved by a dominance of complex I over complex II and the external NADH dehydrogenase of the respiratory chain (for review see Douce, 1985), by the ability of glycine decarboxylase to compete favourably at the level of NAD^+ and by the huge concentration of glycine decarboxylase present in the matrix space. Finally, SHMT (4 x 53 kDa) and the four proteins of the glycine cleavage system including a pyridoxal-P enzyme (P-protein, 2 x 94 kDa), a carrier protein containing covalently bound lipoic acid (H protein, 15 kDa), a protein exhibiting lipoamide dehydrogenase activity (L protein, 2 x 60 kDa) and a tetrahydrofolate dependent enzyme (T-protein, 41 kDa) have been purified to homogeneity from pea leaf mitochondrial extracts (see also Walker and Oliver (1986) in the case of the enzyme complex precipitated by cold acetone).

CONCLUSION

Plant mitochondria prove to be biochemically very flexible organelles. For example the reoxidation of NADH produced in the matrix upon operation of various dehydrogenases including NAD-linked malic enzyme and glycine decarboxylase can be reoxidized equally well by either complex I (in which case substrate oxidation will be linked to the energy status of the mitochondria) or the rotenone-resistant electron pathway (in which case substrate oxidation will not be directly linked to the energy status of the mitochondria). It is clear that the mechanism whereby the rotenone insensitive pathway is engaged and the extent to which it operates are of the utmost importance in the physiological role of this pathway. Another example of this flexibility is the presence of NAD^+-linked malic enzyme present in all the plant mitochondria isolated so far. This enzyme may play a key role in the organic acid metabolism. When stored reserves of malate within the vacuole (the available evidence suggests that malate accumulated in vacuoles is synthesized by a sequence involving phosphoenolpyruvate carboxylase and malate dehydrogenase and rapidly transfered to the vacuole after synthesis) are mobilized, NAD^+-linked malic enzyme allows their complete oxidation via conversion of the malate to pyruvate. Pyruvate is then converted to acetyl-CoA, which in turn can be completely oxidized in the normal reactions of the tricarboxylic acid cycle. In other words malic enzyme allows the conversion of C_4 acids into acetyl-CoA, the normal respiratory substrate, without the necessity of supplying pyruvate from glycolysis (Palmer, 1976).

References

Bahr, J.T., and Bonner, W.D., 1973, Cyanide-insensitive respiration. II Control of the alternate pathway, _J. Biol. Chem_. 248:344.

Bendall, D.S., and Bonner, W.D., 1971, Cyanide-insensitive respiration in plant mitochondria, _Plant Physiol_., 47:236.

Bligny, R., and Douce, R., 1977, Mitochondria of isolated plant cells (_Acer pseudoplatanus_. L.) II. Copper deficiency effects on cytochrome c oxidase and oxygen uptake, _Plant Physiol_., 60:675.

Bonner, W.D., Jr., and Prince, R.C., 1984, The rieske iron-sulfur cluster of plant mitochondria, _FEBS Lett_., 177:47.

Bonner, W.D., Jr., Clarke, S.D., and Rich, P.R., 1986, Partial purification and characterization of the quinol oxidase activity of _Arum_ maculatum mitochondria, _Plant Physiol_., 80:836.

Brouquisse, R., Gaillard, J., and Douce, R., 1986, Electron paramagnetic resonance characterization of membrane bound iron-sulfur clusters and aconitase in plant mitochondria, _Plant Physiol_., 81:247.

Burke, J.J., Siedow, J.N., Moreland, D.E., 1982, Succinate dehydrogenase. A partial purification from mung bean hypocotyls and soybean cotyledons, _Plant Physiol_., 70:1577.

Cammack, R., and Palmer, J.M., 1973, EPR studies of iron-sulphur proteins of plant mitochondria, _Ann. N.Y. acad. Sci_., 222:816.

Cammack, R., and Palmer, J.M., 1977, Iron-sulfur centers in mitochondria from _Arum maculation_ spadix with very high rates of cyanide-resistant respiration, _Biochem. J_., 166:347.

Cottingham, I.R., and Moore, A.L., 1983, Ubiquinone pool behaviour in plant mitochondria, _Biochem. Biophys. Acta_ 724:191.

Crover, S.D., Canellas, P.F., and Wedding, R.T., 1981, Purification of NAD-malic enzyme from potato and investigation of some physical and kinetic properties, _Arch. Biochem. Biophys_, 209:396.

Day, D.A., Neuburger, M., and Douce, R., 1984, Activation of NAD-linked malic enzyme in intact plant mitochondria by exogenous coenzyme A., _Arch. Biochem. Biophys_., 231:233.

Day, D.A., Neuburger, M., and Douce, R., 1985, Biochemical characterization of chlorophyll-free mitochondria from pea leaves, _Aust. J. Plant Physiol_, 12:219.

Degli Espoti, M., Flamini, E., and Zannoni, D., 1985, Functional characterization and partial purification of the ubiquinol cytochrome c oxidoreductase from higher plant mitochondria, Plant Physiol., 77:758.

Denis, M., Richaud, P., 1982, Dynamics of carbon monoxide recombination to fully reduced cytochrome c oxidase in plant mitochondria after low-temperature flash photolysis, Biochem. J., 206:379.

Douce, R., 1985, "Mitochondria in higher plants:structure, function, biogenesis", Academic Press, Orlando.

Douce, R., Neuburger, M., and Douce, R., 1986, Regulation of succinate oxidation by NAD^+ in mitochondria purified from potato tubers, Biochim. Biophys. Acta, 850:64.

Ducet, G., and diano, M., 1978, On the dissociation of the cytochrome bc, of potato mitochondria, Plant Sci. Lett., 11:217.

Ericson, I., Sahlstrom, S., Bergman, A., and Gardestrom, P., 1984, The glycine decarboxylating system in spinach leaf mitochondria, Adv. Photosynth. Res., 3:887.

Gardeström, P., Bergman, A., Sahlstrom, S., Edman, K.A., and Ericson, I., 1983, Comparison of membrane composition of mitochondria isolated from spinach leaves and leaf petioles, Plant Sci. Lett., 31:173.

Gupte, S., Wu, E.S., Hoechli, L., Hoechli, M., Jacobson, K., Sowers, A.E., and Hackenbrock, C.R., 1984, Relationship between lateral diffusion, collision frequency and electron transfer of mitochondrial inner membrane oxidation-reduction components. Proc. Natl. Acad. Sci., USA, 81:2606.

Hatefi, Y., and Galante, Y.M., 1980, Isolation of cytochrome b_{560} from complex II (succinate-ubiquinone oxidoreductase) and its reconstitution with succinate dehydrogenase, J. Biol. Chem., 255:5530.

Hattori, T., and Asahi, T., 1982, Presence of two forms of succinate dehydrogenase in sweet potato root mitochondria, Plant Cell Physiol. 23:515.

Hattori, T., Iwasaki, V., Sakajo, S., and Asahi, T., 1983, Cell-free synthesis of succinate dehydrogenase and mitochondrial adenosine triphosphatase of sweet potato, Biochem. Biophys. Res. Commun., 113:235.

Henry, M.F., and Nyns, E.J., 1975, Cyanide-insensitive respiration. An alternative mitochondrial pathway, Sub-Cell Biochem., 4:1.

Huq, S., and Palmer, J.M., 1978, Isolation of a cyanide-resistant duroquinol oxidase from Arum maculatum mitochondria, FEBS Lett., 95:217.

Kikuchi, G., and Hiraga, K., 1982, The mitochondrial glycine cleavage system unique features of the glycine decarboxylation, Mol. and Cell. Biochem. 45:137.

Kröger , A., and Klingenberg, M., 173a, The kinetics of the redox reactions of ubiquinone related to the electron-transport activity in the respiratory chain, Eur. J. Biochem., 24:358.

Kröger , A., and Klingenberg, M., 173b, Further evidence for the pool function of ubiquinone as derived from the inhibitions of the electron transport by antimycin, Eur. J. Biochem., 39:313.

Lance, C., and Bonner, W.D., Jr., 1968, The respiratory chain components of higher plant mitochondria, Plant Physiol. 43:756.

Lehninger, A.L., 1964, "The mitochondrion", Benjamin, New York.

Lloyd, D., 1974, The mitochondria of microorganisms, Academic Press, London.

Macrae, A.R., 1971, Effect of pH on the oxidation of malate by isolated cauliflower bud mitochondria, Phytochemistry, 10:1453.

Macrae, A.R., and Moorhouse, R., 1970, The oxidation of malate by mitochondria isolated from cauliflower buds, Eur. J. Biochem., 16:96.

Maeshima, M., and Asahi, T., 1978, Purification and characterisation of sweet potato cytochrome c oxidase, Arch. Biochem. Biophys., 187:423.
Maeshima, M., and Asahi, T., 1981, Presence of an inactive protein immunologically analogous to cytochrome oxidase in the inner mitochondrial membrane of sweet potato root mitochondria, J. Biochem., 90:399.
Meeuse, B.J.D., 1975, Thermogenic respiration in Aroids, Ann. Rev. Plant Physiol., 26:117.
Mitchell, P., 1980, Protonmotive cytochrome system of mitochondria, Ann. N.Y. Acad. Sc., 341:564.
Møller , I.M., and Lin, W., 1986, Membrane-bound NAD(P)H dehydrogenases in higher plant cells, Ann. Rev. Plant Physiol. 37:309.
Moore, A.L., and Rich, P.R., 1985, Organization of the respiratory chain and oxidative phosphorylation, in "Higher Plant Cell Respiration", R. Douce and D.A. Day, eds., Springer-Verlag, Berlin.
Moore, A.L., Rich, P.R., Ingledew, W.J., and Bonner, W.D., Jr., 1976, A complex EPR signal in mung bean mitochondria and its possible relation to the alternative pathway, Biochem. Biophys. Res. Commun. 72:1099.
Nakajima, T., Maeshima, M., and Asahi, T., 1984, The subunit composition of sweet-potato Ipomea-Batatas cultival Kokei N° 14 cytochrome b-c_1 complex, Agr. Biol. Chem., 48:3019.
Neuburger, M., and Douce, R., 1980, Effect of bicarbonate and oxaloacetate on malate oxidation by spinach leaf mitochondria, Biochim. Biophys. Acta, 589:176.
Neuburger, M., and Douce, R., 1983, Slow passive diffusion of NAD^+ between intact isolated plant mitochondria and suspending medium, Biochem. J., 216:443.
Neuburger, M., Day, D.A., and Douce, R., 1984, Transport of coenzyme A in plant mitochondria, Arch. Biochem. Biophys. 229:253.
Neuburger, M., Day, D.A., and Douce, R., 1985, Transport of NAD^+ in Percoll-purified potato tuber mitochondria. Inhibition of NAD^+ influx and efflux by N-4-azido-2-nitrophenyl-4-aminobutyryl-3'-NAD^+, Plant Physiol., 78:405.
Neuburger, M., Bourguignon, J., and Douce, R., 1986, Isolation of a large complex from the matrix of pea leaf mitochondria involved in the rapid transformation of glycine into serine, FEBS Lett., in press.
Palmer, J.M., 1976, The organization and regulation of electron transport in plant mitochondria, Ann. Rev. Plant Physiol., 27:133.
Ravanel, P., Tissut, M., and Douce, R., 1986, Platanetin:A Potent natural uncoupler and inhibitor of the exogenous NADH dehydrogenase in intact plant mitochondria, Plant Physiol., 80:500.
Rich, P.R., 1978, Quinol oxidation in Arum maculatum mitochondria and its application to the assay, solubilisation and partial purification of the alternative oxidase, FEBS Lett. 96:252.
Rich, P.R., and Bonner, W.D., Jr., 1978, EPR studies of higher plant mitochondria II. Centre S-3 succinate dehydrogenase and its relation to alternative respiratory oxidations. Biochim. Biophys. Acta, 501:381.
Richaud, P., and Denis, M., 1984, A near-infrared investigation of cytochrome c oxidase in higher plant mitochondria, Arch. Biochem. Biophys., 232:8.
Rieske, J.S., 1986, Experimental observations on the structure and function of mitochondrial complex III that are unresolved by the protonmotive ubiquinone-cycle hypothesis, J. Bioenerg. and Biomemb., 18:235.

Rustin, P., Dupont, J., and Lance, C., 1983, Oxidative interactions between fatty acid peroxy radicals and quinones:possible involvement in cyanide-resistant electron transport in plant mitochondria, Arch. Biochem. Biophys., 225:630.

Schirch, L., and Peterson, D., 1980, Purification and properties of mitochondrial serine hydroxymethyltransferase, J. Biol. Chem., 255:7801.

Singer, T.P., and Johnson, M.K., 1985, The prosthetic groups of succinate dehydrogenase:30 years from discovery to identification FEBS Lett. 190:189.

Stegink, S.J., and Siedow, J.N., 1986a, An apparent inhibitory synergism between antimycin A and salicylhydroxamic acid in plant mitochondria is dependent upon the kinetics of electron transfer through the ubiquinone pool, Plant Science, in press.

Stegink, S.J., and Siedow, J.N., 1986b, Binding of butyl gallate to plant mitochondria II:Relationship to the presence or absence of the alternative pathway, Plant Science, in press.

Storey, B.T., 1976, Respiratory chain of plant mitochondria. XVIII. Point of interaction of the alternate oxidase with the respiratory chain, Plant Physiol., 58:521.

Tobin, A., Djerdjour, B., Journet, E.P., Neuburger, M., and Douce, R., 1980, Effect of NAD^+ on malate oxidation in intact plant mitochondria, Plant Physiol., 66:225.

Tolbert, N.E., 1980, Photorespiration, in "The Biochemistry of Plants. Vol. 2 Metabolism and respiration" D.D. Davies, ed, Academic Press, New York.

Troostembergh, DE, J.C., and Nyns, E.J., 1978, Kinetics of the respiration of cyanide-insensitive mitochondria from the yeast Saccharomycopsis lipolytica, Eur. J. Biochem., 85:423.

Tzagoloff, A., 1982 "Mitochondria", Plenum, New York.

Vanderleyden, J., Van den Eynde, E. and Verachtert, H., 1980, Nature of the effect of AMP on the cyanide-insensitive respiration in mitochondria of Moniliella tomentosa, Biochem.J., 186:309.

Von Jagow, G., and Klingenberg, M., 1970, Pathways of hydrogen in mitochondria of Saccharomyces carlsbergensis, Eur.J. Biochem., 12:583.

Walker, J.L., and Oliver, D.J., 1986, Glycine decarboxylase multienzyme complex purification and partial characterization from pea leaf mitochondria, J. Biol. Chem., 261:2214.

Wikström, M., Krab, K., and Saraste, M. 1981, Proton-translocating cytochrome complexes, Ann. Rev. Biochem., 50:623.

RELATIONSHIPS BETWEEN MEMBRANE POTENTIAL AND OXIDATION RATE IN POTATO MITOCHONDRIA

Philippe Diolez and François Moreau

Laboratoire de Biologie Végétale IV (CNRS, UA 1180)
Université Pierre et Marie Curie
Paris, France

INTRODUCTION

According to the chemiosmotic hypothesis, oxidative phosphorylation is mediated by an electrochemical proton gradient ($\Delta\mu_{H^+}$), or protonmotive force (Δp), created by substrate oxidation and used by the ATP synthase to synthesize ATP (Mitchell, 1961). This $\Delta\mu_{H^+}$ is frequently regarded as the sole factor involved in coupling electron transfer to ATP synthesis (see Boyer et al., 1977) but some observations do not fit into this picture (see Rottenberg, 1979; Westerhoff et al., 1984; Ferguson, 1985). Alternative mechanisms have been proposed in which protons flow within or along the membrane rather than, as in the chemiosmotic theory, via the bulk aqueous phase (Kell, 1979; Haraux and de Kouchkovsky, 1983) or in which a direct relationship exists between respiratory chain and ATP synthase (Padan and Rottenberg, 1973; Tu et al., 1981). By contrast with photosynthetic membranes (Baccarini-Melandri et al., 1977; de Kouchkovsky et al., 1984), relatively little work has been done on plant mitochondria, while some results indicated the interest for a better understanding of mechanisms of phosphorylation in plant cells.

The electrochemical proton gradient comprises two components: a membrane potential (an electrical potential difference $\Delta\psi$) and a pH gradient (ΔpH), which are interconvertible. Previous investigations have shown that in plant mitochondria $\Delta\psi$ is very high (220-240 mV) (Mandolino et al., 1983; Ducet et al., 1983; Diolez and Moreau, 1985) compared to that in animal mitochondria (150-180 mV) (Nicholls, 1982; Akerman and Wikstrom, 1976; Kamo et al., 1979). Furthermore, the value of ΔpH in plant mitochondria appears lower (20-30 hmV) than in animal mitochondria (30-70 mV)(Mitchell and Moyle, 1969; Moore and Rich, 1985; Diolez and Moreau, 1985). Effectively, nigericin, which converts ΔpH to $\Delta\psi$ by exchanging K^+ and H^+, has little effect in plant mitochondria (Mandolino et al., 1983; Ducet et al., 1983; Diolez and Moreau, 1985). The presence of an H^+/K^+ exchanger in plant inner mitochondrial membrane (Hensley and Hanson, 1976; Bernardi and Azzone, 1983; Diolez and Moreau, 1985) could account for this result.

The first set of experiments which is presented in this paper tends to evidence the operation of the H^+/K^+ antiporter in potato mitochondria and to establish a correlation between membrane potential and oxidation rate under non-phosphorylating and phosphorylating steady state conditions. Indeed, multiple relationships have been reported for rat liver mitochondria, a result which has been considered inconsistent with the delocalized chemiosmotic model. In a second set of experiments the double inhibitor titration has been carried out as a means of discriminating between delocalized chemiosmotic mechanism and localized mechanism of energy transduction (see Westerhoff et al., 1984; Ferguson, 1985).

MATERIALS AND METHODS

Extraction and purification of mitochondria from potato tubers (Solanum tuberosum L.) were carried as previously described (Diolez and Moreau, 1983). Purification was performed on 22.5% (v/v) self-generating Percoll gradient. Purified mitochondria were collected in a medium containing 0.3 M mannitol, 1 mM EDTA and 1% (w/v) bovine serum albumin (pH 7,2). Mitochondrial protein was determined by nesslerization.

Oxygen consumption, membrane potential and oxidative phosphorylation were determined simultaneously in a single experiment. Measurements were carried out at 25°C in a glass vessel (final volume, 2 ml) in a medium containing 0.3 M mannitol, 5 mM $MgCl_2$, 30 mM KCl, 1% (w/v) bovine serum albumin and 10 mM potassium phosphate buffer (pH 7.2). About 0.8-1.2 mg protein was added per assay. Oxygen uptake was measured polarographically with a Clark-type electrode (Hansatech Ltd, King's Lynn, U.K.) and membrane potential was measured with a laboratory-made tetraphenyl phosphonium (TPP^+) electrode (Kamo et al., 1979; Diolez and Moreau, 1985). The concentration of TPP^+ in the assay medium was 5 or 10 μM. After calibration mitochondria were incubated in the presence of ATP (100 μM) before addition of 5 mM succinate to create a state 4. The value of the electrode potential after addition of a saturating concentration of valinomycin was used as a reference for all $\Delta\Psi$ determinations (Ducet et al., 1983). An intramitochondrial volume of 1 μl per mg protein (Mandolino et al., 1983) was assumed for the determination of membrane potential by using the equation of Kamo et al. (1979).

RESULTS AND DISCUSSION

Non Phosphorylating (Uncoupling) Conditions

The experiment presented in Fig. 1 was carried out in order to investigate the activity of such an H^+/K^+ antiporter in potato mitochondria. In the absence of valinomycin, a K^+ ionophore, a slow swelling was observed in 0,1 M KCl, the addition of a limiting amount of valinomycin (10 ng per mg protein) induced a rapid swelling. The inner mitochondrial membrane appeared slightly permeable to K^+ in non energized plant mitochondria (Fig. 1A) and freely permeable to Cl^-, which appeared to enter plant mitochondria by passive diffusion (Day and Wiskich, 1984). The energization of mitochondria by adding succinate (Fig. 1B) did not enhance the swelling in the absence of valinomycin but a rapid swelling occurred in its presence. This swelling was followed by a shrinking corresponding to an efflux of K^+ from the mitochondria and a steady state was reached. This shrinking was sensitive to uncouplers, and a slow influx of K^+ occurred after addition of CCCP (carbonyl cyanide-m-chlorophenylhydrazone) as a protonophore (Fig. 1B). The presence of CCCP before the addition of valinomycin caused a very fast swelling and no shrinking occurred (Fig. 1C). In the presence of substrate and of limiting amounts of valinomycin (which do not totally collapse $\Delta\Psi$), a CCCP-sensitive efflux of K^+ occurred. A saturating concentration of valinomycin (20 ng per mg protein) suppressed this effect (result not shown).

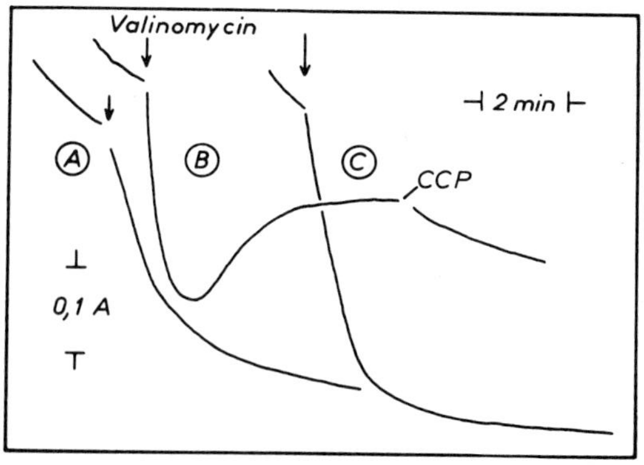

Fig. 1. Swelling of purified potato mitochondria in KCl medium. The same concentration of mitochondria as used for respiratory activities was used in these experiments (about 1 mg protein in 2 ml of KCl medium). Where indicated, 10 ng valinomycin per mg protein and 1 μM CCCP was added. (A) no substrate added; (B) mitochondria were under state 4 (5 mM succinate); (C) substrate (5 mM succinate) was added with CCCP (1 μM) before addition of valinomycin. The swelling of mitochondria was estimated in KCl medium (0.1 M) by measuring the absorption at 520 nm.

This K^+ efflux appeared dependent on membrane energization, since no K^+ efflux could be seen when $\Delta p=0$ (no substrate, saturating valinomycin or CCCP). In consequence, the uncoupling effect of valinomycin in plant mitochondria could be attributed to a $\Delta\Psi$-dependent K^+ cycling mediated by the ionophore and a mitochondrial H^+/K^+ antiporter. In energized mitochondria the effect of valinomycin could thus be equivalent to a proton reentry. By the presence of this H^+/K^+ antiporter, in potato mitochondria $\Delta\Psi$ approaches closely the value of the total electrochemical proton gradient. Furthermore, $\Delta\Psi$, being not affected by electroneutral exchanges including H^+, appeared as a good criterion of ion fluxes (e.g., proton efflux and proton influx) across the inner mitochondrial membrane.

Membrane potential measured with TPP^+-sensitive electrode was plotted against oxidation rate when ionophore concentration was progressively increased (Fig. 2). In this experiment, the effect of CCCP was compared with the effect of valinomycin. As predicted on the basis of the results presented above, the same relationship was obtained between $\Delta\Psi$ and the oxidation rate in these two instances. Furthermore, the presence of nigericin did not affect the curve significantly (Diolez and Moreau, 1985). In state 4, membrane potential was maximal (about 225 mV) while the oxidation rate remained low (about 60 nmol O_2 min^{-1} mg^{-1} protein). The addition of limiting amounts of ionophore (valinomycin or CCCP) caused a concomitant collapse of membrane potential and an increase in the oxidation rate. A linear relationship between $\Delta\Psi$ and the oxidation rate was obtained until the maximal rate of oxidation was reached. Then, no correlation could be found between these two parameters, $\Delta\Psi$ decreased with increasing ionophore concentration but the oxidation rate was then unaffected (Fig. 2). It is noteworthy from this experiment that the maximal oxidation rate was obtained by using an ionophore concentration which lowered membrane potential by only 40-50 mV (approximately 20%). Even though a small variation of $\Delta\Psi$ is involved in the release of the oxidation rate, the above-mentioned results clearly show that a close relationship exists between membrane potential and the oxidation rate.

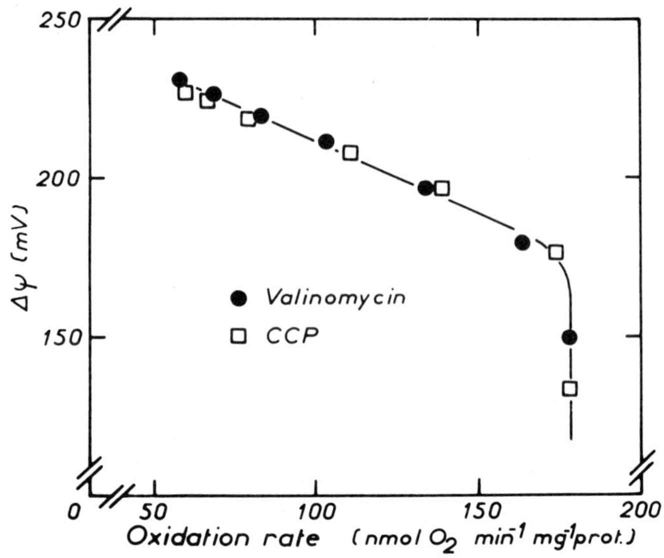

Fig. 2. Correlation between $\Delta\psi$ and oxidation rate in the presence of increasing concentrations of ionophores. These curves were obtained by simultaneously measuring $\Delta\psi$ with a tetraphenylphosphonium-sensitive electrode and the oxidation rates. The membrane potential was progressively collapsed with CCCP (0-1 μM) or with valinomycin (0-20 ng per mg protein) in the presence of 30 mM KCl.

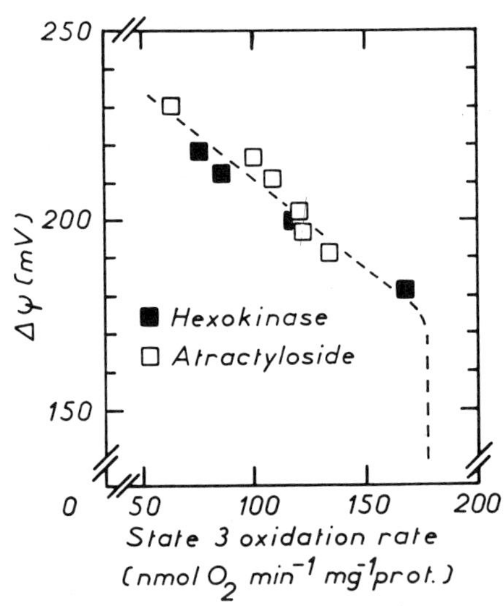

Fig. 3. Relationship between $\Delta\psi$ and the oxidation rates when the phosphorylation rate was limited. The phosphorylation rate was limited by the hexokinase-glucose system or by CAT. Each point (■) represents a single trace, the conditions were: 200 μM ATP, hexokinase (0-17 nkat). Phosphorylating conditions are induced by the addition of 3 mM glucose. These points (□) were obtained when phosphorylation of 200 μM ADP was performed in the presence of increasing concentrations of carboxyatractyloside (0-0.2 μM). The dotted line was obtained in the same experiment by increasing the valinomycin concentration (see Fig. 2).

Phosphorylating Conditions

To test the alternative hypothesises of coupling in potato mitochondria, the relationship between membrane potential and the oxidation rate was investigated during phosphorylation. The simultaneous measurement of $\Delta\psi$ and the oxidation rate in state 3 (ADP present) was performed when phosphorylation was limited. Fig. 3 presents the results obtained with the two methods used: the hexokinase-glucose system (the concentration of hexokinase was the limiting factor of phosphorylation and the reaction was started with the addition of glucose) and the use of carboxyatractyloside (CAT, a specific inhibitor of the adenine nucleotide translocator) which inhibits the rate of phosphorylation by limiting the availability of ADP to the ATP synthase.

Whatever the method used to control the rate of oxidation, the same relationships between $\Delta\psi$ and the oxidation rate were obtained (Fig. 3). The dotted line represents the curve obtained by using increasing valinomycin concentration (see Fig. 2) instead of inhibiting the phosphorylation rates. Such a comparison indicates that identical relationships between $\Delta\psi$ and the oxidation rate were obtained when the proton influx flowed through the ATP synthase (Fig. 3) or directly through the inner mitochondrial membrane under the effect of ionophores (Fig. 2 and dotted line on Fig. 3). These results differ from those previously reported for animal mitochondria (see Rottenberg, 1979), but it should be mentioned that a similar result was obtained with a special type of animal mitochondria : the brown adipose tissue mitochondria (Nicholls and Bernson, 1977). In plant mitochondria, the availability of protons for ATP synthase appears the same as for protonophores or the mitochondrial H^+/K^+ antiporter.

The following experiment, based on the competition between these two ways of proton reentry, was carried out to confirm this observation. Phosphorylation efficiency (ADP/O) was measured with potato mitochondria progressively uncoupled by valinomycin. As already seen in Fig. 2, the

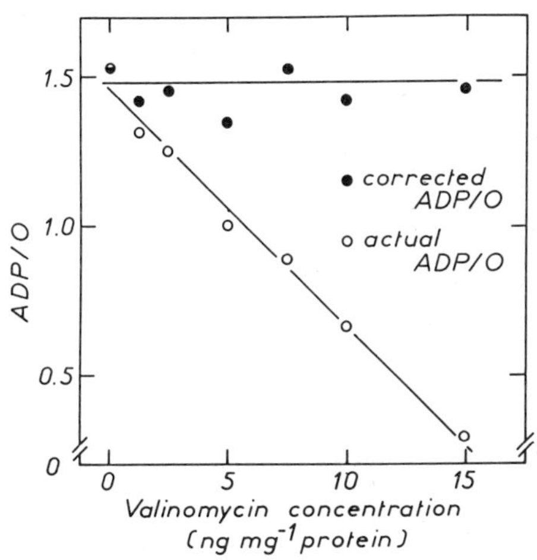

Fig. 4. Phosphorylation in the presence of increasing concentrations of valinomycin. Each actual ADP/O value (O) was measured on a single experiment. ADP/O was calculated from the polarographic traces using the total time of state 4-state 3-state 4 transition measured on the TPP^+ electrode trace. Corrected ADP/O values (●) are calculated as described in the text.

presence of valinomycin decreased $\Delta\psi$ and simultaneously increased the state 4 rate of oxidation. Under these conditions the state 3 rate did not vary and the respiratory control ratio progressively decreased. With limiting concentrations of valinomycin ($\Delta\psi$ decreased from 220 to 150 mV), the ADP/O ratio decreased from 1.5 in the absence of valinomycin to about 0.3 in the presence of 15 ng valinomycin per mg protein (Fig. 4). By substracting the O_2 consumption released by the addition of valinomycin from the total O_2 consumed during the phosphorylation time a 'corrected ADP/O' could be calculated and plotted against the valinomycin concentration (Fig. 4). It can be seen that this 'corrected ADP/O' remained at a quite constant value for valinomycin concentrations increasing from 0 to 15 ng per mg protein. Such a result verifies the hypothetical basis for the correction procedure, namely that the extra-oxygen consumption due to the presence of valinomycin was the same during state 3 as in partly uncoupled state 4. In terms of H^+ fluxes, these results show that the uncoupling effect of valinomycin was linked to a non-phosphorylating H^+ reentry and that this H^+ flux continued at the same rate during phosphorylating conditions. This H^+ flux, linked to the K^+ cycling, appears to be limited only by the valinomycin concentration and was not available for ATP synthesis, suggesting that no preferential proton circuit exists between the respiratory chain and the ATP synthase.

The double-inhibitor titration technique, first used by Baum et al. (1971) with animal submitochondrial particles, was on the basis of some results interpreted as demonstrating the existence of a localized proton gradient in animal mitochondria (Baum et al., 1971; Hitchens and Kell, 1982; 1983; Krasinskaya et al., 1984) or bacteria (Venturoli and Melandri,

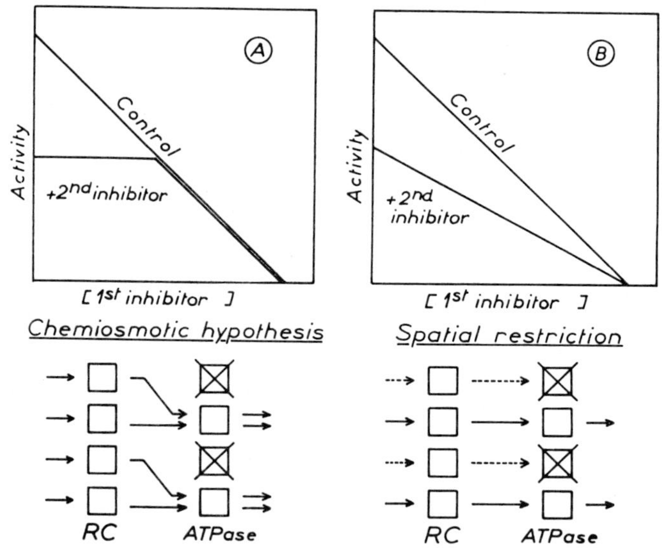

Fig. 5. Rationale of the double-inhibitor titration technique applied to oxidative phosphorylation (after Baum et al., 1971). The double inhibitor titration consits in measuring the stationary state rate of oxidative phosphorylation at different concentrations an inhibitor (1st inhibitor) of respiratory chain (RC) in the absence or in the presence of a constant concentration of an inhibitor (2nd inhibitor) of ATP synthase (ATPase) which inhibits by approx. 50% the activity of this enzyme. Model A: chemiosmotic hypothesis (delocalized); model B: spatial restriction hypothesis (localized chemiosmosis, direct interaction between RC and ATPase).

1982). This interpretation has however been questioned (O'Shea and Thelen, 1984; Pietrobon and Caplan, 1985). The same technique was used here as a means of discriminating between delocalized chemiosmotic and localized mechanism of energy transduction in potato mitochondria.

The rationale of this technique is illustrated on Fig. 5. Two inhibitors are used, each specific for one of the two coupled reactions of the system. In these schemes the expected curves in the case of a strictly delocalized mechanism (Fig. 5A), or when some restriction of the proton flow exists (Fig. 5B)(Baum et al., 1971) are presented. A schematic picture of the proton flow in each case are also presented in this figure. It can be seen that in the presence of only one inhibitor (control), no difference is expected when the activity of the coupled system is plotted against inhibitor concentration. However, when the second inhibitor is added (an inhibitor of phosphorylation in the case of mitochondria) the expected curves are different (Baum et al., 1971).

The results obtained with potato mitochondria are shown in Fig. 6. These experiments were carried out by inhibiting the oxidation rate of succinate by varying malonate concentration (0-2 mM, 1st inhibitor) and the rate of phosphorylation by 0.15 μM CAT (2nd inhibitor) in media with osmolarities varying from 0.3 to 1.0 osM. Indeed, it has been recently reported that a shift from a localized mechanism of coupling to a localized behaviour of oxidative phosphorylation occurred in animal mitochondria whose configuration was modified by increasing the osmolarity

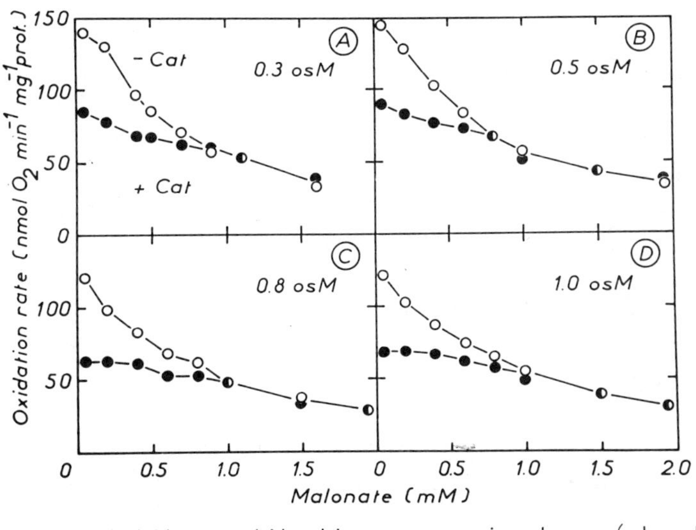

Fig. 6. Double-inhibitor titration experiments (atractyloside-malonate) in media with different osmolarities. The first inhibitor was malonate (0 to 2.0 mM) and the second inhibitor was CAT (0.15 μM). The inhibition of 10 mM succinate oxidation was carried out in state 3 (1 mM ADP), in the presence (●) or in the absence (○) of CAT, by increasing progressively the malonate concentration. The different osmolarities were obtained by varying the mannitol concentration in the assay medium. A, 0.3 osM; B, 0.5 osM; C, 0.8 osM; D, 1.0 osM.

of suspending medium (Krasinskaya et al., 1984). The activity of the system was expressed as the rate of oxidation plotted against malonate concentration (Fig. 6). With potato mitochondria, whatever the osmolarity of the medium in the range 0.3 to 1.0 osM, the double inhibitor titration gave the same results. By contrast with animal mitochondria or bacteria, the curves appeared very close to that expected for a delocalized mechanism, suggesting that no spatial restriction exists in potato mitochondria.

In conclusion, by contrast with animal mitochondria, interestingly no deviation from Mitchell's theory has been observed with potato mitochondria. It appears that no direct relationship exists between respiratory chain and ATP synthase in these mitochondria. A competition between ATP synthase and protonophores against proton flow has been demonstrated and confirmed by the double inhibitor titration technique. In potato mitochondria, a direct intervention of the ATP synthase in the regulation of the oxidation rate can then be disregarded. The primary effector of this regulation appears to be the membrane conductance to protons, whatever the way of protons reentry: by the ATP synthase or through artificial protonophores, or even through natural membrane conductance. The role of the natural and artificial membrane conductance in regulating electron transfer in potato mitochondria needs further investigation.

REFERENCES

Akerman, K.E.O. and Wikstrom, K.F., 1976, Safranine as a probe of the mitochondrial membrane potential, FEBS Lett., 68:191.

Baccarini-Melandri, A., Casadio, R. and Melandri, B.A., 1977, Thermodynamics and kinetics of photophosphorylation in bacterial chromatophores and their relation with the transmembrane electrochemical potential difference of protons, Eur. J. Biochem., 78:389.

Baum, H., Hall, G.S. and Nalder, J., 1971, On the mechanism of oxidative phosphorylation in submitochondrial particles, in: "Energy transduction in respiration and photosynthesis", E. Quagliariello, S. Papa and C.S. Rossi, ed., Adriatica Editrice, Bari.

Bernardi, P. and Azzone, G.F., 1983, Electroneutral H^+/K^+ exchange in liver mitochondria. Regulation by membrane potential, Biochim. Biophys. Acta, 724:212.

Boyer, P.D., Chance, B., Ernster, L., Mitchell, P., Racker, E. and Slater, E.C., 1977, Oxidative phosphorylation and photophosphorylation, Annu. Rev. Biochem., 46:955.

Day, D.A. and Wiskish, J.T., 1984, Transport processes of isolated plant mitochondria, Physiol. Vég., 22:241.

Deléage, G., Penin, F., Godinot, C. and Gautheron, D.C., 1983, Correlations between ATP hydrolysis, ATP synthesis, generation and utilization of pH in mitochondrial ATPase-ATP synthase, Biochim. Biophys. Acta, 725:464.

Diolez, P. and Moreau, F., 1983, Effect of bovine serum albumin on membrane potential in plant mitochondria, Physiol. Plant., 59:177.

Diolez, P. and Moreau, F., 1985, Correlations between ATP synthesis, membrane potential and oxidation rate in plant mitochondria, Biochim. Biophys. Acta, 806:56.

Ducet, G., Gidrol, X. and Richaud, P., 1983, Membrane potential changes in coupled potato mitochondria, Physiol. Vég., 21:385.

Ferguson, S.T., 1985, Fully delocalized chemiosmotic or localized proton flow pathways in energy coupling? A scrutiny of experimental evidence, Biochim. Biophys. Acta, 811:47.

Haraux, F. and de Kouchkovsky, Y., 1983, The energy-transduction theories: a microchemiosmotic approach in thylacoids, Physiol. Vég., 21:563.

Hensley, J.R. and Hanson, J.B., 1975, The action of valinomycin in uncoupling corn mitochondria, Plant Physiol., 56:13.

Hitchens, G.D. and Kell, D.B., 1982, On the extent of localization of the energized membrane state in chromatophores from Rhodopseudomonas capsulata N22, Biochem. J., 206:351.

Hitchens, G.D. and Kell, D.B., 1983, On the functional unit of energy coupling in photophosphorylation by bacterial chromatophores, Biochim. Biophys. Acta, 723:308.

Kamo, N., Muratsugu, M., Hongoh, R. and Kobatake, Y., 1979, Membrane potential of mitochondria measured with an electrode sensitive to tetraphenyl phosphonium and relationship between proton electrochemical potential and phosphorylation potential in steady state, J. Membrane Biol., 49:105.

Kell, D.B., 1979, On the functional proton current pathway of electron transport phosphorylation. An electrodic view, Biochim. Biophys. Acta, 549:45.

Kouchkovsky (de), Y., Haraux, F. and Sigalat, C., 1984, Microchemiosmotic interpretation of energy dependent processes in biomembranes based on the photosynthetic behaviour of thylacoids, Bioelectrochem. Bioenerg., 13:143.

Krasinskaya, I.P., Marshansky, V.N., Dragunova, S.F. and Yaguzhinsky, L.S., 1984, Relationships of respiratory chain and ATP-synthetase in energized mitochondria, FEBS Lett., 167:176.

Mandolino, G., De Santis, A. and Melandri, B.A., 1983, Localized coupling in oxidative phosphorylation by mitochondria from Jerusalem artichoke, Biochim. Biophys. Acta, 723:428.

Mitchell, P., 1961, Coupling of phosphorylation to electron and hydrogen transfer by a chemiosmotic type of mechanism, Nature, 191:144.

Mitchell, P. and Moyle, J., 1969, Estimation of membrane potential and pH difference across the cristae membrane of rat liver mitochondria, Eur. J. Biochem., 7:471.

Moore, A.L. and Rich, P.R., 1985, Organization of the respiratory chain and oxidative phosphorylation, in: "Encyclopedia of plant physiology", Douce R. and Day D.A., Springer-Verlad, Berlin.

Nicholls, D.G., 1982, "Bioenergetics: an introduction to the chemiosmotic theory", Academic Press, London.

Nicholls, D.G. and Bernson, V.S.M., 1977, Inter-relationships between proton electrochemical gradient, adenine nucleotide, phosphorylation potential and respiration, during substrate-level and oxidative phosphorylation by mitochondria from brown adipose tissue of cold adapted Guinea pigs, Eur. J. Biochem., 75:601.

O'Shea, P.S. and Thelen, M., 1984, On the logic of the application of double-inhibitor titrations for the elucidation of the mechanisms of energy coupling, FEBS Lett., 176:79.

Padan, E. and Rottenberg, H., 1973, Respiratory control and the proton electrochemical gradient in mitochondria, Eur. J. Biochem., 40:431.

Pietrobon, D. and Kaplan, S.R., 1985, Further comments on the logic of the application of uncoupler inhibitor titrations for the elucidation of the mechanisms of energy coupling, FEBS Lett., 192:119.

Rottenberg, H., 1979, Non-equilibrium thermodynamics of energy conversion in bioenergetics, Biochim. Biophys. Acta, 549:225.

Tu, S.I., Okazaki, H., Ramirez, F., Lam, E. and Marecek, J.F., 1981, Mutual regulation between mitochondrial ATPase and respiratory chain activities, Arch. Biochem. Biophys., 210:124.

Venturoli, G. and Melandri, B.A., 1982, The localized coupling of bacterial photophosphorylation. Effect of antimycin A and N,N-dicyclohexylcarbodiimide in chromatophores from Rhodopseudomonas sphaeroides, Ga, studied by single turnover event analysis, Biochim. Biophys. Acta, 680:8.

Westerhoff, H.V., Melandri, B.A., Venturoli, G., Azzone, G.F. and Kell, D.B., 1984, Mosaic protonic coupling hypothesis for free energy transduction, FEBS Lett., 165:1.

THE ROLE OF QUINONE POOLS IN REGULATING ELECTRON TRANSPORT IN PLANT MITOCHONDRIA

Christopher J. Kay[a] and John M. Palmer[b]

[a]Department of Biochemistry, University of South Florida Tampa, FL 33612, USA
[b]Department of Pure and Applied Biology, Imperial College London SW7, UK

INTRODUCTION

The plant mitochondrial respiratory chain contains pathways of substrate oxidation not present in the mammalian counterpart. These differences include the ability to oxidise cytosolic NAD(P)H via a dehydrogenase located on the outer face of the inner membrane; the oxidation of glycine by mitochondria from photosynthetic tissues; the oxidation of matrix NADH without energy conservation at site I[1], and the cyanide resistant respiration present in many types of plant mitochondria[2]. The potential of plant mitochondria to oxidise a variety of substrates coupled to different degrees of energy conservation depending on the pathway used, makes it important to understand how these pathways are regulated, to be able to assess their likely significance *in vivo*.

The only component of the respiratory chain common to all pathways is ubiquinone. A range of homologues (UQ-6 to UQ-11) are present in mitochondria from different plants[3]. Electron transfer between dehydrogenases and oxidases is coupled by a stoichiometric excess of ubiquinone (the Q-pool), and much evidence[4] indicates that freely diffusing ubiquinol is the electron carrier, rather than specialized protein-bound quinones. Kroger and Klingenberg[5,6] found that the reduction of ubiquinone (Qo) by dehydrogenase and oxidation of ubiquinol (Qr) by the bc_1 complex were first order with respect to the proportion of the redox active ubiquinone (Qa) in the appropriate redox form (Qo and Qr, respectively). Reduction and oxidation of ubiquinone could therefore be represented by just two rate constants and the steady-state rate of electron transfer (v) be expressed:

$$v = Vr.Vo/(Vr + Vo) \qquad (1)$$

where Vr and Vo are the rate constants for Qo reduction and Qr oxidation, respectively.

Although there have been reports of this "ideal" behaviour in both mammalian and plant mitochondria, in several cases such simple behaviour was not seen. The model represented by equation (1) may be extended to include a second type of dehydrogenase or oxidase activity. However when two dehydrogenases or oxidases are active, the ideal model often does not accurately predict the behaviour of plant mitochondria. Examples of this are the way flux is distributed between cytochrome and alternative pathways[7], the degree of access different substrates have to the alternative pathway[2] and the precedence other substrates, especially glycine[8,9], show over the oxidation of exogenous NADH[10,11]. The inhibition of exogenous NADH oxidation due to succinate oxidation by Jerusalem artichoke tuber mitochondria[11] was chosen for the present study as one example of non-ideal behaviour, since the rates of succinate and exogenous NADH oxidation in these mitochondria tend to be similar, yet when both substrates are present the oxidation of succinate predominates.

MATERIALS AND METHODS

Mitochondria were prepared from Jerusalem artichoke tubers according to the method of Palmer and Kirk[12]. Succinate dehydrogenase was activated by incubation of mitochondria with ATP (1µmol/mg protein), phosphate (1µmol/mg protein) and 5mM $MgCl_2$ for 5min at 25°C.

The ubiquinone content of total lipid extracts was measured by HPLC using a reverse phase silica column (RP-18) developed with methanol:water (97:3 v/v). Stock solutions of UQ-9 were used for calibration and were standardised using an extinction coefficient (280-289nm, oxidised-reduced) of $8.8mM^{-1}cm^{-1}$ [5]. Cytochrome a + a_3 content was measured at 25°C from oxidised and dithionite reduced spectra using an extinction coefficient of $16mM^{-1}cm^{-1}$ [13].

Assays were carried out in a medium containing 0.3M sucrose, 10mM K_2HPO_4, 5mM K^+-TES, 5mM $MgCl_2$, pH 7.2. Oxidation of NADH was measured by dual beam spectroscopy (340-374nm, $4.70mM^{-1}cm^{-1}$). The rate of succinate oxidation in the presence of NADH was determined from parallel measurements of oxygen uptake and NADH oxidation. UQ-1 reduction was followed spectrophotometrically (420nm, $0.465mM^{-1}cm^{-1}$). Concentrations of the inhibitors myxothiazol and antimycin were measured using extinction coefficients of $10.5mM^{-1}cm^{-1}$ (313nm) and $4.8mM^{-1}cm^{-1}$ (320nm), respectively.

RESULTS

The effect of succinate oxidation on the oxidation of exogenous NADH is shown in Fig.1a. Essentially the same inhibition of NADH oxidation was found as reported by Cowley and Palmer[11]. The extent of inhibition was dependent on respiratory state, being greatest under state 4 conditions (61%) and almost absent in uncoupled mitochondria (7%). Inhibition of NADH oxidation required oxidation of added succinate, since addition of sufficient malonate to prevent succinate oxidation also prevented inhibition[11]. Decreased NADH oxidation was not due to reverse electron transfer from succinate to exogenous NAD^+, as no succinate dependent NAD^+ reduction could be detected in the presence of succinate[11]. In addition the electron transport pathway from succinate to exogenous NAD^+ is not energetically favourable since the exogenous NADH dehydrogenase is not energy conserving.

Despite the large inhibition of NADH oxidation by succinate, the influence of NADH oxidation on succinate oxidation (Fig.1b) was slight. It should be noted that the rates of NADH and succinate oxidation tended to be similar to each other in all respiratory states. If the interaction between the substrates were due only to simple competition by the dehydrogenases for Qo, then mutually equal inhibition should have been observed.

The marked dependence of inhibition on respiratory state suggested additional activation of succinate dehydrogenase under assay conditions may have occurred, mediated by NADH, adenine nucleotides or membrane potential as these are known activators of plant succinate dehydrogenase[14]. A role

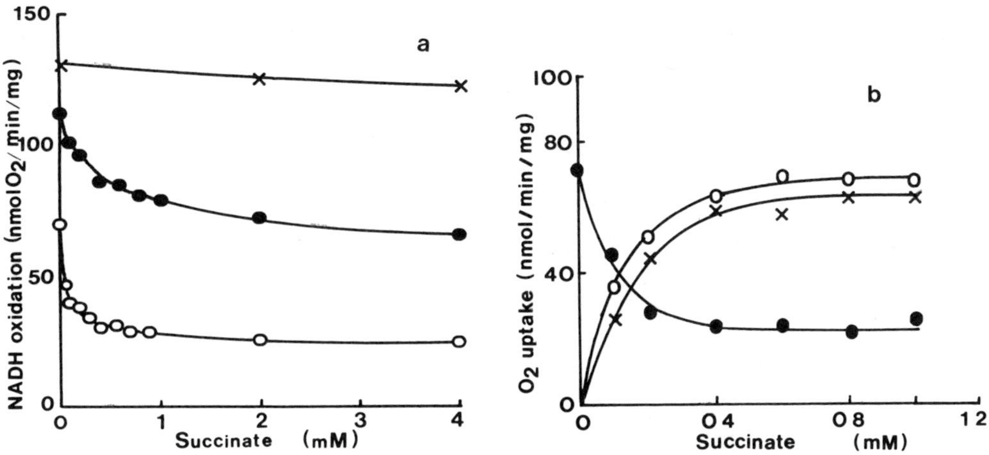

Figure 1a EFFECT OF RESPIRATORY STATE ON THE INHIBITION OF EXOGENOUS NADH OXIDATION BY SUCCINATE.

Rates of NADH oxidation were measured spectrophotometrically in a 1 cm path length cuvette containing 0.22mg mitochondrial protein. Assays were started by the addition of 0.2mM NADH, and after 1 min the appropriate amount of succinate was added.
× +0.2μM FCCP, ● +0.25mM ADP, ○ No addition

Figure 1b INHIBITION OF SUCCINATE OXIDASE BY NADH

Measurements of oxygen uptake and spectrophotometric determinations of NADH oxidation were made under identical conditions in parallel experiments. The rate of succinate oxidation in the presence of NADH was obtained by subtraction. Assays were started by the addition of 0.8mM NADH to the mitochondria (0.16mg protein/ml), followed 1 min later by succinate. ● NADH oxidation, ○ succinate oxidation, × succinate oxidation in the presence of NADH

for adenine nucleotides could be discounted by the use of low concentrations of FCCP to progressively uncouple the cytochrome pathway (Fig.2a). Concentrations of FCCP of up to 70nM caused a threefold increase in the rate of NADH oxidation. The drop in $\Delta\Psi$, estimated using safranine, was small at these FCCP concentrations, presumably because increasing uncoupling was partially compensated by increased NADH oxidation. At higher FCCP concentrations the safranine signal dropped off linearly since NADH oxidation was then maximal. The inhibition of NADH oxidation by succinate still occurred in the absence of added adenine nucleotides (Fig.2a). The degree of inhibition correlated well with the rate of NADH oxidation, but poorly with membrane potential. That a membrane potential was inessential was confirmed using antimycin (Fig.2b) or myxothiazol (not shown) to modulate cytochrome pathway activity in fully uncoupled mitochondria. When the rate of electron transport through the bc_1 complex was decreased using antimycin, the same pattern of asymmetric inhibition of NADH oxidation by succinate was observed as in coupled mitochondria (Fig.2b).

The possibility of interaction between succinate and exogenous NADH dehydrogenase activities was examined by measuring rates of UQ-1 reduction in the presence of cyanide to prevent UQH_2-1 reoxidation. The combined rate of UQ-1 reduction in the presence of both substrates (1474 ± 41 (n=4) nmol UQ/min/mg protein) was very close to the sum of the rates obtained

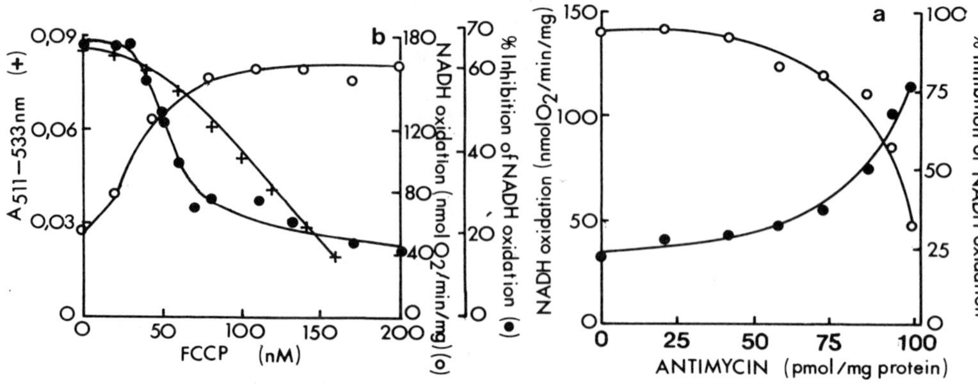

Figure 2a CORRELATION OF MEMBRANE POTENTIAL, NADH OXIDATION AND INHIBITION OF NADH OXIDATION BY SUCCINATE

Mitochondria (0.12mg protein/ml) were suspended in medium containing 5μM safranine and FCCP as indicated. Succinate (10mM) was added 1min after the addition of NADH (0.2mM). In a parallel experiment the membrane potential generated by NADH oxidation was monitored spectrophotometrically (511-533nm).

Figure 2b EFFECT OF ANTIMYCIN ON INHIBITION OF NADH OXIDATION BY SUCCINATE IN UNCOUPLED MITOCHONDRIA

Mitochondria (0.21mg protein/ml) were incubated in medium containing0.2μM FCCP and antimycin as indicated for 2 min prior to addition of NADH (0.2mM). After 1 min succinate was added and the new rate of NADH oxidation recorded. O NADH oxidation, ● % inhibition of NADH oxidation

separately (983 ± 37 (n=4) nmol UQ/min/mg protein for exogenous NADH and 425 ± 28 (n=4) nmol UQ/min/mg protein for succinate). Under identical conditions the rate of NADH oxidation by UQ-1 (995 ± 15 (n=4) nmol NADH/min/mg protein) was unaffected by addition of succinate (962 ± 21 (n=4) nmol NADH/min/mg protein). This suggested that neither succinate nor exogenous NADH dehydrogenase was significantly influenced by the activity of the other when they were decoupled from the rest of the respiratory chain.

Table 1 CONCENTRATIONS OF SOME REDOX COMPONENTS OF JERUSALEM ARTICHOKE MITOCHONDRIA

Ubiquinone content was determined by HPLC. Cytochrome aa_3 was measured spectrophotometrically from oxidised and dithionite reduced spectra. The amount of cytochrome bc_1 complex present was measured from the endpoint of titrations of NADH and succinate oxidase activity with antimycin.
Units are nmol/mg protein.

Ubiquinone-9	2.71 ± 0.36 (3)
Cytochrome aa_3	0.101
bc_1 complex	0.088 ± 0.020 (5)
Ubiquinone-9/bc_1	31
Ubiquinone-9/aa_3	27

These results tended to confirm that interaction was at the level of the Q-pool. One suggestion to explain non-ideal behaviour of the Q-pool has been that Q diffusion is sufficiently slow that the redox state of Q is not homogenous throughout the membrane[15]. The rate of Q turnover is an important factor[4], and will depend inversely on the amount of redox active Q (Qa) present. The major ubiquinone present in Jerusalem artichoke mitochondria was UQ-9 (Table 1). Traces of UQ-10 were also detected, and small amounts of UQ-8 have been reported previously[3]. When related to the amount of bc_1 complex present a ratio of UQ-9/bc_1 of 31 was obtained. The corresponding UQ-9/aa_3 ratio was 27. Similar ratios were obtained using SMP, indicating that the UQ-9 was located in the inner membrane. A ratio of UQ-9/aa_3 of 40 was reported by Schindler et al.[3] in Jerusalem artichoke mitochondria, and estimates combining independent determinations of quinone and cytochrome aa_3 included ratios of 36 for yeast and 7 for _Arum maculatum_[16], although a recent direct determination using Arum gave a higher ratio of 30[3].

As the interaction between succinate and NADH was non-ideal it was of interest to see how closely the oxidation of each substrate separately followed ideal behaviour. Uncoupled mitochondria oxidising NADH (Fig.3a) or succinate (Fig.3b) were titrated with myxothiazol. The degree of saturation with myxothiazol was calculated using a dissociation constant of 2nM, derived from the inhibition by myxothiazol of duroquinol oxidation (not shown), and the bc_1 content (Table 1). When the data were plotted using a linear form of equation (1)[17] it was found that NADH oxidation apparently behaved ideally, corresponding to a ratio of Vo/Vr of 4.4. In contrast succinate oxidation showed significant nonlinearity (non-ideal behaviour). This can be seen in Fig.3b particularly at low inhibitor saturation where inhibition is less than expected, and its possible significance is discussed below.

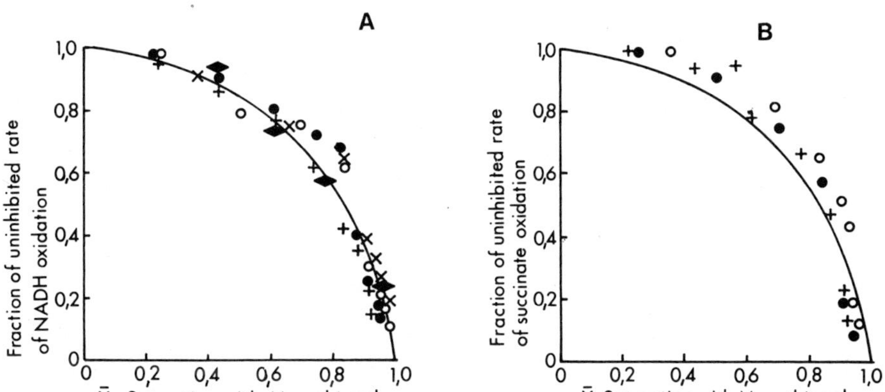

Figure 3 INHIBITION OF EXOGENOUS NADH AND SUCCINATE OXIDATION BY MYXOTHIAZOL.

Mitochondria were incubated with myxothiazol in standard assay medium for 10 min at 25°C. Substrate was added followed 1 min later by 0.2µM FCCP. The rate of oxygen uptake was then recorded. The degree of saturation ($\bar{Y}$) with myxothiazol was calculated as described in the text. (A) Combined results of 5 preparations (0.21-0.51 mg protein/ml) oxidising 1mM NADH. The uninhibited rate was 167 + 40 nmol/min/mg protein, (B) Combined resultsof 3 preparations (0.35-0.51 mg protein/ml) oxidising 20mM succinate. The uninhibited rate was 173 +41 nmol/min/mg protein.

DISCUSSION

One example of apparent non-ideal Q-pool beahviour was chosen for the present study but others may easily be found among the literature of plant mitochondria[2,7-11]. There have been two main kinds of suggestion for this behaviour: either retaining the original assumption of Q-pool homogeneity [5], or proposing heterogeneity within the membrane[4,15]. Implicit in most is a switching of rate limitation from the interaction between Q and Q-redox enzymes to other processes.

The results of the present study are matched to some of these suggestions. The results presented may be summarized as follows. Succinate is better able to compete for access to the respiratory chain than exogenous NADH, despite NADH oxidation being the faster (Fig.1). The degree of inhibition of NADH oxidation by succinate seemed to depend only on the activity of the cytochrome pathway (Fig.2). There seemed no evidence for direct interaction between succinate dehydrogenase and NADH dehydrogenase activities when tested using Q-1 as redox acceptor, nor for additional activation of succinate dehydrogenase in the presence of NADH (Table 1). Interaction appeared to be between sites of Q-1 reduction and the site of myxothiazol inhibition (Q_o of bc_1 complex[18]), that is, at the Q-pool. The asymmetric interaction of these substrates was accompanied by small but consistent differences in myxothiazol inhibition profiles for the two substrates (Fig.3).

(i) Explanations involving heterogeneity

Heterogeneity of the Q-pool has been proposed to explain how mixtures of substrates are oxidised by mammalian SMP[15] and the variable access different substrates have to the alternative pathway of plant mitochondria[2]. Heterogeneity could arise either if Q diffusion were too slow to effectively equalise the redox poise of Q throughout the membrane, or by the more recent suggestion of patching of respiratory components in the membrane, such that electrons from one donor may be channelled to a particular nearby acceptor[4].

The concentration of redox-active Q in the membrane, the diffusability of Q and average distance between sites of Q reduction and oxidation will together determine the extent to which Q diffusion limits rate. The distance Q may travel during steady-state turnover in mammalian SMP has been modelled by Rich[4]. Using a Q:bc_1 ratio of 5, a diffusion coefficient of for Q of $10^{-8}cm^2s^{-1}$, a rate of 200 $e^-s^{-1}.bc_1^{-1}$ complex and a random walk model of Q diffusion, it was calculated that Qr could travel approximately 280nm and potentially interact with 80 bc_1 complex dimers before being oxidised. Applying the same assumptions, Qr in uncoupled Jerusalem artichoke mitochondria would be expected to travel 600nm and potentially interact with 800 bc_1 dimers, comparable to the state 4 situation modelled for mammalian SMP. This difference is mainly due to the larger Q-pool size of Jerusalem artichoke mitochondria[3] (Table 1). These calculations are necessarily approximate, particularly due to continuing uncertainty about the true value for the Q diffusion coefficient. Gupte et al.[19], using the diffusion coefficient of a fluorescent Q-analogue in rat liver mitochondria, calculated that on average 45 collisions between Qr and bc_1 complexes would occur for each turnover. It was concluded that electron transfer was diffusion coupled rather than diffusion limited. Clearly the large Q-pool of Jerusalem artichoke mitochondria (Table 1), compared to mammalian mitochondria decreases the likelihood that Q diffusion is limiting in this case, provided most is redox active[5].

We have observed a similar pattern of inhibition of NADH oxidation by succinate using >90% right-side out SMP[20] as in intact mitochondria (not shown). These SMP lack cristae, consequently if patching of components were responsible for the present observations then the forces driving patching would be intramembrane in origin, in contrast to the stacking of thylakoid membranes, where membrane appression accompanies lateral separation of proteins[21]. In this respect reports of the affinity of succinate dehydrogenase for the bc_1 complex in lipid depleted membranes may be of interest[22] and needs further study.

(ii) Explanations retaining homogeneity

In the original model of Kroger and Klingenberg[5,6], the relationships between Q-pool reduction and the rates of reduction and oxidation of Q were linear. However, all of the results presented above are consistent with a nonlinear relationship between rates and Q-pool reduction. For reasons given below a hyperbolic relationship is used to demonstrate this. Two hypothetical rate constants are required to make the relationship between dehydrogenase activity and Qo/Qa hyperbolic as shown in Fig.4a. The rate constants could, for example, refer to the rate of reduction of dehydrogenase by substrate, and the rate of dehydrogenase reoxidation by Qo, as in the model of Moore and Rich[24]. For simplicity, oxidation of Qr was left as a first order process. The rate of electron transfer is now given by:

$$v = \frac{E.k_1 k_2 (Qo/Qa)}{k_2 + k_1 (Qo/Qa)} = Vo.Qr/Qa \qquad (2)$$

where k_1/k_2 is an index of hyperbolicity and E is the dehydrogenase concentration, assumed to be 0.03 nmol/mg protein by analogy with the ratio of complex I: bc_1 complex in mammalian mitochondria[22].

Equation (2) was used to simulate observed myxothiazol inhibition profiles shown in Fig.3, using a Vo/Vr ratio of 4.4 (Fig.3) and uninhibited rates of oxidation of 195 nmol O_2/min/mg protein. Best fits were found corresponding to k_1/k_2 ratios of 0.01 for NADH oxidation (near ideal behavior) and 5 for succinate oxidation (significantly hyperbolic behaviour). These inhibition profiles are shown in Fig 4b, together with simulations of the interaction between the substrates under these conditions. It can be seen that the pattern of interaction would be very close to that found experimentally, although the uninhibited rates of oxidation of succinate and NADH that were used in the simulation were identical. NADH oxidation would be much more inhibited by succinate oxidation than visa versa. Furthermore, this asymmetry becomes more obvious as Vo is decreased, as found experimentally (Fig.2). A nonlinear relationship between rate of Qo reduction and Qo/Qa would therefore able to explain the experimentally observed characteristics of succinate and NADH oxidation, both separately and together. How might nonlinearity occur?

Nonlinearity, either in the concept of a Km for Qo [23] or by the addition of rate constants not involving Qo in dehydrogenase turnover[24], has already been proposed to explain the substrate dependence of the activity of the alternative pathway in Arum maculatum mitochondria, and the preferential oxidation of glycine, respectively. Nonlinearity may arise in at least two ways, which are experimentally distinguishable according to whether or not nonlinearity is apparent when the dehydrogenase is studied in isolation from the rest of the respiratory chain.

Ragan and Cottingham[25] considered the process of Qo reduction by a dehydrogenase in detail to find how ideal Q-pool behaviour could occur even if the limiting step was not the encounter of a dehydrogenase with Qo, as originally proposed[5]. Assuming equal affinities of Qo and Qr for both oxidised and reduced dehydrogenases, ideal behaviour could result despite Qr and Qo being in rapid equilibrium with Q-binding sites (cases I and II)[25].

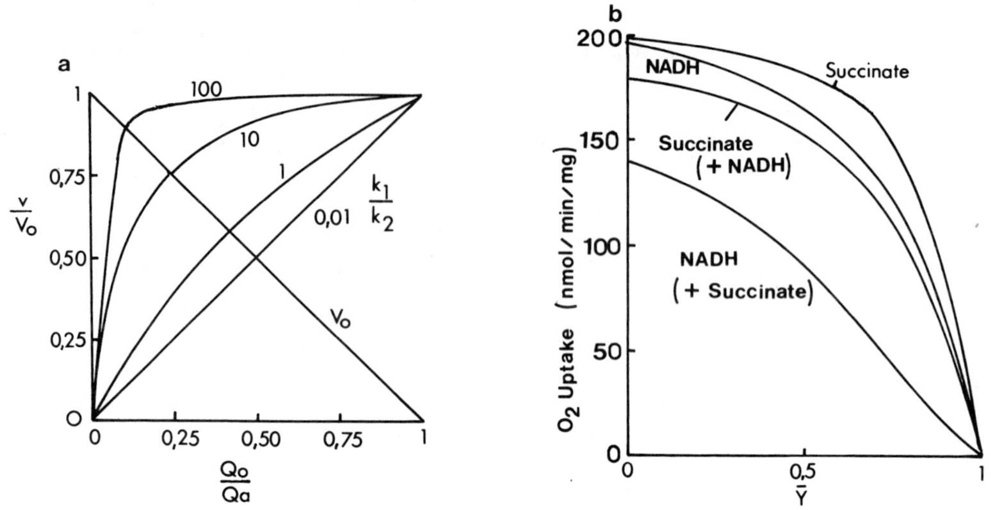

Figure 4a NONLINEAR DEPENDENCE OF DEHYDROGENASE ACTIVITY ON Q-POOL REDUCTION

Lines were calculated using equation (2) for several values of k_1/k_2. They represent the dependence of Qo reduction rate upon Qo/Qa. According to the ideal model[5,6] this would be linear. Ideal ubiquinol oxidase activity is represented by the line labelled Vo.

Figure 4b CONSEQUENCES OF NONLINEAR DEPENDENCE OF DEHYDROGENASE ACTIVITY ON Q-POOL REDUCTION

Rates of succinate and NADH oxidation were calculated using equation (2) at various values of k_1/k_2 to best fit the myxothiazol inhibition profiles shown in Figure 3. These values were then used to simulate the expected rates of oxidation in the presence of the other substrate. The parameters used are given in the text.

Equilibration of the Q-pool is more likely when the pool size is large, as in the present case. The necessary assumptions for ideal behaviour were (i) that the rate constant for dissociation of Qo from dehydrogenase be much greater than that for reduction of bound Qo, and (ii) unless the dehydrogenase was always maintained reduced (saturating substrate) then bound Q must be in equilibrium with a redox centre on the dehydrogenase of identical redox potential to bound Q. Relaxation of these constraints would tend to give a hyperbolic dependence of rate on Qo, and therefore the behaviour shown in Fig.4.

The second way nonlinearity might be generated would not necessarily be detected in studies of isolated dehydrogenase. In general, introduction of any rate control prior to the dehydrogenase might be expected to give rise to a nonlinear relation between rate and Q-pool reduction. Since succinate is oxidized in the matrix, concentrations of succinate, product and effector levels inside the mitochondria are uncertain, and the possible influence of this needs to be considered.

That non-ideal behavior would result from rate limitations prior to the dehydrogenase, despite the dehydrogenase behaving ideally when studied in isolation, is most easily demonstrated using the general theory of metabolic control developed by Kacser and Burns[26]. Ideal Q-pool behavior is easily translated into explicit solutions for control strength (C). Each control strength represents the dependence of electron transfer rate on the level of each enzyme involved. Using equation (1), it may be shown that:

$$Co = \frac{d \ln v}{d \ln Vo} = \frac{Vr}{Vo + Vr} = \frac{Qr}{Qa}$$

where Co represents the control strength for Vo. Similarly the control strength for Vr (represented by Cr) is equal to Qo/Qa. These coefficients represent the linear dependence of rate on Q-pool reduction at any given values of Vr and Vo, and can be visualised as the negative gradient at any point on Fig.3a. The sum of control coefficients for a pathway is equal to one: this is known as the Summation Property[26]:

$$Co + Cr = \frac{Qo + Qr}{Qa} = 1$$

Introduction of rate limitation in addition to Co and Cr, on the substrate side of the dehydrogenase (called Cs) leads to a new Summation condition:

$$Cs + Cr + Co = 1$$

in which the values of Cr and Co, which are determined by the whole system, must both decrease as Cs increases[26]. Consequently for $Cs > 0$:

$$Co < Qr/Qa \quad \text{and} \quad Cr < Qo/Qa$$

So introduction of any rate limitation in substrate supply would be expected to generate nonlinear dependence of rate on Q-pool reduction, as shown in Fig.4, even though the underlying linear dependence of dehydrogenase activity on Qo/Qa (the substrate "elasticity" [26], equal to 1 for ideal behaviour), need not necessarily have changed. One interesting consequence of this would be that the ability of different substrates to compete for access to the respiratory chain _in vivo_ could depend not only on the properties of the respiratory chain itself, but also on the extent of rate control present in the substrate supplying pathway.

In conclusion the characteristics of inhibition of exogenous NADH oxidation by succinate, and the factors determining this have been studied. It is suggested that at present the most likely explanations for this behaviour may be either a nonlinear dependence of rate on Q-pool reduction which would be most pronounced for succinate, or heterogeneity in the distribution of respiratory complexes such that the pathway from succinate to the bc_1 complex is favoured. It is shown how nonlinearity may arise, and what the consequences of this would be. We are currently investigating the relationship between respiratory rate and Q-pool reduction.

ACKNOWLEDGEMENTS

This work was supported by grants from the Science and Engineering Research Council and the Central Research Fund, University of London.

REFERENCES

1. Brunton, C. J. & Palmer, J. M. (1973) Eur. J. Biochem. 39:283-291

2. Lance, C., Chauveau, M. & Dizengremel, P. (1985) Encyclopaedia of Plant Physiology (R.Douce & D.A.Day, eds.), Springer-Verlag, Berlin. pp.202-247
3. Schindler, S., Lichtenthaler, H.K., Dizengremel, P., Rustin, P. & Lance, C. (1984) Structure, Function and Metabolism of Plant Lipids (P.-A.Siegenthaler & W.Eichenberger, eds.), Elsevier, Amsterdam. pp.267-272
4. Rich, P. R. (1984) Biochim. Biophys. Acta 768:53-79
5. Kroger A. & Klingenberg, M. (1973) Eur. J. Biochem. 34:358-368
6. Kroger, A. & Klingenberg, M. (1973) Eur. J. Biochem. 39:313-323
7. Bahr, J. T. & Bonner, W. D. (1973) J. Biol. Chem. 248:3441-3445
8. Gardestrom, P., Bergman, A. & Ericson, I. (1980) Plant Physiol. 65:389-391
9. Dry, I. B., Day, D. A. & Wiskich, J.T. (1983) FEBS Lett. 158:154-158
10. Day, D. A. & Wiskich, J. T. (1977) Phytochem. 16:1499-1502
11. Cowley, R. C. & Palmer, J.M. (1980) J. Exp. Bot. 31:199-207
12. Palmer, J. M. & Kirk, B. I. (1974) Biochem. J. 140:79-86
13. Lance, C. & Bonner, W. D. (1968) Plant Physiol. 43:756-766
14. Oestreicher, G., Hogue, P. & Singer, T. P. (1973) Plant Physiol. 52:622-626
15. Gutman, M. (1980) Biochim. Biophys. Acta 594:53-84
16. Hauska, G. & Hurt, E. (1982) Functions of Quinones in Energy Conserving Systems (B.L.Trumpower, ed.) Academic Press, New York. pp.87-109
17. Zhu, Q. S., Berden, J. A., De Vries, S. & Slater, E. C. (1982) Biochim. Biophys. Acta 680:69-79
18. Meinhardt, S. W. & Crofts, A. R. (1982) FEBS Lett. 149:217-222
19. Gupte. S., Wu, E., Hoechli, L., Jacobson, K., Sowers, A. E. & Hackenbrock, C. R. (1984) Proc. Natl. Acad. Sci. USA 81:2606-2610
20. Kay, C. J., Ericson, I., Gardestrom, P., Palmer, J. M. & Moller, I. M. (1985) FEBS Lett. 193:169-174
21. Williams, R. J. P. (1985) Trends in Biochemical Science 10:341
22. Capaldi, R. A. (1982) Biochim. Biophys. Acta 694:291-306
23. Cottingham, I. R. & Moore, A. L. (1983) Biochim. Biophys. Acta 724:191-200
24. Moore, A. L. & Rich, P. R. (1985) Encyclopaedia of Plant Physiology (R.Douce & D.A.Day, eds.) Springer-Verlag, Berlin. pp.134-172
25. Ragan, C. I. & Cottingham, I. R. (1985) Biochim. Biophys. Acta 811:13-31
26. Kacser, H. & Burns, J. A. (1973) SEB Symposia 27:65-104

THE NATURE OF THE TERMINAL OXIDATION STEP OF THE ALTERNATIVE ELECTRON TRANSPORT PATHWAY

Pierre Rustin

Laboratoire de Biologie Végétale IV (CNRS, UA 1180)
Université Pierre et Marie Curie, Paris, France

INTRODUCTION

The electron transfer system of the inner membrane of plant mitochondria differs from its animal counterpart by its pattern of sensitivity towards electron transfer inhibitors. Inhibitors of complex I of the respiratory chain, such as rotenone and piericidin A, inhibit the oxidation of matrix NADH only partially, whereas inhibitors of complex IV (cytochrome oxidase), such as cyanide, azide or carbon monoxide, or of complex III, such as antimycin A, do not fully inhibit substrate oxidation in isolated mitochondria from many plant species[1].

Resistance to complex I inhibitors is generally ascribed to the existence of an additional NADH dehydrogenase channelling electrons directly from matrix NADH to complex III and therefore by-passing the inhibitor-sensitive site. Alternatively, it can be suggested that the electron transfer pathway through the many sub-units of complex I can be affected by the presence of the inhibitor, and that the electrons reach complex III directly, by-passing again the inhibitor-sensitive site. No evidence is presently available to discriminate between these two hypotheses or to show that such a by-pass of the inhibitor site does indeed occur in the absence of an inhibitor[1].

Resistance to antimycin and cyanide, the most commonly used inhibitors of the cytochrome pathway, is ascribed to the existence of a cyanide-resistant electron transport pathway (the alternative pathway). Since its identification in isolated plant mitochondria, nearly all the components of the respiratory chain have been successively involved in this pathway[2]. Nevertheless, despite the fact that most of these hypotheses have been ruled out, the nature of this pathway is still mysterious.

However, it is now widely admitted that this pathway is branched on the cytochrome chain at the level of ubiquinone. Using CO-poisoned skung cabbage mitochondria, Storey[3] has convincingly demonstrated that a rapid reoxydation of ubiquinol was still possible and could be strongly inhibited by *m*-chlorobenzhydroxamic acid, a hydroxamic acid derivative known to specifically inhibit the functioning of the alternative pathway. It was also shown that a partial extraction by pentane treatment of ubiquinone of *Arum maculatum* spadix mitochondria led to the rapid loss of the alternative pathway activity[4]. Subsequently research was mainly devoted to the

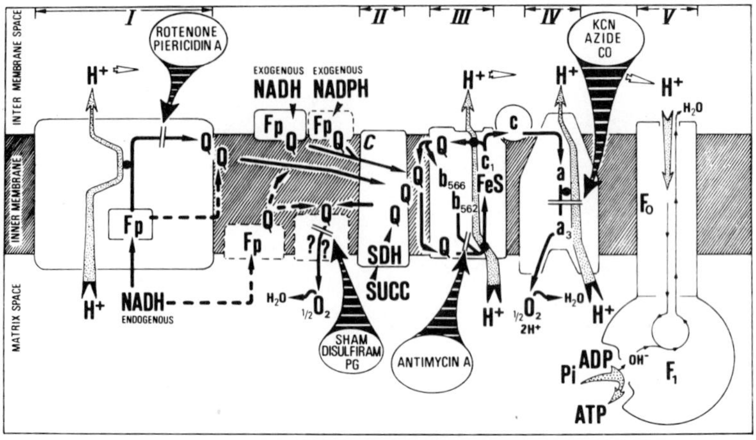

Fig. 1. Organization of the respiratory chain of the inner membrane of plant mitochondria.
Dotted lines correspond to hypothetical components or electron pathways. I, II, III, IV, V, the complexes of the respiratory chain and ATP synthetase; Fp, flavoproteins (various NADH or NADPH dehydrogenases); Q, various micro-domains of ubiquinone; $\underline{a}$, $\underline{a}_3$, $\underline{b}_{552}$, $\underline{b}_{560}$, $\underline{c}$, $\underline{c}_1$, cytochromes; FeS, iron-sulfur protein (Rieske protein in complex III); SDH, succinate dehydrogenase; F_o, F_1, respectively hydrophobic and hydrophilic regions of the ATP synthetase.

mechanisms allowing a cyanide-insensitive reoxidation of ubiquinol. Figure 1, featuring the inner membrane of mitochondria, summarizes the present knowledge on the organization of the respiratory chain in plants.

The aim of this paper is to review some of the arguments used to support or to dismiss the current hypotheses on the nature of the terminal oxidation step of the cyanide-resistant electron transport pathway. As said above, many hypotheses have been proposed in the past, but this paper will mainly focus on the hypotheses that have recently received new support.

1 - QUINOL AUTO-OXIDATION

The first of these hypotheses ascribes to quinol auto-oxidation the electron flux to oxygen which is observed in the presence of cyanide. The basis for such a proposal is that ubiquinone is the only electron carrier that is widely admitted to be involved in the alternative pathway and that its auto-oxidation is sensitive to salicylhydroxamic acid (SHAM) and activated by adenylates as is the alternative pathway in some micro-organisms[5]. However, major arguments can be raised against this hypothesis.

As shown in Figure 2 (trace $\underline{a}$), duroquinol (DQH_2) auto-oxidation readily takes place at pH 7.5 in a medium containing magnesium ions, phosphate buffer and bovine serum albumin (BSA). This auto-oxidation is partially sensitive to SHAM (37 % inhibition). A subsequent addition of

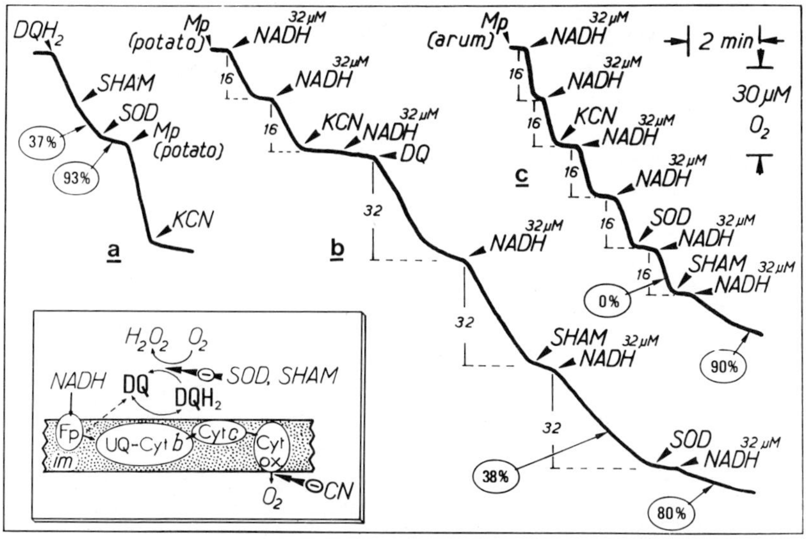

Fig. 2. Discrimination between substrate oxidation through the cyanide-resistant pathway or through quinol auto-oxidation.
a, oxidation of 500 µM DQH_2 through auto-oxidation or by purified potato tuber mitochondria (0.5 mg ml^{-1}). b, oxidation of limiting amounts of NADH through the cytochrome pathway or through duroquinol auto-oxidation in the presence of 50 µM duroquinone. The insert describes these two electron transport pathways (cf. Fig. 1 for abbreviations). c, oxidation of limiting amounts of NADH by purified mitochondria from Arum maculatum spadices in the absence or the presence of cyanide. All these experiments were carried out in a medium consisting of 0.3 M mannitol, 5 mM $MgCl_2$, 10 mM KCl, 10 mM phosphate buffer (pH 7.5) and 1 mg ml^{-1} BSA. Final concentrations were 1 mM SHAM, 500 U ml^{-1} SOD and 1 mM KCN. Numbers along dotted lines correspond to nmol O_2 consumed by the oxidation of the limiting amounts of NADH. Inhibitions are expressed as % of the rate measured before inhibitor additions.

superoxide dismutase (SOD) fully inhibits DQH_2 auto-oxidation. Adding mitochondria causes a new oxygen uptake. Duroquinol is then used as a respiratory substrate whose oxidation is fully cyanide-sensitive.

Figure 2 (trace b) shows the oxidation of limiting amounts of NADH through the cytochrome chain or through the auto-oxidation of duroquinol and allows the comparison of the stoichiometries between NADH oxidized and O_2 consumed according to the system involved in NADH oxidation. On the insert, it is shown that (added) duroquinone is able to divert electrons from the respiratory chain before the antimycin-sensitive site, and that, after its reduction, can be auto-oxidized, leading to H_2O_2 production instead of water. Firstly, it is shown that the stoichiometry of NADH oxidation through the cytochrome pathway is exactly 2 nmol NADH oxidized per nmol oxygen consumed. After cyanide addition, NADH oxidation is no longer oxidized by the mitochondria. However, adding duroquinone allows the electrons to be diverted from the respiratory chain to the added duroquinone, whose reduced form (DQH_2) is oxidizable in this medium. The stoichiometry is now 1 nmol NADH oxidized per nmol oxygen consumed. This oxidation is partially SHAM-sensitive and is fully inhibited by SOD as in

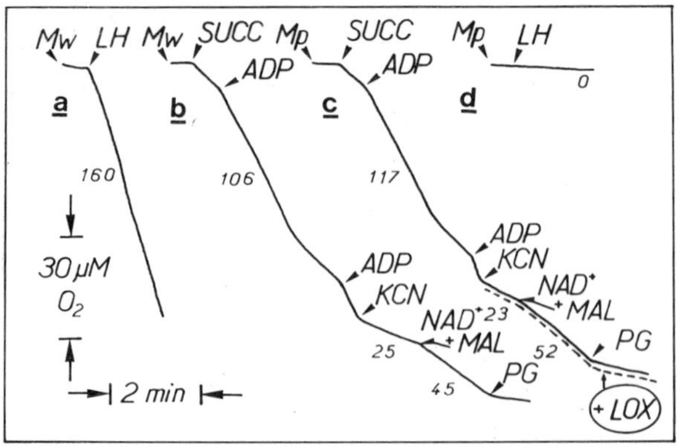

Fig. 3. Cyanide-resistant substrate oxidation in lipoxygenase-contaminated or lipoxygenase-free mitochondria from mung bean.
a, oxidation of linoleic acid by washed mung bean mitochondria. b, succinate oxidation by washed mung bean mitochondria. Effect of cyanide and of a subsequent addition of NAD^+ and malate. c, as in a, but using Percoll-purified mung bean mitochondria. Dotted line, similar experiment carried out in the presence of 1000 U ml^{-1} of added purified lipoxygenase (Type I, Sigma). d, oxidation of linoleic acid by Percoll-purified mung bean mitochondria. Experimental medium as in Fig. 2, but devoid of BSA in a and d. Numbers along the traces are nmol O_2 min^{-1} mg^{-1} protein.

the trace a of Figure 2. Finally, Figure 2 (trace c) shows the oxidation of NADH by the highly-cyanide resistant mitochondria from *Arum maculatum* spadices. On the one hand, in the absence or in the presence of cyanide, the stoichiometry of NADH oxidation is strictly 2 nmol NADH oxidized per nmol oxygen consumed. On the other hand, cyanide-resistant NADH oxidation is not inhibited by SOD, whereas it was fully inhibited by SHAM.

Therefore it clearly appears that a simple quinol auto-oxidation does not account for the stoichiometry between substrate and oxygen. Moreover it does not present a similar response to inhibitors since it is inhibited by SOD whereas cyanide-resistant oxidations are not.

2 - LIPOXYGENASE

A second hypothesis recently suggested is that contaminating lipoxygenase is responsible for cyanide-resistant substrate oxidation[6]. The major basis for such a proposal is a strange similarity of sensitivity towards inhibitors, since the lipoxygenase reaction is cyanide-insensitive, SHAM- and propylgallate-sensitive as is the cyanide-resistant electron transport pathway. In addition, experimental data obtained with wheat mitochondria suggest that mitochondrial fractions rich in lipoxygenase catalyze cyanide-resistant substrate oxidations, whereas after elimination of the contaminating lipoxygenase substrate oxidations would become fully cyanide-sensitive[7].

However, as shown on Figure 3, such data could not be confirmed using mung bean mitochondria. In this experiment, the rates of the cyanide-resistant oxidations of succinate and malate were compared in washed and purified mung bean mitochondria. As shown on trace a, washed mitochondria

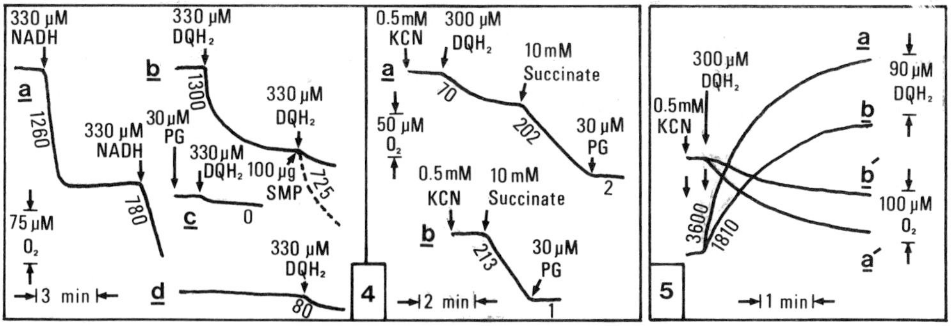

Fig. 4. Discrimination between duroquinol oxidase activity and the terminal oxidation step of the cyanide-resistant pathway using sub-mitochondrial particles from Arum maculatum.
A, oxidation of limiting amounts of NADH (trace a) and of DQH_2 (trace b), inhibition by propylgallate of DQH_2 oxidation (trace c) and effect of 8 min incubation at 25°C on DQH_2 oxidation (trace d). B, effect of 20 min incubation (25°C) of SMP on the successive oxidations of DQH_2 and succinate (trace a) and on the oxidation of succinate (trace b). Numbers along the traces are nmol O_2 consumed min^{-1} mg^{-1} protein. (Adapted from Rustin et al.[9]).

Fig. 5. Effect of antimycin on the cyanide-resistant DQH_2 oxidation by solubilized DQH_2 oxidase from Arum maculatum.
Spectrophotometric (a,b) and polarographic (a',b') recordings of DQH_2 oxidation in the presence of 1 mM KCN. Antimycin (1 μM) was present in a',b'. DQH_2 oxidase activity was solubilized according to Kay and Palmer . Numbers along the traces are nmol DQH_2 oxidized min^{-1} mg^{-1} protein of detergent-treated mitochondria. (Adapted from Rustin et al.[9]).

contained a high lipoxygenase activity which readily oxidized added linoleic acid (LH). Trace b shows that the cyanide-resistant oxidation of succinate by washed mitochondria amounts to 23 nmol min^{-1} mg^{-1} protein, whereas in the presence of malate the rate of cyanide-resistant oxidation can reach 50 nmol. In trace c, a similar recording was made using purified mitochondria. It appears that very similar rates of cyanide-resistant succinate and malate oxidations could be measured, despite the absence of contaminating lipoxygenase, since these mitochondria did not oxidize linoleic acid at a measurable rate, as shown in trace d. It is also shown on trace c that adding purified mung bean lipoxygenase to purified mitochondria did not increase the resistance to cyanide. It follows that there is no correlation between the amount of lipoxygenase present in the mitochondrial preparation and the degree of cyanide resistance. Moreover, purified mitochondria from Arum maculatum spadices do not contain detectable lipoxygenase activity whereas substrate oxidations by these mitochondria are extremely cyanide-resistant[8].

Therefore, one can conclude that lipoxygenase is a contaminating enzyme and that its presence is not the cause of the resistance to cyanide of substrate oxidation.

3 - QUINOL OXIDASE

A third hypothesis suggests the existence of a specific quinol oxidase working as a SHAM-sensitive terminal oxidase. This proposal is based on the isolation by different groups of a protein fraction catalyzing a cyanide-

Table I. Effect of antimycin on cyanide-resistant substrate oxidations by sub-mitochondrial particles of Arum maculatum spadices. KCN (0.5 mM) was present in all conditions. (%) : percent inhibition of the rate measured in the absence of antimycin. (Adapted from Rustin et al.[9])

Substrate	Oxygen uptake (nmol min^{-1} mg^{-1} protein)	
	Control	+ 1 µM antimycin
DQH_2 (500 µM)	1300	765 (51%)
NADH (1 mM)	1420	1432 (0%)
Succinate (10 mM)	570	568 (0%)

insensitive, SHAM- and propylgallate-sensitive quinol oxidation, using mitochondria isolated from the spadices of some Araceae (see Elthon and McIntosh, this Volume).

Using sub-mitochondrial particles (SMP) of Arum maculatum mitochondria, the fate during time of duroquinol oxidase was compared with the ability to catalyze a cyanide-insensitive NADH oxidation (Fig. 4 A). All these experiments were carried out in the presence of 1 mM cyanide. Trace a shows the oxidation of two limiting amounts of NADH, successively added to the SMP. It appears that the rate of oxidation linked to the second addition of NADH was 40% lower than the rate measured after the first addition. A similar experiment, always in the presence of cyanide, was carried out using duroquinol as a substrate. Figure 4 (trace b) shows that duroquinol oxidation tended to decrease rapidly and that this decrease was not overcome by another duroquinol addition. Only the addition of SMP allowed a stimulation of the oxidation rate. Trace c, a control trace, shows that cyanide-insensitive DQH_2 oxidation was totally inhibited by propylgallate. Finally, in trace d, it is shown that incubating SMP for 6 min in the air-saturated electrode medium at 25°C in the absence of DQH_2 led to the disappearance of the ability to oxidize duroquinol.

In Figure 4 B, this comparison was extented to cyanide-resistant succinate oxidation. Trace b shows that in the presence of cyanide succinate oxidation by SMP amounted to 213 nmol O_2 min^{-1} mg^{-1} protein. The next experiment (trace a) was carried out after incubation of SMP during 10 min. Then, the SMP no longer oxidized duroquinol, but they still oxidized succinate without significant change in the rate of oxidation.

Another interesting feature of the duroquinol oxidase activity is shown in Figure 5. Duroquinol oxidation catalyzed by the putative quinol oxidase solubilized from A. maculatum mitochondria was followed in the presence of cyanide through oxygen uptake (a',b') or through spectrophotometric measurement of duroquinone appearance (a,b). This experiment was carried out in the presence (a,a') or in the absence (b,b') of antimycin. It appears that this activity was 50% inhibited by antimycin, whatever the parameter studied: oxygen uptake or duroquinol oxidation. Quite similarly, Table I shows that duroquinol oxidation by SMP was 50% inhibited by antimycin, whereas NADH and succinate oxidations were not affected at all by the presence of the inhibitor. This holds true even for a substrate like NADH that is oxidized at the same rate as duroquinol.

This set of very simple experiments casts serious doubts on the analogy between the quinol oxidase activity that can be solubilized from

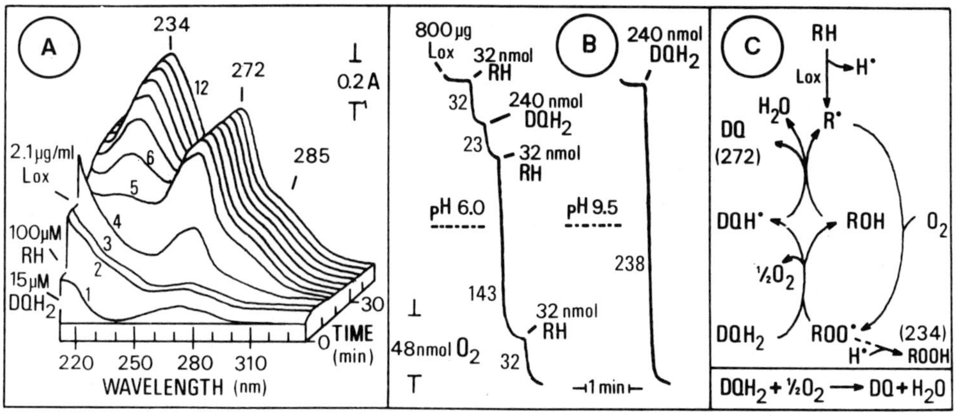

Fig. 6. Oxidative interactions between fatty acid radicals and duroquinol. A, UV spectra showing the rapid oxidation of duroquinol taking place during linoleic acid (RH) peroxidation in 7 mM MES buffer at pH 6.0. B, oxidation of linoleic acid and duroquinol at pH 6.0 catalyzed by lipoxygenase and auto-oxidation of duroquinol at pH 9.5 in 10 mM phosphate buffer. Numbers along the traces refer to nmol O_2 consumed from each substrate addition. C, a model of the cyanide-resistant terminal oxidation step of plant mitochondria (see text for abbreviations)

these mitochondria and the system responsible for the cyanide-resistant oxidations. Moreover, it is noteworthy that such a fraction can not be obtained from other cyanide resistant mitochondria, such as mung bean[10] or *Kalanchoe blossfeldiana* leaf mitochondria (unpublished).

4 - FREE RADICAL MECHANISM

The last hypothesis that will be considered here is the proposal that interactions between various free radicals could constitute a mechanism for the oxidation of quinols of the respiratory chain (Rustin *et al.*[11]). Three major lines of evidence support this proposal.

The first line is the consideration of a model system in which free radicals are generated during the oxidation at pH 6.5 of linoleic acid by lipoxygenase. Figure 6 shows the oxidation of duroquinol (DQH_2) occuring in the presence of linoleic acid (RH) and lipoxygenase (Lox). DQH_2 and RH were added first and the reaction started upon Lox addition. The first event was the rapid oxidation of duroquinol as shown by the increase of absorbance at 272 nm which was followed by the appearance of the hydroperoxides of linoleic acid absorbing at 234 nm. Simultaneously, an accumulation of keto derivatives absorbing at 285 nm could be noticed. These keto derivatives originate from free radical interactions occurring at the pH of 6.5 used in these experiments. It has been checked that duroquinol did not react with lipoxygenase alone, nor with linoleic acid alone or with the hydroperoxides alone (Rustin *et al.*[12]). Therefore, one can possibly conclude that DQH_2 interacts with a transitory product of the reaction. As the reaction takes only place with the type of lipoxygenases that are known to produce free radicals and only under conditions known to favor free radical formation, it can be concluded that duroquinol reacts with free radicals generated during the lipoxygenase catalyzed reaction.

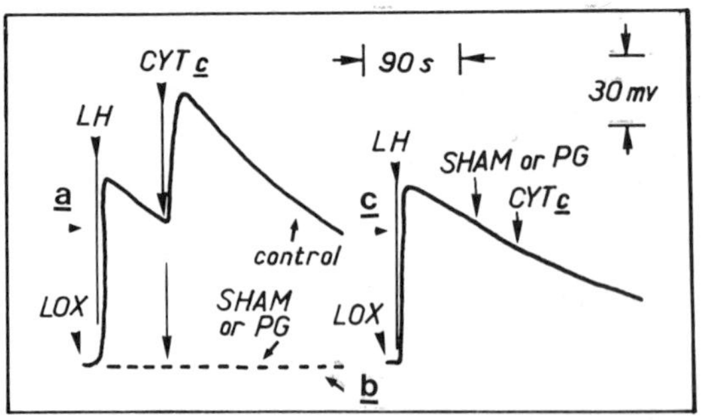

Fig. 7. Radical scavenging properties of SHAM and propylgallate. a, rise of luminescence associated with peroxidation of linoleic acid (LH) by lipoxygenase (LOX) and by the subsequent interaction of linoleic acid hydroperoxides with cyt. c; b, similar experiment as in a, but carried out in the presence of SHAM or PG; c, effect of SHAM or PG on the interaction between linoleic acid hydroperoxides and cyt. c. Measurement of luminescence was carried out in 5 mM phosphate buffer (pH 9.0) in the presence of 25 µM luminol. Final concentrations were 1000 U ml^{-1} mung bean LOX (Type I, Sigma), 20 µM linoleic acid, 1 µM cyt. c, 2 µM PG and 50 µM SHAM.

In a second experiment (Fig. 6 B), the stoichiometry of this reaction was compared with the stoichiometry of duroquinol oxidation induced by an increase of pH. On trace a, limiting amounts of linoleic acid were added to lipoxygenase in the absence and in the presence of duroquinol. The first addition in the absence of duroquinol allows one to measure the oxygen consumed by the oxidation of this limiting amount of linoleic acid. Then linoleic acid was added in the presence of duroquinol. After substraction of the oxygen uptake corresponding to the oxidation of linoleic acid, it is possible to estimate the oxygen consumed by the oxidation of duroquinol and to calculate a stoichiometry of 2 nmol of duroquinol oxidized per nmol O_2 consumed, similar to the stoichiometry measured for the oxidation of ubiquinol through the cyanide-resistant pathway of the mitochondria. In comparison trace b shows the auto-oxidation of the same amount of duroquinol and indicates a stoichiometry for duroquinol auto-oxidation of 1 nmol duroquinol oxidized per nmol O_2 consumed.

In Figure 6 C, a hypothetical scheme represents what could be the reaction between free radicals and duroquinol. The first step would be the production of radicals (R•) and peroxy radicals (ROO•) of linoleic acid. The mechanism of this production remains to be established and could involve traces of superoxide that are known to be produced among numerous radical species during this reaction. Then, peroxy radicals of linoleic acid would successively react with two duroquinol molecules leading to the oxidation of duroquinol to water, and to the regeneration of a free radical. Other schemes could be proposed, but whatever the exact mechanism, the interesting point to be noticed is that quinol oxidation in such a system displays a stoichiometry identical to the stoichiometry measured for the cyanide-resistant system of the mitochondria, in contrast with the auto-oxidation process.

A second line of evidence supporting this proposal is to be found in the properties of the inhibitors of cyanide-resistant substrate oxidations. Radical scavenging properties of SHAM or PG can be illustrated on two successive reactions, respectively producing free radicals through an enzymic and through a chemical system (Fig. 7). The first reaction is the peroxidation of linoleic acid by lipoxygenase, producing linoleic acid hydroperoxides . The second one is a reaction between these hydroperoxides and a hemoprotein , such as cytochrome *c*, that leads to the parallel oxidative destruction of the cytochrome and of the hydroperoxides . As shown in Figure 7 (trace *a*), these reactions lead to the formation of free radicals (producing light that can be measured with a luminometer). Upon addition of lipoxygenase to linoleic acid, a first rise in luminescence is observed, followed by a second rise consecutive to the addition of cytochrome *c*. In the presence of SHAM or propylgallate at non-inhibiting concentrations for the lipoxygenase reaction, no luminescence can be detected when lipoxygenase or cytochrome *c* is added to linoleic acid (trace *b*). Similarly, after the reaction of lipoxygenase on linoleic acid, the presence of the inhibitors suppresses the rise of luminescence caused by the addition of cytochrome *c* (trace *c*).

The situation is slightly more complicated with disulfiram, a third inhibitor of the alternative pathway. Disulfiram was reported not to inhibit the lipoxygenase reaction[13], alhough it is a free radical scavenger[14]. However, the solution came recently when it was shown that its effect depends on the type of lipoxygenase studied[15]. When using the so-called lipoxygenase 1, it was found that no effect could be noticed whereas lipoxygenase 2 was indeed inhibited. The difference between the reactions catalyzed by these two enzymes lies in the fact that lipoxygenase 1 produces free radicals as side products, whereas the reaction mechanism of lipoxygenase 2 requires the production of free radicals. This observation again supports the proposal that a major characteristic of the inhibitor of the alternative pathway could well be their ability to interact with radical species.

A third line of evidences that tends to support the involvement of radicals in cyanide resistance comes from the conditions known to favor the appearance of cyanide-resistant oxidations. These conditions were reviewed recently[16], therefore let's just mention a few of them : (*a*) the necessity of a high level of reduction of the initial part of the respiratory chain, also known to favor free radical production by flavins or quinols; (*b*) the higher degree of cyanide-resistant respiration in senescing tissues corresponding to a slowdown of cytochrome oxidase activity and to an increase in radical content of the cellular membranes; (*c*) the relationship of cyanide resistance with ethylene production which appears to be linked to free radical generation, at least under some conditions.

Thus it appears that a radical mechanism can account for all the known features of the cyanide-resistant substrate oxidations. However, major points remain to be investigated, including the nature of the radicals involved, particularly the potential role of flavin radicals, the mechanisms controlling the appearance of these radicals and the conditions favoring the interactions between these radicals and quinols.

CONCLUSION

As a conclusion one must emphasize the importance for plant mitochondriologists of finally reaching a conclusion on the nature of the mitochondrial altérnative electron transport pathway. This is probably one of the major points that remain to be solved in plant mitochondria, since the activity of such a system will control the efficiency of the ATP synthesis linked to substrate oxidation, and therefore will affect all the energetics of the plant cell. The occurrence of the alternative pathway is naturally mentioned among the first specificities of plant mitochondria, although similar reactions have been described in chloroplasts[17] or in rat liver mitochondria[18], and the concept of cyanide resistance is becoming increasingly used by physiologists. Obviously, a critical examination of the available data leads to the conclusion that more investigations are required before one can conclude on the true nature and physiological significance of these cyanide-resistant oxidations.

REFERENCES

1. R. Douce, Mitochondria in Higher Plants, Academic Press, New York (1985).
2. C. Lance, M. Chauveau and P. Dizengremel, The Cyanide-Resistant Pathway of Plant Mitochondria, in : Higher Plant Cell Respiration, R. Douce and D.A. Day, eds., Encyclopedia of Plant Physiology, new series, vol. 18, Springer Verlag, Berlin (1985).
3. B.T. Storey, The respiratory chain of plant mitochondria. XVIII. Point of interaction of the alternate oxidase with the respiratory chain, Plant Physiol. 58:521 (1976).
4. S. Huq and J.M. Palmer, The involvement and possible role of quinone in cyanide-resistant respiration. in : Plant Mitochondria, G. Ducet and C. Lance, eds., Elsevier/North Holland Biomedical Press, Amsterdam (1978).
5. J. Vanderleyden, E. Van den Eyden and H. Verachtert, Nature of the effect of adenosine 5'-mono-phosphate on the cyanide insensitive respiration in the mitochondria of Moniliella tomentosa, Biochem. J. 186:309 (1980).
6. D.J. Parrish and A.C. Leopold, Confounding of alternate respiration by lipoxygenase activity, Plant PHysiol. 62:470 (1978).
7. A.H. Goldstein, J.O. Anderson and R.G. McDaniel, Cyanide-insensitive and cyanide-sensitive O_2 uptake in wheat. I. Gradient purified mitochondria, Plant Physiol. 66:488 (1980).
8. J. Dupont, P. Rustin and C. Lance, Interaction between mitochondrial cytochromes and linoleic acid hydroperoxide. Possible confusion with lipoxygenase and alternative pathway, Plant Physiol. 69:1308 (1982).
9. P. Rustin, M.-F. Alin and C. Lance, Discrimination between duroquinol oxidase activity and the terminal oxidation step of the cyanide-resistant electron transport pathway of plant mitochondria, Biochem. Biophys. Res. Commun. 135:677 (1986).
10. W.D. Bonner and P.R. Rich, p-Quinol:oxygen oxidoreductase, a new copper oxidase in plant mitochondria, PLant Physiol. 72, 19 (Suppl.).
11. P. Rustin, J. Dupont and C. Lance, A role for fatty acid peroxy radicals in the cyanide-insensitive pathway of plant mitochondria, Trends Biochem. Sci. 8:155 (1983).
12. P. Rustin, J. Dupont and C. Lance, Oxidative interactions between fatty acid peroxy radicals and quinones: possible involvement in cyanide-resistant electron transport in plant mitochondria, Arch. Biochem. Biophys. 225:630 (1983).
13. S.D. Grover and G.G. Laties, Disulfiram inhibition of the alternative respiratory pathway in plant mitochondria, Plant Physiol. 68:393 (1981).
14. U. Kuhnlein, Disulfiram inhibits DNA breakage by hydroxyradical-producing agents. Biochim. Biophys. Acta 609:75 (1980).
15. E.C. Hausknecht and M.O. Funk, The differential effect of disulfiram on lipoxygenases from Glycine Max., Phytochem. 23:1535 (1984).
16. P. Rustin, J. Dupont and C. Lance, On the involvement of lipid peroxy radicals in the cyanide-resistant patway, Physiol. Vég. 22:643 (1984).
17. P. Bennoun, Evidence for a respiratory chain in the chloroplast, Proc. Nat. Acad. Sci. U.S.A. 79:236 (1982).
18. M. Rau and T. Ramasarma, Stimulation of cyanide-insensitive respiration in rat liver mitochondria, Ind. J. Biochem. Biophys. 16:379 (1979).

THE UBIQUINOL/CYTOCHROME C OXIDOREDUCTASE COMPLEX OF PLANT MITOCHONDRIA AND ITS INTERACTION WITH THE ALTERNATIVE CYANIDE RESISTANT OXIDASE: A MEMBRANE COMPLEMENTATION STUDY

Cosima Spinelli and Davide Zannoni

Institute of Botany
University of Bologna
42 Irnerio, I-40126 Bologna

INTRODUCTION

The ubiquinol/cytochrome *c* oxidoreductases (Complex III or bc1 complex) operate a central role in the electron transport and energy transduction of mitochondria, chloroplasts and bacteria. The UH_2/cytochrome *c* oxidoreductase from mitochondria and aerobic bacteria constitutes the middle segment of the respiratory chain; during its catalytic action protons are translocated across the membrane and an electrochemical proton gradient is therefore generated . Similarly, the UH_2/cytochrome *c* oxidoreductase from photosynthetic bacteria is involved in cyclic phosphorylation, and the plastoquinol/plastocyanine (*c*553) oxidoreductase from chloroplasts and cyanobacteria in both cyclic and linear photophosphorylation[1].

The data which are presently available on the molecular nature of Compplex III seem to indicate that the basic content of prosthetic groups of this enzyme is the same in the different organisms, i.e. one 2Fe-2S cluster (Rieske iron-sulphur center), one cytochrome *c* (*c*1 or *f*), and one cytochrome *b*, carrying two heme groups (*b*560 and *b*566) . In addition, the values of the mid-point potentials of the electron transferring carriers are similar in the various *b c*1 complexes, and the polypeptides containing prosthetic groups are also structurally closely related, as indicated by the extensive homology of the primary structure of cytochrome *b* from different organisms[2,3]. These experimental evidences tend to suggest that the molecular mechanism of electron transfer coupled to proton translocation is rather similar in the *b c*1 complexes from the different organisms.

Until recently[4], a remarkable exception to the general concept which predicts a significant conservativity in the molecular composition of the *b c*1 complexes, was offered by the redox chain of plant mitochondria. Indeed, no counterpart of the Rieske iron-sulphur cluster was identified by conventional EPR spectroscopy[5], along with several reports underlying the differences in the composition of the *b* type cytochromes[6,7]. This anomalous situation was also complicated by the fact that mitochondria of many higher plant tissues and fungi possess an alternative, CN^--resistant electron transport pathway[8], which makes the adaptation of the "Q-cycle" scheme, as it is postulated to occur in mammalian mitochondria[9], even more difficult.

Recently, the *bc1* complex from *Helianthus tuberosus* mitochondria has been partially purified and characterized[4]. This paper reviews the current status of the research on the ubiquinol/cytochrome *c* oxidoreductase from *Helianthus tuberosus* mitochondria, along with the most recent studies with specific inhibitors of this multiprotein complex. In addition, the results obtained through the use of the "in vitro" complementation approach to hybrid membranes obtained by fusion of sub-mitochondrial particles and photosynthetic vesicles, will be examined.

MATERIALS AND METHODS

The isolation of the *bc1* complex enriched fractions, S-1 and S-2, from *Helianthus tuberosus* mitochondria has previously been described[4].

SDS-polyacrylamide gel electrophoresis was performed by essentially following the method of Gellefors et al.[10]. The determination of the cytochrome *c1* band was done exploiting the intrinsic fluorescence of the haem in the gels before staining[11].

Arum italicum mitochondria were prepared by sucrose-gradient and sub-mitochondrial particles generated by French-pressure cell treatment in low salt medium[12].

The strain of *Rhodobacter capsulatus* used in this work was ZM6. ZM6 is arsenate resistant (ars^-), capable of photosynthetic and aerobic growth (pho^+, aer^+), and it lacks of the alternative oxidase (cyt.*b*260)[13]. The strain was grown in a malate minimal medium (RCV-medium) with low aeration (2% of oxygen tension) indicated by a VirTis Dissolved Oxygen controller, in the dark. *R.capsulatus*, strain Y142, is streptomycin resistant (str^-), incapable of photosynthetic growth (pho^-) because is deficient in reaction center bacteriochlorophyll (RC^-)[14]. Control strain for light-induced oxygen uptake.

Plasmamembrane vesicles were prepared from cells by disruption in a French pressure cell and differential centrifugation as previously described[13], and resuspended in 10mM $MgCl_2$, 100mM glycylglycine, pH7.5, at 8-10mg protein/ml.

Hybrid membranes were obtained by ultrasonic irradiation of mixtures of ZM6- and SMP-particles (1:1, protein:protein). Sonication was carried out on 1ml samples contained in 5ml cellulose nitrate tubes immersed in a ice bath. The probe of the sonifier (MSE 150Watt Ultrasonic Disintegrator) was immersed about 1cm in the sample, and the sonifier run at full out-put for two 15sec periods with 30sec cooling between. The temperature rose to 14-20°C during sonication. The membranes were enzymatically active for at least 2h.

Cytochrome potentiometric titrations were carried out by following the procedure of Dutton et al.[15], as previously described[4].

EPR spectra were obtained using a Bruker 200 D-EPR spectrometer (Bruker Analyt. Messtechnik, Forchleim, F.R.G.) equipped with a criostat and liquid helium transfer line.

Assay conditions for light-driven oxygen uptake were as previously described[13].

Proteins were assayed by the Lowry method[16].

RESULTS AND DISCUSSION

1. The heme-content of the ubiquinol/cytochrome c oxidoreductase from *Helianthus tuberosus* mitochondria

It has recently been shown that the isolation of mitochondrial *bc1* enriched fractions (named S-1 and S-2) from Helianthus tuberosus can successfully be carried out through the use of the bile salt deoxycholate (DOC) in the presence of KCl [4]. This method produced a complete separation of the bcl complex (Complex III) from the cytochrome c oxidase complex (Cyt.a-a_3 or Complex IV) with a relative enrichment based on ubiquinol/cytochrome c oxidoreductase activity of 7.4 times.

The partial purified complex contains two b type and one c type hemes with mid-point potentials at pH7.0 of +100mV (cyt.b100), -25mV (cyt.b-25) and +240mV (cyt.c240), respectively . The absorption difference spectra at controlled ambient potentials (Zannoni, unpublished) indicated that cyt.b100 has one α band at 562nm, whereas cyt.b-25 presents a shoulder at 558nm and a prominent peak at 566nm. This latter phenomenon, has previously been described in bcl complexes from mammalian mitochondria were the so called cyt.b566/558 corresponds to one of the two identical hemes in which a rhombic distortion of the heme structure plus the negative mid-point potential, seem to split its α band at 562nm into two bands at 558nm and 566nm [17].

The absorption difference spectrum of cyt.c240 (cyt.c1) indicates one major peak in the alfa region at 552nm .

2. The non-heme iron-sulphur content of the bcl complex

The Rieske iron-sulphur cluster is considered an ubiquitous component of all mitochondrial and bacterial bcl complexes, being essential for electron and proton transport[18]. It is therefore surprising that until the application of the ubiquinone analog 5-undecyl-6-hydroxy-4,7-dioxobenzothiazole (UHDBT) as a respiratory inhibitor, several independent and detailed investigations failed to reveil the presence of the Rieske iron-sulphur center in plant mitochondria [5]. Bonner jr and Prince[5], demonstrated that the EPR spectrum of the Rieske-center of mung bean mitochondria becomes clearly detectable by standard EPR analysis in the presence of UHDBT, due to the fact that this inhibitor sharpens the g_y signal of the iron-sulphur cluster. This observation has recently been integrated by a detailed study of the UHDBT effect on electron transport of higher plant mitochondria [19]. It has been concluded that UHDBT is a potent inhibitor of both the CN^--sensitive (Ki=0.5µM) and CN^--insensitive (Ki=0.07µM) pathways.

In mammalian mitochondria, the binding site of UHDBT seems to be at the iron-sulphur center level, as the titer of inhibition is proportional to the content of the Rieske center and the K_d is strongly dependent on the state of its reduction level [20,21]. Oxidation of the Rieske-center decreases the binding of UHDBT by at least two orders of magnitude, therefore binding of this inhibitor changes the observed mid-point potential of the iron-sulphur cluster which is raised by 70mV (from +280mV to +350mV) [18]. These data clearly suggest that the mid-point potential of the Rieske-center in plant mitochondria is markedly lower than the early reported value of +300mV, which originates from the redox titration of the UHDBT-induced g_y=1.89 signal [5]. For this reason, the actual Em at pH7.2 of the Rieske-center in plant mitochondria is probably close to +230mV, this conclusion being supported by the fact that cytochrome c (Em=+240mV) and Rieske iron-sulphur center are usually isopotential [1].

It is noteworthy, that evidence of the presence of the Rieske center in higher plant mitochondria has conclusively been obtained by conventional EPR analysis of the partial purified bcl complex [4]. Indeed, a signal characterized by a g_y =1.89 was seen at 15K with a redox behaviour and temperature dependence similar to those of the Rieske-centers observed in mammalian and bacterial bc1 complexes.

3. Oxidative reactions and effects of inhibitors on the isolated bc1 complex

The partially purified bc1 complex of *Helianthus tuberosus* catalizes the reduction of exogenous horse heart cytochrome *c* by different ubiquinols with a reaction which is fully sensitive to antimycin[4]. Ubiquinol-2 is the most effective electron donor to the bc1 complex with a turnover number of 156 nmoles cyt.*c* per mol cyt.*c*1 per sec and a Km of 2µM. The antimycin-inhibitory pattern gives a Ki=12nM and a maximal binding site content corresponding to 0.4nmol mg protein. According to Rieske[22], this latter value can be used as a reliable indication of the bc1 content.

Another specific inhibitor of the bc1 complex is the antibiotic myxothiazol. Its binding site is close to the cytochrome *b* 558/566 heme environment inducing a 2nm red-shift of the α-absorbance band of this cytochrome[23]. In mammalian bc1 complex shows a K_d of approx. 10^{-10}M and its binding takes place in a linear concentration dependence with a titer of one molecule of myxothiazol per heme. Unexpectedly, the bc1 complex from *Helianthus tuberosus* mitochondria is slightly less sensitive to myxothiazol in comparison to antimycin, with a Ki=31nM[4].

Several quinone analogs have also been reported to inhibit electron flow by the bc1 complex[18]. However, since the K_d values of these compounds are relatively high, they are scarsely useful in elucidating the electron transfer mechanism. Indeed, the ubiquinol/cytochrome *c* reduction by the plant mitochondrial bc1 complex is completely blocked by 2-n-hepthyl-1-hydroxychinoline N-oxide (HQNO) and 2,5-dibromo-3-methyl-6-isopropyl benzoquinone (DBMIB), but with Ki's of 0.1µM and 0.3µM, respectively.

4. The oxidant-induced reduction of cytochromes b

The oxidant-induced reduction of cytochromes *b* in antimycin-inhibited cytochrome *c* reductase is a typical phenomenon of mitochondrial and bacterial bc1 complexes[22,24]. Early electron transfer models to explain this phenomenon, involved a mid-point potential change of the *b* cytochromes modulated by the redox state of either an uncharacterized redox component[25] designated X, or by cytochrome *c*1[26]. Wikstrom and Berden[27] instead proposed a mechanism in which QH_2 transfers one electron to cyt.*b*566 simultaneously with transfer of the second electron to oxygen. Antimycin was postulated to inhibit the reoxidation of cytochromes *b*. This concerted reaction across the bc1 complex was foundation cornerstone for the ideation of the "Q-cycle" hypothesis . In mammalian mitochondria, the general features of the "Q-cycle" scheme can be summarized as follows: (a) Ubiquinol (QH_2) from the ubiquinone pool (Q) is oxidized to ubiquinone (at a reaction site defined as Qo center) in two steps, one electron being transferred to the Rieske-center and then via cytochrome *c*1 to cytochrome *c*, while the second electron is transferred to the low potential heme, *b*558/566 and from there to the high potential heme, *b*562. The oxidation of ubiquinol is linked to a release of two protons to the cytoplasmic side (C-side); (b) At a reaction site defined as Qi center, ubiquinone is re-reduced in a two-step mechanism by the heme *b*562 center,so that half of the electrons which are set free during ubiquinol oxidation return to their point of origin. During the flow of two electrons from ubiquinol to cytochrome *c*1, four protons are released to the outer space while two protons are taken up at the inside; (c) Antimycin blocks regeneration of QH_2 from Q at center _i_ while myxothiazol blocks electron transfer at center _o_. The quinone type inhibitors represented by UHDBT blocks electron transfer at the iron-sulphur center level (see also Fig.1).

Until recently [4]; one major objection to the adaptation of the "Q-cycle" mechanism to plant mitochondria was the lack of evidence for the presence of the Rieske iron-sulphur center, since it is widely accepted that the ubiquinone pool interacts with cytochrome _b_, but not with the _c_ cytochromes. A second difficulty was the involvement of cytochrome _b_566 as an obligatory link in the sequence of electron transport reactions, in the view of the reluctance of this cytochrome to participate in electron transport in intact plant mitochondria[28]. However, the recent key experimental observation that the isolated _bc_1 complex contains a Rieske-like ferredoxin center [4], and also performes the oxidant-induced reduction of cytochromes _b_ which is blocked by myxothiazol [4](as expected by the operativity of a "Q-cycle" mechanism), led to conclude that plant and mammalian ubiquinol/cytochrome _c_ oxidoreductases operate in a similar manner (see below).

5. Polypeptide composition of the _bc_1 complex

The _bc_1 complexes isolated from different sources comprise several sub-units (7 to 11) among which only three carry a prosthetic group [1,22]. Two _b_ type hemes are coordinated to the same hydrophobic apoprotein of ca. 42KDa whereas a _c_ type heme (cyt._c_1) is covalently bound to an amphiphilic protein of 28KDa. The Rieske iron-sulphur center, which is a non-haem electron carrier, is bound to an amphiphilic sub-unit of 25KDa, that can be easily dissociated from the complex.

In SDS-PAGE on a 15% gel of the _bc_1 enriched fraction S-2 from _Helianthus tuberosus_, 8 bands, presumably associated with the ubiquinol/cytochrome _c_ oxidoreductase activity, can be resolved (Degli Esposti and Zannoni, unpublished), namely: the apoprotein carrying the _b_ type heme (36.5KDa), cyt._c_1 (30KDa), the Rieske Fe-S protein (27KDa) and two smaller sub-units (diffuse bands at 15KDa and 8KDa). Interestingly, the two core proteins (48 and 45KDa) althought present, are barely detectable.

6. The branch-point of the respiratory chain in plant mitochondria

Many higher plant tissues and fungi possess an alternative, CN^--resistant electron transport pathway [8]. This pathway is insensitive to antimycin and myxothiazol (see also Tab.1), suggesting that it branches from the main respiratory chain on the substrate side of the _bc_1 complex, i.e. at the ubiquinone pool. The nature of the alternative oxidase is presently obscure. Unfortunately, studies with inhibitors of the alternative oxidase such as hydroxamic acids and ubiquinone analogs, along with EPR analyses, have offered little information concerning the nature of the enzyme. Conflicting results have also been presented on both the iron and copper content of the isolated mitochondrial fractions capable of oxidizing quinols in a cyanide-resistant, SHAM-sensitive manner. On the other hand, investigations involving manipulation of the growth conditions of yeast cultures have definitely demonstrated a role for iron-containing proteins in CN^--insensitive respiration [29,30]. These physiological studies seem to be supported by recent data by Walsh and Moore [31], indicating that the alternative oxidase can readily be isolated as a 30-35KDa protein with a significant iron-content.

An alternative manner of obtaining informations concerning the *in vivo* existence of macromolecular complexes, is the technique of *in vitro* complementation[32,33]. In this technique, membranes from different sources are brought together in an attempt to regain an activity which neither of them possesses alone. For complementation to occur, the "component" which is deficient in one membrane must be replaced by the corresponding functional component from the other membrane. Complementation implies *per se* that the membrane components are free to dissociate and reassemble as needed to regain activity. The lack of complementation might be interpreted as evidence that both membranes are affected at the same level or, alternatively, that the components involved in the complementation-test belong to a single, tightly complexed, functional entity (multiprotein complex).

In this report we have applied the complementation approach to see whether the lack of photo-respiratory activity by membranes from the photosynthetic bacterium *Rhodobacter capsulatus* ZM6, can be restored by the alternative oxidase from higher plant mitochondria. We have analyzed the light-induced oxygen uptake since it has previously been demonstrated that this activity is the result of electron transfer, driven by the photosynthetic reaction center bacteriochlorophyll, from exogenous electron donors, through the alternative pathway of the respiratory chain to oxygen[13]. Indeed, the currently accepted arrangement of electron transport components in membranes of *R.capsulatus* is as follows: NADH- and succinate-dehydrogenases plus the photosynthetic reaction center (RC), each are capable of reducing a ubiquinone pool; ubiquinol is oxidized by a ubiquinol/cytochrome *c* oxidoreductase comprising two *b* type hemes, a Rieske-center and a membrane bound cytochrome *c* (cyt.*c*1). This oxidoreductase complex reduces loosely bound *c* (cyt.*c*2) which may, in turn, reduce either a photo-oxidized reaction center or a cytochrome oxidase of *b* type (Em=+410mV). Similar to plant mitochondria, unidentified redox carriers in the ubiquinone pool region can also reduce an alternative oxidase of *b* type (cyt.o) which also catalises the photo-respiratory activity. In summary, except for the presence of a photochemical reaction center and for the nature of the two respiratory oxidases, the arrangement of the branched respiratory chain of *R.capsulatus* is basically equivalent to that found in plant mitochondria especially in terms of metabolic versatility.

In bacterial chromatophores, two electron transport pathways operate under aerobic illuminated conditions, namely: a cyclic and a linear . They branch at the ubiquinone pool level and the linear electron flow, i.e. photorespiratory activity, is strongly enhanced by antimycin[13], and/or myxothiazol, but it is inhibited by UHDBT and/or carbon monoxide and insensitive to the classic inhibitor of the plant mitochondrial alternative oxidases, SHAM (Zannoni, unpublished). Similar to bacterial alternative oxidases, the mitochondrial counterparts are inhibited by UHDBT[19], so to suggest that this quinone analog may act as a general inhibitor of ubiquinol oxidation in electron transport systems. As UHDBT appears to strongly interact also with the Rieske iron-sulphur center [5], it is likely to conclude that ubiquinol oxidation by plant mitochondria is performed by two distinct sites, operationally defined here as center "r" and center "a". As schematically represented in Fig.1, center "r" would be that at which quinol oxidation occurs through reduction of the "Rieske" iron-sulphur protein (UHDBT low-affinity site), and the center "a" that at which quinol oxidation occurs through the "alternative" pathway (UHDBT high-affinity site). The same concept, can be applied to bacterial chromatophores except for the fact that quinol oxidation

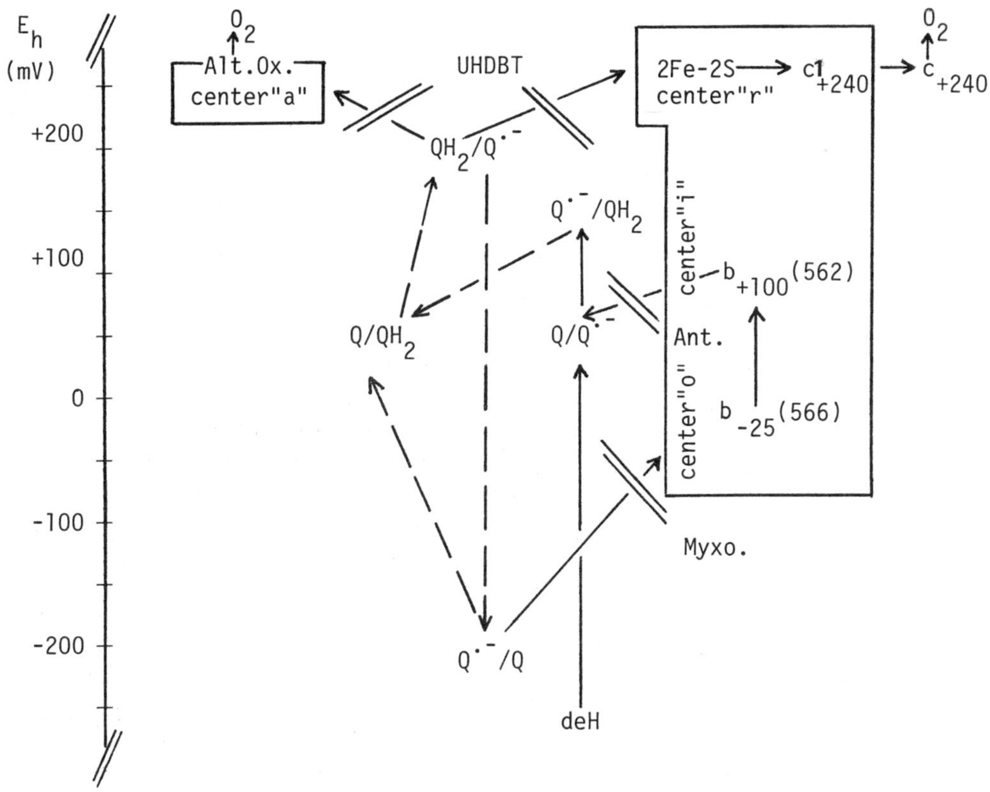

Fig.1. A thermodynamic representation of an hypothetical "Q-cycle" model for electron transfer by plant mitochondria showing the postulated sites of inhibition by antimycin, myxothiazol and UHDBT. The scheme depicts electron transfer in the forward direction, from a dehydrogenase (deH) to either branch, under pre-steady state conditions. Heme- and non-heme redox components are indicated with their $Em_{7.0}$. For consistency with previous literature, site "o" is defined as that at which the reduction of cyt.b566 (cyt.b-25) is blocked by myxothiazol, and site "i" as that at which quinone/semiquinone reduction occurs in an antimycin-sensitive manner. The pathway of protons is omitted. The boxes enclosing the UQ_2/cyt.c oxidoreductase and the alternative oxidase symbolize their multi-protein nature. Abbreviations: Ant.,antimycin; Myxo.,myxothiazol; UHDBT,5-N-undecyl-6-hydroxy-4,7-dioxobenzothiazole; deH,dehydrogenase;Alt.Ox.,alternative CN^--resistant oxidase.
The data supporting the scheme have been taken from von Jagow et al.[37] Look and Kammack[19],Degli Esposti et al.[4].

at center "a" may originate by the input of electrons from either the different respiratory dehydrogenases or the photochemical reaction center.

Similar to higher plant mitochondria in which the proteinaceous nature of the alternative oxidase is still a matter of debate[7], it has been suggested that the photooxidative activity by bacterial chromatophores originates by the oxidation of the photoreduced loosely bound quinone with oxygen[34]. This latter hypothesis is comparable to the free radical theory[35], which would account for mitochondrial CN^--resistant oxidations, although in this case one undefined electron carrier would be necessary to generate the free radicals. In order to discriminate between the different hypotheses involving electron transport by bacterial and mitochondrial "alternative" oxidases, we have therefore initiated a study of the light-driven oxygen uptake in membrane fragments generated by ultrasonic irradiation of a mixed population of R.capsulatus ZM6 chromatophores and sub-mitochondrial particles from Arum italicum and/or Arum maculatum spadices. Membrane fusion caused by co-sonication of mixed membranes has previously been demonstrated to occur between chromatophores isolated from different mutant strains [33]; along with evidence that R.capsulatus ZM6 is deficient in alternative oxidase activity [13]. The preliminary results obtained in the present work are summarized in Tab.1. As expected, ZM6 chromatophores and Arum -SMP do not performe light induced oxygen uptake because they lack of the alternative-oxidase and reaction center bacteriochlorophyll, respectively. A mixture of these two types of membranes show no photorespiratory activity; however, when the mixture is sonicated, a high level of activity appears so to demonstrate in vitro complementation between plant sub-mitochondrial particles and bacterial chromatophores.

In recent years several bacterial respiratory mutant have been isolated and biochemically characterized[36]; for this reason, the preciding results seem to open the door to new avenues in the study of plant mitochondria.

At present, whether complementation involves a process of reassortment of macromolecules such as terminal oxidase proteins or reaction center pro-

Table 1. Reconstitution of the CN^--insensitive photo-respiration in hybrid membranes derived from R.capsulatus ZM6 chromatophores and Arum italicum sub-mitochondrial particles.

Membrane source	Treatment	Substrate	Inhibitor	Experimental conditions	O_2-uptake (neqO_2/min/mg.prot)
ZM6	Sonicated	Asc-TMPD	/	Dark	180
ZM6	Sonicated	Asc-TMPD	KCN(0.1mM)	Dark	0
ZM6	Sonicated	Asc-TMPD	KCN(0.1mM)	Light	20*
ZM6	Sonicated	NADH	/	Dark	2*
Arum	Sonicated	Asc-TMPD	/	Dark	18
Arum	Sonicated	Asc-TMPD	KCN(0.1mM)	Dark	0
Arum	Sonicated	Asc-TMPD	KCN(0.1mM)	Light	0
Arum	Sonicated	NADH	KCN(0.1mM)	Dark	210
ZM6+Arum	Co-sonicated	Asc-TMPD	/	Dark	100
ZM6+Arum	Co-sonicated	Asc-TMPD	KCN(0.1mM)	Dark	0
ZM6+Arum	Co-sonicated	Asc-TMPD	KCN(0.1mM)	Light	124
ZM6+Arum	Co-sonicated	NADH	KCN(0.1mM)	Dark	80
		NADH	Ant./Myxo.	Dark	80

*The light-induced respiration and the NADH-dependent activity in the presence of 0.1mM KCN in wild type membranes of R.capsulatus were 150 and 140 neq O_2/min/mg prot.

Abbreviations: Asc,Na-ascorbate; TMPD,N,N,N'N'-tetramethyl-p-phenylenediamine ; Ant.,antimycin; Myxo.,myxothiazol

Concentrations: Asc,5mM; TMPD,0.250mM; NADH,1mM ; Ant.,1µM;Myxo.,1µM

teins, or the simple diffusion of small molecules like ubiquinones, is not clear (see also Fig.2). Indeed, while the isolation of an alternative-oxidase deficient mutant such as *R.capsulatus* ZM6, indicates *a priori* that photorespiration cannot originate by spontaneous oxidation of loosely bound quinone unless a specific "oxidation-site" is mutated, the role of the mitochondrial ubiquinone pool as a possible source of free radicals cannot be ruled out. Owing to this, we are planning to reconstitute the photorespiratory activity in hybrid membranes derived from ZM6 chromatophores and mitochondria deprived of the ubiquinone pool. Successful reconstitution of light-induced oxygen uptake would indicate that a reduced bacterial ubiquinone pool generated by the reaction center might interact with a specific "component" operating in CN^--resistant respiration of plant mitochondria.

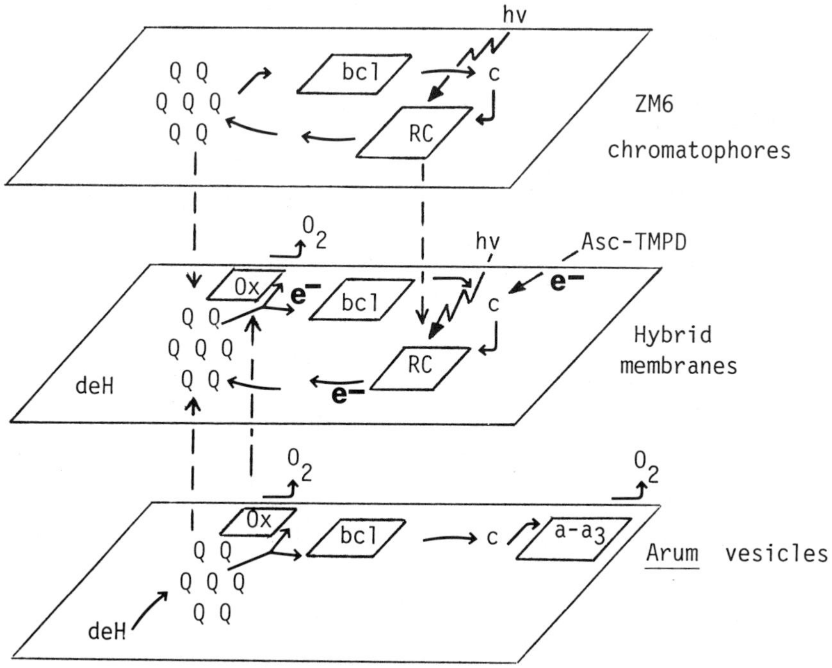

Fig.2. A schematic representation of the membrane reassortment caused by fusion of ZM6- and mitochondrial-vesicles. The different possibilities of diffusion of the redox components giving rise to the complementation observed are indicated by interrupted arrows. Continuous arrows indicate the electron pathways. The rhombic sheets symbolize the membrane fragments.Abbreviations: Q,ubiquinone molecules;RC,reaction center; bc1,complex III; *c*,soluble cyt.*c*; hv, light energy; Ox;,oxidase; Asc-TMPD , exogenous electron donor; deH,dehydrogenase

ACKNOWLEDGEMENTS

D.Z. likes to thank Prof.B.R.Beechey and Dr.A.L.Moore for providing the opportunity to participate in the 2nd International Meeting on Plant Mitochondria (Aberystwyth, 20-24 July, 1986). This work was supported by the Ministry of the Public Instruction of Italy.

REFERENCES

1. G.Hauska, Organization and function of cytochrome b_6f/bcl complexes,in: "Molecular Biology of the Photosynthetic Apparatus",K.E.Steinback, S. Bonitz, C.J.Arntzen, L.Bogorad,eds.,Cold Spring Harbor Laboratory(1985)
2. M.Saraste,Location of haem-binding sites in the mitochondrial cytochrome b, FEBS Lett.166:367(1984)
3. W.R.Widger, W.A.Cramer, R.Hermann, and A.Trebst,Sequence homology and structural similarity between cytochrome b of mitochondrial complex III and the chloroplast b_6f complex:position of the cytochrome b hemes in the membrane, Proc.Natl.Acad.Sci.USA 81:674(1984)
4. M.Degli Esposti,E.Flamini, and D.Zannoni, Functional characterization and partial purification of the ubiquinol/cytochrome c oxidoreductase from higher plant mitochondria(Helianthus tuberosus), Plant Physiol. 77:758(1985)
5. W.D.Bonner jr and R.C.Prince, The Rieske iron-sulphur cluster of plant mitochondria, FEBS Lett. 177:47(1984)
6. A.M.Lambowitz and W.D.Bonner jr, The b cytochromes of plant mitochondria: a spectroscopic and potentiometric study, J.Biol.Chem.249:2428(1974)
7. J.M.Palmer, The organization and regulation of electron transport in plant mitochondria, Ann.Rev.Plant Physiol.27:133(1976)
8. D.A.Day, G.Arnon, and G.G.Laties, Nature and control of respiratory pathways in plants:the interaction of cyanide-resistant respiration with the cyanide-sensitive pathway, in: The Biochemistry of Plants,vol.2, D.D.Davies,ed,Academic Press,New York,London,Toronto,Sydney,San Francisco (1980)
9. P.Mitchell, Proton motive redox mechanism of the cytochrome b/c complex in the respiratory chain: proton motive ubiquinone cycle, FEBS Lett. 56:1(1975)
10. P.Gellefors, T.Johansson, and B.D.Nelson, Isolation of cytochrome bcl complex from rat liver mitochondria, Eur.J.Biochem.115:275(1981)
11. B.B.Katan, Detection of cytochromes on sodium dodecylsulphate-polyacrylamide gels by their intrinsic fluorescence, Anal.Biochem.74:132(1976)
12. I.M.Moller, A.Bergman, P.Gardestrom, I.Ericson, and J.M.Palmer, Characterization and purification of inside out sub-mitochondrial particles obtained from plant mitochondria, FEBS Lett.126:13(1981)
13. D.Zannoni, P.Jasper, and B.L.Marrs, Light induced oxygen reduction as a probe of electron transport between respiratory and photosynthetic components in membranes of Rhodopseudomonas capsulata,Arch.Biochem. Biophys.191:625(1978)
14. P.A.Scolnik,D.Zannoni, and B.L.Marrs, Spectral and functional comparison between the carotenoids of the two antenna complexes of R.capsulata, Biochim.Biophys.Acta 593:230(1980)
15. P.L.Dutton and B.T.Storey, The respiratory chain of plant mitochondria. IX.Oxidation-reduction potentials of the cytochromes of mung bean mitochondria, Plant Physiol.47:283(1971)
16. O.H.Lowry,N.J.Rosebrough,A.L.Farr, and R.J.Randall,Protein measurement with the Folin phenol reagent,J.Biol.Chem.193:265(1951)
17. A.M.Arntjunyan, Y.A.Kamensky, E.Milgrom, S.Surkov, A.A.Konstantinov, and Y.A. Sharonov, Is mitochondrial b566/558 a single hemoprotein or two individual components? , FEBS Lett. 95:40(1978)
18. B.L.Trumpower, Function of the iron-sulphur protein of the cytochrome

b-c1 segment in electron-transfer and energy-conserving reactions of the mitochondrial respiratory chain, Biochim.Biophys.Acta 639:129(1981)
19. N.D.Look and R.Kammack, The inhibition of plant mitochondrial respiration by the synthetic analog of ubiquinone, 5-N-undecyl-6-hydroxy-4,7-dioxobenzothiazole(UHDBT), Arch.Biochem.Biophys. 240:9(1985)
20. J.R.Bowyer, L.A.Edwards, T.Ohnishi, and B.L.Trumpower, An analogue of ubiquinone which inhibits respiration by binding the iron-sulphur protein of the cytochrome b/c1 segment of the mitochondrial respiratory chain, J.Biol.Chem.257:8321(1982)
21. Q.S.Zhu, J.A.Berden, S.de Vries, K.Folkers, T.H.Porter, and E.C.Slater, Identification of two different Q-binding sites in QH_2-cytochrome c oxidoreductase using the Q analogue n-heptadecylmercapto-6-hydroxy-5,8-quinolinequinone, Biochem.Biophys.Acta 682:160(1982)
22. J.S.Rieske, Composition, structure and function of the complex III of the respiratory chain, Biochim.Biophys.Acta 456:195(1976)
23. G.von Jagow and W.D.Engel, Complete inhibition of electron transfer from ubiquinol to cytochrome b by the combined action of antimycin and myxothiazol, FEBS Lett.136:19(1981)
24. N.Gabellini, E.Bowyer, E.Hurt, B.A.Melandri, and G.Hauska, A cytochrome bc1 complex with ubiquinol-cytochrome c oxidoreductase activity from R.sphaeroides GA, Eur.J.Biochem. 126:105(1982)
25. J.S.Rieske, Changes in oxidation-reduction potential of cytochrome b observed in the presence of antimycin A, Arch.Biochem.Biophys.145:179(1971)
26. M.Ericinska and D.F.Wilson, Kinetic studies on cytochrome bc1 interaction in the isolated succinate-cytochrome c reductase, FEBS Lett.24:269(1972)
27. M.K.F.Wikstrom and J.A.Berden, Oxidoreduction of cytochrome b in the presence of antimycin, Biochim.Biophys.Acta 283:403(1972)
28. B.T.Storey, Electron transport and energy coupling in plant mitochondria in: The Biochemistry of Plants, vol.2, D.D.Davies, ed., Academic Press, New York, London, Toronto, Sydney, San Francisco (1980)
29. B.A.Haddock and P.B.Garland, Effect of sulphate-limited growth on mitochondrial electron transfer and energy conservation between reduced nicotinamide-adenine dinucleotide and cytochromes in Torulopsis utilis, Biochem.J. 124:155(1971)
30. M.-F.Henry, W.D.Bonner, and E.J.Nyns, Involvement of iron in the biogenesis of the cyanide-insensitive respiration in the yeast Saccharomycopsis lipolytica, Biochim.Biophys.Acta 460:94(1977)
31. M.C.Walsh and A.L.Moore, Immunological analysis of the alternative oxidase, in: 2nd Int.Meet.on Plant Mitochondria. Structural, Functional and Physiological Aspects, Abstracts A15(1986)
32. R.Sckeckman, A.Weiner, A.Kornberg, Muntienzyme system of DNA replication, Science 186:387(1974)
33. D.Zannoni, S.Peterson, and B.L.Marrs, Recovery of the alternative oxidase dependent electron flow by fusion of membrane vesicles from R.capsulatus mutant strains. An in vitro membrane-complementation study, Arch.Microbiol.144:375(1986)
34. V.G.Remennikov and V.D.Samuilov, Photooxidase activity of isolated chromatophores and intact cells of phototrophic bacteria, Arch.Microbiol. 123:65(1979)
35. P.Rustin, J.Dupont, and C.Lance, Involvement of lipid peroxyradicals in the cyanide-resistant electron transport pathway, Physiol.Veg.22:643(1984)

36. B.L.Marrs, Mutations and genetic manipulations as probes of bacterial photosynthesis, in: Current Topics in Bioenergetics,vol.8, D.R.Sanadi, L.P.Vernon, ed., Academic Press, New York and London (1978)
37. G. von Jagow, T.A.Link, and T.Ohnishi, Organization and function of cytochrome _b_ and ubiquinone in the cristae of beef heart mitochondria, _J.Bioen.Biomemb._ 18:157 (1986)

REGULATION OF ADP-LIMITED RESPIRATION IN ISOLATED PLANT MITOCHONDRIA

D.A. Day[1], J.T. Wiskich, J.H. Bryce and I.B. Dry

Botany Departments, Australian National University[1]
Canberra, A.C.T., 2601 and University of Adelaide
Adelaide, S.A., 5001, Australia

INTRODUCTION

Although the detailed mechanism of oxidative phosphorylation is still open to debate, respiratory control is generally explained in chemiosmotic terms[1]. Electron transport through the respiratory chain is obligatorily coupled to outward H^+ translocation, establishing a proton motive force ($\Delta\mu H^+$) which can drive phosphorylation of ADP. In the absence of ADP, $\Delta\mu H^+$ increases and exerts a negative feedback on the respiratory chain, restricting the rate of electron flow and O_2 consumption. Under any particular set of conditions, the rate of O_2 uptake with a specific substrate in the absence of ADP (state 4) is presumably determined by the $H^+/2e^-$ ratio and the H^+ conductance of the inner membrane. H^+ conductance is dependent on the membrane resistance and $\Delta\mu H^+$. If the conditions or substrate are altered, changes in the redox potential difference (substrate minus O_2) and the $\Delta\mu H^+$ become important.

Isolated plant mitochondria often show relatively rapid rates of state 4 O_2 uptake. This rapid ADP-limited elecron flow has been attributed to several factors, including operation of the cyanide-insensitive alternative path which branches from the cytochrome chain at the level of ubiquinone and which is not linked to H^+ translocation. Operation of this pathway could explain rapid state 4 O_2 uptake with succinate and exogenous NADH, which feed electrons directly to ubiquinone. However, oxidation of NAD-linked substrates should still be restricted by electron flow through the H^+-translocating complex I (endogenous NADH dehydrogenase) of the respiratory chain albeit at a reduced $H^+/2e^-$ ratio. Nonetheless, state 4 O_2 uptake with NAD-linked substrates in isolated plant mitochondria can still be rapid and is remarkably variable, depending on the reaction conditions. We have investigated the nature of variable state 4 O_2 uptake in isolated mitochondria.

MATERIALS AND METHODS

Beetroot (*Beta vulgaris* L.) and potato (*Solanum tuberosum* L.) mitochondria were purified on Percoll gradients[2], while turnip (*Brassica rapa* L.) mitochondria were purified on linear sucrose gradients[3]. Mitochondria from pea (*Pisum sativum*) leaves were purified according to[4].

Oxygen uptake was measured polarographically in a standard reaction medium of 0.3M sorbitol (pea leaves) or 0.25M sucrose (all other tissues), 10mM KH_2PO_4, 10mM TES, 5mM $MgCl_2$ and 0.1% (w/v) BSA. The final pH was 7.2 unless otherwise stated.

The redox level of intramitochondrial NAD^+ was monitored fluorimetrically[5, 6] in a stirred cuvette with a mixing time of less than one second. Traces were corrected for spurious fluorescence or dilution upon additions, in separate experiments without mitochondria.

The electrical potential component of the pmf across the mitochondrial inner membrane was estimated spectrophotometrically using the dye safranin[7]; 15 μM safranin and 0.9 mg mitochondrial protein were used in 3 ml of the standard reaction medium.

Mitochondrial (outer membrane) integrity was estimated as described[2] and was generally found to be better than 90%.

RESULTS AND DISCUSSION

An example of variable rates of state 4 O_2 consumption is shown in Figure 1. At pH6.5, O_2 uptake in state 4 was substantially faster, and respiratory control ratios lower, than at pH7.5 with malate as substrate (Figs 1A and 1B), but ADP/O ratios were unchanged. Succinate oxidation was virtually unaffected by the same change in pH (Figs 1C and 1D), indicating that the pH-mediated increase in O_2 uptake was not due to a direct effect on the activity of the respiratory chain or on the H^+ conductance of the membrane. State 4 O_2 uptake with malate as substrate could be stimulated considerably at pH7.5 by adding thiamin pyrophosphate to the reaction medium. These changes in the rate of malate oxidation have been previously shown to be correlated with changes in the level of oxaloacetate, a product of the malate dehydrogenase reaction, that results in decreased matrix NADH levels[8, 9]. Decreasing external pH tends to favour malate oxidation via malic enzyme, rather than via malate dehydrogenase; matrix NADH levels consequently increase, resulting in

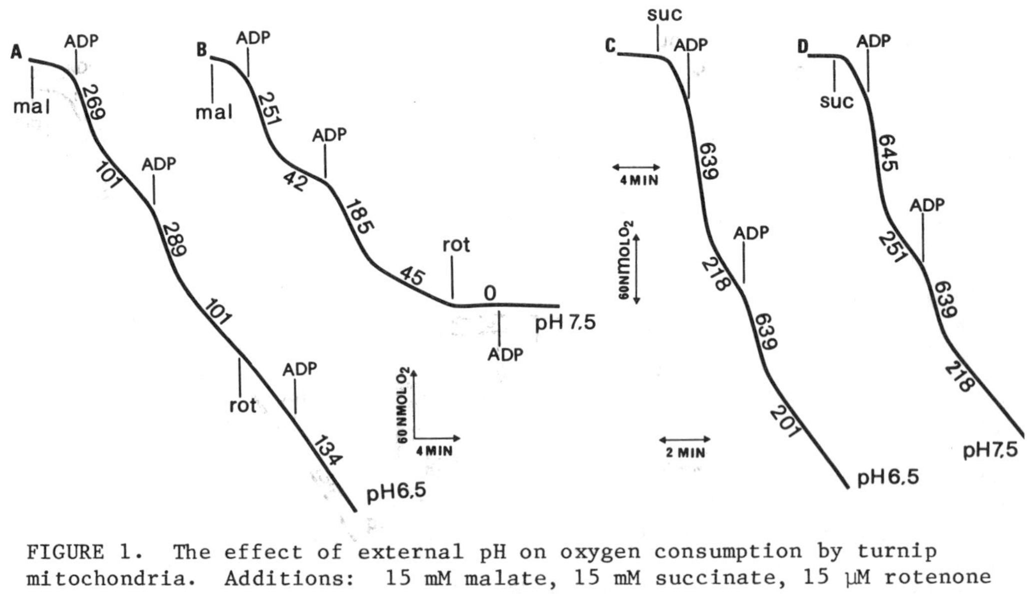

FIGURE 1. The effect of external pH on oxygen consumption by turnip mitochondria. Additions: 15 mM malate, 15 mM succinate, 15 μM rotenone (rot) and 0.3 (traces A and B) or 0.2 (C and D) μmol ADP. Numbers on traces refer to nmol O_2 $min^{-1}mg^{-1}$ protein.

faster state 4 rates (Fig 1). At higher pH, malate dehydrogenase is more active than malic enzyme and oxaloacetate accumulates. Addition of thiamin pyrophosphate activates pyruvate dehydrogenase to produce acetyl coenzyme A which condenses with oxaloacetate, causes intramitochondrial NADH levels to rise, and O_2 uptake to increase.

The effect of providing a second substrate to beetroot mitochondria is shown in Figure 2. Glutamate alone caused some reduction of mitochondrial NAD^+ but O_2 uptake was very slow. Addition of malate caused a much larger reduction of NAD^+ and stimulated O_2 uptake; additions of ADP resulted in the usual state 3/state 4 transitions in O_2 uptake and oxidation/reduction cycles in matrix NAD(H). Similar changes in the redox level of matrix NAD^+ accompanying increases in O_2 uptake have been observed with potato and cauliflower mitochondria oxidizing malate at different pH values[9].

The results presented thus far suggest that increases in state 4 O_2 uptake, with malate as substrate, are the result of more NADH being available to the respiratory chain. Note, however, that state 4 rates need not be strictly correlated with NADH levels. In Fig 2, the level of NADH was higher in state 4 (after ADP addition) than in state 2 (the rate before ADP addition) but O_2 uptake was slower (22 vs 30 nmol.min^{-1}). The $\Delta\psi$ in state 4 was also marginally higher than that before ADP addition (Fig 3: see below). We attribute this difference to the presence of ATP in state 4 which can descrease H^+ conductance through the ATP synthetase; adding ATP (or oligomycin) during state 2 decreases O_2 uptake to state 4 levels (Day, unpublished observations). However, ADP-limited electron flow is generally considered to be a function of $\Delta\mu H^+$. We therefore attempted measurements of the principal component of $\Delta\mu H^+$, the transmembrane electrical potential ($\Delta\psi$), under conditions known to stimulate O_2 uptake. The method used was that of safranin absorbance. Changes in safranin absorbance have been

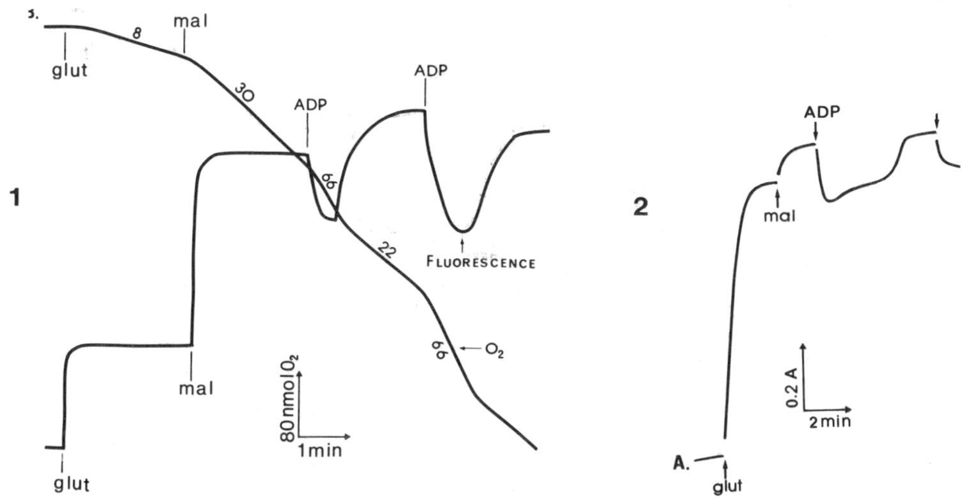

FIGURE 2. Oxygen consumption and endogenous NAD^+ reduction by beetroot mitochondria. O_2 uptake and fluorescence of endogenous pyridine nucleotides were measured simultaneously in parallel experiments and the respective traces superimposed for comparison. ADP added as 0.3 (first addition) or 0.5 (second addition) µmol. An upward deflection of the fluorescence trace indicates reduction of NAD^+.

FIGURE 3. Safranin absorbance changes during respiration by beetroot mitochondria. The external pH was 7.0. An upward deflection of the spectrophotometer trace indicates an increase in membrane potential.

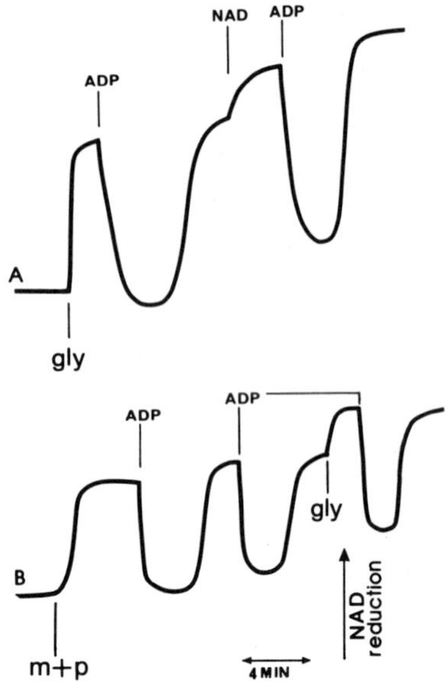

FIGURE 4. Endogenous NAD^+ reduction by pea leaf mitochondria. Mitochondria (0.2 mg protein) were added to 2 ml of standard reaction medium containing 0.2 mM ATP, and 0.3 mM thiamin pyrophosphate. Other additions were 10 mM glycine (gly), 5 mM malate + 5 mM pyruvate (m + p), 0.2 mM NAD and 0.15 mM ADP.

shown to be directly correlated to $\Delta\psi$ in plant mitochondria[7], an increase in $\Delta\psi$ causing an increase in absorbance. Addition of glutamate to beetroot mitochondria caused a large increase in safranin absorbance (Fig 3). Subsequent addition of malate caused a smaller but consistent further rise. Transition to state 3 upon ADP addition was accompanied by a small decrease in absorbance which subsequently recovered in state 4 to a level slightly above that of state 2.

Table 1 depicts other means by which state 4 rates can be stimulated, this time with pea leaf mitochondria. Addition of NAD^+, which is accumulated by plant mitochondria[6] and increases matrix NADH levels (Fig 4), stimulated subsequent state 4 O_2 uptake almost two-fold. Similarly, addition of glycine to leaf mitochondria oxidizing malate plus pyruvate stimulated O_2 uptake (Table 1). Again, the increase in state 4 O_2 uptake was accompanied by an increase in matrix NADH level (Fig 4). The same conditions which stimulated state 4 also stimulated rotenone-

TABLE 1. State 3 and rotenone-insensitive O_2 uptake by pea leaf mitochondria. Assay conditions: 10 mM glycine and pyruvate, 5 mM malate, 0.2 mM NAD, 15 μM rotenone, 0.19 mg protein. Rotenone added in the presence of ADP.

Substrates	Oxygen uptake (nmol O_2 min^{-1} mg^{-1} protein)			
	State 4		Rotenone insensitive	
	−NAD	+NAD	−NAD	+NAD
glycine	47	74	42	79
mal + pyr	47	79	53	90
mal + pyr + gly	74	137	63	137

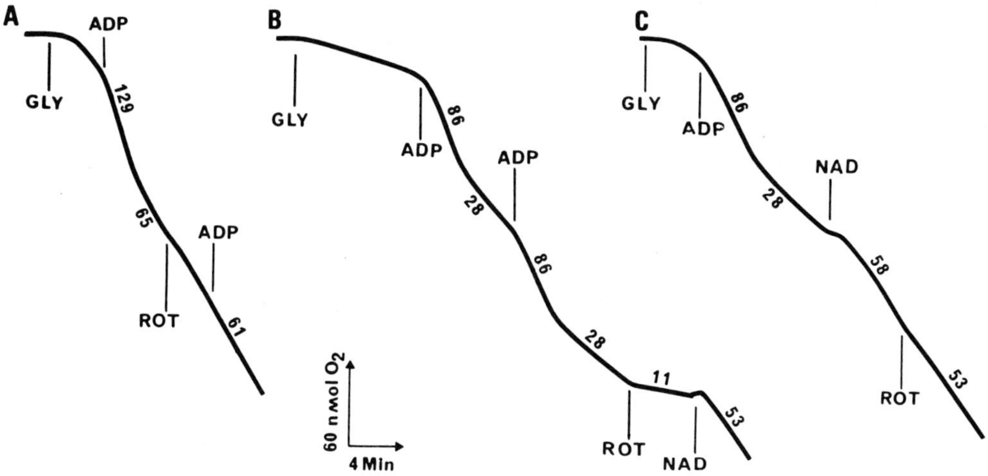

FIGURE 5. Oxygen consumption by pea leaf mitochondria. A: untreated, purified mitochondria (0.33 mg protein). B and C: mitochondria diluted 30-fold with reaction medium and incubated for 4 h at 4 C to deplete them of endogenous NAD^+; the mitochondria were then concentrated by centrifugation and O_2 uptake measured (with 0.27 mg mitochondrial protein). Additions as indicated were, 10 mM glycine, 0.4 µmol ADP, 0.2 mM NAD^+ and 15 µM rotenone (rot). For further details, see figure 1.

insensitive O_2 uptake. Thus O_2 uptake in the presence of rotenone was faster when two substrates were added together and when NAD^+ was included in the reaction medium (Table 1). Note that ADP was present when rotenone was added and hence rates shown are the full capacity of the bypass under the prevailing conditions.

The effect of depleting pea leaf mitochondria of endogenous NAD^+ on O_2 uptake is shown in Fig 5. Both state 3 and 4 rates declined, but the latter more than the former such that RCR increased (Fig 5B), adding NAD^+ to the medium restored O_2 uptake rates (Fig 5C). Inhibition by rotenone was much more severe in NAD^+-depleted mitochondria; adding NAD^+ relieved the inhibition (Fig 5).

Rotenone resistant O_2 uptake is thought to be due to operation of a bypass around complex I, presumably via another NADH dehydrogenase, which does not translocate protons and hence is not restricted by $\Delta\mu H^+$. Affinity of the bypass for NADH is much less than that of complex I[10] and its activity is therefore very sensitive to endogenous NADH levels. In general, any conditions which serve to increase matrix NADH levels can be expected to stimulate operation of the rotenone-insensitive bypass (e.g. rotenone-resistance is greater at pH6.5 than 7.5 with malate as substrate, Fig 1A and 1B, and when a cocktail of substrates is used, Table 1). Since the bypass involves translocation of fewer protons[11], it is tempting to speculate that the increases observed in state 4 respiration rates with NAD-linked substrates, involve engagement of the bypass, which is insensitive to $\Delta\mu H^+$.

Because animal mitochondria do not possess a complex I bypass, they were used to test the above hypothesis. Addition of other NAD-linked substrates to liver mitochondria respiring β-hydroxybutyrate or glutamate, stimulated state 4 O_2 uptake by 25-35% (Table 2). It therefore seems that the presence of a bypass around complex I is not a prerequisite for enhanced state 4 rates in isolated mitochondria. However, it should be

TABLE 2. State 4 O_2 uptake by rat liver mitochondria. Substrates were added at 10 mM to the standard reaction medium.

OH-butyrate	glut	glut + malate	glut + mal + OH-butyrate
		nmol min^{-1} mg^{-1} protein	
5.5	5.5	7.0	7.5

noted that the stimulations observed with plant mitochondria are much more pronounced (up to 200%) than those with rat liver mitochondria.

GENERAL DISCUSSION

Under state 4 conditions the rate of electron transport is such that the rate of H^+ translocation out of the mitochondrion is equal to the rate of H^+ "leakage" back across the inner membrane, maintaining the $\Delta\mu H^+$.[12] Thus state 4 respiration is predominantly due to the inward leak of H^+ (or the H^+ conductance of the inner membrane). Proton conductance is proportional to $\Delta\mu H^+$ and inversely proportional to the membrane resistance. In the key experiments described in this paper we have added NAD^+ or a second substrate and assumed not to have changed the membrane resistance. However, the addition of a second substrate may change the driving force of electron flow (i.e., the redox potential difference between NADH and oxygen). It is obvious from the results presented here that state 4 respiration rates in plant mitochondria are dramatically affected by intramitochondrial NADH level as well as by $\Delta\mu H^+$. The quantitative effect of changing NADH levels is surprising.

Increasing the $NADH/NAD^+$ ratio in the mitochondrial matrix results in an increased driving force for electron flow[13]. Consequently, $\Delta\mu H^+$ is expected to increase, and does so in beetroot mitochondria (Fig.3). This increased $\Delta\mu H^+$ will have two opposing effects: (a) it will increase the backflow of H^+ across the membrane, allowing faster coupled electron flow, and (b) it will oppose some of the increased redox driving force (which, strictly, is [$NADH/NAD^+$ minus $\Delta\mu H^+$]). This mechanism presumably accounts for the stimulation of O_2 uptake in rat liver mitochondria upon addition of a second NAD-linked substrate and for at least some of the results observed with plant mitochondria.

The stimulation of state 4 O_2 uptake in plant mitochondria is much more pronounced than in liver. Every condition that leads to enhanced state 4 rates in plant mitochondria also leads to pronounced rotenone-resistant O_2 uptake. The rotenone-resistant O_2 uptake is due to a non-proton-translocating bypass around site 1, between NADH and ubiquinone (UQ). Engagement of this non-electrogenic bypass lowers the overall $H^+/2e$ ratio and allows faster electron flow at any given $\Delta\mu H^+$. It also removes any thermodynamic effect of $\Delta\mu H^+$ on the equilibrium redox state between NAD(H) and UQ. Increased levels of reduced UQ (that is, increased UQH_2/UQ) will increase the driving force for electron flow to O_2. Clearly, this latter portion of the respiratory chain (UQ to O_2) has the capacity to handle the increase (state 4 rates with succinate are very fast; Figs 1C and 1D) and preliminary experiments with A.L. Moore (University of Sussex) indicate that elevated state 4 rates with NAD-linked substrates are indeed associated with increased levels of reduced ubiquinone.

Leaf mitochondria are also partly cyanide-resistant[14] and concurrent engagement of the non-phosphorylating alternative oxidase with the rotenone-bypass will further serve to increase non-coupled O_2 uptake. Beetroot and turnip mitochondria, on the other hand, lack the alternative oxidase[15] and in these organelles, increased O_2 uptake via the NADH dehydrogenases must be accompanied by an increased rate of coupled electron flow from ubiquinone to cyt. oxidase. The results suggest that $\Delta\mu H^+$ imposes a more severe restriction on electron flow through complex I than on that through the rest of the respiratory chain. This implies that the redox poise between NADH and UQ, through complex I (using NAD-linked substrates), reaches a steady state equilibrium at lower $\Delta\mu H^+$ values than that between UQ and O_2 (using succinate or external NADH). In other words, state 4 rates with these different substrates (Fig 1) are not simply a reflection of their different $H^+/2e^-$ ratios.

Adding rotenone to plant mitochondria during state 4 respiration can have either a marked or a marginal inhibitory effect on O_2 consumption (Fig 5). This effect is dependent on the NAD-levels within the mitochondria[8], which can be manipulated[6] and which affect the activity of the rotenone-insensitive bypass[10]. It is also important to note that state 4 rates themselves are dependent on the level of NAD^+ (Fig 5) and that the slower rates, at limiting levels of NAD^+, show the greatest sensitivity to rotenone. This suggests that if the rotenone-bypass underlies the rapid state 4 rates in plant mitochondria then electron flow through complex I still continues. In other words, we suggest that when intramitochondrial [NADH] is high and state 4 O_2 uptake is rapid, both internal NADH dehydrogenases operate; when matrix [NADH] is low, only the rotenone-sensitive complex I operates and O_2 uptake is slow.

The magnitude of electron flow through complex I may be quite variable. In Fig 5B most of the state 4 rate occurs via complex I and is sensitive to rotenone; the presence of NAD^+ (Fig 5C) produces an increased state 4, most of which appears to be rotenone-insensitive. The extra capacity of the rotenone-insensitive bypass has elevated the level of UQH_2 which in turn increases the rate of electron flow between UQ and O_2; O_2 uptake and $\Delta\mu H^+$ consequently increase. The latter point is illustrated in Figs 2 and 3 - adding a second NAD-linked substrate to beet mitochondra raised $\Delta\psi$ (Fig 3) presumably by increasing matrix NADH (Fig 2) and engaging the bypass. As far as complex I is concerned, increasing matrix NADH may increase the potential driving force for electron transport but it also results in an increased level of UQH_2 and of $\Delta\mu H^+$. The first parameter would tend to increase electron flow through complex I but the latter two parameters would tend to decrease it. Overall, it appears that adding NAD^+ (Fig 5) results in a decreased rate of coupled electron flow through complex I. This is also suggested by the data in Table 1 where the capacity of the rotenone-bypass was less than state 4 rates in the absence of added NAD^+, but greater than state 4 rates when NAD^+ was present.

CONCLUSION

The results presented show that even adenylate-limited respiration is variable and dependent on the supply of reducing power to the mitochondria. This study also suggests that the rotenone-insensitive bypass is a constitutive feature of plant mitochondria which, together with the non-phosphorylating alternative oxidase, will allow rapid respiratory rates in plant cells, even in the presence of high energy charge.

ACKNOWLEGEMENTS

Financial support from the Australian Research Grants Scheme is gratefully acknowledged. We thank Professor Roland Douce and Dr Michel Neuburger for the use of laboratory facilities and assistance with the experiment of Fig 4, and Ms J. Perks for typing the manuscript.

REFERENCES

1. Mitchell, P. (1966). Biol. Rev. 41, 445-502.

2. Neuburger, M., Journet, E-P, Bligny, R., Carde, J-P. and Douce, R. (1982). Arch. Biochem. Biophys. 217, 312-333.

3. Millhouse, J., Wiskich, J.T. and Beevers, H. (1983). Aust. J. Plant Physiol. 10, 167-177.

4. Day, D.A., Neuburger, M. and Douce, R. (1985). Aust. J. Plant Physiol. 12, 219-228.

5. Bryce, J.H. and Wiskich, J.T. (1985). Aust. J. Plant Physiol. 12, 229-239.

6. Neuburger, M., Day, D.A. and Douce, R. (1985). Plant Physiol. 78, 405-410.

7. Moore, A.R. and Bonner, W.D. Jr. (1983). Plant Physiol. 70, 1271-1276.

8. Palmer, J.M., Schwitzguebel, J-P. and Moller, I.M. (1982). Biochem. J. 208, 703-711.

9. Neuburger, M., Day, D.A. and Douce, R. (1984). Physiol. Veg. 22, 517-580.

10. Moller, I.M. and Palmer, J.M. (1982). Physiol. Plant 54, 267-274.

11. Moore, A.L. and Rich, P.R. (1985) in "Higher plant cell respiration", Encyclopedia Plant Physiol (New Series) vol 18, (Douce, R. and Day, D. A. eds) pp 134-72, Springer-Verlag, Berlin.

12. O'Shea, P.S. and Chappell, J.B. (1984). Biochem. J. 219, 401-404.

13. Brown, G.C. and Brand, M. (1985). Biochem. J. 225, 399-405.

14. Day, D.A. and Wiskich, J.T. (1981). Plant Physiol. 68, 425-429.

15. Day, D.A. (1975). Ph.D. Thesis, University of Adelaide.

OHMICITY, PROTICITY AND ELECTRON FLUX IN MUNG BEAN MITOCHONDRIA

Simon T. Dobbs and Anthony L. Moore

Department of Biochemistry, University of Sussex
Falmer Brighton, BN1 9QG, UK

INTRODUCTION

The free energy which is liberated from the redox reactions catalysed by the respiratory chain is conserved as an electrochemical potential difference of protons (Δp) which is directed across the inner mitochondrial membrane[1]. The magnitude of this force has been shown in a number of systems not to be related to the rate of electron flux[2-4], since the rate of respiration can be substantially inhibited by malonate or antimycin resulting in only a minor depression of the protonmotive force. This has been attributed to either the non-Ohmic behaviour of the inner membrane[3,4] or alternatively to variations in the stoichiometry of the redox driven proton pumps (redox slippage)[5,6]. Estimates of the passive conductance of the inner membrane to protons in rat liver mitochondria have shown that this does increase under conditions of a high prevailing protonmotive force when this is imposed as a potassium diffusion potential, in a manner which can at least qualitatively account for the relationship between oxidation rate and protonmotive force, and the occurrence of a finite rate of respiration at static head[4,7]. In the present paper we have investigated the relationship between the electron transport rate and Δp in mung bean mitochondria and show that an increase in the conductance of the inner membrane can quantitatively account for such non-Ohmic behaviour. Furthermore it is shown that an increase in respiratory rate observed under state 4 conditions upon addition of a second substrate is also due to an increase in membrane conductance.

MATERIALS AND METHODS

Mitochondria were isolated from 5-day old etiolated mung bean hypocotyls[8] and approximately 1mg protein was suspended in 2ml of reaction medium (0.3 M mannitol; 10mM KCl; 5mM $MgCl_2$; 10mM phosphate; 10mM MOPS, pH 7.2; 0.5µg/ml nigericin), contained in a cuvette enabling the simultaneous determination of $\Delta\Psi$, the redox poise of ubiquinone and the rate of O_2 consumption. $\Delta\psi$ was measured in the presence of nigericin by monitoring the accumulation of $TPMP^+$ with an ion selective electrode[9]. This was calibrated by successive additions of $TPMP^+$, up to 1µM, and $\Delta\Psi$ calculated using the Nernst equation. The oxygen concentration was measured using a Rank oxygen electrode, and the steady state redox poise of Q-2 added to the reaction medium (1µM) was monitored voltammetrically, using a glassy carbon working electrode, Pt auxiliary electrode and Ag/AgCl reference. The working electrode was set at -360mV relative to the reference.

RESULTS AND DISCUSSION

Titration of succinate oxidation with malonate or exogenous NADH with antimycin yields the familiar hyperbolic relationship which has been observed in other systems. Variations in the inherent conductance of the inner mitochondrial membrane imposed either with uncouplers or by altering the activity of the ATPase with ADP/carboxyatractyloside (Fig 1) or ADP/oligomycin (results not shown) demonstrate that this non-linear behaviour is dependent on that conductance, since the curvature of the relationship decreases as the initial conductance is increased.

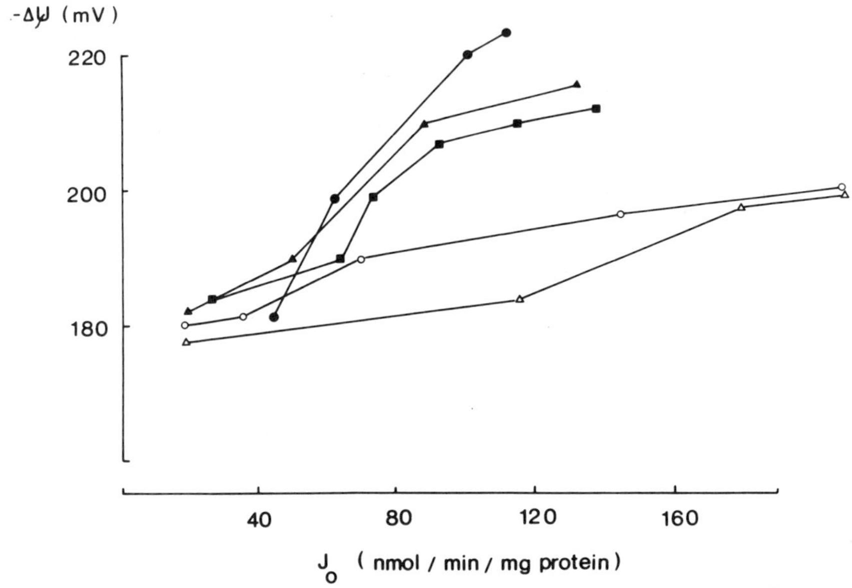

Fig 1. The relationship between membrane potential and the rate of respiration supported by NADH. The reaction medium was supplemented with 1mM NADH and 1mM ADP and the rate of respiration inhibited with antimycin at titres up to 120 pmole. The activity of the ATPase was varied by means of the following concentrations of carboxyatractyloside ● 308; ▲ 231; ■ 154; ○ 115pmol/mg protein

The system therefore becomes Ohmic under conditions of saturating conductance, deviations from linearity becoming apparent under conditions of high prevailing Δp. This supports the observation in rat liver mitochondria that proton fluxes measured with a pH electrode show a hyperbolic dependence on a diffusion potential in native mitochondria, but show a linear dependence in the presence of a protonophore. Titration of NADH or succinate oxidation with valinomycin or FCCP shows that Δp is linearly dependent upon the respiratory rate as the membrane ionic conductance is increased (not shown).

Fig 2. illustrates the dissipation of $\Delta\Psi$ at static head upon addition of a repiratory inhibitor. It can be seen that the rate of dissipation is slower in the presence of antimycin than FCCP suggesting that the response of the electrode is not rate limiting, an observation which is supported by the result that the kinetics of the establishment of a membrane potential and its dissipation are similar to those observed using safranine or radiolabelled lipophilic cations as probes of the membrane potential.

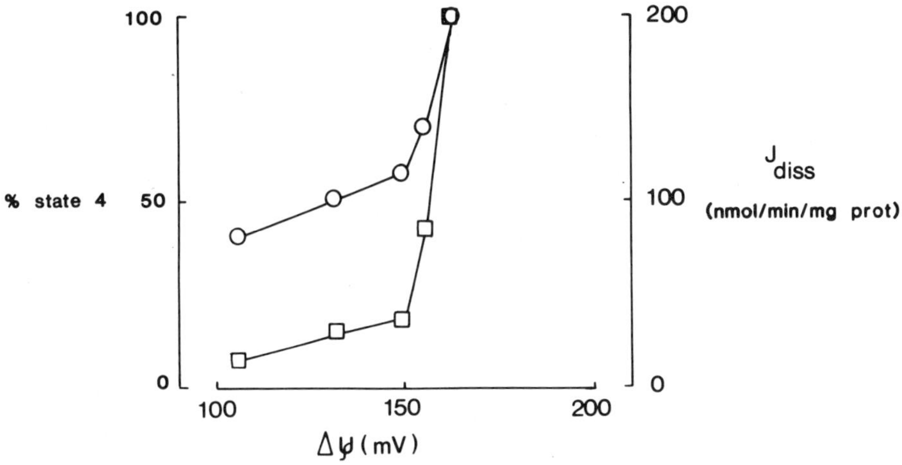

Fig 2. The rate of dissipation of the static head membrane potential upon addition of antimycin or FCCP. Mitochondrial protein (1mg) was suspended in 2ml reaction medium supplemented with 10μM $TPMP^+$ and 10mM succinate. Respiration was then inhibited with 1ng antimycin or 0.1μM FCCP. Note the difference in time base for each inhibitor.

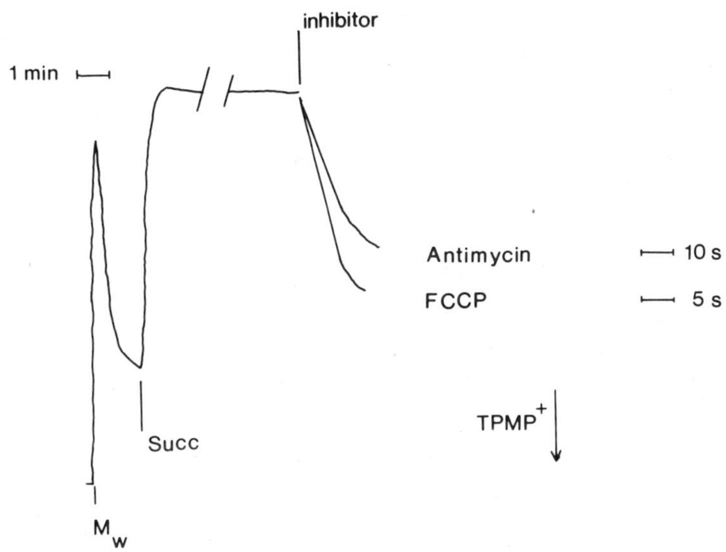

Fig 3. The relationship between the steady state membrane potential and the rate of its dissipation on inhibitor stop. Protocol as Fig 2. The membrane potential was set with 10mM succinate and 0-2mM malonate. The state 4 rate of respiration was 50.9nmol O_2/mg protein/min. Oxidation rate (o), rate of membrane potential dissipation (□).

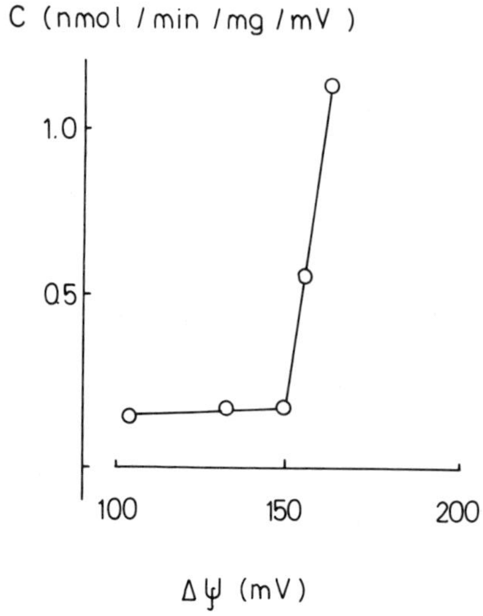

Fig 4. The dependence of ionic conductance of the inner mitochondrial membrane on membrane potential. Data derived from Fig 3.

Variation of the substrate membrane potential with succinate and malonate shows that the ionic current across the membrane is proportional to the potential at low values of ΔΨ (Fig 3) but at 150mV this increases markedly. As can be seen from Fig 4 the conductance of the inner membrane is constant up to a threshold value of membrane potential beyond which it increases. The ionic conductance of the inner mitochondrial membrane estimated in this way is in close agreement with the value obtained using acid pulses in mung bean mitochondria, which yields a value of 2.45nmol H^+ ions/min/mg protein/mV (2.5nmol H^+/s/mg protein/pH unit)[10]. Inhibitor stop experiments, in the presence of a subsaturating concentration of FCCP, (0.1µM) indicated a value of 16nmol/min/mg protein/mV, suggesting that this uncoupler does increase the conductance of the inner mitochondrial membrane. An H^+/O ratio can also be inferred from Fig. 3 based upon the rate of oxygen consumption at static head and the rate of dissipation of ΔΨ. This gives a value of 3.95 which is consistent with oxygen pulse measurements, and is further evidence in favour of the suggestion that there is no change in stoichiometry at static head (ie. no slippage). The shape of the relationship between membrane potential and rate of respiration in state 4 suggests that an increase in respiratory rate could be accommodated by an increase in potential. Such an increase can be elicited by the addition of NADH to mitochondria oxidising succinate. As indicated in Fig 5, the addition of NADH to mitochondria oxidising succinate under state 4 conditions results in a stimulation of respiration, an increase in the steady state reduction of ubiquinone, and a small, but detectable rise in the membrane potential. Similar results were obtained in state 3 (Table 1).

The increase in membrane potential under such circumstances is small, due to the hyperbolic nature of the membrane potential/respiration rate relationship. The much greater increase in respiratory rate may be due in part to either an increase in the membrane conductance or to slippage in the redox driven proton pumps. Recent measurements of the thermodynamic driving force across the bc_1 complex, however, reveal that the H^+/O stoichiometry remains at 2 at static head[11].

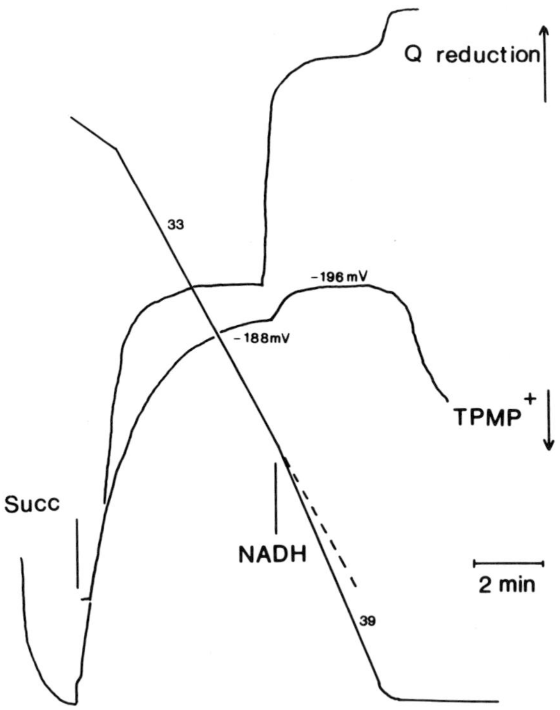

Fig 5. The increase in steady state rate of respiration, membrane potential and redox poise of ubiquinone elicited by the addition of NADH (1mM) to mitochondria oxidising succinate (10mM).

Inhibitor stop experiments suggest that under conditions of the simultaneous oxidation of succinate and NADH, the conductance is 15 nmol/min/mg protein/mV, compared with a value of 12 nmol/min/mg protein/mV for NADH alone. This suggests that the resultant increase in respiratory rate on addition of the second substrate is accommodated in part by an increase in Δp, and in part by an increase in the conductance of the inner membrane.

Table 1. Steady state values of Δp and repiratory rate for mung bean mitochondria oxidising 10mM succinate, 1mM NADH or both. State 3 was initiated by the addition of 1mM ADP.

	Δ p(mV)	J_0(nmol/ min/mg)
STATE 4.		
Succinate	-188	32.5
NADH	-190	34.0
Succinate + NADH	-196	39.0
STATE 3.		
Succinate	-134	93.4
NADH	-143	127.9
Succinate + NADH	-146	158.9

CONCLUSIONS

1. The relationship between the rate of respiration and Δ p in mung bean mitochondria is similar to that observed in other systems. This is due to an increase in the conductance of the inner membrane at high prevailing potentials which permits large changes in respiratory rate with only small associated changes in the potential.
2. Increases in the conductance of the inner membrane imposed with FCCP or by activating the ATPase leads to a decrease in the curvature of the relationship. The system is roughly Ohmic under conditions of high conductance, deviations from linearity becoming more apparent at low conductance.
3. During succinate oxidation, the addition of external NADH leads to an increase in respiratory rate which is accommodated in part by an increase in Δ p and in part by an increase in the conductance of the inner membrane.

ACKNOWLEDGEMENTS

STD gratefully acknowledges the receipt of an SERC studentship.

REFERENCES

1. P. Mitchell, Chemiosmotic coupling in oxidative and photosynthetic phosphorylation. Glynn Research, Bodmin, Cornwall. (1966)
2. D. G. Nicholls, The influence of respiration and ATP hydrolysis on the proton-electrochemical gradient of the inner mitochondrial membrane of rat liver mitochondria as determined by ion distribution, Eur. J. Biochem. 50:305 (1974).
3. N. P. J. Cotton, A. J. Clarke and J. B. Jackson, Changes in membrane ionic conductance, but not changes in slip, can account for the non-linear dependence of the electrochemical proton gradient upon the electron transport rate in chromatophores, Eur. J. Biochem. 142:193 (1984).
4. G. C. Brown and M. D. Brand, Changes in permeability to protons and other cations at high protonmotive force in rat liver mitochondria. Biochem. J. 234:75 (1986).
5. D. Pietrobon, G. F. Azzone, and D. Walz, Effects of funicolosin and antimycin A on the redox driven H^+ pumps in mitochondria-on the nature of "leaks", Eur. J. Biochem. 117:389 (1981).
6. M. Zoratti, M. Favaron, D. Pietrobon, G. F. Azzone, Intrinsic uncoupling of mitochondrial proton pumps. 1. Non-Ohmic conductance cannot account for the nonlinear dependence of static head respiration on Δp, Biochem. 25:760 (1986)
7. G. Krishnamoorthy and P. C. Hinkle, Non-Ohmic proton conduction of mitochondria and liposomes, Biochemistry 23:1640 (1984).
8. A. L. Moore and M. O. Proudlove, Mitochondria and submitochondrial particles, in: "Isolation of membranes and organelles from plant tissues," J. L. Hall and A. L. Moore, eds., Academic Press, London (1983).
9. N. Kamo, M. Muratsugu, R. Hongo, Y. Kobatke, Membrane potential of mitochondria measured with an electrode sensitive to tetraphenyl phosphonium and relationship between proton electrochemical and phosphorylation potential, in steady state, J. Membr. Biol, 49:105 (1979).
10. J. A. Mitchell, D. Phil Thesis. University of Sussex. (1984).
11. S. T. Dobbs and A. L. Moore, The regulation of electron flux in plant mitochondria, Biochem. Soc. Trans., 14:894. (1986).

COMPETITION BETWEEN MALIC ENZYME AND MALATE DEHYDROGENASE IN ISOLATED PLANT MITOCHONDRIA

James H. Bryce and Joseph T. Wiskich

Department of Botany
University of Adelaide
Adelaide
South Australia 5001

INTRODUCTION

It is now sixteen years since Macrae and Moorehouse[1] first described an NAD-dependent enzyme in plant mitochondria which oxidatively decarboxylated malate to pyruvate. Since then it has been established that plant mitochondria normally possess two malate oxidising enzymes; NAD-dependent malate dehydrogenase (EC 1.1.1.37), and NAD-dependent malic enzyme (EC 1.1.1.39). Both enzymes are located in the mitochondrial matrix[2]. Therefore, they may compete for their substrates, and for access to the electron transport chain, in a similar fashion to that described for glycine with TCA* cycle substrates in which glycine is preferentially oxidised[3]. A knowledge of the extent to which these malate oxidising enzymes compete with each other, and how their activity is regulated, is important because malate dehydrogenase produces oxaloacetate which is an essential substrate for the TCA cycle. Therefore, if the activity of these enzymes were unbalanced, the TCA cycle might cease to operate. Furthermore, malic enzyme plays an important role in the decarboxylation of malate in some CAM and C_4 plants.

The equilibrium of malate dehydrogenase for malate oxidation is unfavourable. Therefore, to facilitate its activity, the accumulation of oxaloacetate must be prevented. In studies of competition between malate oxidising enzymes, the activity of malate dehydrogenase may be increased by the inclusion of glutamate in the reaction medium[4], malate dehydrogenase being activated by the removal of oxaloacetate through its transamination by GOT to form aspartate and 2-oxoglutarate. However, in the presence of glutamate, pyruvate, the product of malic enzyme, could be transaminated by GPT to L-alanine and 2-oxoglutarate. Thus, the measurement of pyruvate alone during malate oxidation with glutamate could lead to an underestimation of malic enzyme activity.

*Abbreviations used: BSA, bovine serum albumin; GOT, L-aspartate:2-oxoglutarate aminotransferase, EC 2.6.1.1; GPT, L-alanine:2-oxoglutarate aminotransferase, EC 2.6.1.2; state 3 and state 4, rates of oxygen uptake in the presence and after the depletion of ADP; TCA, tricarboxylic acid; TES, N-tris (hydroxymethyl) methyl-2-aminoethanesulfonic acid; tris, 2-amino-2-hydroxymethylpropane-1, 3-diol.

We therefore set out to determine whether in the presence of glutamate there is significant transamination of pyruvate to alanine during malate oxidation by isolated plant mitochondria, and also whether activation of malate dehydrogenase affects malic enzyme activity.

MATERIALS AND METHODS

Materials

Fresh red beetroot (*Beta vulgaris* L.) and turnip (*Brassica rapa* L.) tissues were obtained from local markets. Chemicals were obtained from Sigma Chemical Co. (St. Louis, Missouri) and Calbiochem (San Diego, California).

Isolation of Mitochondria

Beetroot and turnip mitochondria were isolated according to Bryce and Wiskich[4]. Protein content of mitochondria was estimated by the method of Lowry[5] using BSA as the standard.

Oxygen consumption

Oxygen consumption was measured polarographically at 25°C using a Clark oxygen electrode. The standard reaction medium was 0.25 M sucrose, 10 mM KH_2PO_4, 5 mM $MgCl_2$, and 10 mM TES-KOH (pH 6.8 or 7.3). Metabolism of malate was studied under state 3 conditions using glucose (9.5 mM) and hexokinase (3 units; Sigma H-5250) to continuously regenerate ADP (191 µM).

Measurement of Pyruvate and Alanine formed during Malate Oxidation

Pyruvate was measured spectrophotometrically by determining the oxidation of NADH at 340 nm following deproteinisation of samples[4]. For measurement of L-alanine, deproteinisation was modified to include heating of the acidified samples at 95°C for 5 min before they were neutralised. This ensured that there was no residual malate dehydrogenase activity. L-alanine was measured in a 2 ml reaction mixture which consisted of 2.8 mM EDTA, 667 mM hydrazine hydrate, 27 mM tris (pH 10.0), 0.8 mM NAD, and neutralised sample[6]. The reaction was started by the addition of L-alanine dehydrogenase (0.2 units; Sigma A-7653), and NAD reduction determined spectrophotometrically at 340 nm against an identical assay mixture to which no enzyme was added.

RESULTS AND DISCUSSION

Transamination of Pyruvate to Alanine

Our aim was to investigate a possible non-oxidative loss of pyruvate as a consequence of having glutamate in the reaction mixture during malate oxidation. Beetroot mitochondria were isolated, and pyruvate production measured during their oxidation of malate in the presence of arsenite with and without glutamate. Previously, it has been established that deproteinisation of samples from such reactions does not cause any significant loss of pyruvate[4], and furthermore, that with arsenite neither pyruvate nor 2-oxoglutarate are oxidised. Duplicate reactions were run concurrently in oxygen electrode chambers, except that to one duplicate a known amount of pyruvate (approximately the amount produced by malic enzyme in 10 min) was added. Oxygen uptake was initiated by 19 mM malate with a further addition after 6 min of 29 mM glutamate. Pyruvate was determined in the duplicate ractions at 2 min intervals. During the oxidation of malate alone, the pyruvate content of the duplicate reaction with additional pyruvate remained higher by an amount close to that added (recovery > 95%). However, within 4 min of adding glutamate to the reactions the pyruvate recovery fell to 79%, and within 6 min it was down to 63%. There was no difference in oxygen uptake between the pairs of reactions. Therefore the inclusion of glutamate to facilitate the removal of oxaloacetate and stimulate malate dehydrogenase activity resulted in a loss of pyruvate from the reaction mixture.

The data for oxidation of malate in the presence of glutamate by turnip mitochondria (Table 1) show that the loss of pyruvate was due to its transamination to alanine. Where 388 nmol of pyruvate were initially added to the reaction mixture, the total pyruvate plus alanine after 9 min exceeded that in the reaction mixture where no pyruvate had been added by 366 nmol, a 94% recovery. Similarly, alanine was recovered when it was added rather than pyruvate (data not shown). During malate oxidation with glutamate, some pyruvate produced by malic enzyme is clearly transaminated to alanine. Therefore, determination of malic enzyme activity in the presence of glutomate requires the measurement of both pyruvate and alanine.

Effect of Malate Dehydrogenase on Malic Enzyme

During malate oxidation by freshly isolated beetroot and turnip mitochondria, the activity of malic enzyme was estimated from measurements of pyruvate production with malate alone, but when glutamate was included in the reaction mixture both pyruvate and alanine were measured. Malate dehydrogenase activity was calculated from oxygen uptake not accounted for by malic enzyme.

Malic enzyme was much more active at pH 6.8 in turnip mitochondria than in beetroot mitochondria (Table 2). The percentage of malic enzyme activity accounted for by alanine was 38%, 20%, and 30% for beetroot pH 6.8, turnip pH 6.8, and turnip pH 7.3 respectively. Where malic enzyme was extremely active in turnip mitochondria at pH 6.8, activation of malate dehydrogenase with glutamate had no effect on malic enzyme. On the contrary, with beetroot mitochondria at pH 6.8, where malic enzyme activity was already low, activation of malate dehydrogenase caused a 41% reduction in malic enzyme. With turnip mitochondria at pH 7.3, where malic enzyme activity was lower and glutamate stimulation of malate dehydrogenase greater than at pH 6.8, activation of malate dehydrogenase reduced malic enzyme activity by 18% (Table 2). These data therefore demonstrate that there is competition between malate dehydrogenase and malic enzyme, and that the extent of the competition may be affected by pH.

Table 1. Production of Pyruvate and Alanine by Turnip Mitochondria during the Oxidation of Malate[a] (pH 7.3) for 9 min with Glutamate[a]

	Product (nmōl)		
	Pyruvate	Alanine	Pyruvate + Alanine
Control plus:	1236	219	1455
Pyruvate (388 nmol)	1565	256	1821

[a]Concentrations used were 19 mM malate and 6 mM glutamate.

Table 2. Effect of Glutamate[a] and pH on Malate[a] Oxidation by freshly isolated Turnip and Beetroot Mitochondria

	Oxygen uptake (nmol. mg protein^{-1} min^{-1})					
	Beetroot		Turnip			
	pH 6.8		pH 6.8		pH 7.3	
	ME[b]	MDH[b]	ME	MDH	ME	MDH
Malate	7	7	94	4	68	3
+ Glutamate	4	21	96	26	56	52

[a]Concentrations used were 26 mM glutamate and 18 mM malate.
[b]ME, malic enzyme; MDH, malate dehydrogenase.

ACKNOWLEDGEMENTS

We gratefully acknowledge financial support from the Australian Research Grants Scheme.

REFERENCES

1. A.R. Macrae and R. Moorehouse, The oxidation of malate by mitochondria isolated from cauliflower buds, Eur. J. Biochem. 16:96 (1970).
2. J.T. Wiskich, Control of the Krebs Cycle, in: "The Biochemistry of Plants Vol. 2. Metabolism and Respiration," D.D. Davies, ed., Academic Press, London (1980).
3. I.B. Dry, D.A. Day, and J.T. Wiskich, Preferential oxidation of glycine by the respiratory chain of pea leaf mitochondria, FEBS Lett., 158: 154 (1983).
4. J.H. Bryce and J.T. Wiskich, Effect of NAD and rotenone on the partitioning of malate oxidation between malate dehydrogenase and malic enzyme in isolated plant mitochondria, Aust. J. Plant Physiol., 12:229 (1985).
5. O.H. Lowry, N.J. Rosebrough, A.L. Farr, and R.J. Randall, Protein measurement with the Folin phenol reagent, J. Biol. Chem., 193:265 (1951).
6. D.H. Williamson, L-alanine: determination with alanine dehydrogenase, in: "Methods of Enzymatic Analysis," H.U. Bergmeyer, ed., Academic Press, New York and London (1974).

EFFECTS OF ADENINE DERIVATIVES ON PERMEABILITY AND DEHYDROGENASE ACTIVITIES OF POTATO MITOCHONDRIA

Michèle Chauveau, Isabelle Funes and Jean Roussaux

Laboratoire de Biologie Végétale IV (CNRS, UA 1180)
Université Pierre et Marie Curie, Paris, France

INTRODUCTON

Some substituted adenines with or without cytokinin activity are able to inhibit mitochondrial oxidations (Miller, 1979, 1982; Chauveau et al., 1983). In particular Roussaux et al. (1986) have reported on a selective action by some adenine derivatives on the cyanide-sensitive cytochrome pathway. Because these compounds are rather lipophilic this effect could result, at least for a part, from interactions with the mitochondrial membranes. To check this hypothesis, the variations of mitochondrial permeability to K+ or to substrates were studied by measuring the mitochondrial swelling in the presence or in the absence of 6-benzylaminopurine (BA) or 6-benzoylaminopurine (RD 1). The effects observed were compared with those of two auxin-related compounds : 2,3,5-triiodobenzoic acid (TIBA) and 2,4-dichlorophenoxyacetic acid (2,4-D). In order to define the conditions of specificity more accurately, structural modifications have been taken into account.

MATERIALS AND METHODS

Mitochondria were isolated from fresh potato tubers (*Solanum tuberosum* L. var. Bintje) by differential centrifugation and purification on a Percoll gradient (Moreau and Romani, 1982). O2 uptake was measured at 25°C with a Clark oxygen electrode in a medium containing 300 mM mannitol, 5 mM MgCl2, 10 mM KCl, 10 mM phosphate buffer and 1 mg/ml bovine serum albumin. pH was 6.5 for malate oxidation *via* ME and 7.9 for malate oxidation *via* MDH (Chauveau et al., 1983). pH was 7.2 for other substrates. Mitochondrial swelling was followed spectrophotometrically at 520 nm according to Yoshida and Sato, (1968). Dehydrogenase activities were measured at 600 nm using phenazine methosulfate-dichlorophenolindophenol (PMS-DCPIP) as an electron acceptor in 5 mM phosphate buffer (pH 7.2 or 7.9, depending on substrate used) containing 1 uM antimycin and 1 mM KCN. Mitochondrial proteins were estimated by mineralization and nesslerization. BA, TIBA and 2,4-D were from Serva. Other compounds were synthesized by Dr Rolando.

RESULTS

Effects on mitochondrial oxidations.

As shown in table I, BA specifically inhibited malate oxidation

Table I. Effects of adenine derivatives and auxin-related compounds on the oxidation of various substrates. (% inhibition).

	nmol O2/min.mg protein				
	Succinate (10 mM)	Malate (30 mM)		NADH (1 mM)	Duroquinol (1 mM)
		via MDH	*via* ME		
Control	336	62	114	264	504
BA (0.75 mM)	336 (0%)	18 (71%)	107 (6%)	264 (0%)	504 (0%)
RD1 (0.75 mM)	45 (86%)	62 (0%)	114 (0%)	264 (0%)	504 (0%)
TIBA (1.5 mM)	37 (89%)	6 (90%)	0 (99%)	72 (73%)	252 (50%)
2,4-D (3 mM)	100 (70%)	21 (66%)	9 (92%)	124 (53%)	180 (64%)

via MDH and RD 1 specifically inhibited succinate oxidation. Both compounds were ineffective on malate oxidation *via* ME as well as on NADH and duroquinol oxidation. These results suggest that the compounds act on the initial segment of the respiratory chain, before the ubiquinone branch point. By contrast, auxin-related compounds (TIBA and 2,4-D) were inhibitors of all oxidations to various extents.

Effects on mitochondrial swelling

In an iso-osmotic KCl medium, mitochondria underwent passive swelling due to K+ penetration in the matrix space. At concentrations which inhibit succinate or malate oxidation RD 1 had strickly no effect on membrane permeability to K+, but BA increased the swelling slightly. By contrast, TIBA and 2,4-D increased the permeability to K + markedly (Fig. 1A).

Mitochondria energized by substrates exhibited a weak contraction. In the presence of succinate, RD 1 inhibited this contraction and restored passive swelling as in the control. BA had only a slight effect (Fig. 1B). On the contrary, with malate (+ glutamate), at pH 7.9 to make MDH

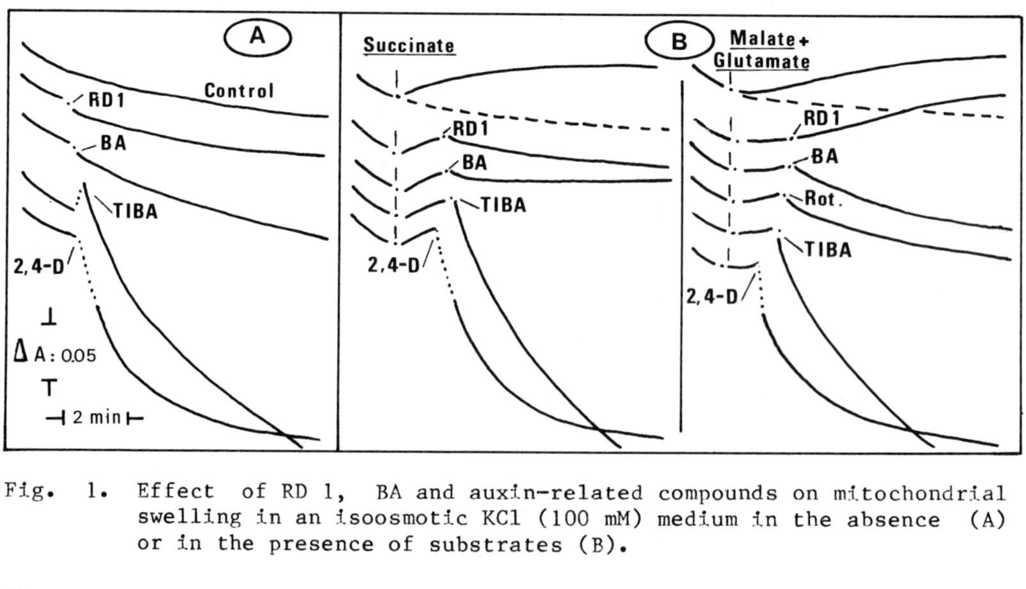

Fig. 1. Effect of RD 1, BA and auxin-related compounds on mitochondrial swelling in an isoosmotic KCl (100 mM) medium in the absence (A) or in the presence of substrates (B).

operative, no effect of RD 1 and a potent inhibitory effect of BA were observed on mitochondrial contraction. Moreover, BA induced a swelling more important than in the control, just as did rotenone (Fig. 1B). Thus, BA and RD 1 have the same specificity on mitochondrial contraction as on mitochondrial oxidations. Auxin-related compounds inhibited membrane contraction whatever the substrate used and brought about a fast swelling.

In order to test a possible action of the adenine derivatives on the substrate transporting systems, the mitochondrial swelling was measured in iso-osmotic ammonium salt media. In solution of ammonium succinate or malate, the light swelling observed (Fig. 2) appeared to be a consequence of dicarboxylate penetration accross the inner mitochondrial membrane (Phillips and Williams, 1973). RD 1 and BA had no effect on this transport, even in the presence of small amounts of ammonium phosphate which increased the dicarboxylate transport. TIBA and 2,4-D induced a large swelling only in the absence of ammonium phosphate.

Effects on dehydrogenase activities

RD 1 and BA were assayed on succinate- or malate-(PMS-DCPIP) reductase activity. Under this condition only a short segment of the respiratory chain was functioning. Fig. 3 shows that BA was ineffective on dehydrogenase activities whatever the substrate used. By contrast, RD 1 inhibited succinate-(PMS-DCPIP) reductase to 70%.

Structural requirements for inhibition

Table II summarizes the cytokinin activity and the inhibitory effect on the oxidative pathway of ten adenine derivatives. As BA, alkyl derivatives are all specific inhibitors of malate oxidation and have a good cytokinin activity. Acyl derivatives exhibit a weaker cytokinin activity. Only RD 1, RD 4 (6-thenoylaminopurine) and the fluoroderivative m-F-RD 1 were specific inhibitors of succinate oxidation. On the contrary, RD 2 (6-cyclohexoylaminopurine) and the fluoroderivative p-F-RD 1 had no inhibitory action.

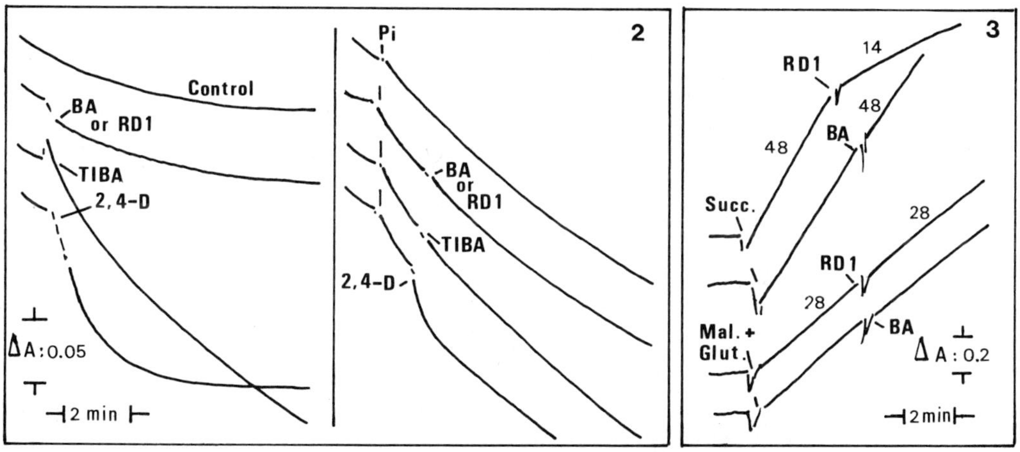

Fig. 2. Effect of RD 1, BA and auxin-related compounds on mitochondrial swelling in an iso-osmotic ammonium succinate or ammonium malate medium. Ammonium salts : 100 mM; RD 1, BA : 0.75 mM; TIBA : 1.5 mM; 2,4-D : 3 mM.

Fig. 3. Effect of RD 1 and BA on PMS-DCPIP reductase. RD 1, BA : 0.75 mM.

Table II. Inhibition of substrate oxidation and cytokinin activity of some adenine derivatives.

		Inhibition (%)		Cytokinin
		Succinate	Malate	activity
Alkyl aminopurine	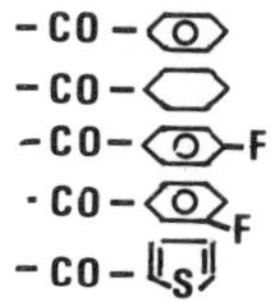			
BA		0	71	+++
RD 12		0	76	++
p-F-BA		0	68	+++
m-F-BA		0	70	+++
RD 14		0	71	++
Acyl aminopurine	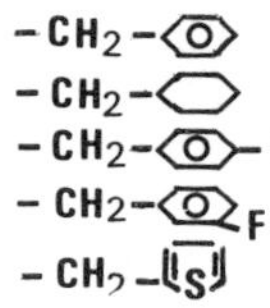			
RD 1		86	0	+
RD 2		0	0	0
p-F-RD 1		0	0	+
m-F-RD 1		79	0	0
RD 4		95	0	+

CONCLUSIONS

Whereas auxin-related compounds have no specific effect on electron transport, adenine derivatives do present a specific effect : acyl aminopurines inhibit succinate oxidation and alkyl aminopurines inhibit malate oxidation only when MDH is operative. Adenine derivatives do not inhibit the substrate carriers in mitochondria and do not modify membrane permeability appreciably. Thus their actions are specifically located on complex I or complex II. BA and other alkyl derivatives are not active on the NADH-linked dehydrogenases themselves but on the rotenone-sensitive site as shown previously (Chauveau et al, 1973). On the opposite RD 1 and some other acyl derivatives seem to act at the succinate dehydrogenase level.

Replacing -CO- by -CH2- seems to be responsible for the specificity. Whatever the substitution, alkyl compounds are active on malate oxidation whereas there are more stringent requirements for the effect of acyl compounds on succinate oxidation.

REFERENCES

Chauveau M., Dizengremel P. and Roussaux J., 1983, Interaction of benzylaminopurine with electron transport in plant mitochondria during malate oxidation. Plant Physiol. 73 : 945.

Miller C.O., 1979, Cytokinin inhibition of respiration by cells and mitochondria of soybean, Glycine max (L.) Merrill. Planta, 146 : 503.

Miller C.O., 1982, Cytokinin modification of mitochondrial function. Plant Physiol., 69 : 1274.

Moreau F. and Romani R., 1982, Preparation of avocado mitochondria using self-generated Percoll density gradient and changes in buoyant density during ripening. Plant Physiol., 70 : 1380.

Phillips M.L. and Williams G.R., 1973, Anion transporters in plant mitochondria. Plant Physiol., 51 : 667.

Roussaux J., Dizengremel P. and Chauveau M., 1986, Inhibition of mitochondrial oxidations by adenine derivatives and auxin related compounds. J. Plant Physiol., 123 : 55.

Yoshida K. and Sato S., 1968, Swelling and contraction of isolated plant mitochondria. I. Passive swelling in sugar and electrolyte solutions. J. Fac. Sci. Univ. Tokyo III, 10 : 49.

KINETICS OF ELECTRON TRANSPORT IN COMPLEX III OF PLANT MITOCHONDRIA DURING AGING *IN VITRO*

Anne-Catherine Fricaud and Jacques Dupont

Laboratoire de Biologie Végétale IV (CNRS, UA 1180)
Université Pierre et Marie Curie, Paris, France

INTRODUCTION

During storage at 0°C, isolated organelles such as mitochondria undergo progressive degradations. The rates of substrate oxidations or the resistance toward inhibitors are changed. At the end of the process, the membrane components (proteins, lipids) are destroyed. These phenomena are strongly increased by aging of the organelles *in vitro* at 25°C[1,2].

The aim of this work was to characterize the effect of aging *in vitro* on the complex III (*b*-c_1) of the respiratory chain of the mitochondria from potato tubers. The electron flow through this complex has been shown to be strongly inhibited by antimycin [3]. The action of this inhibitor is not total and the changes in the degrees of resistance to antimycin were followed during the aging process. On the other hand, the number of cytochromes *b* (cyt. *b*) in plant mitochondria is still a matter of debate [3]. The study of the different kinetics of degradation of the cyt. *b* could provide a means to characterize the various species of cyt. *b*.

MATERIALS AND METHODS

Mitochondria were prepared from potato tubers (*Solanum tuberosum L.*) as previously described [4]. Mitochondria were aged at 25°C in the absence of bovine serum albumin and samples were taken at various times for the determination of electron transport activities. Oxygen uptake was measured at 22°C with a Clark-type electrode. Cyanide and propylgallate were added after antimycin to measure the electron flow by-passing the antimycin site. The reduction of cyt. *b* and *c* were studied at 560 *minus* 580 nm and 550 *minus* 580 nm respectively with an Aminco DW 2A dual-wavelength spectrophotometer. All assays were made in the following medium: 300 mM mannitol, 10 mM $MgCl_2$, 5 mM KCl and 10 mM phosphate buffer (pH 7.20).

RESULTS

Effect of aging on resistance to antimycin

Figure 1 shows that after 7 h of aging the rate of oxidation of NADH in the absence of antimycin was only 40% of the rate measured initially

(t_0). In the presence of antimycin, the rate of oxidation of NADH was only 20% of the initial inhibited rate. A parallel decrease of the rate of NADH oxidation and of the resistance to antimycin was therefore observed.

Figure 2 A shows the optical changes of the baseline (trace _a_) and the reductions of cyt. _c_ (trace _b_) and cyt. _b_ (trace _c_) which result from the addition of NADH to freshly isolated mitochondria. The addition of antimycin brought about an increased reduction of cyt. _b_ (trace _c_) and a full oxidation of cyt. _c_ (trace _b_). A further addition of cyanide led to the complete reduction of cyt. _c_ (trace _b_). During aging, the reduction of cyt. _b_ upon antimycin addition increased (fig. 2 B, trace _c_) and then decreased (fig. 2 C, trace _c_). In contrast, the oxidation of cyt. _c_ upon antimycin addition decreased only during aging (fig. 2 B, C, traces _b_), once corrected for the baseline changes. The addition of cyanide led to a reduction of cyt. _c_ that became biphasic (fig. 2 B, trace _b_) with a slow phase a lower reduction of cyt. _c_ than in the control mitochondria. This latter situation appeared more marked as aging progressed (fig. 2 A, B, C, traces _b_).

Obviously, these results suggest deep alterations in the respiratory chain expressed by lower levels of reduction.

Cytochromes b variations during aging

Figure 3 A, spectrum _a_ is a classical low temperature spectrum of NADH-antimycin-aerobic _minus_ aerobic potato mitochondria. A 3-peaked α - band (554, 559 and 563 nm) and a 2-peaked β -band + 1 shoulder (536 nm, 527 nm and 543 nm respectively) clearly identify these peaks as belonging to the _b_ group of cytochromes. Cytochrome oxidase is detected by the absorbance at 595 nm. During the first hours of aging, the absorbance at 563 nm decreased whereas the absorbances at 554 nm and 559 nm increased

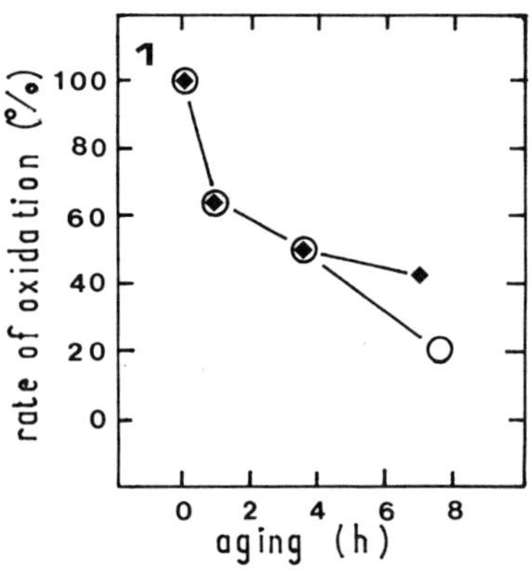

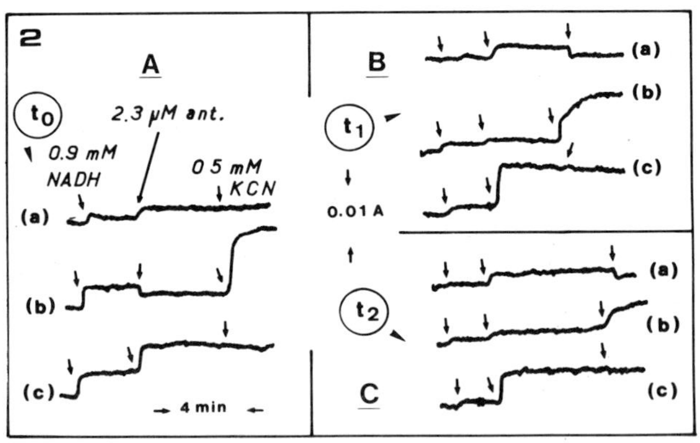

Fig. 1. Changes in the rate of exogenous NADH oxidation during aging in the presence or in absence of antimycin. NADH was 1 mM and antimycin 0.9 μM. ◆ , no antimycin, initial rate: 100%= 270 nmol min^{-1} mg^{-1} prot. ○ , with antimycin, initial rate: 100%= 11 nmol min^{-1} mg^{-1} prot.

Fig. 2. Absorbance changes at 540 _minus_ 580 nm (A), 550 _minus_ 580 nm (B) and 560 _minus_ 580 nm (C) observed during aging of washed mitochondria upon addition of 0.9 mM NADH followed by 2.3 μM antimycin and 0.5 mM KCN. Bandpath: 2.5 nm. Protein concentration: 2.42 mg ml^{-1}. t_0, t_1 and t_2 =0, 48, 258 min aging respectively.

(spectra b and c). After 520 min aging, the absorbance of the 3 cyt. b and of cytochrome oxidase decreased (spectrum d).

A similar experiment was carried out in figure 3 B, spectrum a. As soon as the absorbance at 596 disappeared (after 265 min of aging), a peak at 548 nm and a shoulder at 554 nm were detected (fig. 3 B, spectra b). It is characteristic of the band of cyt. c which is reduced, even in the absence of KCN. The various cyt. b and c have been detected before aging after addition of cyanide to the sample cell (fig. 3 C, spectrum a). After 590 min aging, a general decrease in absorbance of all the cytochromes was observed (fig. 3 C, spectrum b). This phenomenon was observed in the absence of cyanide (fig. 3 A, spectrum d) and likely resulted from the breakdown of the cytochromes rather than from variations of their midpoint potentials.

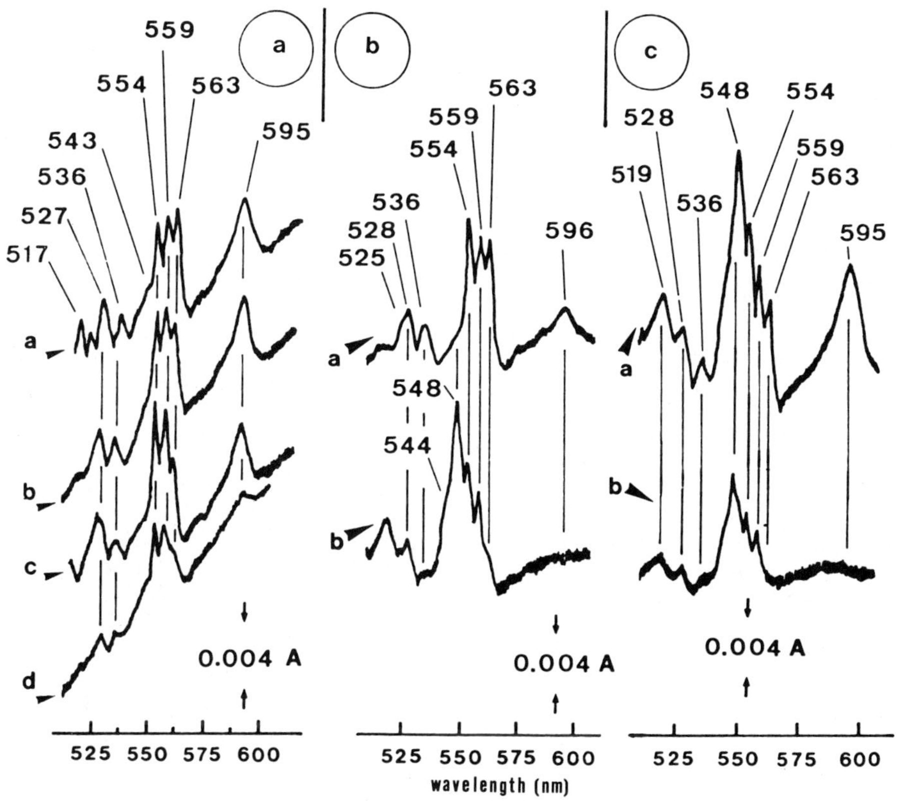

Fig. 3. Difference spectra of purified mitochondria at liquid nitrogen temperature. A: spectra recorded as follows: NADH (2.15 mM) + antimycin (10 µM) aerobic minus aerobic. Bandpath: 0.95 nm. Mitochondrial protein: 0.96 mg ml^{-1}. Spectrum a: control; spectra b to d correspond 267 min, 400 min and 520 min aging respectively. B: spectra were recorded as follows: NADH (2.35 mM) + antimycin (4.7 µM) aerobic minus aerobic (O_2 bubbling). Bandpath: 0.8 nm. Mitochondrial protein: 1.5 mg ml^{-1}. Spectrum a: control; spectrum b: 265 min aging. C: The conditions were NADH (2.35 mM) + antimycin (4.7 µM) + cyanide (9.4 mM) aerobic minus aerobic. All other conditions as in B. Spectrum a: control; spectrum b: 425 min aging.

CONCLUSIONS

The present study describes some aspects of the kinetics of electron transport in complex III when isolated potato mitochondria are submitted to aging *in vitro*.

Antimycin does not completely inhibits electron transfer in complex III in freshly isolated mitochondria when the electron flow is delivered from NADH. During aging the net electron flow by-passing the antimycin binding site decreases as the pressure of electrons delivered by NADH decreases (fig. 1).

The transiently enhanced increase in antimycin-induced reductibility of *b* type cytochromes (fig. 2 B, C, traces *c*) is attributable to the species characterized by bands at 554 nm and 559 nm. The progressive decrease in the height of the peak at 563 nm likely results from a specific breakdown of this compound, possibly as a consequence of a local formation of radicals which could attack the heme of the cytochrome [1,5]. This is evidenced by the impossibility to increase the level of reducibility of cyt. *b*-563 by addition of cyanide (fig. 3 C). The differential breakdown of the various spectral species of cyt. *b* support the conclusion of the existence of at least two different *b*-type heme compounds.

In aged mitochondria, the destruction of cytochrome oxidase, mimicking the effect of cyanide, decreases the oxidizing power of the respiratory chain. This leads to a partial reduction of cyt. *c* in the presence of antimycin alone. Such a cyt. *c* reduction can be obtained even if the electron flow by-passing the antimycin site is decreased (fig. 1).

REFERENCES

1. J. Dupont, P. Rustin, and C. Lance, Effect of lipid peroxidation on cytochrome composition during the aging of plant mitochondria, *in*: "Stucture, Function and Metabolism of Plant Lipids", P.-A. Siegenthaler and W. Eichenberger, ed., Elsevier Science Publisher, Amsterdam (1984).
2. R. Douce, and C. Lance, Alteration des activités oxydatives et phosphorylantes de mitochondries de Chou-fleur sous l'action de phospholipases et du vieillissement, *Physiol. Vég.*, 10: 181 (1972).
3. Y. Hatefi, The mitochondrial electron transport and oxidative phosphorylation system, *Annu. Rev. Biochem.* 54: 1015 (1985).
4. P. Diolez, and F. Moreau, Effect of bovine serum albumin on membrane potential in plant mitochondria, *Physiol. Plant.*, 59: 177 (1983).
5. J. Dupont, P. Rustin, and C. Lance, Interaction between mitochondrial cytochromes and linoleic acid hydroperoxide. Possible confusion with lipoxygenase and alternative pathway, *Plant Physiol.*, 69: 1308 (1982).

THE EFFECTS OF INHIBITORS OF THE B - C1 COMPLEX ON THE RESPIRATION OF MITOCHONDRIA FROM AGED POTATO DISCS

João Daniel Arrabaça and Ana Maria Tenreiro

Department of Plant Biology, Faculty of Science
University of Lisbon, Portugal

INTRODUCTION

The nature and function of the alternative, cyanide insensitive terminal oxidase, found in most plant tissues and in some microorganisms, has long been puzzling plant physiologists and biochemists.

The search for the alternative terminal oxidase resulted in claims of its partial purification (Kay and Palmer, 1985; Bonner et al., 1986) from mitochondria of Arum maculatum, in the form of a quinol oxidase, insensitive to cyanide and sensitive to inhibitors of the alternative pathway, like salicyl hydroxamate (SHAM) and propyl gallate. However, some doubts have been cast (Rustin et al., 1986) as to the correspondence of the isolated enzymes to the alternative terminal oxidase. Alternatively, a provocative proposal has been put forward (Rustin et al., 1984a, b) assigning the cyanide insensitive respiration to the formation and scavenging of fatty acid peroxy radicals inside the inner mitochondrial membrane. This hypothesis could imply some interactions between the proton-motive Q-cycle (Mitchell and Moyle, 1985), known to work in plant mitochondria (Rich and Moore, 1976) and the cyanide insensitive respiration.

We tried to take advantage of mixothyazol, a recently described inhibitor of the reduction of the Rieske iron-sulfur center of the b-c1 complex (Thierbach and Reichenbach, 1983) and of DTNB (5-5´-dithiobis(2-dinitrobenzoate), known to act on the same site (Marres et al., 1982).

MATERIAL AND METHODS

Discs were obtained from potato tubers purchased in the local market. They were incubated for 24 hours, at room temperature and with intense aeration, in buffer containing 0.1 mM $CaSO_4$ and 10 µM Fe-EDTA, at about 4 l of buffer per kg of discs. It was found that the presence of iron in the medium would generally lead to higher rates of cyanide insensitive respiration in isolated mitochondria.

Mitochondria were isolated as described by Dizengremel and Lance (1976) and purified with self-generating gradients of 24% Percoll (Jackson et al., 1979).

Respiration rates of purified mitochondria were measured with a Clark-type oxygen electrode (Hansatech), maintained at 25 ^oC, using the buffers described by Dizengremel and Lance (1976).

RESULTS AND DISCUSSION

In this system, the respiration of succinate can be up to 60% insensitive to cyanide, in state 3. As shown in Figure 1, the addition of 5 mM ferricyanide, in the presence of rotenone, in order to avoid interaction with complex I, resulted in the total inhibition of oxygen consumption, while the electron acceptor was being reduced. After reduction of ferricyanide, the cyanide insensitive respiration showed its initial rate. The addition of mixothyazol inhibited the reduction of the dye, and induced a renewal of oxygen consumption, indicating that it is accepting electrons from cytochrome c1 or c, and not from complex II.

DTNB inhibited the respiration of succinate (Figure 2), in a way similar to mixothyazol. Subsequent additions of antimycin resulted in further reduction of the respiratory rate and SHAM completely stops respiration. However, ascorbate (plus TMPD) could still be respired at high rates, though with a very low degree of coupling. It seems that the effects of DTNB and antimycin could be partially additive. Figure 3 shows that the inhibition of succinate respiration by KCN and by mixothyazol also show some degree of additivity. Again, the addition of SHAM lead to rate of respiration near zero.

The respiration of externally added NADH showed a quite different pattern towards the action of inhibitors. In fact, either cyanide, antimycin and mixothyazol inhibited almost completely the rate of respiration and SHAM had no effects on the residual respiration.

The effect of ferricyanide on the respiration of succinate could well be explained by competition towards reducing equivalents by the electron acceptor and a non-enzymatic mechanism. On the other hand, given the rates of ferricyanide reduction, the activity of a terminal oxidase should still be visible. SHAM had negligible effect on the oxidation of ferricyanide.

The additivity of the effects of antimycin and mixothyazol, present at concentrations that should fully inhibit their respective sites of action, could indicate that indeed there could be an interplay between the electron transport at the level of the Q-cycle and the oxido-reduction of ubiquinone in the cyanide insensitive respiration, probably the quinol or the semiquinone. Thus, the cyanide insensitive respiration may be somawhat more complicated than a reaction exclusively catalyzed by a simple terminal oxidation.

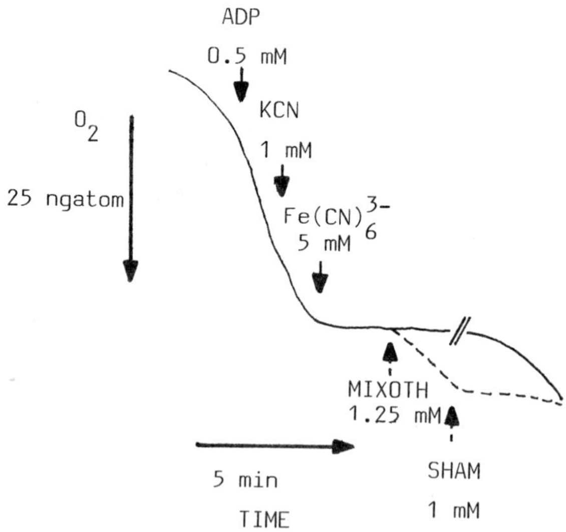

Fig.1. THE EFFECT OF FERRICYANIDE ON THE RESPIRATION OF SUCCINATE
5 mM succinate, 2.5 μM rotenone
0.2 mM ATP, 0.2 mg prot.ml^{-1}

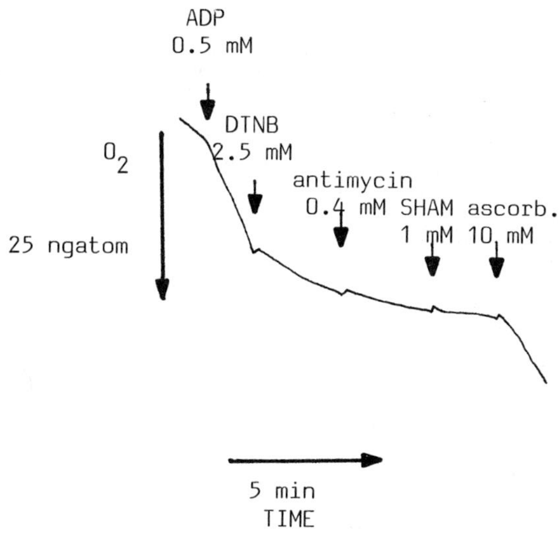

Fig. 2. THE EFFECT OF DTNB ON SUCCINATE RESPIRATION
conditions as in Fig. 1.

We are fully aware that other techniques, to follow the oxidation of ubiquinol and the oxido-reduction of the Rieske center are needed to complete the present data. Also, the use of a different biological system, like Arum, mitochondria where the physiological function of the cyanide insensitive respiration may be different, could prove important.

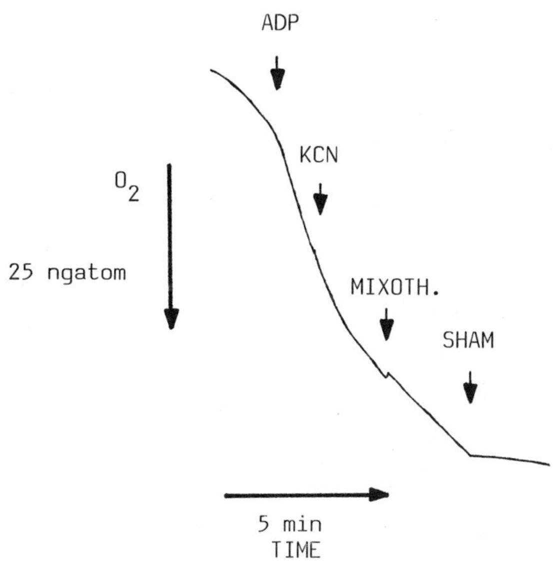

Fig. 3. THE EFFECT OF MIXOTHYAZOL ON SUCCINATE RESPIRATION
conditions as in Fig. 1.

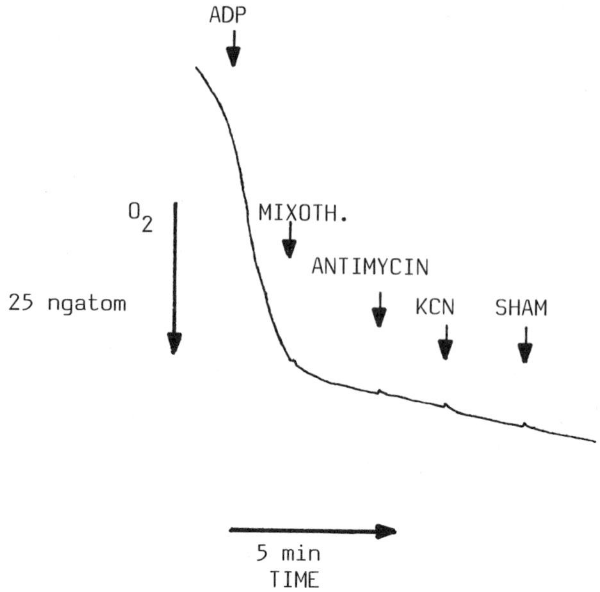

Fig. 4. INHIBITION OF EXTERNAL NADH RESPIRATION
ATP ommited, NADH concentration 10 mM

REFERENCES

Bonner Jr., W., Clarke, S. D. and Rich, P., 1986, Partial purification and characterization of the quinol oxidase activity of Arum maculatum mitochondria, Plant Physiol. 80:838

Dizengremel, P. and Lance, C., 1976, Control of changes in mitochondrial activities during ageing of potato slices, Plant Physiol., 58:147

Jackson, C., Dench, J. E., Hall, D. O. and Moore, A. L., 19879, Separation of mitochondria from contaminating subcellular structures utilizing sili silica sol gradient centrifugation, Plant Physiol., 64:150

Kay, C. J. and Palmer, J. M., 1985, Solubilization of the alternative oxidase of cuckoo-pint (Arum maculatum) mitochondria, Biochem. J. 228:309

Marres, C. A. M., DeVries, S. and Slater, E. C., 1982, The site of inhibition by 5,5'-dithiobis(2-nitrobenzoate) in ubiquinol:cytochrome c oxidoreductase, Biochim. Biophys. Acta 681:323

Mitchell, P. and Moyle, J. 1985, The role of ubiquinone and plastoquinone in chemiosmotic coupling between electron transfer and proton translocation, in: "Coenzyme Q", G. Lenaz , ed.; WileyChichester

Rich, P. R. and Moore, A. L., 1976, The involvement of the protonmotive ubiquinone cycle in the respiratory chain of higher plants and its relation to the branch point of the alternate pathway, FEBS Lett. 65:339

Rustin, P., Alin, M.-F. and Lance, C., 1986, Discrimination between duroquinol oxidase activity and the terminal oxidation step of the cyanide -resistant electron transport pathway of plant mitochondria, Biochem. Biophys. Res. Commun., 135:677

Rustin, P., Dupont, J. and Lance, C., 1984a, Free radicals and cyanide resistance in plant mitochondria, in:"Structure, Function and Metabolism of Plant Lipids", P.-A. Siegenthaler and W. Eichenberger, eds., Elsevier,

Rustin, P., Dupont, J. and Lance, C.,1984b, Involvement of lipid peroxy radicals in the cyanide-resistant electron transport pathway, Physiol. Vég. 22:643

Thierbach, G. and Reichenbach, H., 1983, The effect of the new antibiotic myxothiazol on the respiration of Paracoccus denitrificans, Arch. Microbiol., 134:104

THE REDOX STATE OF MITOCHONDRIAL PYRIDINE NUCLEOTIDES VERSUS RATE OF SUBSTRATE OXIDATION

Pierre Rustin*, Michel Neuburger**, Roland Douce** and Claude Lance*

*Laboratoire de Biologie Végétale IV (CNRS, UA 1180) Université Pierre et Marie Curie, Paris, France and

**Département de Recherche Fondamentale, Laboratoire de Physiologie Cellulaire Végétale (CNRS, UA 576), Centre d'Etudes Nucléaires et Université Scientifique et Médicale de Grenoble, Grenoble, France

INTRODUCTION

In addition to the tricarboxylic cycle (TCA) enzymes, the matrix of plant mitochondria is endowed with enzymes involved in specific metabolic pathways. Thus, mitochondria of C_3 plants are equipped with a glycine decarboxylase, present in very large amount, while those of CAM (Crassulacean Acid Metabolism) plants are especially rich in NAD^+-malic enzyme. These metabolic pathways are bound to interact at the level of the matrix pyridine nucleotides (PN). Such interactions are experimentally supported by the effect of oxaloacetate (OAA) on O_2 uptake linked to the oxidation of substrates generating NADH. In the presence of malate dehydrogenase (MDH), OAA constitutes a powerful NADH-reoxidizing system whatever the enzyme involved in the production of NADH, and successfully competes with the respiratory chain[1].

However, a constant equilibration of the different metabolic pathways through the redox equilibration of the PN and NAD^+-linked dehydrogenases is difficult to admit. For example, the added reducing powers of the NADH-generating TCA enzymes, of malic enzyme and of glycine decarboxylase would maintain OAA at such low levels that citrate synthase would not be able to work, and the TCA as well. It has been suggested that proteins are associated according to their involvement in metabolic pathways[2]. Such associations (called metabolons) could determine the existence of PN micro-domains that would not necessarily equilibrate. In addition, it must be noticed that PN are distributed between free and bound forms, and that their amounts in the matrix are highly variable, due to their ability to be transported accross the inner membrane[3].

The aim of this work, carried out on Percoll-purified mitochondria from potato tuber, was to simultaneously study the rate of O_2 uptake and the redox state of the matrix PN associated to the oxidation of various substrates, in order to characterize the modalities of the PN functioning. The different methods of PN reduction measurement presently available have been used, since each of these presents specific drawbacks, <u>i.e.</u> UV measurements are complicated by the unspecific absorption of ketoacids in the UV region, spectrofluorometric signals of PN depend on the ratio of bound and free PN, whereas enzymic measurements only give punctual values.

MATERIALS AND METHODS

Mitochondria were extracted from potato (*Solanum tuberosum* L.) tubers and purified on two successive self-generating Percoll gradients. Pea (*Pisum sativum* L.) leaf mitochondria were isolated and purified according to Day et al.[4]. The simultaneous recording of PN redox state (340 *minus* 375 nm) and of O_2 uptake were carried out according to Rustin and Valat[5], whereas spectrofluorometric measurements of the redox state of PN were carried out according to Neuburger and Douce[6], and bioluminescent assays according to Karp et al.[7]. All experiments were carried out in a medium consisting of 0.3 M mannitol, 10 mM KCl, 5 mM $MgCl_2$, 10 mM phosphate buffer and 1 mg ml^{-1} BSA. Unless otherwise indicated, the pH of the medium was adjusted at pH 7.3.

RESULTS

Table I presents the values measured for the O_2 consumed and the PN redox state (studied by UV spectrophotometry) during the oxidation of various substrates. Under state 3 conditions, succinate is the substrate the more actively oxidized. Since the activity of the respiratory chain is not limiting, this allows one to estimate the reducing power (*i.e.* NADH) generated by the various enzymes. It appears that isocitrate is slowly oxidized when compared with other substrates and therefore gives the lowest reducing power. Under these state 3 conditions, PN are maintained oxidized, supporting the conclusion that the respiratory chain activity is not the limiting factor of the oxidation rates in potato tuber mitochondria.

As shown in table 1, in pea leaf mitochondria oxidizing glycine, matrix PN, strongly reduced under state 4 conditions, remain slightly reduced under state 3 conditions : the activity of the glycine decarboxylase complex is so high that the respiratory chain becomes the limiting factor of the oxidation. It is noteworthy that the concentration

Table 1. Simultaneous measurement of O_2 uptake and PN redox state during the oxidation of various substrates by potato tuber and pea leaf mitochondria.

Substrates	State 3		State 4	
	O_2 uptake (nmol min^{-1} mg^{-1}prot)	NAD reduction (%)	O_2 uptake (nmol min^{-1} mg^{-1}prot)	NAD reduction (%)
Potato tuber				
Succinate	180	0	60	100[a]
Malate	130	0	30	26
α-ketoglutarate	175	0	43	40
Isocitrate	75	0	20	44
Pea leaf				
Glycine	186	8	83	92

[a] The value of NAD^+ reduction caused by reversed electron flow from succinate under state 4 condition was taken as a 100% reduction of the NAD pool .

of matrix PN will play a major role by controlling the activity of the NAD-dependent enzymes according to their Km towards NAD^+ and thus affecting the ratio between the activities of the enzymes and that of the respiratory chain. Under state 4 conditions, succinate is also oxidized more rapidly that the NAD-linked substrates. This again suggests that the overall rate of oxidation of these substrates is mainly controlled by the activity of the concerned dehydrogenases rather than by the activity of the respiratory chain. However, the comparison of these oxidation rates with the redox states of matrix PN clearly indicates that no relationship actually exists between these two parameters. In the presence of malate, which is rapidly oxidized under state 4 conditions, only 26% of the PN are reduced. In contrast isocitrate, which gives a very low rate of O_2 uptake, reduces about half of the matrix PN, a value similar to the measurement obtained with α-ketoglutarate which presents gives a significantly higher rate of O_2 uptake.

Such an observation was confirmed by spectrofluorometric measurement of PN redox changes during malate and citrate oxidation (Fig. 1A). In order to stimulate malic enzyme activity, malate oxidation was studied at pH 6.5. It is shown that the rate of malate oxidation under state 4 conditions is significantly higher than the rate of citrate oxidation under similar conditions. Again, when comparing the level of PN reduction, it appears that PN were less reduced with malate than with citrate.

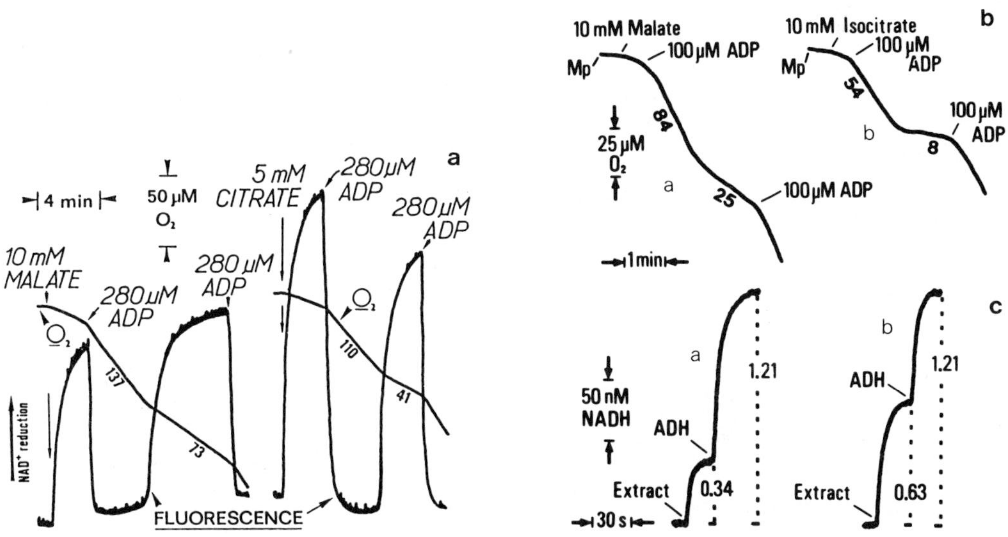

Figure 1. Oxygen uptake and redox state of matrix PN during the oxidation of malate, citrate and isocitrate by potato tuber mitochondria. A, Spectrofluorometric measurement of the redox changes of PN and oxygen uptake associated with malate and citrate oxidation. The experiment using malate as substrate was carried out at pH 6.5. B, O_2 uptake during malate (a) and isocitrate (b) oxidation. Number along the traces are nmol O_2 min^{-1} mg^{-1}protein. Protein concentration was 0.3 mg ml^{-1}. C, Luminescent assay of NADH present in the mitochondrial matrix during malate (a) and isocitrate (b) oxidation under state 4 conditions. Malate oxidation was studied at pH 7.3 in the presence of 100 μM coenzyme A. NADH was extracted from mitochondria by ethanol treatment. After measurement of the NADH present in the extract, the addition of alcohol dehydrogenase (ADH) causes the full reduction of the NAD^+ present in the extract (0.15 mg protein ml^{-1}). Dotted lines are nmol NADH mg^{-1} protein. Numbers along the polarographic recordings are nmol O_2 consumed min^{-1} mg^{-1} protein.

Finally, the redox state of PN during malate and isocitrate oxidation under state 4 conditions was measured by using the luminescent assay. In order to stimulate malic enzyme activity, malate oxidation (at pH 7.3) was studied in the presence of coenzyme A[8]. NADH reduced by malate or isocitrate during state 4 was detected by the luminescent reagent upon addition of the extract. It appears that 28% (0.34 nmol) of the PN were reduced by malate, whereas 52% (0.63 nmol) were reduced by isocitrate (Fig 1C). Compared with the respective rates of oxygen uptake linked to the oxidation of the two substrates under similar conditions (Fig. 1B), it again appears that a higher rate of oxygen uptake was measured using malate as substrate, despite the fact that less NADH was present in the mitochondrial matrix.

CONCLUSION

Whatever the method used to measure the redox state of PN, it appears that in potato tuber mitochondria no quantitative relationship exists between the level of reduction of the PN and (a) the rate of O_2 uptake under state 4 conditions or (b) the reducing power (as estimated under state 3 conditions) represented by the activity of the various NAD-dependent enzymes. Such an observation leads to the conclusion that the respiratory chain does not react at an identical rate with all the NADH present in the matrix space. On the other hand, a spatial compartmentation of the matrix PN, leading to the existence of several distinct pools, has also to be ruled out on the basis of the OAA effect on the oxidation of NADH for all NAD-dependent enzymes[1]. Therefore, one has to admit that the amount of PN present in the matrix space, the respective amounts of free and loosely bound forms of PN, the amount, activity and affinity for PN of each enzyme, the diffusion rate of PN, the matrix or membrane location and the association of the enzymes (metabolons) determine the existence of a kinetic compartmentation within the pool of PN shared among the different dehydrogenases, as it has also been suggested for the pool of ubiquinone of the inner mitochondrial membrane[9]. These various micro-domains, characterized by their specific kinetic properties, do not react homogeneously with the respiratory chain under state 4 conditions.

REFERENCES

1. R. Douce and W.D. Bonner, Oxalacetate control of Krebs cycle oxidations in purified plant mitochondria, Biochem. Biophys. Res. Commun. 47:619 (1972).
2. J.B. Robinson and P.A. Srere, Organization of the tricarboxylic acid cycle enzymes in mitochondria, J. Biol. Chem. 260:10800 (1985).
3. M. Neuburger and R. Douce, Transport of NAD^+ through the inner membrane of plant mitochondria, in: Plant Mitochondria, G. Ducet and C. Lance, eds., Elsevier/North Holland Biomedical Press, Amsterdam (1978).
4. D.A. Day, M. Neuburger and R. Douce, Interactions between glycine decarboxylase, the tricarboxylic acid cycle and the respiratory chain in pea leaf mitochondria, Aust. J. Plant Physiol. 12:219 (1985).
5. P. Rustin and M. Valat, The control of malate dehydrogenase activity in purified potato tuber (Solanum tuberosum L.) mitochondria, Arch. Biochem. Biophys. 247:26 (1986).
6. M. Neuburger and R. Douce, Slow passive diffusion of NAD^+ between intact isolated plant mitochondria and suspensing medium, Biochem. J. 216:443 (1983).
7. M.T. Karp, R.P. Raunio and T.N.E. Lövgren, Simultaneous extraction and combined bioluminescent assay of NAD^+ and NADH, Analyt. Biochem. 128:175 (1983).
8. M. Neuburger and R. Douce, Transport of coenzyme A in plant mitochondria, Arch. Biochem. Biophys. 229:253 (1984).
9. M. Gutman, Functional compartmentation of the mitochondrial quinone during the oxidation of two substrates. in Bioenergetics of Membranes, L. Packer, G.C. Pappageorgiou and A. Trebs, eds., Elsevier/North Holland Biomedical Press, Amsterdam (1977).

EFFECTS OF INHIBITORS ON THE MEMBRANE POTENTIAL (ΔΨ) OF MUNG BEAN MITOCHONDRIA

Donald E. Moreland and William P. Novitzky

U. S. Department of Agriculture, Agricultural Research Service, and Departments of Crop Science and Botany, North Carolina State University, Raleigh, NC 27695-7620 (USA)

ABSTRACT

Safranine O was used to measure changes imposed on the membrane potential (ΔΨ) of mung bean mitochondria by different metabolic conditions and inhibitors. Changes in ΔΨ were correlated with simultaneously made measurements of O_2 uptake. ΔΨ values of 129 ± 3 mV and 144 ± 4 mV were obtained under state 3 and state 4 conditions, respectively, during the oxidation of malate. At concentrations of electron transport inhibitors that maximally inhibited O_2 utilization, ΔΨ was only partially collapsed. Under state 4 conditions, the collapse of ΔΨ to the Donnan potential by uncouplers correlated with the concentration at which state 4 respiration was stimulated maximally. Inhibition of O_2 utilization by F_0 inhibitors correlated with an increase in ΔΨ from the state 3 to the state 4 potential.

INTRODUCTION

The synthesis of ATP in mitochondria, according to the chemiosmotic hypothesis, is driven by an electrochemical gradient (pmf or $\Delta\mu H^+$) generated by the electron transport chain[1]. The gradient has two components, an electrical component (ΔΨ) and a chemical component (ΔpH). ΔΨ has been estimated to comprise approximately 80% of the total pmf of plant mitochondria. Actual values reported in the literature vary with the source from which the mitochondria were isolated and the measurement technique used. For example, state 4 ΔΨ values of 220 to 230 mV have been measured for potato tuber mitochondria with the TPP^+ electrode[2], whereas a value of 135 mV was reported for mung bean mitochondria with safranine O as the probe[3]. Only limited information is available on how ΔΨ is affected by the metabolic state of plant mitochondria and how standard inhibitors affect the generation and maintenance of ΔΨ. The characterization of effects on ΔΨ produced by inhibitors may provide an insight into how the various types interact with the mitochondrial inner membrane. Consequently, the objectives of this study were to identify and correlate the effects of metabolic factors, electron transport inhibitors, uncouplers, and energy transfer inhibitors on the transmembrane potential (ΔΨ) and O_2 uptake of mung bean (Vigna radiata L.) hypocotyl mitochondria.

MATERIALS AND METHODS

Mitochondria (MBM) were isolated from 3-day-old etiolated mung bean hypocotyls[4]. Assays were conducted in a medium that consisted of 0.3 M mannitol, 10 mM KCl, 0.5 mM $MgCl_2$, 10 mM phosphate buffer (pH 7.2), 5 μM safranine O, and 0.4 mg mitochondrial protein in a 2.0 mL cuvette (25 C). Respiratory substrates and ADP were added at the concentrations indicated elsewhere. Oxygen uptake was measured with a Clark-type electrode and ΔΨ was measured simultaneously as safranine absorbance changes monitored at the wavelength pair 510-532 nm with a diode-array spectrophotometer. Absorbance changes were converted to mV with K^+ diffusion gradients according to the Nernst equation[3].

RESULTS AND DISCUSSION

Shown in Fig. 1 are representative traces of safranine absorbance changes, converted to mV, and the concomitant oxygen consumption traces during state 3/state 4 transitions for the oxidation of malate by MBM. The MBM exhibited an initial apparent ΔΨ of 84 ± 6 mV. The addition of malate caused an increase of ΔΨ to 144 ± 4 mV (state 4) and a small increase in O_2 uptake. The addition of ADP (0.15 mM) caused ΔΨ to drop to 129 ± 3 mV and O_2 uptake to accelerate (state 3). Upon depletion of the ADP, ΔΨ increased back to the state 4 level and O_2 uptake slowed. Additional aliquots of ADP repeated the cycles. When an uncoupler (FCCP, 0.02 μM) was added, ΔΨ collapsed to about 60 mV (the Donnan potential of the MBM) and O_2 uptake increased to about the state 3 rate.

The addition of ADP to MBM lowered ΔΨ (Fig. 1). The extent of the decrease was related to the concentration of ADP added as shown in Fig. 2 for the oxidation of malate. The decrease of ΔΨ was 5, 10, 12, 14, 16, and 20 mV respectively when 0.05, 0.15, 0.25, 0.5, 1, and 2 mM ADP was added. The state 4 level of ΔΨ (145 mV) was found to be the same for all respiratory substrates tested (malate, succinate, NADH, and TMPD/asc). However, the state 3 level of ΔΨ was inversely proportional to the number of "coupling sites" associated with the oxidation of different substrates.

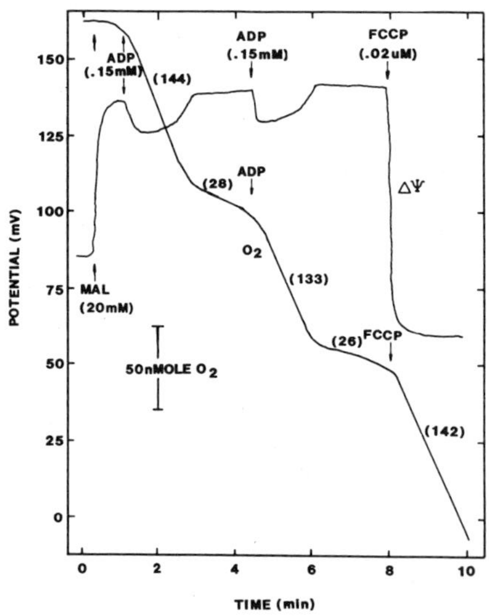

Figure 1. Representative traces that depict the responses of ΔΨ and O_2 utilization during state 3/state 4 transitions in mung bean mitochondria with malate as substrate. Oxygen uptake rates expressed as nmoles/mg protein/min are indicated parenthetically.

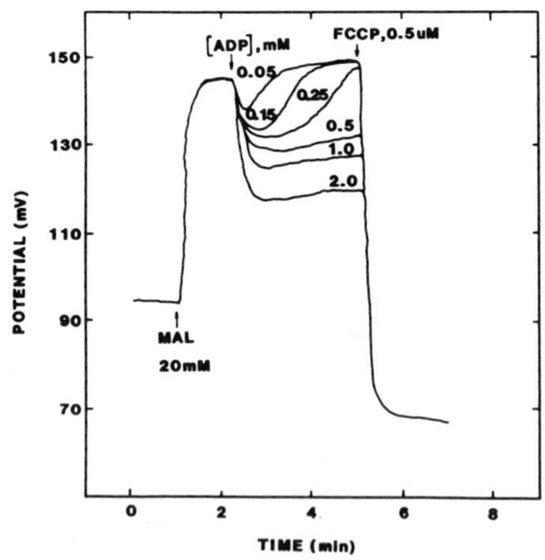

Figure 2. Representative traces of safranine O absorbance changes, converted to mV, that show the effect of ADP concentration on the state 3 decrease of ΔΨ by mung bean mitochondria during the oxidation of malate.

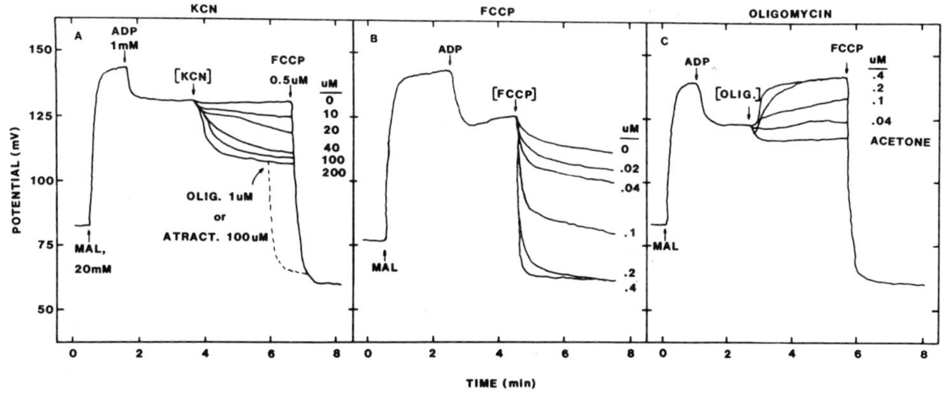

Figure 3. Representative traces of absorbance changes of safranine O, converted to mV, that show the effects of different concentrations of (A) KCN, (B) FCCP, and (C) oligomycin on ΔΨ for the oxidation of malate by mung bean mitochondria under state 3 conditions.

The addition of 1 mM ADP decreased ΔΨ from 145 mV to 129 mV, 127 mV, 127 mV, and 121 mV for malate, succinate, NADH, and TMPD/asc, respectively. The substrates are linked respectively to 3, 2, 2, and 1 coupling site(s).

The effects of reference inhibitors of electron transport and phosphorylation on ΔΨ during state 3 respiration were related to the type of inhibitory action expressed. Shown in Fig. 3 are representative traces of absorbance changes of safranine, that depict these differential effects. Electron transport inhibitors such as KCN (Fig. 3A), antimycin a, and azide caused a concentration-dependent collapse of ΔΨ to between 100 and 110 mV. The further addition of oligomycin or atractyloside to maximally inhibited MBM (e.g., 200 μM KCN) or FCCP, at any concentration of KCN, caused ΔΨ to collapse to the Donnan potential. The results indicate that ΔΨ was apparently being maintained by ATPase activity arising from the utilization of ATP synthesized prior to the addition of the inhibitor. Under state 4 conditions, only if no adenylate had been added, ΔΨ was collapsed to the Donnan potential by electron transport inhibitors without the addition of FCCP, oligomycin, or atractyloside (data not shown). The state 4 data support an ATPase-related maintenance of ΔΨ. Electron transport inhibitors prevent the generation of ΔΨ by inhibiting the proton translocating steps of the chain. Uncouplers of oxidative phosphorylation such as FCCP caused a concentration-related collapse of ΔΨ to the Donnan potential under both state 3 (Fig. 3B) and state 4 (not shown) conditions. The collapse of ΔΨ under state 4 conditions was marginally more sensitive to FCCP than under state 3 conditions. Energy transfer inhibitors that interact with the F_0 portion of the coupling factor complex, such as oligomycin, DCCD, and TBT cause a concentration-dependent increase of ΔΨ from the state 3 level back to the state 4 level, as shown in Fig. 3C. The addition of FCCP following oligomycin treatment collapsed ΔΨ to the Donnan level. Under state 4 conditions the energy transfer inhibitors had no effect on ΔΨ (data not shown).

Dose-response curves for effects on ΔΨ were developed from the traces depicted in Fig. 3. and were compared to the concomitant effects produced by the reference standards on state 3 O_2 utilization (Fig. 4). The relationship between effects on ΔΨ and O_2 uptake was related to the type of inhibitory action expressed. Under state 3 conditions, electron transport inhibitors such as KCN (Fig. 4A) caused ΔΨ to collapse to about 110 mV at concentrations that caused near-maximum inhibition of O_2 uptake (70 to 80%). The addition of SHAM did not appreciably change the level of ΔΨ in the presence of high concentrations of KCN (data not shown). Uncouplers

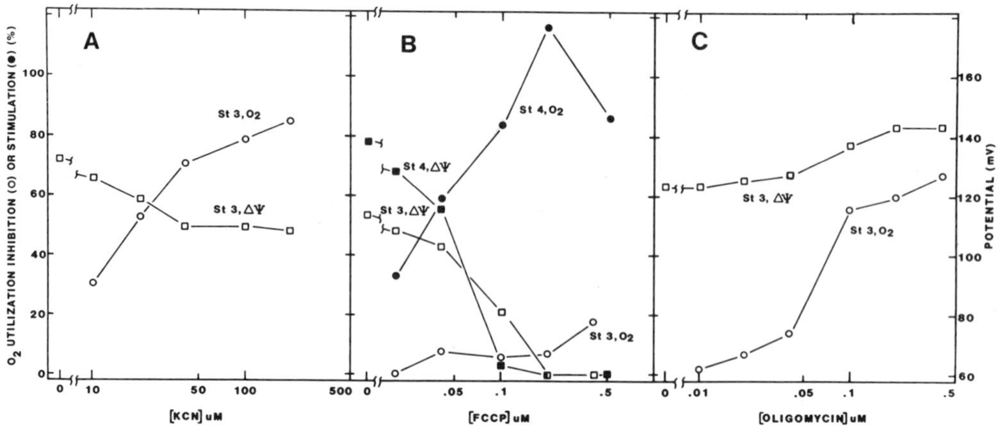

Figure 4. Dose-response curves plotted semilogarithmically that show effects imposed by (A) KCN, (B) FCCP, and (C) oligomycin on O_2 utilization expressed as percent inhibition or stimulation and on ΔΨ expressed as mV under state 3 and state 4 (Fig. 4B only) conditions with mung bean mitochondria during the oxidation of malate.

such as FCCP (Fig. 4B) caused state 3 ΔΨ to collapse to the Donnan potential at concentrations that did not strongly inhibit O_2 uptake. At higher FCCP concentrations, O_2 uptake began to be inhibited. Under state 4 conditions, ΔΨ began to decrease at FCCP concentrations that stimulated the rate of O_2 uptake. The concentration of FCCP that caused maximum stimulation of state 4 respiration also caused ΔΨ to collapse completely. Energy transfer inhibitors such as oligomycin (Fig. 4C) caused state 3 ΔΨ to increase from the state 3 level to the state 4 level over the same concentration range that inhibited O_2 uptake. The concentration of oligomycin that produced maximal inhibition of O_2 uptake (>65%) coincided with ΔΨ attaining the state 4 level. Oligomycin had no effect on either ΔΨ or O_2 uptake under state 4 conditions. The data demonstrate that the action of different types of inhibitors can readily be ascertained by the effects on ΔΨ.

ACKNOWLEDGEMENTS

Cooperative investigations of the United States Department of Agriculture, Agricultural Research Service, and the North Carolina Agricultural Research Service, Raleigh, North Carolina. Paper No. 10709 of the Journal Series of the North Carolina Agricultural Research Service, Raleigh, North Carolina 27695-7601. This investigation was supported in part by Grant No. PHS ES-00044 from the National Institute of Health. Appreciation is extended to F. S. Farmer for technical assistance.

REFERENCES

1. P. Mitchell, Coupling of phosphorylation to electron and hydrogen transfer by a chemi-osmotic type mechanism, Nature (London). 191:144 (1961).
2. P. Diolez and F. Moreau, Correlation between ATP synthesis, membrane potential and oxidation rate in potato mitochondria, Biochim. Biophys. Acta 806:56 (1985).
3. A. L. Moore and W. D. Bonner, Jr., Measurements of membrane potentials in plant mitochondria with the safranine method, Plant Physiol. 70:1271 (1982).
4. D. E. Moreland, Interaction of perfluidone with mitochondrial, thylakoid, and liposome membranes, Pestic. Biochem. Physiol. 15:21 (1981).

EFFECT OF THE ANTIMALARIAL QUINONE BW 58C ON CYANIDE-SENSITIVE AND CYANIDE-INSENSITIVE PLANT RESPIRATION

D.C. Holt*, J.B. Hyndman, and R. Cammack

Department of Biochemistry, King's College
Campden Hill Rd, London W8 7AH
*Present address: Imperial Chemical Industries
The Heath, Runcorn, WA7 4QF

Introduction

The ubiquinone analogue 5-n-undecyl-6-hydroxy-4,7 -dioxobenzothiazole (UHDBT) is a potent inhibitor of the cyanide sensitive pathway, acting at the level of complex III. Cook & Cammack (1984) have found that it also inhibits cyanide -insensitive respiration at similar concentrations, in the order of 10^{-7}M.

A number of synthetic naphthoquinones have been developed for use as drugs against parasitic protozoa, such as 2-(4-butyl cyclohexyl) 1,3-hydroxy-1,4-naphthoquinone (BW 58C), which inhibits the growth of the malarial parasite *Plasmodium falciparum* at concentrations of 10^{-11} M (Fry, Hudson and Randall, 1984). The erythrocytic stage of these parasites displays a very low level of mitochondrial respiration, for which the only known function is the oxidation of dihydroorotate, and which is partially inhibited by both cyanide and hydroxamic acids. BW 58C and similar naphthoquinones are known to be inhibitors of respiration in mitochondria from mammals and protozoa. They are thought to inhibit at the level of ubiquinone.

Here we report the effects of BW 58C on cyanide-resistant respiration in mitochondria isolated from *A. maculatum*. These have been compared with the effects that the same naphthoquinones have on cyanide-sensitive respiration, in mitochondria isolated from the tubers of *Helianthus tuberosus* (Jerusalem Artichoke).

Methods

Mitochondria were prepared from *Arum maculatum* inflorescence spadices as described by Cammack and Palmer (1977). Mitochondria from *Helianthus tuberosus* tubers were prepared according to the method of Palmer and Kirk (1974). All assays were performed in a Clark-type oxygen electrode according to the method of Douce et al. (1973), in reaction medium containing 0.3M mannitol, 5mM $MgCl_2$, 10mM phosphate buffer (pH 7.05), at 25°C. The concentration of mitochondria used was 0.1 to 0.5 mg protein/ml.

The inhibitor BW 58C was kindly provided by Dr. D.J. Linstead, Wellcome Laboratories, Langley Court, Beckenham, Kent. Solutions of concentrations ranging from 0.003 to 3.0 mg/ml were prepared in dimethylformamide (DMF) immediately prior to experimentation, and kept shielded from light. Rates were corrected for a small (less than 10%) inhibition by DMF.

UHDBT samples were kind gifts from Dr. B.L. Trumpower, Department of Biochemistry, Dartmouth Medical School, Hanover, USA and Dr. K. Folkers, Institute for Biomedical research, Austin, Texas, USA. The purity of the samples was confirmed by mass spectrometry.

The measurement of succinate oxidation in mitochondria was preceded by pre-incubation with ATP (0.2 mM) for two minutes prior to the addition of substrate. The reaction was initiated by an addition of succinate to a concentration of 33mM. In the measurement of NADH oxidation, the reaction was started by the addition of 0.83mM NADH. Where necessary, mitochondria were pre-incubated with the inhibitor for up to five minutes at 25°C. In order to prevent oxidation by the cyanide-sensitive respiratory pathway, cyanide was present in the reaction medium at a concentration of 100uM. Respiratory rates were corrected for the small rate of non-mitochondrial substrate oxidation; measured after the addition of 100um KCN and 1mM salicylhydroxamic acid (SHAM) to *A. maculatum* mitochondria, or 100uM KCN to *H. tuberosus* mitochondria.

Results

Table 1 (below) shows a comparison of the effects of BW 58C and UHDBT on plant, protozoan, and vertebrate mitochondria. The concentrations shown (in nanomoles) are those required for a 50 % inhibition of NADH and succinate oxidase activity. The published values for the effects of BW 58C on mitochondria from the coccidian protozoan *Eimeria tenella* and from rat and chicken liver (Fry et al., 1984) are added for comparison.

Table 1

Inhibitor	Dehydrogenase substrate	Source of mitochondria: *Helianthus tuberosus*	*Arum maculatum*	*Eimeria tenella*	Chick liver	Rat liver
UHDBT	NADH	145	75			
	succinate	70	300			
BW 58C	NADH	45	24500	0.053	540	530
	succinate	58	56000	0.062	580	640

Figs. 1a and 1b show the inhibition of respiration in A. maculatum and H. tuberosus mitochondria respectivèly, by UHDBT and BW 58C.

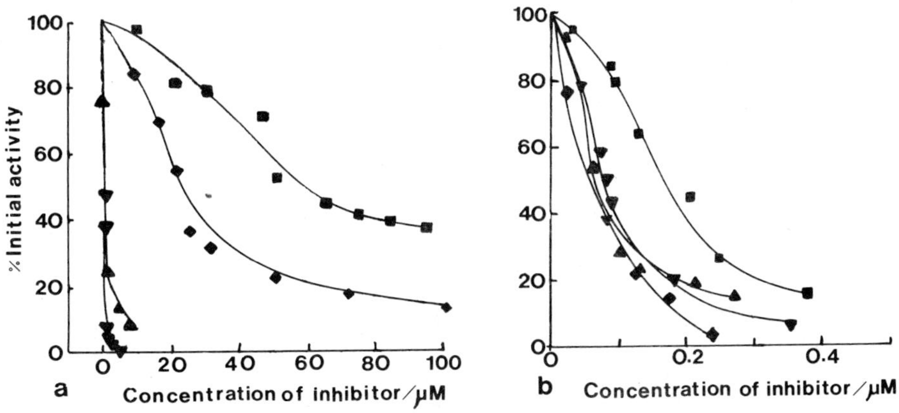

Substrate	Inhibitor
succinate	UHDBT ▲
NADH	UHDBT ▼

Substrate	Inhibitor
succinate	BW 58C ■
NADH	BW 58C ◆

Figure 1a: Relative effects of UHDBT and BW 58C on succinate and NADH oxidation in Arum maculatum mitochondria.

Figure 1b: Relative effects of UHDBT and BW 58C on succinate and NADH oxidation in Helianthus tuberosus mitochondria.

Conclusions

In the normal, cyanide-sensitive respiratory chain of plant mitochondria, BW 58C is an effective inhibitor. The site of inhibiton is presumably complex III. The concentration required for inhibition was approximately 10-fold less than for inhibition of vertebrate mitochondria.

However this concentration is considerably greater than that required to inhibit protozoal respiration. This is in part due to a decreased level of binding sites in the protozoa, but the shape of the binding curves (fig. 1) indicates also that there is a less high affinity. This indicates a difference between the plant and protozoal respiratory chains.

In the cyanide-resistant pathway of A. maculatum mitochondria, BW 58C is a considerably less effective inhibitor than it is of the cyanide-sensitive pathway. If BW 58C inhibits the malarial parasite by acting on a cyanide-resistant respiratory chain, the results indicate a difference between the plant and protozoal alternative oxidases.

UHDBT specifically inhibited the cyanide-resistant respiration of A. maculatum mitochondria. BW 58C was only effective at higher concentrations, similar to low-affinity inhibitors such as SHAM. Among the quinone analogues, UHDBT appears to be unique in its effect on the cyanide-resistant oxidase.

Acknowledgements

This work was supported by a grant from the Agriculture and Food Research Council.

References

Cammack, R., and Palmer, J.M. ,1977, Iron sulphur centres in mitochondria from *Arum maculatum* spadix with very high rates of cyanide resistant respiration:
Biochem. J., 166:347-355.

Cook, N.D., and Cammack, R., 1984, The inhibition of plant mitochondrial respiration by the synthetic analog of ubiquinone, 5-n-undedyl-6-hydroxy-4,7-dioxobenzothiazole (UHDBT):
Arch. Biochem. Biophys., 240:9-14.

Douce, R., Manella, C.A., and Bonner, W.D. Jr., 1973, The external NADH dehydrogenases of intact plant mitochondria:
Biochim. Biophys. Acta, 292:105-116.

Fry, M., Hudson, A.T., Randall, A.W., Williams, R.B., 1984, Potent and selective hydroxynaphthoquinone inhibition of mitochondrial electron transport in the coccidian *Eimeria tenella*:
Biochem. Pharm., 33:2115-2122.

Palmer, J.M., and Kirk, B.I., 1974, The influence of osmolarity on reduction of exogenous cytochrome c and permeability of the inner membrane of Jerusalem artichoke mitochondria:
Biochem. J., 140:79-86.

REDOX PROPERTIES OF INHIBITORS OF THE ALTERNATIVE PATHWAY OF PLANT MITOCHONDRIA

Jacques Dupont and Pierre Rustin

Laboratoire de Biologie Végétale IV (CNRS, UA 1180)
Université Pierre et Marie Curie, Paris, France

INTRODUCTION

The mechanisms of inhibition of the alternative pathway of plant mitochondria by salicylhydroxamic acid (SHAM), propylgallate (PG), 8-OH quinoline or disulfiram are still a subject of controversy. It has been suggested that these inhibitors could bind to a specific protein (Siedow, 1982) or could act as free radical scavengers (Rustin et al.,1984). In this work, it is shown that these inhibitors also display redox properties and could act as electron donors to the respiratory and photosynthetic chains.

MATERIALS AND METHODS

Mitochondria extracted from potato tubers (*Solanum tuberosum* L.) were purified on two successive self-generating Percoll gradients. Chloroplasts extracted from spinach leaves (*Spinacia oleracea* L.) were purified on a self-generated Percoll gradient. Oxygen uptake and spectrophotometric measurements were simultaneously carried out in a 2.1 ml quartz-cell, thermostated at 27°C and magnetically stirred. Inhibitors were dissolved in dimethyl sulfoxide as 1 M solutions.

RESULTS

Redox interactions with electron acceptors in solution

As shown in figure 1a and b, PG and SHAM readily reduce cytochrome *c* (cyt *c*) in solution. The rate of cyt *c* reduction, monitored at 550 nm, was 50-fold faster in the presence of PG than in the presence of SHAM. It was also found that dichlorophenolindophenol (DCPIP) or tetramethylphenyldiamine (TMPD) could be used as electron acceptors using PG or SHAM as reductants (not shown). Disulfiram and 8-OH quinoline (two other inhibitors of the cyanide-resistant electron pathway) were also able to reduce cyt *c*, although the rate of reduction of cyt *c* by the former was relatively slow (not shown). In the presence of SHAM (Fig. 1b), an additional absorption peak (571 nm) was observed. Its importance was dependent on the relative concentrations of SHAM and cyt *c*. It was observed immediately upon mixing of the two compounds and was unchanged during the course of the redox reaction. It could be linked to the formation of a complex between the

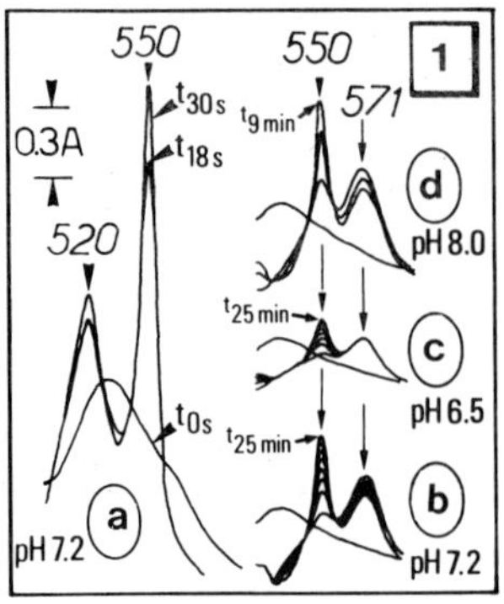

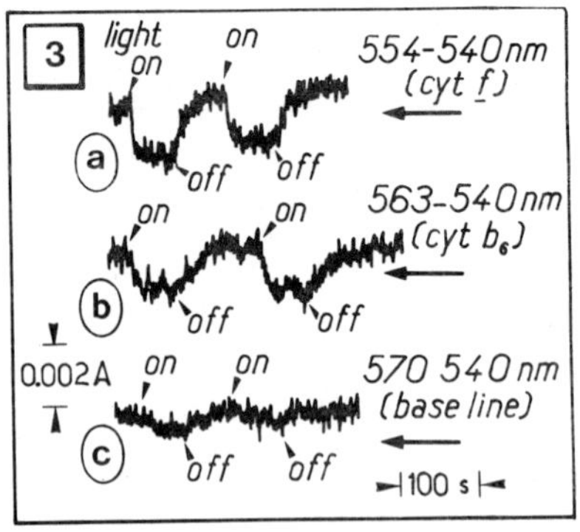

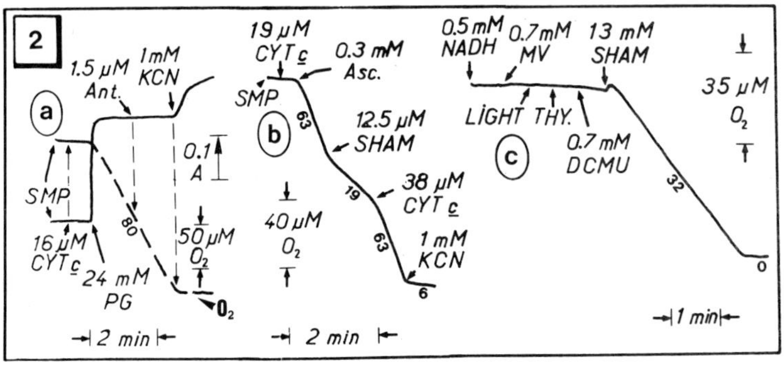

Fig. 1. Cytochrome c reduction by PG (a) or SHAM (b-d). a: Reduction of 25 µM cyt c by 20 mM PG. Scanning was run at 20 nm s^{-1}; b-d: Effect of pH on the reduction of 21 µM cyt c by 33 mM SHAM. Scanning was run at 1 nm s^{-1}. All the experiments were carried out at 27°C in 50 mM phosphate buffer adjusted for each at the pH value indicated on the figure.

Fig. 2. Interactions of PG or SHAM with mitochondrial (a-b) or thylakoid membranes (c). a: cyt c-mediated oxidation of PG by sub-mitochondrial particles (SMP) (obtained by osmotic shock) in 20 mM phosphate buffer (pH 7.3). Continuous line represents the changes of redox state of cyt c, whereas dotted line corresponds to the O_2 uptake. b: Effect of SHAM on the oxidation of ascorbate-reduced cyt c by SMP in 20 mM phosphate buffer (pH 7.3). c: light-dependent O_2 uptake by thylakoids in the presence of SHAM. The experiment was carried out in 100 mM sorbitol and 20 mM phosphate buffer (pH 7.3). Chlorophyll was 35 µg in c. Mitochondrial protein in a and b was 0.5 mg. Numbers along the traces are nmol O_2 min^{-1} mg^{-1} protein.

Fig. 3. Reduction of thylakoid cytochromes by SHAM. The reaction medium contained 50 mM tricine-NaOH (pH 7.8), 10 mM NaCl, 5 mM $MgCl_2$, 1 uM methylviologen, 2 mM DCMU, 16 mM SHAM and 0.8 mM NADH. Chlorophyll was 64 µg.

inhibitor and the hemoproteins, as previously observed with other hemoproteins (Rich et al., 1978). The reduction of cyt c by SHAM was strongly pH-dependent and faster at alcaline pH (237 nmol cyt c reduced min^{-1} ml^{-1} at pH 8.2) than at acidic pH (46 nmol cyt c reduced min^{-1} ml^{-1}

at pH 6.5). The rate of reduction of cyt _c_ by SHAM could be greatly increased by the addition of NADH. Under conditions similar to those of figure 1 b, _i.e._ 25 μM cyt _c_, 33 mM SHAM, at pH 6.5, the rate of cyt _c_ reduction measured in the presence of 400 uM NADH was increased 10-fold and reached values higher than 500 nmol cyt _c_ reduced min^{-1} ml^{-1}. Then, lowering SHAM concentration decreased the rate of the reaction, but it was still significant with 1 mM SHAM, since a value of 100 nmol cyt _c_ reduced min^{-1} ml^{-1} could still be measured. It is noteworthy that a shift in the pH to 8.0 led to a very fast reaction, since a rate of about 1400 nmol cyt _c_ reduced min^{-1} ml^{-1} could then be measured.

This set of experiments clearly establishes that (_a_) the widely used inhibitors of the alternative pathway can act as very potent reductants for electron acceptors, such as cyt _c_, DCPIP or TMPD and (_b_), as shown using SHAM, cyt _c_ and NADH, can act as electron mediators between electron carriers (NADH and cyt _c_ in this instance)

Redox interactions with biological membranes

Mitochondria. As ascorbate (a hydrophilic compound) and duroquinol (a lipophilic compound), SHAM and PG are thus able to rapidly reduce cyt _c_ in solution (Fig. 1). As these compounds, SHAM and PG are also unable to rapidly react with the membranous cyt _c_ in the absence of an electron mediator, despite the peripherical location of cyt _c_ in the inner mitochondrial membrane. However, in the presence of a trace of added exogenous cyt _c_ acting as an electron mediator, sub-mitochondrial particles (SMP) readily oxidized PG (Fig. 2a). The O_2 uptake associated with the reaction and the redox state of the cyt _c_ were simultaneously recorded. Upon PG addition to the mitochondria incubated with 16 uM cyt _c_, an O_2 consumption was rapidly observed, whereas a large part of the cyt _c_ was reduced (about 70% of the added cyt _c_). This level of reduction was controlled by the respective activities of PG, acting as a reductant, and of cyt oxidase, acting as an oxydant. Adding cyanide, caused a full inhibition of the O_2 uptake and an increase of cyt _c_ reduction. Under these conditions, more than 95% of the added cyt _c_ was reduced. Identical results were obtained using a mixture of PG and TMPD. A similar experiment carried out using SHAM instead of PG indicated that a significant reduction of the added cyt _c_ did occur but that it only corresponded to a very low O_2 uptake (not shown). As previously quoted, SHAM could form a complex with the hemoprotein and it can be suggested that the formation of this complex could prevent the further reaction of reduced cyt _c_ with the site of cyt _c_ oxidation in the inner mitochondrial membrane. This hypothesis was investigated in the next experiment (Fig. 2b). In this experiment, added cyt _c_ was reduced by ascorbate allowing to measure a cyt oxidase activity with SMP. The addition of SHAM (12.5 mM) caused a significant inhibition of the O_2 uptake (about 70%). This inhibition could be released by a new addition of cyt _c_, whereas the addition of cyanide totally blocked the activity of cyt oxidase, and therefore the O_2 uptake. Such an experiment supports the above hypothesis that the formation of a complex between SHAM and cyt _c_ strongly affects the further reactivity of cyt _c_ with its site of oxidation.

From these experiments, it appears that PG is able to rapidly feed electrons to the respiratory chain in the presence of an adequate electron mediator, such as cyt _c_ or TMPD. Curiously, it is then even possible to measure the activity of the cyt oxidase using PG as electron donor, or to inhibit this activity, catalysed by a mixture of ascorbate and cyt _c_, by SHAM.

Chloroplasts. SHAM can also be used as electron donor to the photosynthetic electron transport chain (Fig. 2c). In the presence of SHAM (and NADH), an active light-dependent O_2 uptake can be measured. Since this O_2 uptake was measured in the presence of DCMU (an inhibitor of PS II) and methylviologen (which accepts the electrons PS I), it involved an electron transfer through PS I. In such an experiment, the uptake of O_2 was linked to the auto-oxidation of methyl viologen, reduced at the level of ferredoxin (after PS I) by the electrons arising from SHAM. In order to establish the site of interaction of SHAM with the photosynthetic chain, the redox changes of the $\underline{b}_6/\underline{f}$ complex induced by SHAM were investigated. As shown in figure 3, cyt $\underline{f}$ (554-540 nm) was reduced upon addition of SHAM to the thylakoid suspension in the dark. When light was switched on, cyt $\underline{f}$ was fully reoxidized. Subsequent dark-light transitions lead to redox transitions of cyt $\underline{f}$. It is also shown (Fig. 3) that cyt $\underline{b}_6$ (563-540 nm) was similarly reduced by SHAM, and reoxidized upon activation of PS I by light. A control experiment has established that the baseline (570-540 nm) was not significantly affected during the dark-light transitions. Similar results were obtained using PG instead of SHAM (not shown).

This study establishes that SHAM and PG are able to feed electrons to the photosynthetic chain somewhere between the DCMU-sensitive site and the $\underline{b}_6$-$\underline{f}$ complex of the chain. The specificity of this reaction allows the reduction of the chloroplast cytochromes without the need of an activation of the PS II by light (680 nm) to generate electrons.

CONCLUSION

The above results establish that the inhibitors of the mitochondrial alternative pathway, such as SHAM, PG, 8-OH quinoline and disulfiram are redox-active compounds. They are able to feed electrons to the respiratory and photosynthetic chains under suitable conditions.

As a first consequence, great care has to be taken when using these compounds as inhibitors in the presence of redox active compounds <u>in solution</u> or of membranous electron carriers. In particular, the measurement of activities involving soluble cyt $\underline{c}$ (<u>i.e.</u> cyt $\underline{c}$ reductase or oxidase activities) can be strongly affected by the presence of these compounds. On the other hand, the study of the binding of these inhibitors to the mitochondrial membranes, in order to characterize an eventual oxidase involved in the alternative pathway, might take into account their possible binding with hemoproteins, at least for SHAM.

The nature of the chemical groups involved in the redox properties (phenols, hydroxamic groups, etc.) and the eventual relationship between the redox and inhibitory properties of these compounds are presently under investigation. Are their anti-oxidant properties affected by their redox state ? Are the values of their Ki for the cyanide-resistant electron pathway or lipoxygenase dependent on their redox states ? At any rate, the demonstration of the redox properties of these inhibitors fits well with the previous suggestion that these compounds could act as potent free radical scavengers (Rustin et al., 1984).

<u>Acknowledgement</u> – This work was supported by a grant from PIRSEM (N°3293).

REFERENCES

Rich, P.R., Wiegand, N.K., Blum, H., Moore, A.L. and Bonner, W.D., Studies on the mechanism of inhibition of redox enzymes by substituted hydroxamic acids, Biochim. Biophys. Acta, 525: 325 (1978).

Rustin, P., Dupont, J. and Lance, C., Involvement of lipid peroxy radicals in the cyanide-resistant electron transport pathway, Physiol. Vég., 22: 643 (1984).

Siedow, J.N., The nature of the cyanide-resistant pathway in plant mitochondria, Rec. Adv. Phytochem., 16: 47 (1982).

HYDROXAMATE-ACTIVATED PEROXIDASES :

INTERFERENCE WITH DETERMINATION OF RESPIRATORY PATHWAY ACTIVITY ?

Linus H.W. van der Plas, Henk Gude and Marijke Wagner

Dept.of Plant Physiology,Vrije Universiteit
De Boelelaan 1087,1081 HV Amsterdam
The Netherlands

INTRODUCTION

In many studies on plant respiration, hydroxamates are an important tool in the determination of the flow through the "efficient" cytochrome path and the "inefficient" alternative path (Laties,1982). A complete and specific inhibition of the alternative pathway by these hydroxamates is a prerequisite for such an use of this type of inhibitors. In this paper the properties of a peroxidase will be described,that causes an increased oxygen uptake after addition of hydroxamates.

MATERIAL AND METHODS

Callus tissue cultivation,isolation of mitochondria and measurement of oxygen uptake was previously described by Van der Plas and Wagner (1983) and Hemrika-Wagner et al. (1986).

RESULTS AND DISCUSSION

1.Hydroxamate stimulated oxygen uptake

Addition of an hydroxamate (salicylhydroxamate,SHAM or benzohydroxamate,BHAM) to NADH-consuming mitochondria often caused an enormous stimulation of the oxygen uptake; such a stimulation was not found with succinate.This stimulation made determination of the activity of the cytochrome and alternative pathway activity with NADH as the substrate impossible.

Experiments on the subcellular localization of the process that caused this hydroxamate stimulated oxygen uptake, revealed that only a small part was localized in the mitochondria; more than 95 % was localized in the soluble fraction.

The rate of this hydroxamate stimulated oxygen uptake in cell homogenates of potato tuber callus was huge : activities of 10,000 to 100,000 nmol O_2-uptake.min $^{-1}$.g FW.$^{-1}$ were often observed(for comparison: respiratory activity was generally less than 100 nmol O_2-uptake.min $^{-1}$.g FW.$^{-1}$, Van der Plas and Wagner, 1983,Hemrika-Wagner et al.,1986).

2. Properties of the hydroxamate-stimulated oxygen uptake process

Low concentrations hydroxamate (2 - 4 mM) gave maximal stimulation of the oxygen uptake reaction. High concentrations (especially of BHAM) were less stimulatory. The hydroxamate was not consumed during the reaction.

NADH or NADPH are consumed during the reaction and serve as the electron donor for the hydroxamate-stimulated oxygen uptake. The affinity for NAD(P)H was very low, resulting in a Km of about 15 mM (the Km for NADH of the mitochondrial respiratory oxygen uptake was about 1000 times lower,).
The hydroxamate-stimulated oxygen uptake was completely inhibited by cyanide, but insensitive to azide and antimycin-A.
By sephadex column chromatography at least two protein fractions could be isolated ,that were responsible for the stimulatory effect.One fraction had a low molecular weight (about 15 kD), the other a much higher one (more than 50 kD).

3. The effect of catalase and the supposed reaction scheme

The addition of (excess , 5000 U) catalase at various moments after start of the reaction (Fig. 1) had two effects :
1- Instantaneous production of oxygen upon addition of catalase, indicating the accumulation of hydrogenperoxide.

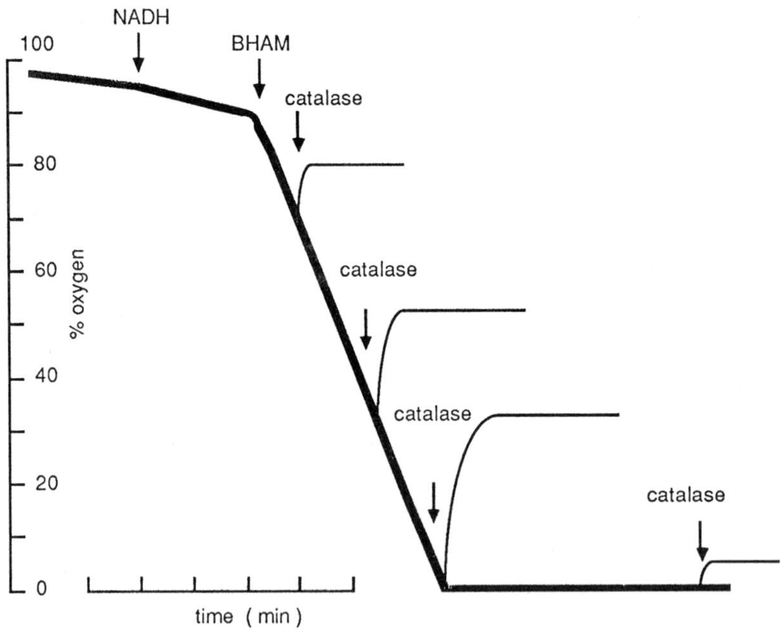

Fig. 1 The effect of catalase on the hydroxamate stimulated oxygen uptake .The reaction mixture contained 50 µl of light fraction peroxidase, purified by sephadex column chromatography.NADH (2 mM),BHAM (1.5 mM) and catalase (10 µl, 5000 U) were added at the indicated time.Each catalase addition refers to a separate, consecutive experiment.

The amount of accumulated hydrogenperoxide increased with the reaction time and was maximal at the moment,that all the oxygen was consumed. The hydrogenperoxide disappeared again during the succeeding period of anaerobiosis.

2- A complete inhibition of the hydroxamate stimulated oxygen uptake (and of the concomittant NADH-uptake). Superoxide dismutase had the same inhibitory effect.

Apparently two reactions occur :

$$NADH + H^+ + O_2 \;\; -> \;\; NAD^+ + H_2O_2 \qquad (1)$$

$$NADH + H^+ + H_2O_2 \;\; -> \;\; NAD^+ + 2H_2O \qquad (2)$$

The expected stoichiometry of 2 NADH : 1 O_2 (after completion of the reaction) was indeed experimentally observed.The enzyme, catalyzing both reactions, apparently is an hydroxamate-stimulated peroxidase.

Endogenous catalase (when present in small, non-inhibiting amounts) may also contribute to the disappearance of the hydrogenperoxide, via the reaction :

$$H_2O_2 \;\; -> \;\; H_2O + 1/2\ O_2 \ . \qquad (3)$$

<u>4.Does the hydroxamate-stimulated peroxidase contribute to the oxygen uptake, during the measurement of *in vivo* respiration ?</u>

Hydroxamates did not stimulate the oxygen uptake by intact callus tissue (Fig. 2).Only when exogenous NADH was added , a peroxidase mediated oxygen uptake (sensitive to catalase and CN, but insensitive to azide) was observed. Apparently, in tissue respiration experiments, peroxidase activity is limited by NADH-availability. Electron transfer from endogenous, NADH or NADPH is probably prevented by a combination of the low affinity of the peroxidase for NAD(P)H and the different localization of the

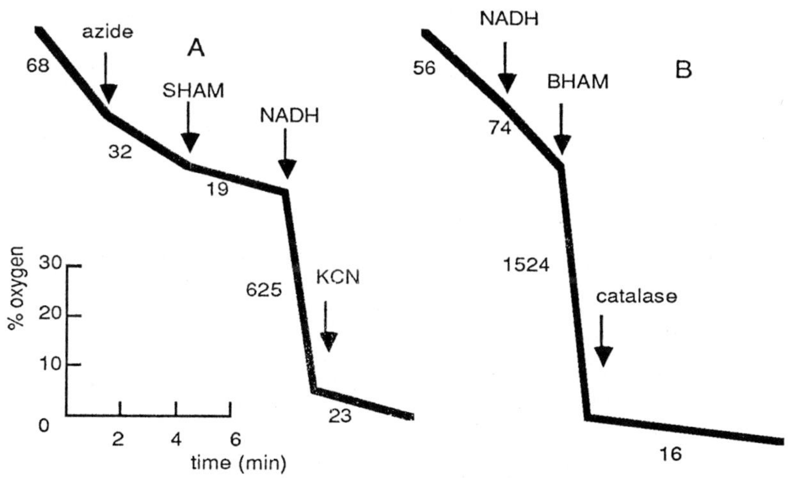

Fig.2. Effect of hydroxamates on the oxygen uptake by potato tuber callus pieces.BHAM (2.5 mM),SHAM (2.5 mM), NADH (0.5 mM),azide (5 mM),KCN (2 mM) and catalase (10 µl, 5000 U) were added at the indicated time.The values refer to the resulting O_2 uptake in nmol $.min^{-1}.g\ FW^{-1}$.

chief pool of endogenous NAD(P)H (the cytoplasm) and the peroxidase.

As exogenous NADH is necessary for the peroxidase reaction and added catalase inhibits this reaction, at least part of the peroxidase is located outside the plasmamembrane barrier, presumably the cell wall.This also explains, why the peroxidase activity is easily washed out from the callus tissue.

When also some peroxidase is present in the cytoplasm, it has to compete for NAD(P)H with other NAD(P)H consuming processes (especially the mitochondrial respiration); the extermination of the high affinity respiratory NADH-consumption without affecting the low affinity peroxidase with the aid of azide ,then must lead to an increase of the involvement of the peroxidase. However, when the effect of azide plus SHAM (inhibiting only the mitochondrial NADH-consumption) was compared with that of cyanide (inhibiting both mitochondrial and peroxidase mediated NADH-consumption) , no significant differences were observed. Also the addition of cyanide after azide (plus hydroxamate) did not result in a substantial extra inhibition, indicating that the peroxidase is not involved in 'normal' oxygen uptake by the callus tissue. However, in roots a stimulation of the oxygen uptake by hydroxamates was described by Spreen-Brouwer et al.(1986).

GENERAL CONCLUSION

Hydroxamate-stimulated peroxidases may interfere with the determination of the activity of the cytochrome and the alternative pathway in plant tissue.In experiments with mitochondria these peroxidases can be specifically inhibited by including catalase in the reaction mixture, without affecting respiratory oxygen uptake.In experiments with intact tissue,the use of high concentrations of hydroxamate may be recommended (Spreen-Brouwer et al,1986). However, in some tissues (including potato tuber callus, Van der Plas and Wagner 1980), such high concentrations may have side-effects on the cytochrome pathway. In that case, the use of normal (low) concentrations of hydroxamates, only is allowed when first is ascertained that the peroxidase is not involved in *in vivo* oxygen uptake, as is the case in potato tuber callus.

Hydroxamates never must be used as respiratory inhibitor, without investigating the presence and the role of hydroxamate stimulated peroxidases.

REFERENCES

Hemrika-Wagner,A.M.,Gude,H.,Marissen,N.,Van der Plas,L.H.W., and Verleur,J.D.,1986,Compartmentation of alternative oxidase in plant mitochondria,Plant Cell Physiol. 27 :499.

Laties,G.G.,1982,The cyanide-resistant,alternative path in higher plant respiration,Annu.Rev.Plant Physiol.,33:519.

Spreen-Brouwer,K.,Van Valen ,T.,Day,D.,and Lambers,H.,1986, Hydroxamate-stimulated O2 uptake in roots of Pisum sativum and Zea mays,mediated by a peroxidase.,Plant Physiol., in press

Van der Plas,L.H.W. and Wagner,M.J.,1980,Influence of ethanol on alternative oxidase in mitochondria from callusforming potato tuber discs,Physiol.Plant. 49:121.

Van der Plas ,L.H.W. and Wagner,M.J.,1983,Regulation of the activity of the alternative oxidase in callusforming discs from potato tubers,Physiol.Plant. 58 : 311.

CYTOKININS AND THE INDUCEMENT OF THE CYANIDE-RESISTANT ALTERNATIVE PATHWAY

[*]Annick Hourmant and [**]Pierre Dizengremel

[*]Laboratoire de Physiologie Végétale, Université de Bretagne Occidentale, Brest. [**]Laboratoire de Physiologie Cellulaire Végétale, Université de Nancy I, France

INTRODUCTION

In potato tuber slices, ageing induces a large increase in tissue respiration and the development of an alternative electron transport pathway[1]. Cytoplasmic protein synthesis was found to be required for this process[1]. In a previous publication, the addition of cytokinins to the ageing medium was shown to prevent the increase in the rates of respiration and phosphate uptake[2]. Moreover, cytokinins are well known to promote many cellular events and especially protein synthesis. In this respect, Keim et al.[3] reported a cytokinin-binding site in plant mitochondria.

This paper describes the effect of cytokinins, present in the ageing medium, on the properties of mitochondria isolated from potato tuber slices.

MATERIAL AND METHODS

Mitochondria from fresh and (24 h) aged slices of potato tubers (Solanum tuberosum L.) were isolated and purified according to classical techniques[1]. Oxygen uptake was measured polarographically (Hansatech O_2 electrode) at 25°C in a standard assay medium[1]. Mitochondrial proteins were estimated with the procedure of Lowry et al[4]. Substrates were used at the following concentrations : 20 mM Succinate, 1 mM NADH and 30 mM malate. Glutamate (2 mM) was added to malate to prevent oxaloacetate accumulation. NAD (0.6 mM) was added to obtain a maximal rate of malate oxidation. The two-dimensional analysis of mitochondrial membrane proteins was carried out according to O'Farrell[5] using isoelectric focusing in the presence of ampholines. Separation of the electrofocused proteins by molecular sieving in the second dimension was carried out on a slab containing 13 % polyacrylamide gel. The second dimension gel was stained with the silver staining procedure.

RESULTS

Effect of benzylmaminopurine (BAP) on succinate oxidation

Purified mitochondria from aged slices oxidized succinate in a cyanide-resistant way (Fig. 1). The inducement of cyanide insensitivity in potato

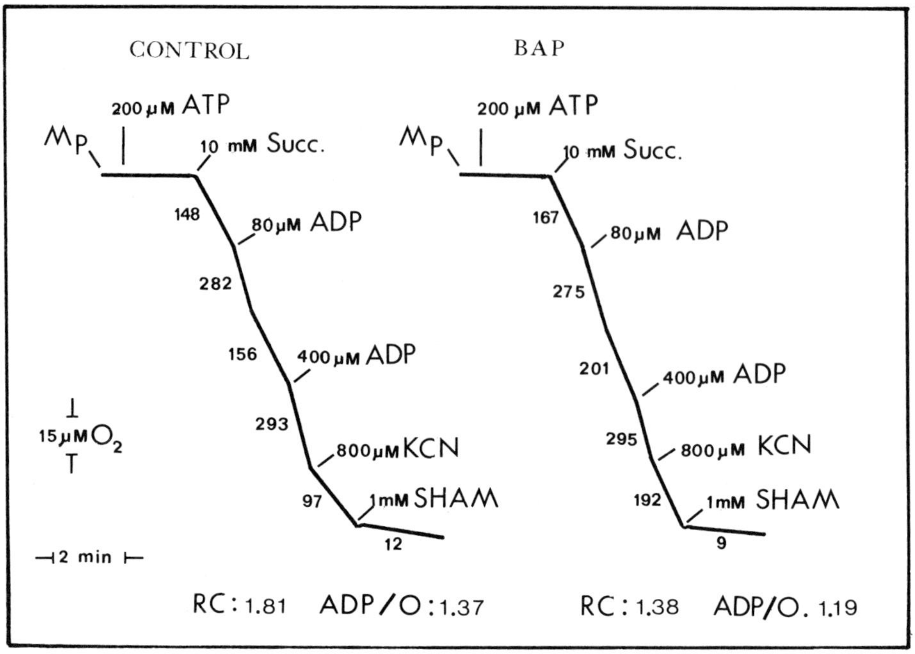

Fig. 1. Effect of the presence of benzylaminopurine (BAP, 44 µM) during ageing on the oxidation of succinate by mitochondria from potato slices. Numbers along the traces indicate O_2 uptake in nmol. min^{-1}. mg^{-1} protein. RC:respiratory control.

mitochondria (35 %) was enhanced up to 65 % by the presence of BAP in the ageing medium (Fig. 1). Although the rate of oxidation remained unchanged, the efficiency of oxidative phosphorylation was severely decreased. This

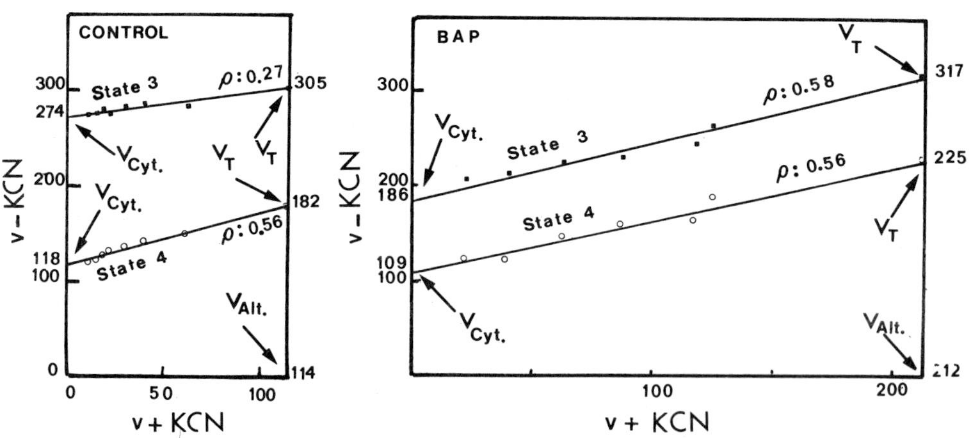

Fig. 2. Titration of respiration of mitochondria with SHAM and KCN with succinate as substrate. V_T : total electron transport rate ; V_{cyt} : contribution by the cytochrome pathway ; V_{alt}: maximal capacity of the alternative pathway ; ρ: fraction of the full alternative pathway which is operative.

effect of BAP was observed only when given in the first 8 to 10 hours following the cutting of potato tuber slices.

An attempt was made to determine the respective participation of the two pathways by using the method of Bahr and Bonner[6]. Figure 2 shows that BAP did not modify the participation of the alternative pathway in state 4. By contrast, the engagement of the cyanide-resistant pathway was promoted by BAP in state 3.

Effect of BAP on exogenous NADH and malate oxidations

The marked increase in the rate of NADH oxidation commonly observed after ageing was not modified by the presence of BAP. The oxidative phosphorylation staid unchanged (RC : 2.45 ; ADP/O : 1.25). The poor cyanide resistance observed with this substrate (10 %) was not enhanced.

Malate oxidation preferentially occurs through either malic enzyme (pH 6.5) or malate dehydrogenase (pH 7.8)[7]. The rate of malate oxidation was slightly higher at pH 6.5. In the presence of NAD, whatever the pH of the assay medium, the rates were increased and tended to reach a similar level (Table 1). Under optimal conditions for the measurement of the oxidative activity, the cyanide resistance in mitochondria from control tissues (-BAP) increased when the pH decreased (Table 1). The presence of BAP induced a very strong cyanide resistance, which was identical (≃ 76 %) whatever the pH.

Table 1. Effect of the presence of benzylaminopurine during ageing on malate oxidation measured at different pH's

Parameter		pH 6.5		pH 7.2		pH 7.8	
		Control	BAP	Control	BAP	Control	BAP
State 3 rate[a]	- NAD	67	60	50	45	44	45
	+ NAD	101	95	97	94	73	88
R.C.[b]		1.51	1.36	2.03	1.80	2.11	1.95
ADP/O[b]		2.10	2.12	1.98	1.54	1.88	1.57
CN^- Resistance (%)[b]		69	74	54	77	49	78

[a]The rates are expressed in nmol O_2. $min^{-1}.mg^{-1}$ protein.

[b]In the presence of NAD.

Effect of other phytohormones

Several phytohormones were tested in the ageing medium. Kinetin (45 µM), like BAP, accentuated the development of the alternative pathway during ageing but the effect was slightly less important (47 % instead of 65 %). By contrast, GA_3 (100 µM), ABA (40 µM) and 2,4-D (50 µM) had no effect.

Relationship between cyanide resistance and protein synthesis

Cycloheximide (0.8 µg/ml) blocked the inducement of cyanide resistance during ageing with both normal and BAP-supplemented media. It could be inferred that BAP increased the participation of a cytoplasmic-synthesizing system to the observed change. A two-dimensional electrophoresis of the mitochondrial membrane proteins showed a few new polypeptides (Fig. 3).

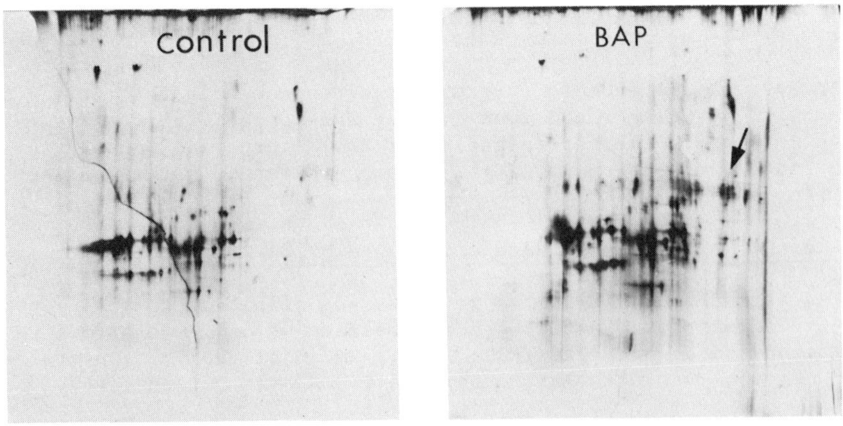

Fig. 3. Two dimensional isoelectrofocusing-SDS-polyacrylamide gels of mitochondrial membrane proteins. (↗: new proteins).

DISCUSSION

Cytokinins promote the development of the alternative pathway during ageing of potato tuber slices resulting in an increase of 30 % cyanide resistance associated to a strong decrease in oxidative phosphorylation. A gradation in cyanide resistance of malate oxidation in function of pH takes places in aged mitochondria. Ageing with BAP sets up a high and identical cyanide resistance (≃ 75 %) whatever the pH. This promoting effect of BAP on the alternative pathway which occurs at low concentrations (20-40 µM) is probably totally different from the inhibitory effect of this cytokinin (1mM) on the electron transport mediated *in vitro* by mitochondria[8]. The stimulating effect is likely related to protein synthesis from which alternative pathway development is dependent. The study of ^{14}C leucine incorporation in aged potato slices does not show any quantitave effect of BAP on protein synthesis (unpublished data). However, it cannot be excluded the possibility that cytokinins bring about qualitative changes in the spectrum of proteins as shown by Jouanneau[9] and Fosket and Tepfer[10]. The electrophoresis of mitochondrial membrane proteins supports such an hypothesis. It is possible that a new protein might be implied in the operation of the alternative pathway. This latter possibility was raised by Stegink and Siedow[11] who suggest that an engaging factor could be induced during ageing to couple the alternative oxidase to the cytochrome pathway.

REFERENCES

1. P. Dizengremel and C. Lance, Plant Physiol., 58, 147-151 (1976).
2. A. Hourmant, A. Pradet and M. Penot, Physiol. vég., 17, 483-499 (1979).
3. P. Keim, J. Erion and J.E. Fox, J. Guern and C. Péaud-Lenoël, ed., Springer-Verlag, New York (1981).
4. O.H. Lowry, N.J. Rosebrough, A.L. Farr and R.J. Randall, J. Biol. Chem., 193, 265-275 (1951).
5. P.H. O'Farrel, J. Biol. Chem., 250, 4009-4021 (1975).
6. J.T. Bahr and W.D. Bonner, J. Biol. Chem., 248, 3441-3445 (1973).
7. A.R. MacRae, Phytochemistry, 10, 1453-1458 (1971).
8. P. Dizengremel, M. Chauveau and J. Roussaux, Plant Physiol., 70, 585-589 (1982).
9. J.P. Jouanneau, Physiol. Plant., 23, 232-244 (1970).
10. D.E. Fosket and D.A. Tepfer, In vitro, 14, 63-75 (1978).
11. S.T. Stegink and J.N. Siedow, Plant Physiol., 80, 196-201 (1986).

A DETERMINATION OF THE MOLECULAR WEIGHT OF THE AROID ALTERNATIVE OXIDASE BY RADIATION INACTIVATION ANALYSIS

Deborah A. Berthold,[1] Donald J. Fluke,[2] and James N. Siedow[1]

Depts. of [1]Botany and [2]Zoology
Duke University
Durham, North Carolina 27706

INTRODUCTION

Cyanide-resistant respiration occurs in a variety of plants and fungi and has been shown to be due to an "alternative" electron transfer pathway located in the mitochondrion. The alternative pathway accepts electrons from the ubiquinone pool of the main pathway and, like cytochrome oxidase in the main pathway, uses these reducing equivalents to reduce oxygen to water.[1,2] The components of the alternative pathway are unknown, although it has been proposed that the pathway consists of a single complex, termed the "alternative oxidase," which functions as a ubiquinol:oxygen oxidoreductase. In aroid spadix mitochondria, duroquinol can be used as a water-soluble ubiquinol analogue to donate electrons to this oxidase in a cyanide-resistant, SHAM-sensitive manner. Because the detergent-solubilized duroquinol oxidase activity has proven difficult to fractionate, we decided to investigate this enzyme using radiation inactivation analysis.

Radiation inactivation is based on the premise that the larger the mass of an enzyme, the more likely it is to intercept the ionizing radiation being directed at it.[3,4] Each interaction deposits an average of 900 kcal/mol, causing a complete inactivation of the enzyme. Plots of inactivation versus dose of radiation yield an exponential dose at 37% survival.

Radiation inactivation analysis has been invoked in recent years for molecular weight determinations of membrane-bound proteins. Because loss of enzymic activity with increasing dose of radiation is directly examined, the resulting molecular weight is independent of other proteins in the sample; that is, the protein of interest does not need to be purified. Another advantage this technique has over more conventional methods is that the molecular weight thus determined is a "functional" molecular weight: it is a molecular weight relevant to the activity of the enzyme. Depending on how independently the subunit polypeptides act, the functional molecular weight may represent the molecular weight of one or more subunits of an enzyme, or it may represent the molecular weight of an entire enzyme complex.

MATERIALS AND METHODS

Mitochondria were isolated from skunk cabbage (*Symplocarpus foetidus*) and voodoo lily (*Sauromatum guttatum*) spadix tissue and purified using sucrose gradient centrifugation. Alternative oxidase activity was assayed in a citrate-containing mannitol reaction buffer (0.7 *M* citrate, 0.3 *M* mannitol, 10 m*M* KCl, 5 m*M* $MgCl_2$, 1 m*M* EDTA, 0.25 m*M* KCN, 0.1% BSA, pH 6.8) with a Clark-type oxygen electrode using 0.5 m*M* duroquinol as a substrate.

Cytochrome oxidase was assayed in bovine heart mitochondria as the cytochrome *c*-dependent, cyanide-sensitive oxygen uptake in a medium containing 8 m*M* ascorbate, 0.5 m*M* TMPD, 0.1 m*M* EDTA, 1 m*M* lauryl maltoside, 0.1% BSA, and 50 m*M* K phosphate, pH 6.5. Heart mitochondria were provided by Dr. G. T. Babcock, E. Lansing, MI.

Glucose-6-phosphate (G-6-P) dehydrogenase (from *Leuconostoc mesenteroides*) served as an internal standard in some experiments and was assayed spectrophotometrically as NAD reduction at 340 nm. The assay medium contained 3 m*M* NAD, 3 m*M* G-6-P, 3 m*M* $MgCl_2$, and 50 m*M* TRIS, pH 7.8.

Mitochondrial samples were prepared for irradiation as follows. A 25 µl sample was placed on a mylar film in good thermal contact with an aluminum block at 2°C. The pressure over the sample was incrementally decreased over an hour to allow for degassing, and then a final decrease in pressure permitted the sample to dry. The dry samples were then transferred to an irradiation chamber, maintained at 25 ± 1°C over the course of the irradiation. Resuspension of unirradiated samples yielded a 100% recovery of cytochrome oxidase and G-6-P dehydrogenase activities, and an 85% recovery of duroquinol oxidase activity.

RESULTS AND DISCUSSION

Inactivation of duroquinol oxidase activity from skunk cabbage mitochondria using high-energy ionizing radiation yielded an exponential function (Fig. 1). The functional molecular weight derived from these data is 38,000. Upon repeating this experiment, we have obtained functional molecular weights that range from 36,000-40,000.

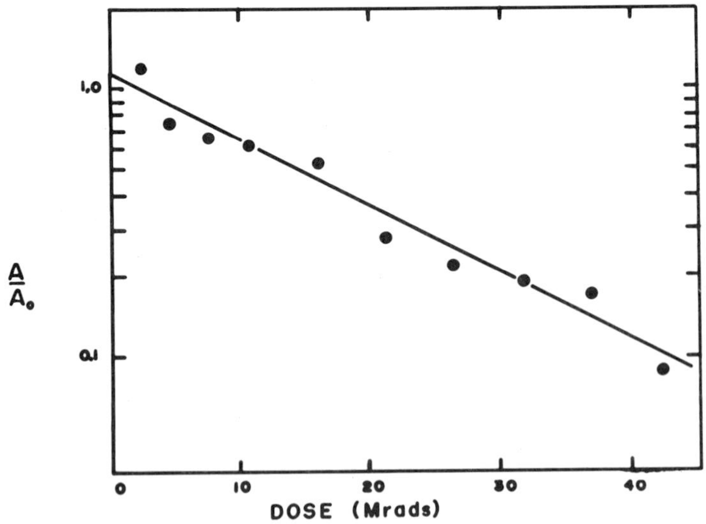

Fig. 1. Radiation inactivation of skunk cabbage alternative oxidase. The dose for 37% survival is 17.1 Mrads.

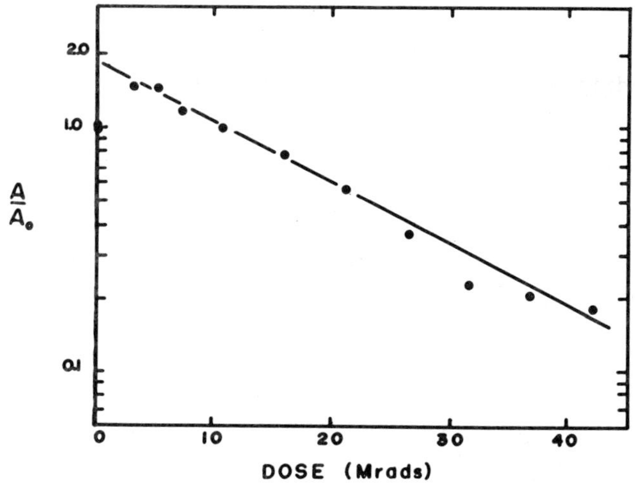

Fig. 2. Radiation inactivation of voodoo lily alternative oxidase. The dose for 37% survival was found to be 17.0 Mrads.

Radiation inactivation of duroquinol oxidase activity in voodoo lily mitochondria produced an exponential function with a slope also giving rise to a functional molecular weight of 38,000 (Fig. 2). However, the line extrapolated to an intercept of 190% initial activity. There are several possible explanations for this behavior which are currently being investigated.

When glucose-6-phosphate dehydrogenase was added as an internal standard to skunk cabbage mitochondria prior to irradiation, an exponential inactivation of the enzyme was observed over three orders of magnitude and gave a functional molecular weight of 111,000. This value compares well with the molecular weight of 104,000 obtained by sedimentation analysis.[6]

Skunk cabbage spadix cytochrome oxidase and bovine heart cytochrome oxidase in purified mitochondria were each inactivated exponentially with dose of high-energy radiation. In both cases a molecular weight of around 77,000 was obtained. Bovine cytochrome oxidase is an enzyme composed of 8-12 subunits with a molecular weight between 160,000 and 200,000. The prosthetic groups are believed to reside on subunits I and II and the functional molecular weight obtained here may correspond to the sum of these two subunits (83 kDa). Previous workers have carried out radiation inactivation using purified bovine heart cytochrome oxidase and have found functional molecular weights in the range 70,000-115,000, also much less than the molecular weight of the entire complex.[7]

A functional molecular weight determined by radiation inactivation analysis may correspond to the molecular weight of a single polypeptide, the sum of the molecular weights of two or more component subunits, or in rare instances, it may represent a somewhat isolated domain within a single polypeptide. There is reason to believe that a functional molecular weight of 38,000 for the duroquinol oxidase activity corresponds to a single polypeptide. Bertrand et al.[8] correlated the appearance of a 35 kDa polypeptide on an SDS gel with the appearance of cyanide-resistant respiration in mutants to <u>Neurospora crassa</u>. Further, a partial purification of duroquinol oxidase activity from <u>Arum maculatum</u> has yielded a major constituent in this same range (30-35 kDa).[9] Hence the value of 38,000 obtained in the present study may correspond to the same polypeptide other groups have implicated. However, we cannot rule out the possibility of a pathway consisting of several diffusible proteins of which 38,000 is the molecular weight of the largest component.

ACKNOWLEDGEMENT

This work was supported by NSF grant DMB85-166951 to J. N. S.

REFERENCES

1. C. Lance, M. Chauveau, and P. Dizengremel, The cyanide-resistant pathway of plant mitochondria, in: "Higher Plant Cell Respiration," R. Douce and D. A. Day, eds., Springer-Verlag, New York (1985).
2. J. N. Siedow and D. A. Berthold, The alternative oxidase: a cyanide-resistant pathway in higher plants, Physiol. Plant. 66:569 (1986).
3. D. E. Lea, "The Actions of Radiations on Living Cells," Cambridge Press, Cambridge, U. K. (1946).
4. C. Y. Jung, Molecular weight determinations by radiation inactivation, in: "Receptor Biochemistry and Methodology," vol. 3, J. C. Venter and L. C. Harrison, eds., Alan R. Liss, New York (1984).
5. G. R. Kepner and R. I. Macey, Membrane enzyme systems: molecular size determination by radiation inactivation, Biochim. Biophys. Acta 163: 188 (1968).
6. C. Olive and H. R. Levy, Glucose 6-phosphate dehydrogenase from Leuconostoc mesenteroides: physical studies, J. Biol. Chem. 246: 2043 (1971).
7. M. D. Suarez, A. Revzin, R. Narlock, E. S. Kempner, D. A. Thompson, and S. Ferguson-Miller, The functional and physical form of mammalian cytochrome c oxidase determine by gel filtration, radiation inactivation, and sedimentation equilibrium analysis, J. Biol. Chem. 259: 13791 (1984).
8. H. Bertrand, C. A. Argan, J. Vanderleyden, and N. A. Szakacs, Genetic control of inducible cyanide-insensitive respiratory activity in Neurospora crassa, Genetics 91s:9 (1979).
9. W. D. Bonner, Jr., S. D. Clarke, and P. R. Rich, Partial purification and characterization of the quinol oxidase activity of Arum maculatum mitochondria, Plant Phys. 80:838 (1986).

PARTIAL PURIFICATION, RECONSTITUTION, AND ANTIBODY PRODUCTION TO THE ALTERNATIVE OXIDASE OF *SAUROMATUM GUTTATUM* Mitochondria

Thomas E. Elthon and Lee McIntosh

MSU-DOE Plant Research Laboratory and Biochemistry Department
Michigan State University, East Lansing, Michigan 48824

ABSTRACT

The alternative oxidase of *Sauromatum guttatum* appendix mitochondria has been highly purified using a combination of cation exchange and hydrophobic interaction chromatography. Nickel stained SDS-PAGE (sodium dodecyl sulfate-polyacrylamide gel electrophoresis) of the purified preparation indicates a cluster of proteins near 35 kd. The partially purified alternative oxidase was found to require lipids for optimal activity, with phospholipids being the most effective. Polyclonal antibodies have been raised in mice to the alternative oxidase preparation. These antibodies are proficient in immunoprecipitating the protein(s) responsible for alternative oxidase activity.

INTRODUCTION

Although considerable research has been done on the alternative pathway at the physiological level, relatively little has been done at the biochemical or protein levels. The alternative oxidase has been solubilized with detergents from mitochondria of the spadices of *Arum maculatum* (3,7,8,9) and *Sauromatum guttatum* (5). The resulting solubilized preparations show inhibition characteristics and other properties similar to the alternative pathway activity *in situ*. However, the protein profiles of these solubilized alternative oxidase preparations are still complex (3,5). The research presented here is directed towards further purification of the alternative oxidase using solubilized preparations from *Sauromatum guttatum* appendix mitochondria.

MATERIALS AND METHODS

Voodoo lilies (*Sauromatum guttatum* Schott) were maintained in a glasshouse as previously described (5). Isolation of mitochondria, solubilization of the alternative oxidase, assay of the alternative oxidase, and protein assays were also done as previously described (5). Experiments were conducted with mitochondria isolated from appendix tissue the day before anthesis (D-1). All results are representative of three or more experiments.

CM-Sepharose CL-6B fractionation. A 2 ml (2 cm high x 1 cm in diameter) column was used for the initial fractionation of proteins in BIGCHAP (N,N-*bis*-3-D-glucoamidopropyl deoxycholamide) solubilized alternative oxidase preparations. The column was poured and equilibrated in BIGCHAP buffer (30 mM MOPS, 1 mM EDTA, and 0.5% BIGCHAP at pH 6.8). Two ml of solubilized alternative oxidase (approximately 2.0 mg total protein) was thawed and applied to the column. The column was washed with BIGCHAP buffer and the alternative oxidase eluted with BIGCHAP buffer plus 100 mM NaCl. The alternative oxidase activity elutes immediately, followed by a broad peak which consists primarily of lipids.

Phenyl-Sepharose CL-4B fractionation. The appropriate CM-Sepharose fractions were then passed through a 200 µl Phenyl-Sepharose column (11 mm high x 5 mm in diameter). The Phenyl-Sepharose column was also poured and equilibrated with BIGCHAP buffer. After application of the CM-Sepharose fractions, the column was washed with BIGCHAP buffer. Following this wash, the alternative oxidase activity remains associated with the resin.

Preparation of Asolectin vesicles. Asolectin is a commercial preparation of phospholipids (Associated Concentrates, Woodside, New York). Asolectin was dispersed in BIGCHAP buffer at a concentration of 50 mg/ml. Upon formation of an even suspension, the preparation was sonicated to effect formation of small vesicles. The vesicles were stored for periods of up to 1 week at 4 C.

Silicic acid chromatography of lipids. Total lipids were extracted from various fractions with the method of Bligh and Dyer (2). The resulting chloroform-lipid solution was applied to a silicic acid column (Biorad Bio-Sil A 100-200 mesh - 3 cm high x 0.5 cm in diameter) equilibrated with chloroform. Elution of the neutral lipid, glycolipid, and phospholipid fractions was carried out according to Carroll (4). Each fraction was then evaporated to dryness with a stream of nitrogen, resuspended in BIGCHAP buffer, and sonicated to vesicularize the lipids.

Antibody production. Antibodies were raised in 8-week old female Balb/c mice. Approximately 75 µg of protein was used per injection. Proteins were injected in PBS (10 mM NaH_2PO_4 and 150 mM NaCl at pH 7.2) and Freund's incomplete (1:3 ratio). Injections were made into the peritoneal cavity at 2 week intervals.

Gel electrophoresis, Westerns, and antibody probing of Westerns. SDS-PAGE was carried out as before (5), except that proteins were often concentrated for use with ammonium sulfate. Westerns were run using the protocol of Blake et al. (1). The nitrocellulose was then blocked by washing twice for 15 min in 100 ml of PBS-Tween (0.3% Tween 20) and stored dry. Antibody probing was done in a manner similar to the alkaline phosphatase method of Blake et al. (1), except that antibody incubations were for 1 hr at room temperature. When desirable, total proteins were stained on nitrocellulose sheets using India ink according to Hancock and Tsang (6).

Immunoprecipitation. Serum (40 µl) was added to BIGCHAP solubilized alternative oxidase (200 µg protein), and incubated for 4 hours at 4 C to allow for formation of precipitin complexes. Following centrifugation at 13,000 g for 5 min, the supernatants were assayed for alternative oxidase activity.

RESULTS AND DISCUSSION

Partial purification of the alternative oxidase. In a recent paper (5) we described a method for optimal solubilization of the alternative oxidase from voodoo lily appendix mitochondria. This procedure yields a highly active preparation (BIGCHAP solubilized preparation) and some degree of purification. The first step used for further purification of the alternative oxidase from the BIGCHAP solubilized preparation was a CM-Sepharose column. About 39% of the protein passes through the column while most of the alternative oxidase activity (73%) is retained. The alternative oxidase activity that is not retained on the CM column is believed to be within incompletely solubilized membrane fragments. The alternative oxidase activity is eluted from the resin with BIGCHAP buffer supplemented with 100 mM NaCl. The appropriate CM-Sepharose fractions were then passed through a Phenyl-Sepharose column, and the column washed with BIGCHAP buffer. The alternative oxidase sticks tenaciously (100% of the recoverable activity is retained), while most proteins (85%) pass through. So far, we have been unable to remove the alternative oxidase from this resin and retain activity. However, the proteins can be removed with 0.1% SDS. The effectiveness of this purification scheme is represented in Figure 1. The Phenyl-Sepharose protein profile suggests that the protein(s) responsible for alternative oxidase activity are associated with the cluster of proteins near 35 kd.

Reconstitution of alternative oxidase activity. During the purification process, considerable loss of alternative oxidase activity occurs. Part of this loss in activity results from the separation of the alternative oxidase protein(s) from essential lipids. This is readily demonstrated through ammonium sulfate fractionation which effectively separates the proteins (alternative oxidase precipitates between 45 and 55% saturation) from the lipids (which precipitate between 60 and 70%) saturation). Ammonium sulfate precipitation causes an 85% loss of alternative oxidase activity. When the separated lipids were fractionated into the broad classes of neutral lipids, phospholipids, and glycolipids (2,4), the phospholipid fraction was the most effective at restoring activity. The commercial phospholipid preparation "Asolectin" was found to be particularly effective in that it restores 48% of the lost activity.

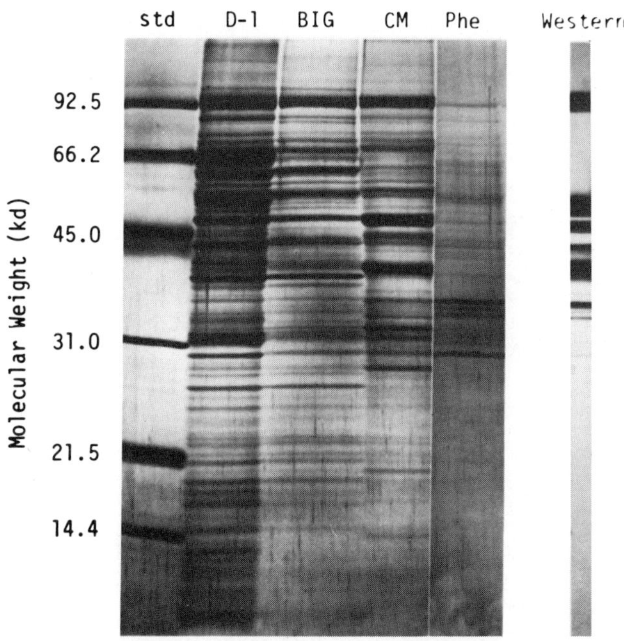

Figure 1. SDS-PAGE of various fractions during purification of the alternative oxidase. The total mitochondrial protein lane (D-1) was loaded at 20 μg, and the other lanes as appropriately as possible. The Western lane shows the results of antibody probing of a D-1 lane.

Table I. Activity of Various Fractions during Purification of the Alternative Oxidase. Asolectin vesicles (50 μl) were added to an appropriate aliquot of each fraction prior to assay for alternative oxidase activity. Alternative oxidase activity was assayed as SHAM (salicylhydroxamic acid)-sensitive DHQ (durohydroquinone)-dependent oxygen uptake in 1.0 ml of BIGCHAP buffer.

fraction	alt oxidase activity	protein	specific activity
	natoms O/min	μg	natoms O/min/mg protein
D-1 mitochondria	634	2000	317
BIGCHAP preparation	251	164	1530
CM-Sepharose	67.8	100	678
Phenyl-Sepharose	74.6	14.8	5041

The ability to reconstitute alternative oxidase activity with Asolectin allows for a useful evaluation of the specific activities of various fractions during purification (Table I). The final Phenyl-Sepharose fraction retains 12% of the initial activity and represents a 16-fold purification. However, only 15% of the initial activity is recoverable in the final fractions. Thus 85% of the activity is lost and cannot be reconstituted with Asolectin. If this activity could be reconstituted, the specific activity of the Phenyl-Sepharose fraction would represent a 133-fold purification. The activity of the CM-Sepharose fraction was low because of the presence of NaCl which is inhibitory.

Antibody production. Polyclonal antibodies have been raised in mice to the CM-Sepharose fraction. Of the proteins in this fraction, only 13 are antigenic (refer to the Western lane in Figure 1). Four of these are within the cluster of proteins retained on Phenyl-Sepharose. Sera (12 week) obtained from three different mice have been found to be effective in immunoprecipitating the alternative oxidase activity (an average of 85% of the activity is precipitated). Pre-immune and other control sera precipitate on the average only 21% of the activity.

Acknowledgements. Thanks to Bastiaan Meeuse for his expert advice, and for providing us with voodoo lily corms. We also thank Roxy Nickels and Cindy Garbus-Gooch for their assistance with polyclonal antibodies. Supported by NSF Fellowship (TEE) Grant No. DMB-8508782 and DOE Contract DE-AC02-76ERO-1338.

REFERENCES

1. BLAKE MS, KH JOHNSTON, GJ RUSSELL-JONES, EC GOTSCHLICH 1984 Anal Biochem 136:175-179
2. BLIGH EG, WJ DYER 1959 Can J Biochem and Physiol 37:911-917
3. BONNER WD JR, SD CLARKE, PR RICH 1986 Plant Physiol 80:838-842
4. CARROLL KK 1976 In GV Marinetti, ed, Lipid Chromatographic Analysis, Marrell Dekker, Inc, New York, pp 173-214
5. ELTHON TE, L MCINTOSH 1986 Plant Physiol:in press
6. HANCOCK K, VCW TSANG 1983 Anal Biochem 133:157-162
7. HUQ S, JM PALMER 1978 FEBS Lett 95:217-220
8. KAY CJ, JM PALMER 1985 Biochem J 228:309-318
9. RICH PR 1978 FEBS Lett 96:252-256

IMMUNOLOGICAL AND ELECTROPHORETIC ANALYSIS OF THE ALTERNATIVE OXIDASE OF *Arum Maculatum* SUB-MITOCHONDRIAL PARTICLES

Michael C. Walsh and Anthony L. Moore

Department of Biochemistry, University of Sussex, Falmer
Brighton BN1 9QG, U.K.

INTRODUCTION

Solubilization and partial purification of the alternative (quinol) oxidase of mitochondria isolated from spadices of Arum maculatum has been described several times[1-3]. Failure to further purify the oxidase has led to other suggestions regarding the nature of SHAM sensitive quinol oxidation in plant mitochondria, such as non-enzymic lipid peroxidation[4]. Recently a comprehensive purification using ammonium acetate precipitation and sucrose density gradient centrifugation resulted in a preparation of increased activity, similar inhibitor sensitivities and significant iron content. SDS-PAGE of the preparation reveals a major band of molecular weight 34kD however a number of other bands are still present[5].

The main problems in further characterizing the oxidase are the poor yields obtained using conventional protein purification techniques with plant tissues and identification and assay of the enzyme. The only diagnostic feature of alternative oxidase activity is its sensitivity to inhibitors such as SHAM and n-propyl gallate[6,7].

Using immunological techniques, the isolation of the alternative oxidase may be achieved by identifying the specific band on a gel and using these small amounts of protein to raise monospecific antibodies. As a first step polyspecific antisera have been raised to Arum maculatum sub-mitochondrial particles and to a fraction of the partially purified alternative oxidase described above.

METHODS AND MATERIALS

Spadices of *Arum maculatum* were collected from local woods, mitochondria were isolated *immediately* and stored in liquid nitrogen. Isolation of sub-mitochondrial particles, solubilization and assay of alternative oxidase activity, polyspecific antisera production and immunoprecipitation of alternative oxidase activity have all been previously described[8].

Crossed-immunoelectrophoresis. A 1% (w/v) agarose gel was cast onto a 50 x 50mm sheet of Gel Bond (Pharmacia) and a sample of solubilized Arum SMP (20-30ug of protein) was loaded into a well cut into a corner of the gel. Electrophoresis was carried out at 5V/cm for 4 h in the first dimension and the same voltage for 18 h in the second dimension. All experiments were performed at 4°C to maintain enzyme activity, and using purified IgG fractions[9].

Non-dissociating PAGE. Two systems were used both comprising 3-25% gradient polyacrylamide gels (125 x 80 x 1mm) containing:

1. 50mM Tris-glycine buffer pH 9.0
   0.05% (w/v) deoxycholate
   0.05% (w/v) octyl-B-D-thioglucoside

2. 50mM Tris-citrate buffer pH 7.6
   0.05% (w/v) deoxycholate
   0.05% (w/v) octyl-B-D-thioglucoside

Electrophoresis, in 100mM gel buffer containing 0.05% (w/v) deoxycholate and 0.05% (w/v) octyl-B-D-thioglucoside, was run at 10mA and 4°C for 4-5 h at the higher pH and overnight at the lower pH. Samples of solubilized Arum SMP were prepared in dilute buffer and were diluted twofold in running buffer before loading.

Staining. Prior to staining, agarose immunoelectrophoresis plates were washed in phosphate buffered saline and pressed dry using chromatography paper, polyacrylamide gels were rinsed in running buffer. Gels were stained for protein and for alternative oxidase activity, protein bands were visualized using Coomassie Brilliant Blue R. Alternative oxidase activity was detected by staining with 1mM duroquinol, 5mg/ml neo-tetrazolium chloride in 10mM MOPS pH 7.2 containing 0.05% (w/v) deoxycholate and 0.05% (w/v) octyl-B-D-thioglucoside. Colour development takes place within about 2 h. In addition gels were stained for other enzyme activities such as NADH and succinate dehydrogenases, cytochromes and the ATPase.

Western blotting and antibody probing. Electrophoretic transfer of proteins to nitrocellulose paper was carried out after non-dissociating PAGE. Transfer was carried out, in 100mM Tris-glycine pH 9.0 containing 0.05% (w/v) deoxycholate and 0.05% (w/v) octyl-B-D-thioglucoside, at 150mA and 4°C overnight. All subsequent steps were carried out at room temperature in phosphate buffered saline containing 5% (w/v) Marvel (Cadbury's), 0.3% (v/v) Tween 20 and 0.1% (v/v) antifoam. Unoccupied protein binding sites were blocked by washing three times for 20 mins each, nitrocellulose sheets were then probed with the polyspecific antisera (1:100 dilution) or with purified IgG fraction (1:200 dilution). Bound antibodies were then visualized using a goat anti-rabbit peroxidase conjugate system.

RESULTS AND DISCUSSION

Crossed-immunoelectrophoretic analysis of the antibody content of the two antisera are shown in Fig.1. The antisera raised to Arum SMP has precipitated at least 6 antigenically distinct proteins (A) while the antisera raised to the partially purified alternative oxidase precipated at least 8 antigenically distinct proteins (B). The large number of precipitin peaks present on plate B confirms that there are still a number of contaminants in the alternative oxidase preparation, however the increased antigenicity (as compared to the anti-Arum SMP antisera) may be a result of the increased specific activity of the preparation.

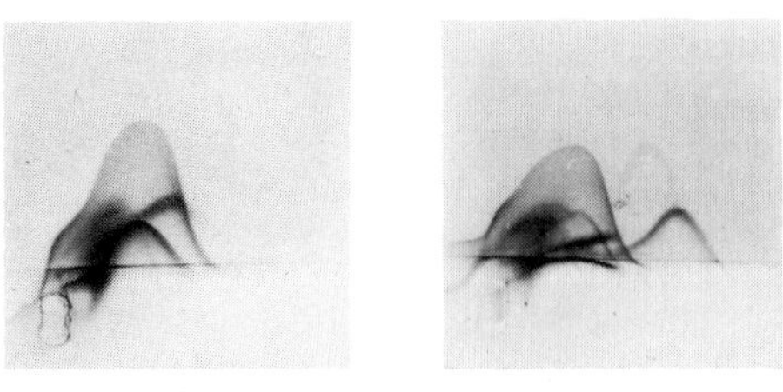

A B

Fig.1. Crossed immunoelectrophoresis of Triton X-100 solubilized Arum SMP.
Crossed immunoelectrophoresis of Triton X-100 solubilized Arum SMP was performed using antisera raised to Arum SMP (A) and to the partially purified enzyme (B), both plates were stained for protein with Coomassie Blue.

The presence of antibodies to the alternative oxidase is shown in Fig. 2. The antisera raised to the partially purified preparation of the alternative oxidase immunoprecipitates over 70% of the alternative oxidase activity while similar concentrations of the anti-Arum SMP antisera removes no activity by immunoprecipitation. The failure of the anti-Arum SMP antisera to precipitate any activity acts as a control showing the effect is not due to non-specific IgG binding.

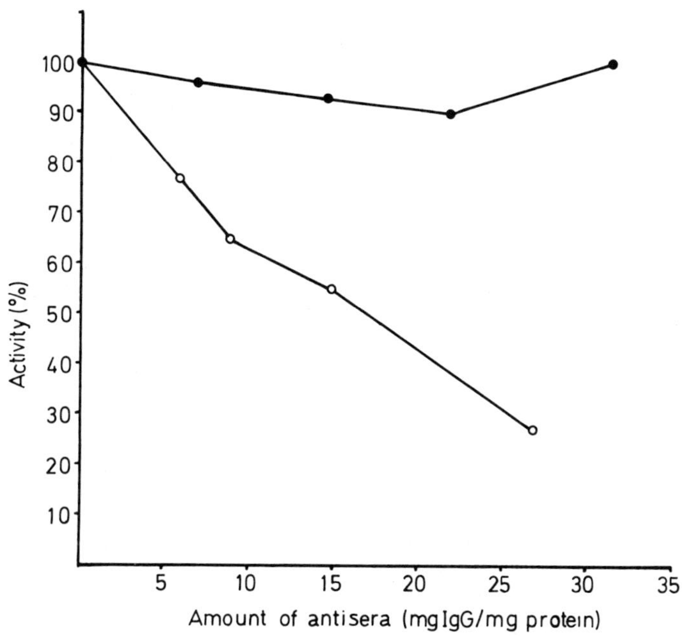

Fig.2. Immunoprecipitation of alternative oxidase activity. Activity was measured in the supernatant, after incubating solubilized Arum SMP with antisera raised to Arum SMP (●) or to a partially purified alternative oxidase fraction (O) and removing the immunoprecipitate by centrifugation. Duroquinol oxidation was followed in a Rank oxygen electrode containing 2ml 10mM MOPS (pH 7.4); 1ug/ml antimycin A; 0.5mM KCN; 1mM duroquinol. All rates were 90-100% inhibited by 1mM SHAM and are expressed as percentage activity of the alternative oxidase activity in the original solubilized preparation.

Analysis of solubilized Arum SMP on the high pH non-dissociating PAGE system reveals up to 16 protein staining bands. Staining for alternative oxidase activity on this system reveals relatively little despite the fact that NADH dehydrogenase, succinate dehydrogenase and cytochrome staining bands are all revealed (Fig.3.). Similar attempts to stain for alternative oxidase activity on agarose immunoelectrophoresis plates were also unsuccessful. The failure of the alternative oxidase activity stain, in both systems, is explained by the pH profile of the enzyme (data not shown). This shows that solubilized alternative oxidase activity is completely inactivated above pH 8.0 however the non-dissociating PAGE and agarose immunoelectrophoresis systems are run at pH 9.0 and pH 8.5 respectively.

The lower pH non-dissociating PAGE system gives much poorer resolution of protein stained bands due to the very slow migration of the solubilised protein complexes under these conditions. Staining for alternative oxidase activity, however, reveals two bands, one that corresponds to a heavily staining protein band and a very weak staining band (Fig. 4.).

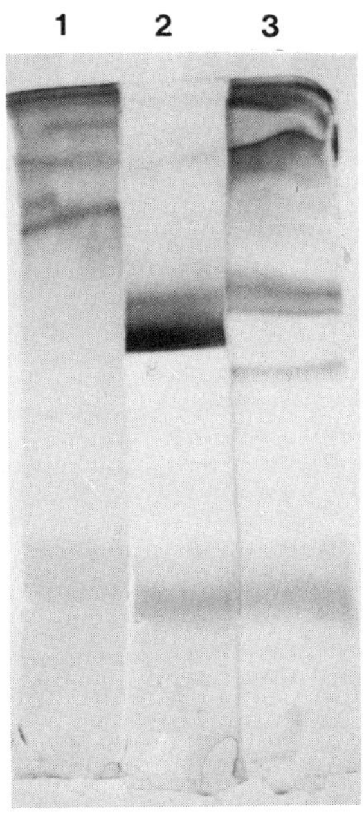

Fig.3. Non-dissociating PAGE of solubilized Arum SMP at pH 9.0.
Gel electrophoresis was performed as described in the text and subsequently stained for:-
1. cytochrome
2. succinate dehydrogenase
3. NADH dehydrogenase

Fig.4. Non-dissociating PAGE of solubilized Arum SMP at pH 7.6.
Gel electrophoresis was performed as described in the text and subsequently stained for:-
1. protein
2. alternative oxidase

Western -immunoblotting of the higher pH gel with the antisera to the partially purified alternative oxidase shows 8 protein bands binding antibodies, of these, the 4 higher molecular weight proteins have been identified as two NADH dehydrogenases, succinate dehydrogenase and cytochrome staining bands. The other 4 antibody binding bands are close together on the gel and have molecular weights less than succinate dehydrogenase, any of these may be the alternative oxidase (Fig. 5.).

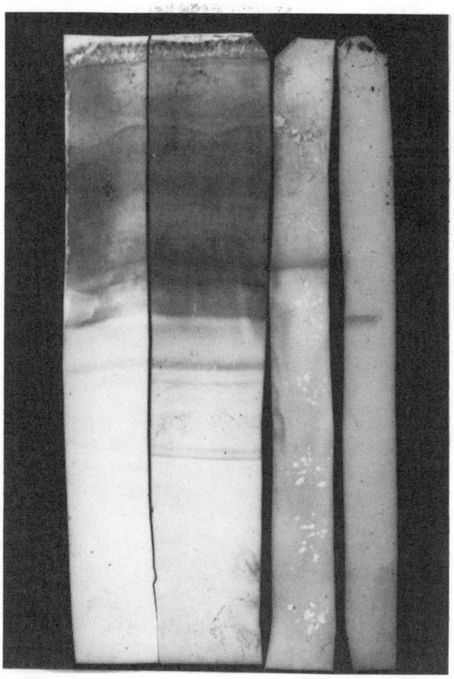

Fig.5. Western immuno-blotting of solubilized Arum SMP from higher pH non-dissociating gel with antisera raised to the partially purified alternative oxidase.
Protein bands were electrophoretically transferred from the non-dissociating gel to nitrocellulose paper, the paper was then probed with:-
1. raw serum (1:100)
2. purified IgG fraction (1:200)
3. NADH dehydrogenase stain
4. succinate dehydrogenase stain

Now that the alternative oxidase stain has identified complexes on the low pH gel system. Immunoblotting from these gels or band excision followed by line immunoelectrophoresis will allow isolation of the specific precipitin complexes, from which a monospecific antiserum can be raised. In addition, two dimensional electrophoresis using the non-dissociating system for first dimensional separation followed by SDS-PAGE in the second dimension should reveal a lot of data concerning the polypeptide profiles of all the protein complexes present in solubilized Arum SMP.

In conclusion, the antisera raised to the partially purified alternative oxidase is capable of immunoprecipitating SHAM sensitive and antimycin A insensitive quinol oxidation in solubilized Arum SMP, strongly suggesting this activity is associated with a protein. The number of antibody binding proteins has been determined and half of these have been identified, of the remaining 4 unidentified antibody binding proteins, one or more may be the alternative oxidase enzyme.

ACKNOWLEDGMENTS

We would like to thank Dr. P.R.Rich (Glynn Research Institute) for the gift of the partially purified oxidase fraction. This work was supported by an S.E.R.C. grant awarded to A.L.M.

REFERENCES

1. Rich, P.R. (1978) FEBS Lett. 96, 252-256
2. Huq, S. & Palmer, J.M. (1978) FEBS Lett. 95, 217-220
3. Kay, C.J. & Palmer, J.M. (1985) Biochem. J. 228, 309-318

4. Rustin, P., Dupont, J. & Lance, C. (1983) Arch. Biochem. Biophys. 225, 630-639
5. Bonner, W.D., Clarke, S.D. & Rich, P.R. (1986) Plant Physiol. 80, 838-842
6. Schonbaum, G.R., Bonner, W.D., Storey, B.T. & Bahr, J.T. (1971) Plant Physiol. 47, 124-128
7. Siedow, J.N. & Girvin, M.E. (1980) Plant Physiol. 65, 669-674
8. Walsh, M.C., Cottingham, I.R. & Moore, A.L. (1986) Biochem. Soc. Trans. 14, 895
9. Owen, P. & Salton, M.R.J. (1975) Proc. Natl. Acad. Sci. U.S.A. 48, 1657-1663

THE EFFECTS OF DIFFERENT IONIC-CONDITIONS ON THE ACTIVITY OF CYTOCHROME C-OXIDASE IN PURIFIED PLANT MITOCHONDRIA

Bo Wigge and Per Gardeström

Department of Plant Physiology
University of Umeå
S-901 87 Sweden

INTRODUCTION

Cytochrome C-oxidase is one of the most commonly used marker enzymes for mitochondria. Since the intact outer mitochondrial membrane is not permeable to the substrate (reduced Cytochrome C) and as Cyt C interacts with the outer surface of the inner membrane, the assay has also been used as an intactness assay of isolated mitochondria[1] and for determination of the polartiy of submitochondrial particles (SMP)[2].

Here we report on the effects of different ionic conditions on the activity of Cytochrome C-oxidase in purified plant mitochondria. From these studys we have composed a medium for the Cyt c-oxidase assay which give a high activity and the same intactness and latency as previously used media.

METHODS

Isolation of mitochondria

Pea leaf mitochondria and mitochondria from potato tubers were isolated[4,5] and purified on a 28% selfgenerating Percoll-gradient (40000 g, 45 min.). Both types of mitochondria were suspended in: 0.3 M sucrose,10 mM Hepes, 2mM Glycin, 0.1% w/v BSA, 1 mM K_2HPO_4, pH 7.2, except when SMP were prepared.

Isolation of SMP

SMP was isolated in low salt + EDTA medium by French press treatment and in high salt medium by sonication[2].

Assays

Cytochrome C oxidase was assayed at pH 7.2 $\pm$ 0.02% Triton X-100. The assay was started with 50 µM reduced Cytochrome C and followed at 550 nm with 540 nm as reference wavelength. Activity is expressed as % of the activity in basic medium (0.3 M sucrose, 10 mM KPB pH 7.2, 0.02% Triton X-100). The activity in basic medium varied from 800 to 2000 nmol Cyt C * mg prot^{-1} * min^{-1} between different preparations.
Intactness was calculated as:

Intactness=(1-(rate without triton / rate with triton))* 100

The latency of the SMP:s was determined and calculated in in the same way as intactness and then expressed as " % right side out"(%RO).

%RO=(rate without triton / rate with triton) * 100

Protein determination

The sample was solubilized in 0.1 M KOH at 80°C for 1.5 h. The protein was then determined spectrophotometrically with Comassie Brilliant Blue using BSA for calibration.

RESULTS AND DISCUSSION

Figure 1 shows that the highest activity is obtained with 100 mM KCl. Also $MgCl_2$ up to 25 mM stimulates the activity while $CaCl_2$ seeems to inhibit.

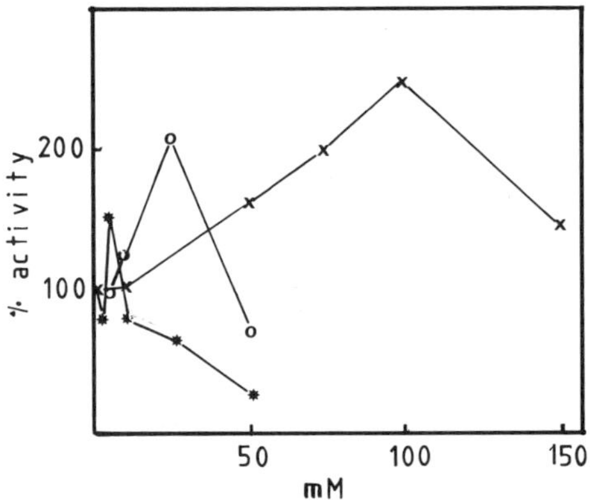

Fig. 1. Activity of Cyt C-oxidase at different concentrations of KCl (x-x), $MgCl_2$ (o-o) and $CaCl_2$ (∗-∗).

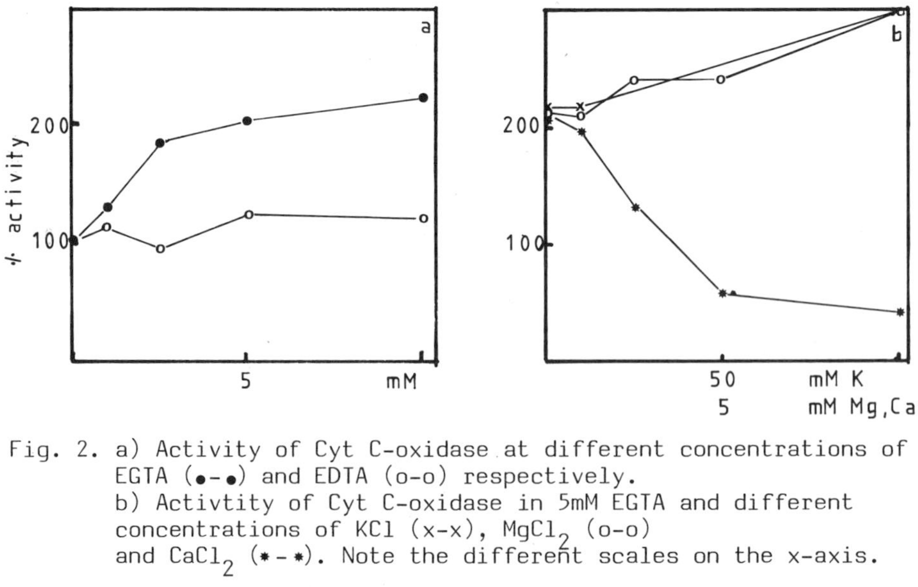

Fig. 2. a) Activity of Cyt C-oxidase at different concentrations of EGTA (•-•) and EDTA (o-o) respectively.
b) Activtity of Cyt C-oxidase in 5mM EGTA and different concentrations of KCl (x-x), $MgCl_2$ (o-o) and $CaCl_2$ (∗-∗). Note the different scales on the x-axis.

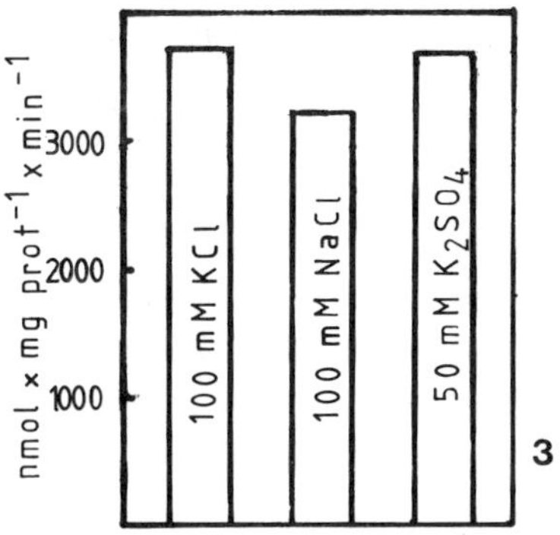

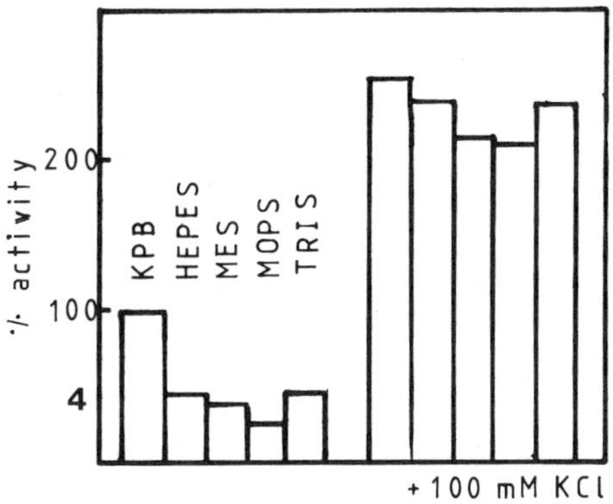

Fig. 3. Activity of Cyt C-oxidase with 100 mM KCl, 100 mM NaCl and 50 mM K_2SO_4 respectively.
Fig. 4. Activity of Cyt C-oxidase in different buffers with and without 100 mM KCl.

The inhibition by Ca^{2+} is also shown on addition of EGTA, which complexes Ca^{2+} (Fig. 4a). The additon of EDTA does not cause any stimulation. The comparison of KCl, NaCl and K_2SO_4 (Fig.2) shows that it is probably the cations that cause the effect.

The addition of $(DM)Br_2$ up to 10 mM and $(TEC)Cl_3$ up to 1 mM stimulated the activity very little (data not shown). This indicates that the effect of K^+ Na^+ and Mg^{2+} is not a charge-screening effect .

Figure 4 shows that the choice of organic buffer has very little effect on the activity of Cytochrome C-oxidase. The higher activity in KPB is probably due to the higher K^+ concentraion in this buffer.

From the data above a medium for the assay of Cyt C-oxidase was composed: 0.3 M sucrose, 100 mM KCl, 10 mM KPB pH 7.2

Table 1. Four different assay-media for Cyt C-oxidase were compared. Medium A is a medium designed from the results in this study, B was used by Møller et al[2], C by Møller et al[3] and D is a commonly used medium for respiration studies.
The activity is expressed as nmol Cyt C * mg protein^{-1} * min^{-1}.
Medium A: 0.3 M sucrose, 100 mM KCl, 10 mM KPB pH 7.2.
B: 0.3 M sucrose, 5 mM $MgCl_2$, 10 mM MOPS pH 7.2.
C: 0.3 M sucrose, 10 mM K_2HPO_4, 2 mM $MgCl_2$, 10 mM MOPS pH 7.2.
D: 0.3 M sucrose, 10mM KCl, 5 mM $MgCl_2$, 10 mM KPB pH 7.2.

	PEA MITOCHONDRIA Assay medium				POTATO MITOCHONDRIA Assay medium			
	A	B	C	D	A	B	C	D
- Triton	52	14	21	26	27	9	15	15
+ 0.02 % Triton	5090	720	1930	2105	2616	744	1375	1516
Intactness	99	99	99	99	99	99	99	99

Table 2. % RO in SMP from Pea leaf and Potato tuber mitochondria measured in four different media (see legend to Table 1).

	PEA SMP Assay medium				POTATO SMP Assay medium			
	A	B	C	D	A	B	C	D
Low salt SMP	43	58	51	60	55	47	58	50
High salt SMP	11	18	15	14	19	26	21	20

The medium composed from the present study was compared with other previous used media with respect to the activity of Cyt C-oxidase and intactness of mitochondria from pea and potato tubers. Table 1 shows that the new medium (Medium A) gives the highest activity and the same intactness as the other media.

The media was also tested on SMP from Pea and Potato tuber mitochondria with respect to polarity.In both cases medium A gave the the highest activity and the same polarity as the other media used.

REFERENCES

1. M. Neuburger, E-P. Journet, R Bligny, R Carde, R Douce, Purification of Plant Mitochondria by Isopycnic Centrifugation in Density Gradients of Percoll, Arch. Biochem. Biophys. 217:313-323 (1982).
2. I.M. Møller, I. Ericson, P. Gardeström, Isolation of submitochondrial particles with different polarities, in: "Methods of Enzymology Series. Plant Cell Membranes," L.Packer and R.Douce, eds., Academic Press, New York (1986) In press.
3. I.M. Møller, A Bergman, P. Gardeström, I. Ericson, J. Palmer, Characterisation and Purification of Inside-out Submitochondrial Particles Obtained from Plant Mitochondria, FEBS Lett. 126:13-17 (1981).
4. D.A Day, M. Neuburger, R. Douce, Biochemical Characterisation of Chlorophyll-free Mitochondria from Pea Leaves Aust. J. Plant Physiol. 12:219-228 (1985).
5. P.J. Petit, K.A. Edman, P. Gardeström, I. Ericson, Some properties of mitochondria, mitoplasts and submitochondrial particles of different polarities from plant tissues, Biochim. Biophys. Acta In press.

PURIFICATION OF JERUSALEM ARTICHOKE MITOCHONDRIA ON A CONTINUOUS PERCOLL GRADIENT

Ian M. Møller and Annika Lidén

Department of Plant Physiology
University of Lund
Box 7007, S-220 07 Lund
Sweden

INTRODUCTION

Mitochondria from a number of plant species, mainly from green leaves, have been purified on continuous Percoll gradients. In non-green tissues the main contaminants are etioplasts and microbodies and they are removed almost completely through this purification step (e.g. 1). Since we wished to investigate whether there is a homogeneous or a heterogeneous lateral distribution of proteins in the inner mitochondrial membrane (2), it was essential to start with uncontaminated mitochondria. We have, therefore, optimized Percoll purification for Jerusalem artichoke mitochondria.

MATERIALS AND METHODS

Crude mitochondria were isolated from Jerusalem artichoke (Helianthus tuberosus L.) tubers as described in (3). Three ml of crude mitochondria (30-50 mg protein) were layered on top of 36 ml of a medium containing 21, 22, 25, or 28% (v/v) Percoll. The gradient was developed by centrifugation at 40 000 g x 30 min in an angle rotor. For the characterization of the gradient, it was pumped from the bottom and 20-30 fractions were collected and assayed for protein content and enzyme activities. For the isolation of purified mitochondria, the top 15 ml were sucked off and the next 23 ml collected (the bottom 1 ml was discarded), diluted and the mitochondria pelleted at 48 000 g x 3 min.

Cytochrome c oxidase (CCO, EC 1.3.9.1) and NAD-dependent malate dehydrogenase (MDH, EC 1.1.1.37) were measured according to (3). Catalase (EC 1.11.1.6) was assayed at 25°C in a medium containing 20 mM MOPS-KOH, pH 7.0 and 1 mM H_2O_2. The respiration of the mitochondria was measured as oxygen consumption in a medium containing 0.3 M sucrose, 5 mM KPi, 5 mM MOPS-KOH and 2.5 mM $MgCl_2$, pH 7.2 The latency of CCO and MDH was estimated by measuring the activity ± 0.025% (w/v) Triton X-100 and then calculating the latency as:

$$\frac{(\text{rate +Triton}) - (\text{rate -Triton})}{(\text{rate +Triton})} \times 100\%$$

RESULTS

Initially we used 28% Percoll as in the purification of potato mitochondria (1). However, few mitochondria entered the gradient under these conditions (not shown). As the concentration of Percoll was gradually decreased via 25% and 22% to 21% the mitochondrial peak moved towards the bottom of the gradient and it grew in size until at 21% Percoll about 2/3 of the CCO activity was found there (Fig. 1). When this band was taken out, diluted and the purified mitochondria pelleted some loss of protein and activity occurred. Incomplete pelletation of mitochondria due to the formation of a Percoll cushion may have been the cause.

Assays for catalase showed that even at 21% Percoll the microbodies stayed at the top of the gradient (Fig. 1); only about 5% of the catalase was found in the mitochondrial band subsequently used as purified mitochondria (1-24 ml; Fig. 1, Table 1). The result was an almost 8-fold purification of mitochondria relative to the microbodies. The mitochondrial band also had higher specific activities of CCO than crude mitochondria. The latencies of CCO and MDH indicated that 98% of the outer membranes and 96% of the inner membranes were intact in the purified mitochondria, a marked improvement over the crude mitochondria (Table 1). Damaged mitochondria banded just below the microbodies on the gradient. They showed CCO latencies of 50-90% (Fig. 1). The distribution and latency of MDH on the Percoll gradient showed that the microbodies contained substantial amounts of latent MDH (Fig. 1, Table 1).

Crude mitochondria showed reasonable oxidation rates, respiratory controls (RC) and ADP/O ratios with NADH and succinate (Table 1). The oxidation rate with malate alone was slow and decreased progressively in successive State 3/State 4 cycles (not shown), but it was markedly improved by a preincubation with NAD^+ and ATP. The purified mitochondria showed significantly improved respiratory properties with all substrates except with malate alone (Table 1). The very high RC observed with malate + NAD^+ was the result of an initial slow State 4 (see ref. 4).

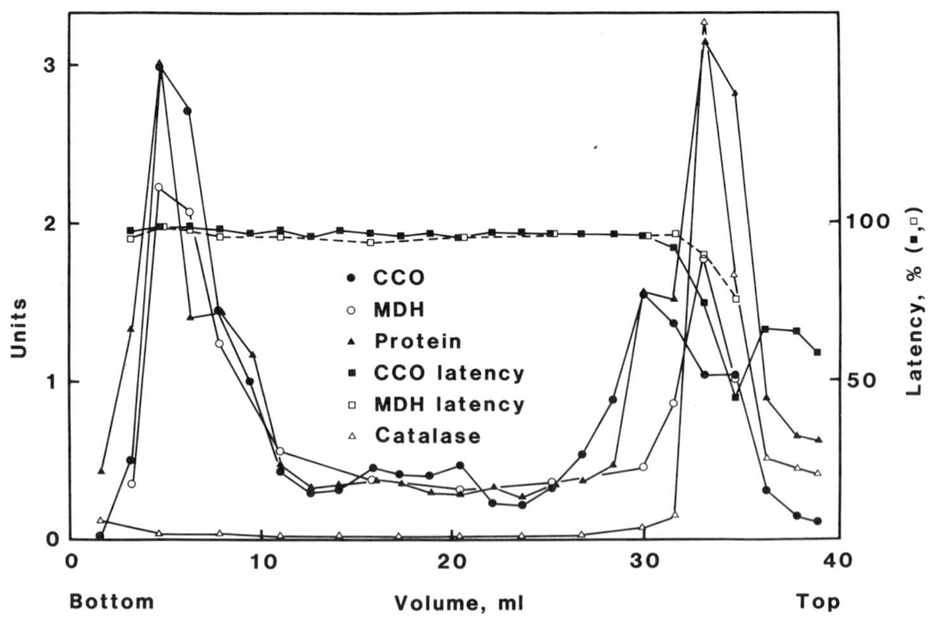

Fig. 1. Purification of crude mitochondria on a 21% Percoll gradient. One unit (left axis) represents 1 µmol/min x ml CCO activity, 1 mg/ml protein, 5 µmol/min x ml MDH activity or 10 µmol/min x ml catalase activity.

Table 1. Comparison of the properties of crude and purified Jerusalem artichoke mitochondria. All data from the same preparation.

Parameter	Unit[g)]	Mitochondria	
		Crude	Purified
Protein	mg	30	11
	%	100	36
CCO latency	%	90	98
specific activity[a)]	μmol $mg^{-1}min^{-1}$	1.17	1.83
recovery	%	100	56
MDH latency	%	83	96
specific activity[b)]	μmol $mg^{-1}min^{-1}$	5.0	4.2
recovery	%	100	31
Catalase specific activity[c)]	μmol $mg^{-1}min^{-1}$	1.91	0.40[h)]
recovery	%	100	8
CCO specific activity / Catalase specific activity	-	0.61	4.6[h)]
Respiratory properties			
NADH[d)] +FCCP	nmol $mg^{-1}min^{-1}$	160	225
RC	-	3.7	5.3
ADP/O	-	1.25	1.54
Succinate[e)] State 3	nmol $mg^{-1}min^{-1}$	118	245
RC	-	3.4	5.6
ADP/O	-	1.25	1.54
Malate[f)] State 3	nmol $mg^{-1}min^{-1}$	51	113
RC	-	5.3	11.2
ADP/O	-	2.07	2.50

a) Measured as µmol reduced cytochrome c oxidized (3)
b) Measured as µmol NADH oxidized (3)
c) Measured as µmol oxygen released
d) 1 mM NADH; 0.4 µM carbonylcyanide p-trifluoromethoxyphenylhydrazone; 3rd State 3/State 4 cycle
e) 10 mM succinate + 0.12 mM ATP (to activate succinate dehydrogenase); 3rd State 3/State 4 cycle
f) Preincubation with 0.5 mM NAD^+ + 0.12 mM ATP; 25 mM malate added after 3 min; 2nd State 3/State 4 cycle (the State 4 measured is the initial slow phase - see ref. 4)
g) All specific activities are per mg mitochondrial protein
h) Normally considerably better removal of catalase (see Table 1 in ref. 2)

DISCUSSION

We have reduced a major microbody contamination in crude Jerusalem artichoke mitochondria by ca 90% through the use of a continuous Percoll gradient (Fig. 1, Table 1). This was a major objective of the present study as the mitochondria were to be used in studies on inner membrane polypeptide composition (2). In addition to the increased purity, damaged mitochondria were also removed and the purified mitochondria showed better integrities, higher specific activity of CCO and higher ADP/O ratios, RC and respiratory rates with NADH, succinate and malate (plus NAD^+) than crude mitochondria (Fig. 1, Table 1). Similar improvements in quality after purification of mitochondria on continuous Percoll gradients have been reported for potato tubers (1) and pea leaves (5).

Jerusalem artichoke mitochondria, both crude and purified, appeared to have suboptimal contents of NAD^+ as the uptake of this coenzyme, induced through energization of the mitochondria with ATP (6), greatly improved their rate of malate oxidation. The presence of NAD^+ also prevented the gradual decrease in the State 3 with successive State 3/State 4 cycles (not shown). This is achieved through a displacement of the equilibrium of MDH by NAD^+ so that it tolerates better the accumulation of oxaloacetate (4).

The microbodies of Jerusalem artichoke tubers are less dense than intact mitochondria (Fig. 1). This is also true for acetate-grown Neurospora crassa (7) whereas microbodies are more dense than mitochondria for potato tubers (1) and pea leaves (5). Thus there does not appear to be a fixed relationship between the "true" densities of these two organelles as determined on a Percoll gradient. An optimal purification procedure must be worked out specifically for each new tissue.

ACKNOWLEDGEMENTS

We are grateful to Eva Olsson for drawing the figure. This study was supported by a grant from the Swedish Natural Sciences Research Council to IMM.

REFERENCES

1. M. Neuburger, E.-P. Journet, R. Bligny, J.-P. Carde, and R. Douce, Purification of plant mitochondria by isopycnic centrifugation in density gradients of Percoll, Arch. Biochem. Biophys. 217:312 (1982).
2. A. Lidén, M. Sommarin, and I.M. Møller, Lateral heterogeneity in the inner mitochondrial membrane, These proceedings.
3. I.M. Møller, I. Ericson, and P. Gardeström, Isolation of submitochondrial particles with different polarities, Methods Enzymol. (In press).
4. J.M. Palmer, J.-P. Schwitzguébel, and I.M. Møller, Regulation of malate oxidation in plant mitochondria. Response to rotenone and exogenous NAD^+, Biochem. J. 208:703 (1982).
5. D.A. Day, M. Neuburger, and R. Douce, Biochemical characterization of chlorophyll-free mitochondria from pea leaves, Aust. J. Plant Physiol. 12:219 (1985).
6. A. Tobin, B. Djerdjour, E. Journet, M. Neuburger, and R. Douce, Effect of NAD^+ on malate oxidation in intact plant mitochondria, Plant Physiol. 66:225 (1980).
7. J.-P. Schwitzguébel, I.M. Møller, and J.M. Palmer, Changes in density of mitochondria and glyoxysomes from Neurospora crassa: a re-evaluation utilizing silica sol gradient centrifugation, J. Gen. Microbiol. 126: 289 (1981).

GENERATION AND CHARACTERIZATION OF SUB-MITOCHONDRIAL PARTICLES: ISOELECTRIC POINTS AND CHARGE SURFACE DENSITY OF PLANT MITOCHONDRIAL MEMBRANES

Patrice Petit*, Kjell Edman**, Per Gardeström*** and Ingemar Ericson**

*Laboratoire de Biologie Végétale IV (CNRS, UA 1180), Université Pierre et Marie Curie, Paris, France.
**Department of Biochemistry and ***Department of Plant Physiology, University of Umeå, Umeå, Sweden

INTRODUCTION

Mitochondrial research is witnessing a fundamental change which is beginning to influence our view on how mitochondria work and how they interact with other compartments of the eukaryotic cell (Yaffe and Schatz, 1984).

Most of the phenomena actually discussed as protein import, interaction of mitochondria with each other or with the cytoskeleton, and mitochondrial DNA control of the expression of surface antigenes are probably mediated by the outer and may be by the inner mitochondrial membrane proteins.

Tranverse asymmetry is a basic feature of biological membranes which is of particular functional importance for energy transducing membranes. The classical experimental tools for testing this asymmetry, i.e. chemical labels or antibodies, can give conclusive results only when both right-sided and inverted vesicles of a membrane are available. Biological membranes can also show a heterogeneity in the lateral distribution of membrane components. In mitochondria as in the chloroplasts (Andersson et al., 1985), the isolation of right side out and inside-out vesicles may lead to the understanding of the transbilayer and lateral heterogeneity of the membranes (Møller et al., 1984).

The knowledge of some physical properties of the different surfaces of the mitochondrial membranes as isoelectric point and charge surface density may lead to some rapid progress in the studies of interaction mechamisms.

MATERIALS AND METHODS

Potato tuber (Solanum tuberosum L.) mitochondria were prepared and purified (MP) as respectively described by Diolez and Moreau (1983) and Petit et al. (1986). Either Percoll- and sucrose-purified mitochondria were used for obtaining mitoplasts by osmotic shock. Sub-mitochondrial particles (SMP) were generated as described by Kay et al. (1985). The sub-mitochondrial particles were pelleted at 300,000 g and were rinsed twice with either 0.3 M sucrose and 2 mM MOPS (pH 7.0) for charge density measurements (to reduce ionic strength) or with 0.3 M sucrose and 50 mM K^+-phosphate (pH 7.2) for cross-partition (to reduce chloride content). The isoelectric point of MP, mitoplasts, right side-out (RO-SMP) or inside-out (IO-SMP) sub-mitochondrial particles were determined by cross-partition. A relatively large electrostatic potential difference between the bulk phases was produced by including either a cationic (trimethylamino-polyethylene glycol) (TMA-PEG) or an anionic (polyethylene glycol sulfonate) (S-PEG) derivative in the phase systems (Westrin _et al._, 1983). The measurement of the charge surface density was performed as described by Johnston _et al._ (1979) with the modifications of Møller _et al._ (1984).

RESULTS AND DISCUSSION

Irrespective of the starting material (washed mitochondria, MP , mitoplasts) SMP of well-defined polarities can be generated by French press treatement in low salt medium and by sonication in high salt medium. These

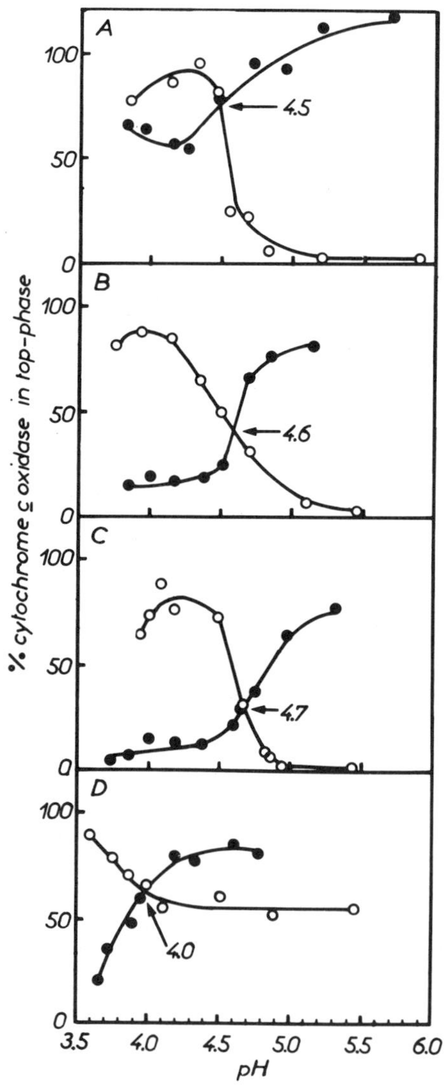

Figure 1. Cross-partition partition of intact mitochondria (A), mitoplasts (B), RO-SMP (C) and IO-SMP (D). Closed circles represent the series of phase systems with TMA-PEG and open circles the series of phase systems with S-PEG. The cross-point values indicate the isoelectric point of each vesicles. The number with the arrow is the measured isoelectric point in the system.

treatments will result in SMP preparations which are more than 80% right side-out or inside-out respectively. The yield of potato tuber SMP, both on a protein and a cytochrome oxidase basis, is about two fold lower for IO-SMP than for RO-SMP. The yield of SMP production may be correlated with the proposed mechanisms (Kay _et al._, 1985) for the formation of both RO-SMP and IO-SMP. If the IO-SMP would primarily derive from tightly stacked cristae regions and the RO-SMP from the inner boundary membrane, the yield of the two types of vesicles might be different.

The differences in isoelectric points between intact mitochondria and mitoplasts reflect differences in lipid (Møller _et al._, 1984) and polypeptide compositions between the outer and inner mitochondrial membranes. The observed differences in the isoelectric points between RO-SMP and IO-SMP demonstrate differences between the outer and inner surfaces of the inner membrane (Figure 1).

The difference between the various mitochondrial membranes in the content of acidic and basic groups demonstrated by cross-partition is also seen when the charge density at pH 7.0 is measured by the 9-aminoacridine method (Table 1) The outer surface of the outer membrane has only half the value of those for the outer and the inner surfaces of the inner membrane. Even though the charge densities for the two surfaces of the inner membrane are similar, IO-SMP are much more negatively charged since the IO-SMP are much larger, having a surface about 9 times larger than for RO-SMP.

Table 1 - Surface charge density of plant mitochondrial membranes. The net surface charge densities were measured by using the 9-aminoacridine fluorescence. The area per net charge is also calculated for each charge surface density.

Membrane	Surface charge density ($mC\ m^{-2}$)	Area per charge unit ($A^{-2}\ e^{-1}$)
Mitochondria	- 27	588
Mitoplasts	- 35	432
RO-SMP	- 51	313
IO-SMP	- 51	313

The results presented here clearly demonstrate the possibility to obtain sub-mitochondrial particles of various polarities and sizes. As stated before (Kay _et al._, 1985; Møller _et al._, 1985) RO-SMP and IO-SMP will be useful and available tools in the studies of transverse asymmetry of the inner mitochondrial membrane. The two populations of SMP may also prove to be of importance to ascertain any possible lateral molecular heterogeneity along the inner membrane.

REFERENCES

Andersson, B., Sundby, C.,Akerlund, H.-E., and Albertsson, P.-A., 1985, Inside-out thylakoid vesicles. An important tool for the characterization of the photosynthetic membrane, _Physiol. Plant._, 65: 322-330.

Barber, J., 1982, Influence of surface charges on thylakoid structure and function, _Annu. Rev. Plant Physiol._, 33: 261-295.

Diolez, P. and Moreau F., 1983, Effect of bovine serum albumin on membrane potential in plant mitochondria, _Physiol. Plant._, 59: 177-182.

Johnston, S.P., Møller, I.M. and Palmer, J.M., The stimulation of exogenous NADH oxidation in Jerusamem artichoke mitochondria by screening of charges on the membranes, FEBS Lett., 108: 28-32.

Kay, C.J., Ericson I., Gardeström P., Palmer, J.M. and Møller, I.M., Generation and purification of sub-mitochondrial particles of different polarities from plant mitochondria, FEBS Lett., 193: 169-174.

Møller, I.M., Lundborg, T. and Bérczi, A., 1984, The negative surface charge density of plasmalemma vesicles from wheat and oat roots, FEBS Lett., 167: 181-185.

Møller, I.M., Ericson, I. and Gardeström, P., 1986, Isolation of sub-mitochondrial particles with different polarities, Methods Enzymol. (in press).

Petit, P., Diolez, P., Müller P. and Brown, S.C., 1986, Binding of concanavalin A to the outer membrane of potato tuber mitochondria detected by flow cytometry, FEBS Lett., 196: 65-70.

Westrin, H., Shanbhag, V.P. and Albertsson P.-A., 1983, Isoelectric point of the membrane surfaces of three spinach chloroplast classes determined by cross-partition, Biochim. Biophys. Acta, 732: 83-91.

Yaffe, M. and Schatz, G., 1984, The future of mitochondrial research, Trends Bioch. Sci., 9: 179-181.

LATERAL HETEROGENEITY IN THE INNER MITOCHONDRIAL MEMBRANE

Annika Lidén, Marianne Sommarin and Ian M. Møller

Department of Plant Physiology
University of Lund
Box 7007, S-220 07 Lund
Sweden

INTRODUCTION

When grana stacks are disrupted stacked or appressed membranes yield inside-out (IO) vesicles whereas non-stacked or non-appressed membranes give right side-out (RO) vesicles. By separating RO and IO vesicles it has been possible to show a marked lateral and transverse heterogeneity in the thylakoid membrane (1).

We thought it was possible that cristae membranes might be analogous to stacked thylakoid membranes and thus produce IO submitochondrial particles (SMP) upon disruption of the mitochondria. Furthermore, since the activity of the entire electron transport chain is electrostatically regulated (2,3) it appeared possible that cristae formation was similarly affected. This was shown to be the case: IO-SMP are the primary product when mitochondria are disrupted under high cation conditions whereas mainly RO-SMP are produced in a (low cation + EDTA)-medium (4,5). In this study we present evidence for a lateral heterogeneity in the inner mitochondrial membrane - cytochrome c oxidase (CCO) activity as well as a number of polypeptides are concentrated in the cristae regions.

MATERIALS AND METHODS

Purified mitochondria from potato and Jerusalem artichoke tubers were isolated as described elsewhere (6,7). IO-SMP were produced by sonication in a medium containing 50 mM $MgCl_2$. RO-SMP were produced by French Press treatment of mitochondria in a low-salt medium containing 5 mM EDTA. In both cases the SMP were pelleted at 105 000 g x 60 min after mitochondria and larger fragments had been removed at 48 000 g x 10 min.

Protein content determination, enzyme activity measurements and SDS-PAGE were performed according to standard methods. The latency of CCO and MDH was measured as described in (7).

RESULTS AND DISCUSSION

The CCO activity was high and latent in IO-SMP whereas it was low and almost completely unaffected by Triton X-100 in RO-SMP. The opposite pattern was observed for MDH. The results with Jerusalem artichoke mitochondria (Table 1) and potato mitochondria (not shown) were similar.

The results are consistent with our model for SMP formation where RO-SMP derive from the non-cristae regions and reseal around matrix proteins (here represented by MDH) whereas IO-SMP derive from the cristae regions and reseal around intermembrane space (4,5). The results would then point at a higher activity of CCO in the cristae regions (IO-SMP) of the inner membrane. This result is not caused by inactivation of CCO by French Press treatment (used to produce RO-SMP) as IO- and RO-SMP purified from a mixture of SMP produced by sonication showed a similar trend - more CCO and succinate dehydrogenase activity in IO-SMP than in RO-SMP (4 and unpublished results).

SDS-PAGE showed marked differences in the polypeptide pattern of RO- and IO-SMP (Fig. 1). For both species many bands were found in RO-SMP, particularly in the 35-50 kDa regions, which were absent or much less pronounced in IO-SMP. At least some of these probably represent matrix proteins. However, some bands were more marked in IO-SMP than in RO-SMP (Fig. 1). Two of the very strong bands between 52 and 55 kDa probably represent the α and β subunits of the F_1-ATPase. Some of the other bands enriched in the IO-SMP may represent subunits of CCO.

CONCLUSIONS

a) IO-SMP deriving from the stacked (cristae) regions of the inner mitochondrial membrane contain more CCO activity than RO-SMP deriving from non-cristae regions (Table 1). IO-SMP are also enriched in several polypeptides compared to RO-SMP (Fig. 1).

b) This points at a lateral heterogeneity in the inner mitochondrial membrane where CCO (and other e^- transport components such as succinate dehydrogenase - see ref. 4) are concentrated in the cristae. Proteins preferentially located to the non-cristae part of the inner membrane would reasonably include transport proteins.

c) The Vmax of e^- transport is strongly enhanced when cations are added to the assay medium (2,3). This can now be explained by a lateral redistri-

Table 1. Properties of submitochondrial particles from purified Jerusalem artichoke mitochondria.

Parameter	Mitochondria	Submitochondrial particles		
		IO	RO	IO/RO
Protein, mg	9.2	0.91	1.25	0.73
%	100	9.9	13.5	-
CCO, latency, %	98	88	6	15
spec. act.[a)]	1.45	1.25	0.29	4.3
recovery, %	100	8.6	2.7	3.2
MDH, latency, %	92	6	65	0.09
spec. act.[a)]	3.43	0.066	0.65	0.10
recovery, %	100	0.2	2.6	0.08
Catalase, spec. act.[a)]	0.15	N.D.	N.D.	-

a) µmol (mg protein)$^{-1}$min^{-1}
b) N.D., not detectable

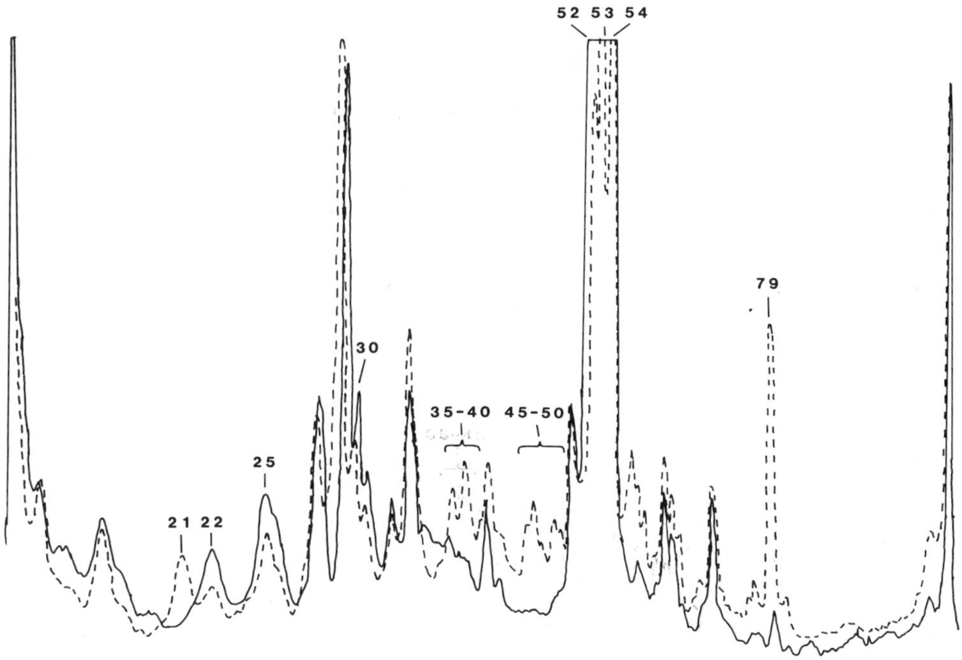

A. Jerusalem artichoke tubers

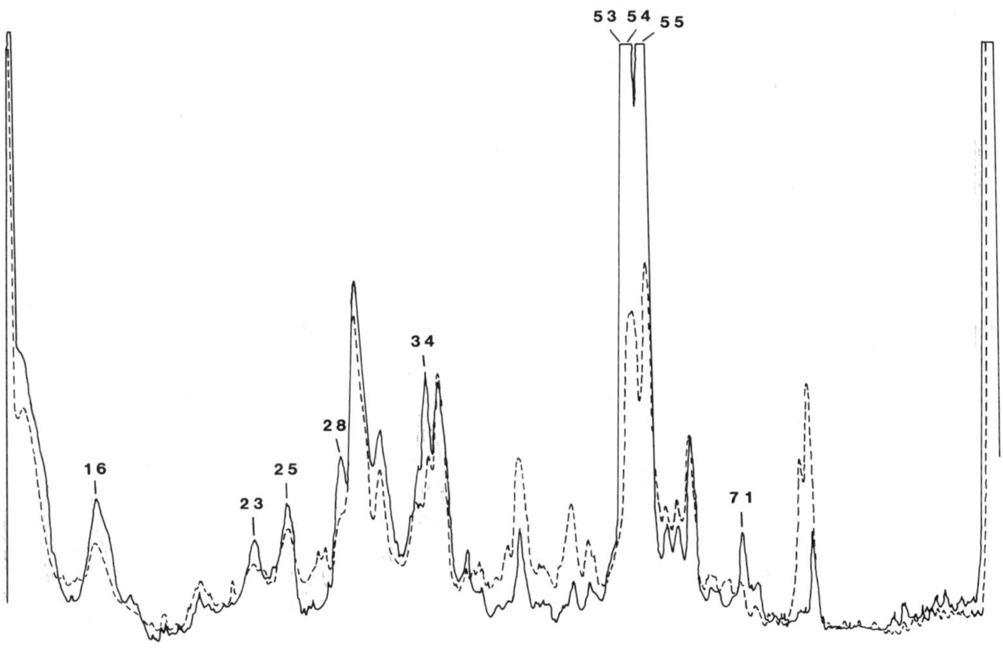

B. Potato tubers

Fig. 1. Scans of SDS-PAGE runs with IO-SMP (continuous line) and RO-SMP (broken line) from Jerusalem artichoke (A) and potato tubers (B). The gels were stained with Coomassie Brilliant Blue. Numbers indicate the molecular weights of polypeptides in kDa.

bution of e^- transport components as a result of inner membrane stacking (= cristae formation). The concentration of e^- transport components is increased and the collision frequency, which is rate limiting (8) is increased.

d) The following model illustrates the above points:

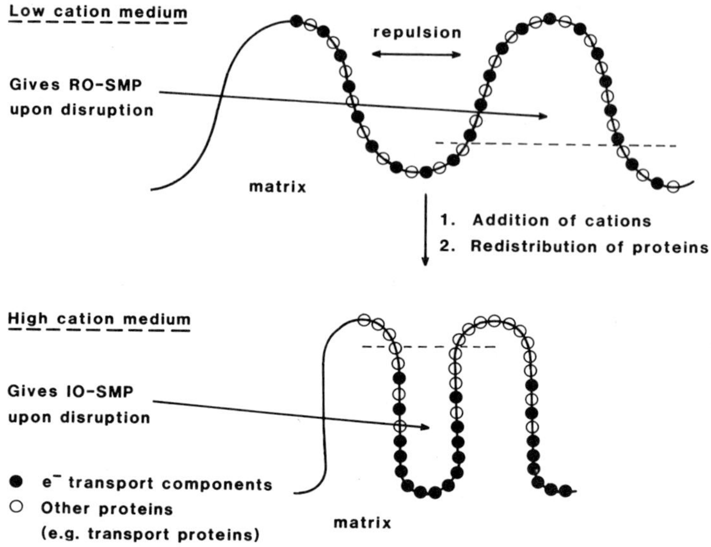

e) Regulation of cristae frequency and size could be important in the overall function of the mitochondrion. Factors which may affect cristae formation are energization/de-energization, protein phosphorylation and the concentration, valency and distribution of cations.

REFERENCES

1. B. Andersson, C. Sundby, H.-E. Åkerlund, and P.-Å. Albertsson, Inside-out thylakoid vesicles. An important tool for the characterization of the photosynthetic membrane, Physiol. Plant. 65:322 (1985).
2. I.M. Møller, C.J. Kay, and J.M. Palmer, Electrostatic screening stimulates rate-limiting steps in mitochondrial electron transport, Biochem. J. 223:761 (1984).
3. I.M. Møller and J.M. Palmer, Charge screening by cations affects the conformation of the mitochondrial inner membrane. A study of exogenous NAD(P)H oxidation in plant mitochondria, Biochem. J. 195:583 (1981).
4. C.J. Kay, I. Ericson, P. Gardeström, J.M. Palmer, and I.M. Møller, Generation and purification of submitochondrial particles of different polarities from plant mitochondria, FEBS Lett. 193:169 (1985).
5. I.M. Møller, I. Ericson, and P. Gardeström, Isolation of submitochondrial particles with different polarities, Methods Enzymol. (In press).
6. M. Neuburger, E.-P. Journet, R. Bligny, J.-P. Carde, and R. Douce, Purification of plant mitochondria by isopycnic centrifugation in density gradients of Percoll, Arch. Biochem. Biophys. 217:312 (1982).
7. I.M. Møller and A. Lidén, Purification of Jerusalem artichoke mitochondria on a continuous Percoll gradient, These proceedings.
8. S. Gupte, E.-S. Wu, L. Hoechli, M. Hoeckli, K. Jacobson, A.E. Sowers, and C.R. Hackenbrock, Relationship between lateral diffusion, collision frequency, and electron transfer of mitochondrial inner membrane oxidation-reduction components, Proc. Natl. Acad. Sci. USA 81:2606 (1984).

RECONSTITUTION OF OXIDATIVE PHOSPHORYLATION IN PEA COTYLEDON SUBMITOCHONDRIAL PARTICLES

Arnost Horak, Helena Horak and Mary Packer

Department of Plant Science, Faculty of Agriculture and Forestry, University of Alberta, Edmonton, Alberta T6G 2P5, Canada

INTRODUCTION

We have described the purification and subunit composition of the mitochondrial F_1-ATPase from pea cotyledons[1]. The purified F_1-ATPase stimulated ATP formation in CF_1-depleted pea chloroplast membranes[1]. In this report, we describe the preparation of mitochondrial membranes which have low rates of residual oxidative phosphorylation and show that the ATP formation in these membranes is stimulated on incubation with the purified mitochondrial F_1-ATPase.

PREPARATION OF F_1-DEPLETED MITOCHONDRIAL MEMBRANES

Submitochondrial particles were prepared from pea cotyledon mitochondria by sonication in 250 mM sucrose/50 mM TES/5 mM $MgCl_2$ (pH 7.0). The resulting particles (Mg^{2+}-SMP) were extracted with various concentrations of urea to remove the F_1-ATPase. The urea extraction was done by a modification of the procedure described by Pedersen and Hullihen[2] for rat liver mitochondria. To 14 mg of Mg^{2+}-SMP in 0.5 ml of 250 mM sucrose, 0.1 ml of 12 mM TES/0.6% fatty acid poor bovine serum albumin (pH 7.4) was added, followed by addition of 0.6 ml of urea in water to obtain a final urea concentration between 1.0 M and 4.0 M. The suspension was incubated on ice for 5 min and the extraction stopped by addition of 20 ml of cold 250 mM sucrose/2 mM TES/0.1% fatty acid poor bovine serum albumin (pH 7.4). Final suspension of urea extracted membranes was made in 250 mM sucrose at 1 mg/ml. Freshly prepared particles were used for ATPase (30 - 60 μg protein) or oxidative phosphorylation (100 - 200 μg protein) assays. Oxidative phosphorylation was measured in a final volume of 1 ml with 0.88 mM NADH as substrate. Conditions described by Grubmeyer et al.[3] were followed, except that 0.2% fatty acid poor bovine serum albumin was also included in the assay medium.

The ATPase activity of SMP or of urea particles was measured in the same way as described for the ATPase activity of F_1[1]. When oligomycin was included, a final concentration of 1 μg/ml was used.

The results are shown in Fig. 1. At a urea concentration higher than 1.0 M, ATP synthesis is rapidly decreasing, while the decline in

Table 1. Reconstitution of oxidative phosphorylation in urea particles by purified F_1. Urea particles (UP) were prepared as described above. For reconstitution, freshly prepared urea particles (400 μg protein) were incubated with 100 μg of purified F_1 in a final volume of 0.1 ml. The incubations were carried out in the presence of 1% fatty acid poor bovine serum albumin in "reconstitution buffer" consisting of 100 mM sucrose/100 mM K_2HPO_4/2 mM EDTA/1 mM TES (pH 7.5). After 1 hour on ice, 40 μl aliquots were taken for the oxidative phosphorylation assay. Unextracted SMP or urea extracted membranes alone (for measurements of residual oxidative phosphorylation) were subjected to the same reconstitution procedure prior to the oxidative phosphorylation assay.

Membrane preparation	ATP formed (nmol/min/mg protein)
Mg^{2+}-SMP	120
1.7 M UP	35
1.7 M UP + F_1	133
2.0 M UP	11
2.0 M UP + F_1	54
2.3 M UP	2
2.3 M UP + F_1	17
3.0 M UP	0
3.0 M UP + F_1	0

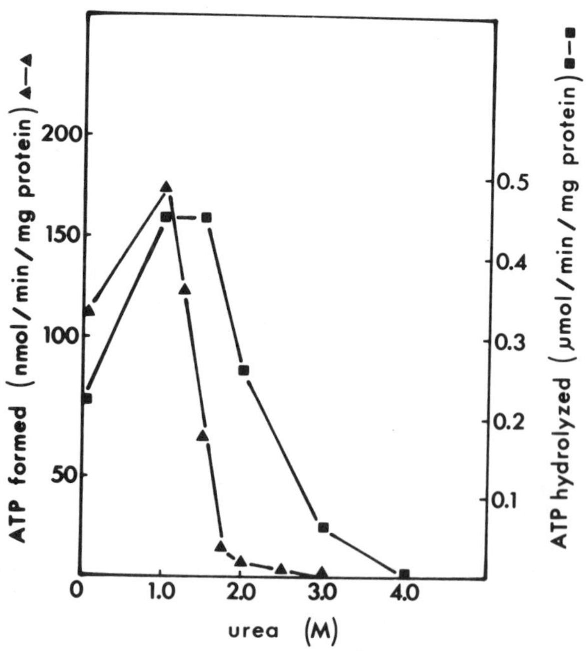

Fig. 1. Oxidative phosphorylation and oligomycin sensitive ATPase in urea extracted Mg^{2+}-SMP.

ATPase activity is more gradual. 3.0 M urea treatment completely abolishes oxidative phosphorylation and decreases the ATPase activity to a low level.

RECONSTITUTION OF OXIDATIVE PHOSPHORYLATION IN UREA TREATED MEMBRANES

Results in Table 1 show that ATP formation is reconstituted in 1.7 - 2.3 M urea-extracted membranes on incubation with a saturating amount of purified mitochondrial F_1-ATPase. In contrast to the urea-extracted membranes from rat liver mitochondria[2], treatment of pea cotyledon Mg^{2+}-SMP with 3.0 M urea resulted in membranes that were not capable of reconstitution of oxidative phosphorylation on preincubation with purified F_1-ATPase. We have used 1.9 M urea for large scale preparation of stable urea particles (1.9 M UP) that could be stored in liquid N_2 without any loss of reconstitution capacity. These particles were used in subsequent experiments.

Results in Fig. 2 show that purified mitochondrial F_1-ATPase could readily reconstitute oxidative phosphorylation in 1.9 M UP reaching saturating levels when 100 μg of F_1 were incubated with 400 μg of urea particles. In contrast to the purified mitochondrial enzyme, purified activated pea chloroplast CF_1 was not capable of increasing ATP synthesis in 1.9 M UP from pea cotyledon mitochondria (Fig. 2). This result is interesting in view of the fact that both chloroplast CF_1, as well as mitochondrial F_1 could reconstitute photophosphorylation in CF_1-depleted pea chloroplast membranes[1].

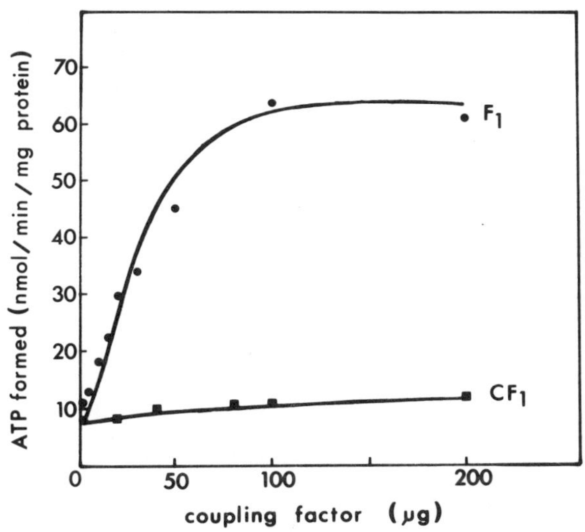

Fig. 2. ATP-synthetic capacity of 1.9 M urea particles reconstituted with various concentrations of purified F_1 or CF_1. Mg^{2+}-SMP were extracted with 1.9 M urea and stored in liquid nitrogen. For reconstitution, 400 μg of urea particles were incubated with the indicated amount of purified mitochondrial F_1 or purified activated chloroplast CF_1 and subsequently 40 μl aliquots were used in the oxidative phosphorylation assay. The activation of chloroplast CF_1 was achieved by incubation of CF_1 with 50 mM dithiothreitol in 20 mM TES (pH 8.0).

Table II. Stimulation of oxidative phosphorylation in 1.9 M urea particles by DCCD and by purified F_1. Mg^{2+}-submitochondrial particles were extracted with 1.9 M urea and stored in liquid nitrogen. For reconstitution, 400 μg of urea particles were incubated with the indicated amount of DCCD (in 5 μl of ethanol) or with purified mitochondrial F_1 under conditions described in Table 1.

Addition to 1.9 M urea particles		ATP formed (nmol/min/mg protein)
A. DCCD (nmoles)		
	0	6
	0.1	8
	0.5	14
	1	25
	2	44
	3	48
	5	27
	10	3
B. F_1	200 μg	61

The urea particles used in our reconstitution experiments were partially deficient in, but not devoid of endogenous ATPase activity. In the experiment shown in Table II, we tried to distinguish whether the added F_1 was fulfilling its structural role only or whether it was catalytically active as well. The structural role of coupling factors (blocking of proton pores in the membranes) can be simulated by incubation of F_1-depleted membranes with low concentrations of DCCD. We have preincubated the 1.9 M urea particles with low concentrations of DCCD in an experiment analogous to that of McCarty and Racker[4]. As shown in Table II, the phosphorylation rates measured after preincubation of 1.9 M UP with F_1-ATPase were about 25% higher than the highest rates observed after preincubation of 1.9 M UP with DCCD. This result suggests that the added F_1-ATPase, besides fulfilling its structural role, was also catalytically active in ATP synthesis. However, more resolved mitochondrial membranes containing low amounts of residual F_1-ATPase will be required for unequivocal proof of catalytic activity of the added enzyme.

REFERENCES

1. A. Horak and M. Packer, Coupling factor activity of the purified pea mitochondrial F_1-ATPase, Biochim. Biophys. Acta 810:310 (1985).
2. P.L. Pedersen and J. Hullihen, Adenosine triphosphatase of rat liver mitochondria, J. Biol. Chem. 253:2176 (1978).
3. C. Grubmeyer, D. Melanson, I. Duncan, and M. Spencer, Oxidative phosphorylation in pea cotyledon submitochondrial particles, Plant Physiol. 64:757 (1979).
4. R.E. McCarty and E. Racker, Partial resolution of the enzymes catalyzing photophosphorylation, J. Biol. Chem. 243:3435 (1967).

A PLANT MITOCHONDRIAL ATPASE/SYNTHASE

John O'Rourke and S. Brian Wilson

Department of Biochemistry
University of Aberdeen
Marischal College
Aberdeen
AB9 1AS
Scotland

INTRODUCTION

The membrane-bound proton transporting ATPase/ ATP-synthase of bacteria, mitochondria and chloroplasts catalyses the reversible synthesis/hydrolysis of ATP from ADP and Pi. The mitochondrial enzyme is the major source of cellular ATP in the absence of photosynthesis. It is believed that the electro-chemical gradient produced during electron transport is utilised by the ATPase during ATP synthesis in a process involving proton translocation. The enzyme has a requirement for a divalent cation for activity. Normally this requirement is fulfilled by magnesium ions although the chloroplast (Vambutas and Racker 1965) and pea mitochondrial enzymes are also activated by calcium ions (Grubmeyer and Spencer 1979).

The structure and function of the ATPases from a variety of sources have been widely studied and extensively reviewed. The mitochondrial enzyme consists of two multi-subunit parts, an intrinsic membrane component; the "F_o" portion and a peripheral protein component; the "F_1" complex. The F_1 complex can be readily released from the membrane and has the ATPase/ATP-synthase activity. The F_1 component was first purified from beef-heart mitochondria by Serrano *et al* (1976) by a multistep process. More recently a single step procedure (Beechey *et al* 1975) involving release of the ATPase from the membrane by chloroform has been used to purify ATPases from various sources. The F_1 component is usually isolated containing five subunits α, β, γ, δ and ϵ, which are arranged in a 3:3:1:1:1 stoichiometry. However there are a number of reports of different stoichiometries and of six subunit F_1-ATPases being purified. While the α, β and γ subunits appear similar in all species, there is much more variation in the δ and ϵ subunits, e.g. between beef heart and *E. coli* (Walker *et al* 1982)

While the F_1-ATPase from animal mitochondria has been widely studied that from higher plants has not been extensively investigated, although recently there have been several reports of the purification of plant F_1-ATPase (Hack and Leaver [1983], Iwasaki and Asahi [1983], Boutry *et al* [1983], Dunn *et at* [1985], Randall *et al* [1985] and

Spitzberg *et al* [1985]). Some of of these report the occasional appearance of six subunits although it has been suggested that the sixth subunit might be a proteolytic artifact (Iwasaki and Asahi 1983). The F_1-ATPase of higher plants appears to differ from that of animals in a several important respects: (a) - the α subunit is encoded by the mitochondrial rather than the nuclear DNA (Boutry *et al* 1983. Hack and Leaver 1983), (b) - uncouplers do not usually release significant ATPase activity when applied to higher plant mitochondria, whereas with mammalian mitochondria, there is a large uncoupler induced increase in ATPase activity, (c) - there may be six subunits in the plant enzyme, (d) - the enzyme is reported to be activated by calcium as well as magnesium ions (Grubmeyer and Spencer 1979).

This paper describes the release from the membrane and subsequent purification of the F_1-ATPase from Turnip *(Brassica napus L.)* mitochondria. Conditions for the stable storage of the enzyme are described.

MATERIALS AND METHODS

Mitochondria were isolated and purified from Turnip *(Brassica napus L.)* purchased from local sources as described by Gauvrit and Wilson (1983) except that bovine serum albumin was omitted from the medium. Mitochondria were only used for ATPase isolation if they met the usually accepted criteria for intact pure plant mitochondria i.e. high ADP:O ratios and a high degree of membrane integrity.

All chemicals were obtained from commercial sources (BDH Ltd., Poole, Dorset, U.K., Sigma Chemical Co. Ltd., Poole, Dorset, U.K. and BCL London Ltd., Lewes, Sussex, U.K.) and were of AR or the best available grade. Chromatography media were obtained from Pharmacia Ltd., London, U.K. Electrophoresis materials and standard proteins for SDS gel electrophoresis and column chromatography were from BDH Ltd.

ATPase activity was measured by colourimetric assay of phosphate released from ATP or spectrophotometrically. Incubations were carried out in the presence of 2 mM Phosphoenol-pyruvate, 10 units pyruvate kinase, 5 mM ATP in Tris-HCl buffer pH 8.5, 10 mM KCl and either 3mM $MgCl_2$, 0.3%w/v Na_2SO_3 (Magnesium-ATPase), or 10 mM $CaCl_2$ (Calcium-ATPase). The ATPase reaction was stopped by adding equal volumes of 10%w/v TCA and 10%w/v SDS prior to the determination of phosphate. Phosphate was measured using a modification of the method of Peterson (1978) using 1%w/v p-methylaminophenol in 3%w/v sodium metabisulphite as reducing reagent.

Protein was determined as described by Bradford (1976) using bovine serum albumin as standard.

Polyacrylamide gel electrophoresis was carried out with or without SDS as described by Laemmli (1970). Gels were stained for protein either with 0.1%w/v coomassie blue in methanol/acetic acid or silver stained. In some cases non-denaturing gels were stained for ATPase activity (Horak and Hill 1972).

RESULTS

Purification of the Turnip F_1 -ATPase

The ATPase was readily released from membrane fragments by solvent treatment in the presence of an inhibitor of proteolysis. Recovery of activity was found to be greater with dichloromethane than with other chlorinated hydrocarbon solvents. Gel electrophoresis of the released protein showed that unlike the chloroform released beef-heart mitochondrial ATPase, the solublised turnip ATPase was not pure and further purification was required.

Following release of the ATPase from the membrane it was found to be cold-labile and all subsequent operations were carried out at room temperature. The solvent released ATPase was treated with ammonium sulphate and the resulting fraction loaded onto a glycerol density gradient. The ATPase active band from the density gradient was finally subjected to ion-exchange chromatography on DEAE Sephacel using a KCl gradient. The ATPase active peak from the ion-exchange column was electrophoretically pure and was concentrated for subsequent determination of the properties of the ATPase

Properties of the Turnip F_1 -ATPase

The purified protein appeared as a single band after polyacrylamide gel electrophoresis with either coomassie blue or silver stain. The protein band was found to correspond to the band detectable by either calcium-ATPase or magnesium-ATPase activity stain showing that the both activities resided in the same protein. Under optimal conditions the specific activity of the preparation was 135 μMol min^{-1} mg^{-1} protein.

The molecular weight estimated by chromatography on Sepharose 6B in comparison with standard proteins was 340kDa. Using SDS polyacrylamide gel electrophoresis and comparison with molecular weight markers the turnip ATPase was found to contain six subunits with relative molecular masses of 55 kDa, 52 kDa, 31 kDa, 21.8 kDa, 20.4 kDa and 12.5 kDa respectively. However delays in the isolation of the ATPase resulted in preparations which lacked the sixth subunit.

The enzyme was found to be cold-labile so the conditions for stable storage were studied. It was found possible to store the turnip ATPase at 20°C in 30% glycerol buffer for at least 120 hr.

DISCUSSION

The purified turnip ATPase has a specific activity higher than previously reported for a higher plant F_1-ATPase. In many other respects the preparation resembles those reported for other higher plant mitochondrial ATPases. The subunit molecular weights are more like those reported for the *E. coli* ATPase than the mammalian mitochondrial enzyme, in particular the δ and ϵ (20.4 and 12.5 kDa respectively) subunits. The presence of a sixth subunit (21.8 kDa) in samples of ATPase processed rapidly in the presence of an inhibitor of proteolysis suggests that six is the normal number of subunits in the plant F_1-ATPase rather than five. Six subunits have also been reported for a solvent released preparation of the beef-heart mitochondrial ATPase (Fisher *et al* 1981) when it was suggested that the additional subunit might be involved in the binding of the ATPase to the membrane. The role of the turnip sixth subunit is or its relationship to the other subunits is not known, however the sixth subunit is clearly not required for activity since both forms of the enzyme are active.

ACKNOWLEDGEMENT

The authors are grateful to the AFRC for financial support, to Mrs. A. Flett for expert technical assistance and to Professor J.E. Fothergill and Dr. L.A. Fothergill-Gilmore for extensive and helpful discussion.

REFERENCES

Beechey, R.B., Hubbard, S.A., Linnett, P.E., Mitchell, A.D. and Munn, E.A. (1975), A simple and rapid method for the preparation of adenosine triphosphatase from submitochondrial particles, Biochem. J., 148: 533-537.

Boutry, M., Briquet, M. and Gaffeau, A. (1983), The α subunit of plant mitochondrial F_1-ATPase is translated in mitochondria, J. Biol. Chem., 258: 8524-8526.

Bradford, M.M. (1976), A rapid and sensitive method for the quantitation of microgram quantities of protein utilising the principle of protein-dye binding, Anal. Biochem., 72: 248-254.

Dunn, P.P.J., Slabas, A.R. and Moore A.L. (1985), Purification of the F_1-ATPase from cockoo-pint (Arum maculatum) mitochondria, Biochem. J., 225: 821-824.

Fisher, R.J., Liang, A.M. and Sundstrom, G.C. (1981), Selective disaggregation of the H^+-translocating ATPase; isolation of two discrete complexes of the rutamycin-insensitive ATPase differing in mitochondrial membrane-binding properties, J.Biol. Chem., 256: 707-715.

Gauvrit, C. and Wilson, S.B. (1983), Uncoupling properties of the herbicide N-n-Pentyl-N-Methyl-N'-(3,4-Dichlorophenyl)-Urea, J.Exp. Bot., 34: 367-380.

Grubmeyer, C. and Spencer, M. (1979), Effects of anions on a soluble ATPase from mitochondria of pea cotyledons, Plant and Cell Physiol., 20: 83-91.

Hack, E. and Leaver C.J. (1983), The α subunit of the maize F_1-ATPase is synthesised in the mitochondrion, EMBO J., 2: 1983-1989.

Horak, A. and Hill, R.D. (1972), Adenosine triphosphatase of bean plastids, its properties and site of formation, Plant Physiol., 49: 365-370.

Iwasaki, Y. and Asahi, T. (1983), Purification and characterisation of the soluble mitochondrial adenosine triphosphatase from sweet potato, Arch. Biochem. and Biophys., 227: 164-173.

Laemmli, U.K. (1970), Cleavage of structural proteins during the assembly of the head of bacteriophage T_4, Nature, 227: 680-685.

Peterson, G.L. (1978), A simplified method for the analysis of inorganic phosphate in the presence of interfering substances, Anal. Biochem., 84: 164-172.

Randall, S.K., Wang, Y. and Sze, H. (1985) Purification and Characterisation of the soluble F_1-ATPase of oat root mitochondria, Plant Physiol., 79: 957-962.

Serrano, R., Kanner, B.I. and Racker, E. (1976), Purification of the proton-translocating adenosine triphosphatase complex of bovine heart mitochondria, J. Biol. Chem., 251: 2453-2461.

Spitzberg, V.L., Pfeiffer, N.E., Partridge, B., Wylie, D.E. and Schuster, S.M. (1985), Isolation and antigenic characterisation of corn mitochondrial ATPase, Plant Physiol., 77: 339-345.

Walker J.E., Runswick, M.J. and Saraste, M. (1982), Subunit equivalence in Escherichia coli and bovine heart mitochondrial F_1F_0 ATPases, FEBS Lett., 146: 393-396.

Vambutas, V. and Racker, E. (1965),Partial resolution of the enzymes catalysing photophosphorylation. I, Stimulation of photophosphorylation by a preparation of a latent Ca^{++} dependent adenosine triphosphatase from chloroplasts, J. Biol. Chem., 240: 2660-2667.

INTERACTIONS OF MITOCHONDRIA WITH OTHER METABOLIC PROCESSES — AN OVERVIEW

Ian B. Dry and Joseph T. Wiskich

Botany Department
The University of Adelaide
Adelaide 5001, Australia

Mitochondria have an important role in the supply of ATP and carbon skeletons for various biosynthetic, transport and other processes in cells. However, there are also a number of other metabolic processes, in plant cells, which are known, or thought, to interact with mitochondria for reasons other than this basic need for respiratory products. These include photosynthesis/chloroplast-based reactions, decarboxylation processes in C_4 and crassulacean acid metabolism (CAM), lipid metabolism in germinating fatty seeds, nitrate assimilation and photorespiration. Our current understanding of the basis of the interaction of mitochondria with these various processes will be discussed.

PHOTOSYNTHESIS/CHLOROPLAST-BASED REACTIONS

It is well established that photosynthesis has an *indirect* influence on the mitochondrial activity of photosynthetic or leaf cells (in the post-illumination period) by influencing the level of available substrate[1]. For example, Azcon-Bieto and Osmond[2] demonstrated that the rate of dark respiration in wheat leaves, immediately after the light period, was correlated with accumulated net CO_2 fixation in the preceding light period.

It has also often been suggested that photosynthesis or chloroplast-based reactions have a *direct* effect on mitochondrial activity causing both tricarboxylic acid cycle (TCA) and/or electron transport chain (ETC) activity to be depressed during the light period. This suggestion is based on the premise that the energy needs of the cell, in the light, can be derived from chloroplast-based reactions i.e. photophosphorylation, making mitochondrial metabolism redundant. Graham[3] reviewed the literature on the effect of light on dark (mitochondrial) respiration and concluded the evidence strongly favoured the continued operation of the TCA cycle in the light (in some cases at rates approaching those in the dark). The situation is clearly not resolved, however, given the range of estimates of the relative rates of TCA cycle activity in light vs. dark which have been published since that review (i.e. 18-100%[2,4,5,6]). One should not, of course, overlook the role of the TCA cycle in the supply of carbon skeletons such as 2-oxoglutarate, for light-dependent biosynthetic reactions (e.g. synthesis of amino acids, porphyrins etc.). This would necessarily require some TCA turnover during the light period, however, the rate required to satisfy these cellular requirements is unknown.

Quite separate from the question of TCA cycle turnover in the light is the question of whether the mitochondrial ETC operates during photosynthesis. Many conflicting results have been obtained using various photosynthetic tissues and experimental techniques ranging from a complete inhibition of respiratory O_2 uptake in the light, to little or no effect[7,8,9,10,11].

In summary there does appear to be some support for a continuation of both mitochondrial ETC and TCA cycle activity in the light, but at what rate compared to "dark" rates is yet to be resolved.

DECARBOXYLATION PROCESSES IN C_4 AND CAM METABOLISM

The C_4 and CAM photosynthetic pathways involve a two phase CO_2 fixation process - initial fixation by PEP carboxylase to form a C_4 acid followed by decarboxylation of that C_4 acid and refixation of the CO_2 in the normal C_3 cycle. In C_4 photosynthesis these two phases are spatially separate i.e. in the mesophyll and bundle sheath cells; in CAM photosynthesis they occur within the same cell, but are separated in time i.e. night and day.

The major interaction between mitochondria and the C_4/CAM pathways arises during the decarboxylation phase. However, the degree of interaction varies markedly depending on the type of decarboxylation enzyme utilised i.e. NADP-malic enzyme, PEP-carboxykinase or NAD-malic enzyme (for a more detailed review see Gardeström and Edwards[12]). The most significant interactions occur in those C_4 and CAM species in which decarboxylation occurs via the mitochondrial enzyme, NAD-malic enzyme.

Decarboxylation in NAD-malic enzyme type C_4 species

Malate decarboxylation via mitochondrial NAD-malic enzyme may potentially be supported by NADH reoxidation via the ETC. However, studies with mitochondria isolated from these C_4 types, indicates that the capacity for reoxidation via this pathway, when expressed on a whole tissue basis (approx. 50 μmol/mg Chl/h[13]), is insufficient to account for *in vivo* rates of malate decarboxylation (i.e. 180-300 μmol/mg Chl/h[14]).

Kagawa and Hatch[14] have proposed that NADH may be recycled, within the mitochondrion, through the coupled action of aspartate aminotransferase (AsAT), malate dehydrogenase (MDH) and NAD-malic enzyme (Fig. 1). This system would utilise the extremely high levels of AsAT present in the bundle sheath mitochondria of these C_4 types (i.e. 20-40 x greater than in C_3 leaf mitochondria[15]). Certain aspects of this decarboxylation scheme are still to be resolved. First, by what mechanism do the various metabolites cross the inner mitochondrial membrane i.e. via exchange (and if so, with what) or via diffusion? Second, what regulates pyruvate dehydrogenase to ensure that pyruvate produced during malate decarboxylation is conserved within the C_4 cycle and is not lost through oxidation to acetyl-CoA.

Decarboxylation in Malic enzyme (NADP and NAD) type CAM species

In contrast to C_4 species, CAM species may contain high levels of both NADP- and NAD- malic enzyme and, thus, decarboxylation may occur via both enzymes within the same cell. Decarboxylation of malate via cytoplasmic NADP-ME will result in the production of NADPH which has been shown to be rapidly oxidised by isolated mitochondria from ME-type CAM species[16,17]. The NADH produced during the decarboxylation of malate via mitochondrial NAD-ME may be reoxidised via the ETC or by some form of substrate shuttle [e.g. malate/oxaloacetate (OAA) or malate/aspartate] which would result in the export of NADH to the cytoplasm for use in the gluconeogenic conversion of pyruvate back to carbohydrate. In contrast to NAD-ME type C_4 mitochondria,

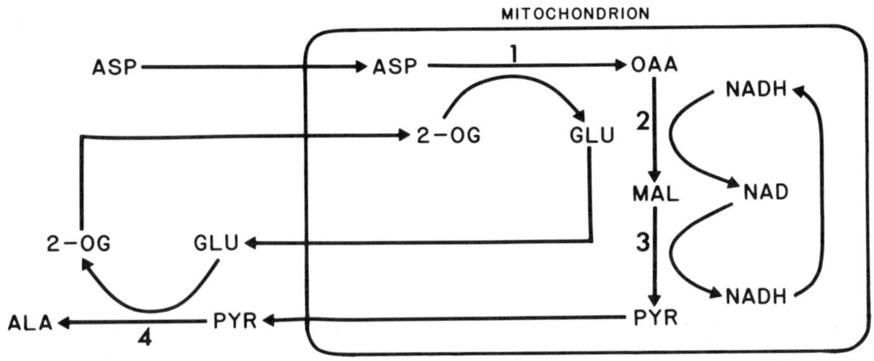

Fig. 1. Proposed scheme for malate decarboxylation in bundle sheath cells of NAD-malic enzyme type C_4 leaves. Enzymes: (1) aspartate aminotransferase (2) malate dehydrogenase (3) NAD-malic enzyme (4) alanine aminotransferase.

rates of malate oxidation via the ETC, in isolated CAM leaf mitochondria [17,18] appear to be sufficient to account for *in vivo* rates of malate decarboxylation (i.e. 60-100 μmol/mg Chl/h[17,18]). These rates are, of course, extrapolated from state 3 rates with isolated mitochondria, whereas ETC activity *in vivo* may be restricted by ADP supply (ATP turnover). The presence of a particularly active rotenone-insensitive bypass[16,17] and the possible involvement of the alternative pathway[19], may ensure that NADH reoxidation during the decarboxylation phase is not restricted by ADP supply.

Finally, evidence obtained by Holtum and Osmond[20] indicates that the bulk of the pyruvate produced during the decarboxylation process is converted back to storage carbohydrate and is not metabolised via pyruvate dehydrogenase and the TCA cycle. Thus, as in NAD-ME type C_4 mitochondria, pyruvate dehydrogenase and/or the TCA cycle enzymes appear to be under some form of control to restrict pyruvate oxidation.

LIPID METABOLISM IN FATTY SEEDS

During the germination of fatty seeds, such as castor beans, the lipid stored in the endosperm is converted to sucrose for use by the growing embryo. This conversion involves the integrated action of two organelles - the glyoxysome and the mitochondrion[21].

Lipids are broken down via the β-oxidation pathway in the glyoxysome producing acetyl-CoA and NADH. The acetyl-CoA is converted to succinate in the glyoxylate cycle and is then transported to mitochondria for the final conversion to OAA which is the initial precursor for the gluconeogenic reactions in the cytoplasm. The NADH produced in the glyoxysome must also be transported to mitochondria for reoxidation, as glyoxysomes do not possess an electron transport system capable of reoxidising this NADH. In order to facilitate the movement of both NADH and carbon (i.e. succinate) between glyoxysomes and mitochondria, a shuttle system has been proposed [22,23] which utilises (as in the NAD-ME type C_4 case) the extremely high levels of AsAT present in mitochondria and glyoxysomes of castor bean endosperm (i.e. 30-50 x higher than in C_3 leaf mitochondria[22]). The basic component of the shuttle scheme is a malate/aspartate shuttle for the movement of reducing equivalents from the glyoxysome to the mitochondrion, upon which is superimposed the movement of carbon into (in the form of succinate) and out of (in the form of 2-oxoglutarate) the mitochondrion. The operation of such a shuttle system is supported by the results of transport studies with purified castor bean mitochondria[24].

One of the major questions yet to be resolved regarding the conversion of lipid to carbohydrate in the germinating castor bean, is the fate of the so called "excess" ATP produced during the coupled oxidation of NADH and $FADH_2$ via the mitochondrial ETC. There is no evidence of alternative pathway activity in castor bean endosperm mitochondria and therefore all oxidation via the ETC will result in the production of ATP. The conversion of one 16 carbon fatty acid molecule to one 12 carbon sucrose molecule will result in the net production (taking into account the requirements for gluconeogenesis) of 30 ATP. Thus, the question remains as to how this ATP is recycled to ensure that NADH reoxidation, via the mitochondrial ETC, is not restricted by ADP supply. One possibility is that while the endosperm tissue itself is not growing during fat breakdown, the biosynthesis of molecules within the endosperm and their subsequent transport to the growing embryo may represent a significant sink for ATP.

NITRATE ASSIMILATION IN LEAF TISSUE

There is still much debate as to the potential interaction of mitochondrial metabolism with nitrate assimilation in leaf tissue. Nitrate reductase is located in the cytoplasm and catalyses the reduction of nitrate to nitrite using NADH (Fig. 2). The nitrite is subsequently reduced to ammonia, in the chloroplast, in a light-dependent reaction via ferredoxin-dependent nitrite reductase. Various sources of NADH for nitrate reduction have been proposed including glycolysis[25], chloroplasts (via substrate shuttles[26]) and mitochondria[27].

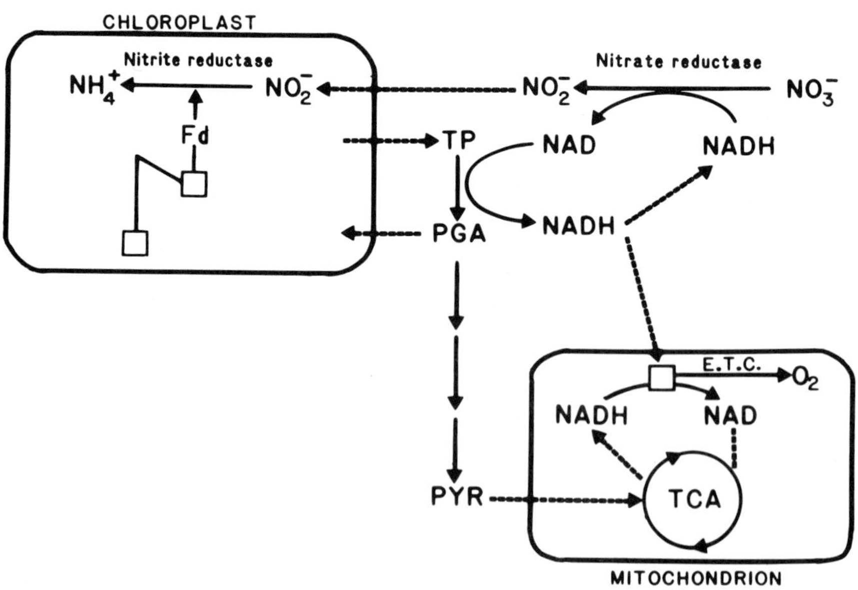

Fig. 2. Interactions of mitochondrial metabolism with nitrate assimilation in leaf tissue.

Much of the data implicating mitochondrial involvement in nitrate reduction has been obtained using the "classical" in vivo nitrate reductase assay which involves measurement of nitrite accumulation in leaf slices incubated in the dark. It has been observed that under these experimental conditions, nitrate reduction is significantly increased by an inhibition of the mitochondrial ETC[28]. This suggests that nitrate reductase and the mitochondrial ETC may compete for cytoplasmic NADH. However, it has also been proposed that this observed competition may be for intramitochondrial NADH[27], and that under conditions of suppressed ETC activity (as has been proposed to occur during photosynthesis) intramitochondrial NADH is shuttled out to the cytoplasm resulting in rapid rates of nitrate assimilation. The applicability of these results to the true in vivo situation (i.e. light, aerobic conditions), remains unclear since the "classical" in vivo assay system necessarily eliminates the chloroplast as a potential source of reducing equivalents for nitrate reduction. Thus, an alternative proposal[29] is that nitrate reductase and the mitochondrial ETC do compete for cytoplasmic NADH, but that in the light there is a sufficiently high level of NADH generated as a result of chloroplast metabolism, either directly through the operation of substrate shuttles[26] or indirectly through the production of glycolytic intermediates (i.e. triose phosphates[25]), to alleviate the competition between them and allow for maximal rates of nitrate reduction.

PHOTORESPIRATION

Photorespiration results from the binding of O_2 by ribulose-1,5P_2 carboxylase to produce a two carbon compound, phosphoglycolate. Phosphoglycolate is subsequently recycled back to 3-phosphoglycerate, via the photorespiratory cycle, which involves both peroxisomes and mitochondria (Fig. 3). Within the mitochondria, two molecules of glycine are converted to serine by the action of the enzymes, glycine decarboxylase and serine hydroxymethyltransferase. These reactions also produce NADH, NH_3 and CO_2. It is well accepted that the NH_3 is refixed via the glutamine synthetase/glutamate synthase pathway within the chloroplast[30], however the ultimate fate of the NADH is yet to be fully resolved.

There are two possible fates for the NADH generated during glycine oxidation: (a) reoxidation via the ETC or (b) reoxidation via a substrate shuttle and transport of the reducing equivalents to the peroxisome for use by hydroxypyruvate reductase (Fig. 3). Photorespiratory glycine oxidation alone, almost certainly produces reducing equivalents at a rate greater than that produced by the whole TCA cycle during normal dark respiration*. This raises a number of questions regarding the capacity of the mitochondria to oxidise this large amount of NADH, be it via the ETC or a substrate shuttle.

Let us assume, for example, that the ETC does operate in the light. If so, would there be sufficient mitochondrial ETC capacity to reoxidise all of the NADH produced during photorespiratory glycine oxidation? Maximal rates of glycine oxidation via the ETC of isolated leaf mitochondria (expressed on a whole tissue basis), give rates of 25-46 µmol CO_2/mg Chl/h[33] which are just sufficient to account for in vivo rates of glycine oxidation. However, this would also require that the ETC operate under state 3 conditions in vivo - thereby necessitating rapid rates of cellular ATP turnover. Alternative pathway activity, at least in pea and spinach

---

*Photorespiratory rate (based on 25-35% of net photosynthetic rate[31] of 150 µmol/mg Chl/h = 38-52 µmol CO_2 (NADH)/mg Chl/h. Dark respiration rate of leaf tissue[32] = 7-14 µmol O_2/mg Chl/h ≡ 14-28 µmol (NADH + $FADH_2$) / mg Chl/h.

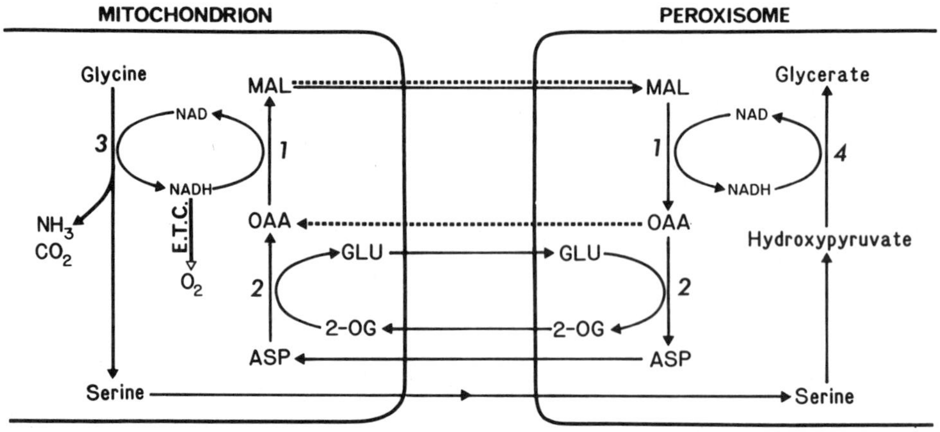

Fig. 3. Potential mechanisms of reoxidation of NADH produced during photorespiratory glycine oxidation. Enzymes: (1) malate dehydrogenase (2) aspartate aminotransferase (3) glycine decarboxylase/serine hydroxymethyltransferase (4) hydroxypyruvate reductase. Symbols: malate/OAA shuttle (---), malate/aspartate shuttle (——). ETC, electron transport chain.

leaf mitochondria[33,34] are not sufficient to account for the high rates of glycine oxidation _in vivo_. It should also be noted that if all of the NADH produced by glycine oxidation were reoxidised via the ETC, rates of respiratory O_2 consumption would be significantly higher in the light than in the dark - a result which is yet to be observed experimentally. Another important question is what influence would the generation of reducing equivalents, by the TCA cycle dehydrogenases in the light, have on the potential oxidation of glycine via the ETC? The results of a number of studies[33,35,36] have shown that in a competitive situation in which the capacity of the individual dehydrogenases exceeds the reoxidation capacity of the ETC, glycine is preferentially oxidised at the expense of the other respiratory substrates. Indeed, Day et al.[37] have suggested that this competition between glycine and TCA cycle dehydrogenases, for access to the ETC, may explain some of the observed decreases in TCA cycle activity in leaf tissue in the light.

As mentioned previously, the alternative fate for NADH generated during glycine oxidation involves its transport to the peroxisome to support hydroxypyruvate reduction. Two types of substrate shuttle have been proposed to mediate this transfer - a malate/OAA shuttle and a malate/aspartate shuttle (Fig. 3). The effective operation of either of these shuttles, in supporting rapid rates of glycine oxidation, is obviously dependent on the presence of sufficiently high levels of MDH and AsAT in both the mitochondria and the peroxisomes and on the sufficiently rapid movement of the shuttle metabolites across the mitochondrial and peroxisomal membranes.

Malate/oxaloacetate shuttle

In the case of the malate/OAA shuttle (Fig. 3) there appear to be sufficiently high levels of MDH in both the mitochondria and peroxisomes to support hydroxypyruvate reduction. Movement of OAA across the mitochondrial membrane, at least, is also extremely rapid[38,39]. Day and Wiskich[33], however, have suggested that the overall activity of this shuttle may be limited, not by the rate of OAA uptake, but by the rate of malate efflux which was found to be much slower than the rate of OAA uptake. This also raises the question of the transport mechanism involved in the movement of OAA and malate across the inner mitochondrial membrane. Possible exchange mechanisms are (a) on the same carrier in a 1:1 exchange[39] or (b) on separate carriers via an OAA-specific carrier and the dicarboxylate (malate)/phosphate exchange carrier[33].

Malate/aspartate shuttle

Malate/aspartate shuttle activities have been estimated previously, with isolated mitochondria, from the degree of inhibition of O_2 uptake in the combined presence of aspartate and 2-oxoglutarate (Fig. 4)[2]. However, it appears that part of the observed inhibition may result from the uptake of OAA produced by AsAT located external to the inner membrane, as demonstrated by the partial reversal of the inhibition by a low concentration of phthalonate (a specific inhibitor of the OAA transporter). Allowing for this correction we estimate that maximum rates of NADH export from the mitochondria, via this shuttle, (expressed on a whole tissue basis) will be of the order of 7-11 µmol/mg Chl/h, which is well below that proposed to be produced during glycine oxidation *in vivo*. This, together with the observation that there are only low levels of AsAT activity in leaf peroxisomes [41], raises serious doubts regarding the viability of the malate/aspartate shuttle as the sole means of transfer of NADH between the mitochondria and peroxisomes during photorespiration.

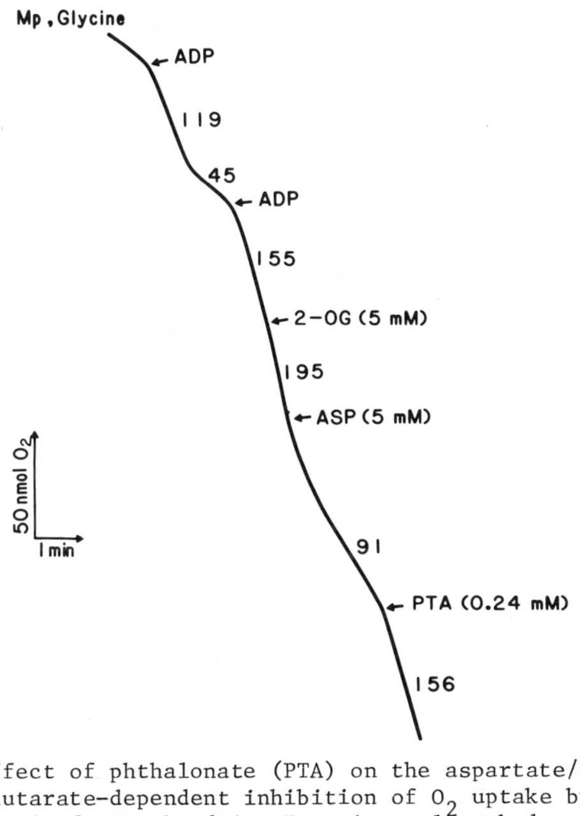

Fig. 4. Effect of phthalonate (PTA) on the aspartate/2-oxoglutarate-dependent inhibition of O_2 uptake by purified pea leaf mitochondria. Experimental methods as given in Dry and Wiskich[42]. Additions: 0.4 mg mitochondrial protein, 10 mM glycine. Rates expressed as nmol O_2 consumed/mg protein/min.

SUMMARY

Mitochondria have an important role, in plant cells, not only in the production of respiratory products such as ATP and carbon skeletons, but also in the reoxidation (or in some cases supply) of reducing power required to maintain the operation of certain key metabolic processes.

ACKNOWLEDGEMENTS

We thank Carol Robinson for typing this manuscript and gratefully acknowledge the financial support of the Australian Research Grants Scheme.

REFERENCES

1. H. Lambers, Respiration in intact plants and tissues: its regulation and dependence on environmental factors, metabolism and invaded organs, in: "Encyclopaedia of Plant Physiology, New Series, vol XVIII", R. Douce and D.A. Day, eds, Springer-Verlag, Berlin (1985).
2. J. Azcon-Bieto and C.B. Osmond, Relationship between photosynthesis and respiration. The effect of carbohydrate status on the rate of CO_2 production by respiring in darkened and illuminated wheat leaves, Plant Physiol. 71:574 (1983).
3. D. Graham, Effects of light on dark respiration, in: "The biochemistry of plants. A Comprehensive treatise, vol. II", P.K. Stumpf and E.E. Conn, eds, Academic Press, New York (1980).
4. M. Peisker and P. Apel, Dark respiration and the effect of oxygen on CO_2 compensation concentration in wheat leaves, Z. Pflanzenphysiol. 100:398 (1980).
5. R.E. Sharp, M.A. Matthews and J.S. Boyer, Kok effect and the quantum yield of photosynthesis, Plant Physiol. 75:95 (1984).
6. A. Brooks and G.D. Farquhar, Effects of temperature on the CO_2/O_2 specificity of ribulose-1,5-bisphosphate carboxylase/oxygenase and the rate of respiration in the light, Planta 165:397 (1985).
7. D.T. Canvin, J.A. Berry, M.R. Badger, H. Fock and C.B. Osmond, Oxygen exchange in leaves in the light, Plant Physiol. 66:302 (1980).
8. A. Gerbaud and M. Andre, Effect of CO_2, O_2 and light on photosynthesis and photorespiration in wheat, Plant Physiol. 66:1032 (1980).
9. D. Chevallier and R. Douce, Interactions between mitochondria and chloroplasts in cells. I. Action of cyanide and of 3-(3,4-Dichlorophenyl)-1,1-dimethylurea on the spore of Funaria hygrometrica, Plant Physiol. 57:400 (1976).
10. G. Peltier and P. Thibault, O_2 uptake in the light in Chlamydomonas. Evidence for persistent mitochondrial respiration, Plant Physiol. 79:225 (1985).
11. K.A. Walker, C.V. Givan and A.J. Keys, Glutamic acid metabolism and the photorespiratory nitrogen cycle in wheat leaves. Metabolic consequences of elevated ammonia concentrations and of blocking ammonia assimilation, Plant Physiol. 75:60 (1984).
12. P. Gardeström and G.E. Edwards, Leaf mitochondria (C_3 + C_4 + CAM), in "Encyclopaedia of Plant Physiology, New Series, vol. XVIII", R. Douce and D.A. Day eds, Springer-Verlag, Berlin (1985).
13. P. Gardeström and G.E. Edwards, Isolation of mitochondria from leaf tissue of Panicum miliaceum, a NAD-malic enzyme type C_4 plant, Plant Physiol. 71:24 (1983).
14. T. Kagawa and M.D. Hatch, Mitochondria as a site of C_4-pathway photosynthesis, Arch. Biochem. Biophys. 167:687 (1975).
15. M.D. Hatch and S. Mau, Activity, location and role of aspartate aminotransferase and alanine aminotransferase isoenzymes in leaves with C_4 pathway photosynthesis, Arch. Biochem. Biophys. 156:195 (1973).

16. G.P. Arron, M.H. Spalding and G.E. Edwards, Isolation and oxidative properties of intact mitochondria from leaves of Sedum praealtum. A crassulacean acid metabolism plant, Plant Physiol. 64:182 (1979).
17. D.A. Day, Malate decarboxylation by Kalanchoe diagremontiana mitochondria and its role in crassulacean acid metabolism, Plant Physiol. 65:675 (1980).
18. M.H. Spalding, G.P. Arron and G.E. Edwards, Malate decarboxylation in isolated mitochondria from the crassulacean acid metabolism plant Sedum praealtum, Arch. Biochem. Biophys. 199:448 (1980).
19. P. Rustin and C. Queiroz-Claret, Changes in oxidative properties of Kalanchoe blossfeldiana leaf mitochondria during development of crassulacean acid metabolism, Planta 164:415 (1985).
20. J.A.M. Holtum and C.B. Osmond, The Gluconeogenic metabolism of pyruvate during deacidification in plants with crassulacean acid metabolism, Aust. J. Plant Physiol. 8:31 (1981).
21. H. Beevers, The role of the glyoxylate cycle, in: "The biochemistry of plants, vol. IV," P.K. Stumpf, ed., Academic Press, New York (1980).
22. I.J. Mettler and H. Beevers, Oxidation of NADH in glyoxysomes by a malate-aspartate shuttle, Plant Physiol. 66:555 (1980).
23. J.T. Wiskich, Control of the Krebs cycle, in: "The biochemistry of plants, vol. II," D.D. Davies, ed. Academic Press, New York (1980).
24. J. Millhouse, J.T. Wiskich and H. Beevers, Metabolite oxidation and transport in mitochondria of endosperm from germinating castor bean, Aust. J. Plant Physiol. 10:167 (1983).
25. L. Klepper, D. Flesher and R.H. Hageman, Generation of reduced nicotinamide adenine dinucleotide for nitrate reduction in green leaves, Plant Physiol. 48:580 (1971).
26. C.K.M. Rathnam, Malate and dihydroxyacetone phosphate-dependent nitrate reduction in spinach leaf protoplasts, Plant Physiol. 62:220 (1978).
27. M.S. Naik and D.J.D. Nicholas, Malate metabolism and its relation to nitrate assimilation in plants, Phytochem. 25:571 (1986).
28. S.K. Sawhney, M.S. Naik and D.J.D. Nicholas, Regulation of NADH supply for nitrate reduction in green plants via photosynthesis and mitochondrial respiration, Biochem. Biophys. Res. Comm, 81:1209 (1978).
29. A.J. Reed and D.T. Canvin, Light and dark controls of nitrate reduction in wheat (Triticum aestivum L.) protoplasts. Plant Physiol. 69:508 (1982).
30. R.M. Wallsgrove, A.J. Keys, P.J. Lea and B.J. Miflin, Photosynthesis, photorespiration and nitrogen metabolism, Plant Cell Envir. 6:301 (1983).
31. S.C. Somerville and C.P. Somerville, Effect of oxygen and carbon dioxide on photorespiratory flux determined from glycine accumulation in a mutant of Arabidopsis thaliana, J. Exp. Bot. 34:415 (1983).
32. H. Lambers, D.A. Day and J. Azcon-Bieto, Cyanide-resistant respiration in roots and leaves. Measurements with intact tissues and isolated mitochondria, Plant Physiol. 58:148 (1983).
33. D.A. Day and J.T. Wiskich, Glycine metabolism and oxaloacetate transport by pea leaf mitochondria, Plant Physiol. 68:425 (1981).
34. R. Douce, A.L. Moore and M. Neuburger, Isolation and oxidative properties of intact mitochondria isolated from spinach leaves, Plant Physiol. 60:625 (1977).
35. I.B. Dry, D.A. Day and J.T. Wiskich, Preferential oxidation of glycine by the respiratory chain of pea leaf mitochondria, FEBS Lett. 158:154 (1983).
36. A. Bergman and I. Ericson, Effects of pH, NADH, succinate and malate on the oxidation of glycine in spinach leaf mitochondria, Physiol. Plant. 59:421 (1983).
37. D.A. Day, M. Neuburger and R. Douce, Interactions between glycine decarboxylase, the tricarboxylic acid cycle and the respiratory chain in pea leaf mitochondria, Aust. J. Plant Physiol. 12:119 (1985).

38. D.J. Oliver and G.H. Walker, Characterization of the transport of oxaloacetate by pea leaf mitochondria, Plant Physiol. 76:409 (1984).
39. H. Ebbinghausen, Chen Jia and H.W. Heldt, Oxaloacetate translocator in plant mitochondria, Biochim. Biophys. Acta 810:184 (1985)
40. E.P. Journet, M. Neuburger and R. Douce, Role of glutamate-oxaloacetate transaminase and malate dehydrogenase in the regeneration of NAD^+ for glycine oxidation by spinach leaf mitochondria, Plant Physiol. 67:467 (1981).
41. M.R. Schmitt and G.E. Edwards, Provisions of reductant for the hydroxypyruvate to glycerate conversion in leaf peroxisomes. A critical evaluation of the proposed malate/aspartate shuttle, Plant Physiol. 72:728 (1983).
42. I.B. Dry and J.T. Wiskich, Comparative aspects of aminooxyacetate inhibition of glycine oxidation and aminotransferase activity of pea leaf mitochondria, Plant Sci. 44:23 (1986).

THE EFFECT OF LIGHT ON MITOCHONDRIAL RESPIRATION STUDIED BY RAPID FRACTIONATION OF PROTOPLASTS

Per Gardeström

Department of Plant Physiology
University of Umeå
S-901 87 Umeå, Sweden

ABSTRACT

Rapid fractionation by membrane filtration was used to determine ATP/ADP ratios in chloroplasts, mitochondria and cytosol of barley protoplasts incubated under different conditions.

In darkness the ATP/ADP ratio was higher in the cytosol than in chloroplasts and mitochondria. In the light at saturating CO_2 the ATP/ADP ratio increased in chloroplasts and mitochondria compared to darkness whereas a slight decrease was observed in the cytosol. In the light at limiting CO_2 the ATP/ADP ratio increased in all compartments compared to saturating CO_2. The cytosolic ATP/ADP ratio is higher in the light at limiting CO_2 compared to darkness. As CO_2 is limiting for photosynthesis in C_3 plants at atmospheric conditions this has important implications for the regulation of the metabolism in the cytosol in the light.

When photosynthesis was reduced by lack of CO_2 the cytosolic ATP/ADP ratio was much higher than when photosynthesis was limited by inhibition of the Calvin cycle. From this it is suggested that, under certain conditions, photorespiration can function to shuttle ATP from the chloroplasts into the cytosol.

INTRODUCTION

Adenylates (ATP, ADP, AMP) participate in many reactions in the cell, and are also very important regulators for many metabolic reactions. Therefore the phosphorylation level of the adenylate system has been studied to gain information about the metabolic state of the cell. The phosphorylation level can be expressed as energy charge, as phosphorylation potential or as ATP/ADP ratio. For a review see Pradet and Raymond (1983).

The net transport of adenylates over the chloroplast and the mitochondrial membranes are relatively slow. Therefore the adenylates will be present in different pools in the cytosol, chloroplast stroma and the mitochondrial matrix. The ATP/ADP ratios differ but are not independent in these pools since the phosphorylation level is communicated via shuttles and exchange translocators over the organell membranes. The most important systems for this communication are regarded to be the triosephosphate-phosphoglycerate shuttle in the chloroplast envelope (Flügge and Heldt 1984) and the ATP/ADP exchange translocator in the inner mitochondrial membrane (Day and Wiskich 1984). It is thus not sufficient to determine the overall ATP/ADP ratio of the cell, but the ratio has to be determined in each compartment. The turn-over of the adenylate system is very rapid - often in the order of one second (Stitt et al. 1982). Thus, conventional techniques for organelle separation can not be used for the determination of ATP/ADP ratios in different cell compartments. Non-aqueous fractionation has been used to determine the concentration of metabolites in the chloroplast and the rest of the cell (Santarius and Heber 1965). Hampp (1981) has developed a technique for fractionation of protoplasts using silicon-oil centrifugation. Lilley et al. (1982) developed a method to fractionate protoplasts very quickly by membrane filtration. In this method organelles (chloroplasts and mitochondria) are removed from ruptured protoplasts by membrane filtration. The metabolite level in the different cell compartments is then obtained by subtraction.

In the present study I have used the method of membrane filtration to study the ATP/ADP ratios in chloroplasts, mitochondria and cytosol of barley protoplasts. Before fractionation the protoplasts were incubated in darkness, in light with saturating CO_2 or in light with limiting CO_2. Also the effect of the Calvin cycle inhibitor glycolaldehyde was studied. The results are discussed with reference to the function and regulation of mitochondrial respiration in the light.

METHODS

Isolation of protoplasts

Protoplasts were isolated from 6 to 10 days old barley leaves (*Hordeum vulgare* var Gunilla) grown in a green house with supplementing artificial light. After cutting into small pieces the tissue was digested for about 2 h at 28°C in: 1 % cellulase (Cellulysin from Calbiochem), 0.3 % pectinase (Macerase from Calbiochem), 0.5 M sorbitol, 0.2 % BSA, 0.05 % PVP, pH=5.0. After removal of the digestion medium protoplasts were released by gentle shaking in: 0.5 M sorbitol, 1 mM $CaCl_2$, 0.05 % BSA, 0.05 % PVP, 5 mM Hepes (pH=7.0) and centrifuged at 300 g for 3 min. The pellet was suspended in: 0.5 M sucrose, 1 mM $MgCl_2$, 5 mM Hepes (pH=7.0). This suspension was transferred to centrifuge tubes and overlayered with: 0.4 M sucrose, 0.1 M sorbitol, 1 mM $MgCl_2$, 5 mM Hepes (pH=7.0) and 0.5 M sorbitol, 1 mM $MgCl_2$, 5 mM Hepes (pH=7.0). After centrifugation at 500 g for 5 min purified protoplasts were collected from the interface between 0.4 M sucrose/0.1 M sorbitol and 0.5 M sorbitol. The composition of the assay medium was: 0.25 M sucrose, 0.25 M sorbitol, 10 mM KCl, 0.5 mM $MgCl_2$, 0.06 % BSA, 0.2 % PVP, 10 mM Hepes (pH=7.2). This medium has the same density as the protoplasts. Boiled water was used to prepare a CO_2-free assay medium. The KOH used for neutralization was also washed in boiled water before it was dissolved.

The protoplasts were more than 80% intact as determined by the latency of glycolate oxidase (Nishimura et al. 1984). The rate of photosynthetic oxygen evolution was 100-160 µmol O_2 * mg Chl^{-1} * h^{-1} with saturating CO_2 (10 mM $NaHCO_3$) in assay medium.

Fractionation of protoplasts

Protoplasts (20 µg $Chl*ml^{-1}$) were incubated in three glass syringes fitted to filter holders with different filter combinations giving three different filtrates according to Table I. The protoplasts are ruptured by passing through the 15 µm nylon net. Chloroplasts and mitochondria are collected on the 8 µm and 0.45 µm membrane filter respectively. The syringes

Table I. The filter combinations used to obtain the three different filtrates. The nylon net was obtained from Henry Simon Ltd. (Stockport, England) and membrane filters from Millipore. F0 = total extract, F1 = total extract - chloroplasts, F2 = total extract - chloroplasts - mitochondria.

Filter	Filtrate		
	F0	F1	F2
15 µm nylon net	X	X	X
8 µm membrane filter		X	X
0.45 µm membrane filter			X

were closed simultaneously in a specially constructed apparatus using the principle developed by Lilley et al. (1982), with two important improvements: 1. A system was developed where the acid used to denaturate proteins was injected into the filtrate after the filters. This was obtained by using small plastic syringes that were closed simultaneously as the protoplast syringe. The advantage with this system is that it is quicker and gives a constant acid concentration during acidification. 2. Each filtrate was divided into three parts (instead of two) and collected in eppendorf tubes. One part was mixed with buffer and used for assays of marker enzymes. The second part of each filtrate was mixed with acid and used for measurements of ATP and ADP. The third part will allow measurements of NADH and NADPH after mixing with KOH. (Measurements of NADH and NADPH was not a part of the present study.)

At the rate of filtration used (3.6 ml/sec) the protoplasts were mixed with trichloroacetic acid (TCA) about 0.1 s after entering the filter holder. The final concentration of TCA in the filtrates was about 5 %.

Enzyme assays

Fumarase was used as marker for mitochondria and was assayed by coupling to NADP-Malic enzyme basically as described by Hatch (1978). As marker for the cytosol, PEP carboxylase was measured according to Gardeström and Edwards (1983). NADP-triophosphate dehydrogenase (NADP-TPD) was assayed according to Winter et al. (1982) and used as marker for chloroplasts. Chlorophyll was determined according to Bruinsma (1961).

Determination of ATP and ADP

Adenylates were determined by the firefly lusiferase method using an ATP monitoring kit (LKB Wallac, Turku, Finland) and a LKB 1250 Luminometer. Assays were run in a final volume of 285 µl in: 70 mM Hepes pH=7.75, 3.5 mM $MgCl_2$, 1.4 mM EDTA, 6 mM PEP. The sample was clearified by centrifugation and 7 µl added without neutralization. ADP was converted to ATP by addition of 1 unit Pyruvate kinase (Serva). ATP and ADP was assayed in the same sample. Each assay was standardized by the addition of 7 µl 1 µm ATP. At least triplicate assays were run for each sample.

Table II. The distribution of adenylates between the different filtrates. The values ± S.D. are based on 40 filtration experiments.

	Filtrate		
	FO	F1	F2
		%	
ATP+ADP	100	34 ± 4	29 ± 4

Calculations

The separation of each filtrate into three parts was not fully reproducable. Therefore the eppendorf tubes used to collect the filtrates were weighed before and after filtration. As the amount of acid mixed into each tube was the same between experiments the dilution could be determined for the individual samples. Using this, the distribution of adenylates (ATP+ADP) between the three filtrates was determined. The average distribution

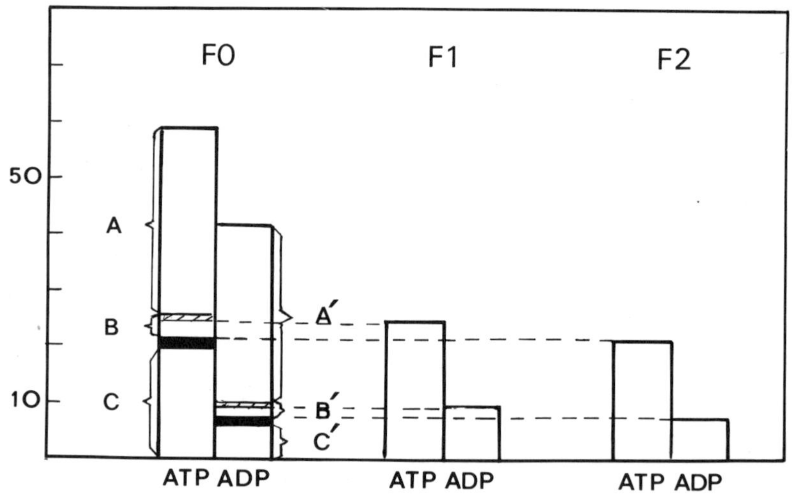

Figure 1. Example of the calculations of ATP/ADP ratios in chloroplasts, mitochondria and cytosol at high CO_2 in the light. ATP/ADP ratios in the filtrates were: FO=1.4, F1=2.6, F2=2.8. According to Table II (ATP+ADP) in FO=100%, F1=34%, F2=29%. The amounts of ATP and ADP in F1 and F2 is projected on the FO histogram. The hatched area represents mitochondria filtered off on the 8 µm filter (Table III). A and A' represents ATP and ADP in chloroplasts. B and B' represents ATP and ADP in mitochondria. The cytosol values for ATP and ADP (C and C') are corrected for 5% breakage of chloroplasts during filtration (black area).

Table III. The distribution of marker enzymes between the different filtrates. The activity in F1 and F2 is expressed as % of the activity in F0. The specific activity in the total extract (F0) was: fumarase 27±4, PEP carboxylase 36±8 and NADP-TPD 589±164 µmol/mg Chl*h. Values ± S.D. are the mean of 30 fractionation experiments.

Filtrate	Fumarase	PEPcase	NADP-TPD
F0	100	100	100
F1	74 ± 7	91 ± 7	12 ± 7
F2	7 ± 2	81 ± 7	9 ± 3

obtained (Table II) was used in calculations of the ATP/ADP ratios in the different compartments. The amounts of ATP and ADP in chloroplasts, mitochondria and cytosol is obtained by subtraction between the different filtrates. The results were corrected for the removal of mitochondria on the 8 µm filter and the breakage of chloroplasts during fractionation. No correction was made for broken protoplasts. An example of the calculation is shown in Figure 1. In this method of calculation the ATP/ADP ratio measured in the different filtrates is used directly. This should avoid that errors in the determination of the dilution for each fraction directly affect the results as is the case if the calculations are made from the amounts of ATP and ADP found in the filtrates.

RESULTS

As shown in Table III the 8 µm membrane filter is very effective in removing chloroplasts as only 12 % of the chloroplast marker NADP-TPD appears in the filtrate. This high retention also proves that the chloroplast breakage is very low in the filtration. By comparing the activity of NADP-TPD in the F1 filtrate to a F1 filtrate obtained at lower speed the chloroplast breakage during filtration was estimated to 5 % (results not shown). Some mitochondria (26 %) were lost on the 8 µm filter. The 0.45 µm filter removed all of the intact mitochondria remaining since the fumarase activity in the F2 filtrate was not stimulated by addition of Triton X-100 (results not shown).

The ATP/ADP ratio in the total extract (F0) was somewhat higher in the light at saturating CO_2 compared to darkness (Table IV). The increase was confined to the chloroplasts and the mitochondria. In the cytosol the ATP/ADP ratio was somewhat lower in the light compared to darkness. This observation is similar as reported for wheat protoplasts under similar conditions (Stitt et al. 1982).

A situation with limiting CO_2 supply to the protoplasts was obtained by adding bicarbonate to a CO_2-free reaction medium. The bicarbonate was continuously injected at a constant rate during the incubation. After a lag of a couple of minutes the protoplasts adjusted to a steady-state rate of photosynthesis which was dependent on the rate of bicarbonate addition. When a constant rate was obtained the protoplasts were transferred to syringes and fractionated by membrane filtration as described above.

Table IV. ATP/ADP ratio in chloroplasts, mitochondria and cytosol of barley protoplasts incubated in darkness, in light at high CO_2 (10 mM $NaHCO_3$) and in light at low CO_2 (obtained as described in "Methods"). Saturating light was used (500 µE * m^{-2} * s^{-1}). The rate of photosynthetic oxygen evolution at low CO_2 was about 20 % of the rate at saturating CO_2. The mean ± S.D. is based on nine experiments in dark, nine at high CO_2 and four at low CO_2.

Incubation	Total extract	Chloroplasts	Mitochondria	Cytosol
Dark	1.1 ± 0.3	0.8 ± 0.3	0.6 ± 0.4	4.3 ± 1.3
Light -high CO_2	1.5 ± 0.6	1.2 ± 0.7	1.2 ± 0.5	3.1 ± 0.5
Light -low CO_2	4.0 ± 0.5	3.2 ± 0.7	4.2 ± 0.8	9.0 ± 3.0

When CO_2 was limiting the ATP/ADP ratio was drastically increased in the total extract (Table IV). The increase was seen in all compartments. The big increase in the cytosolic ATP/ADP ratio observed is especially interesting since this means that the cytosolic ATP/ADP ratio is higher in the light at limiting CO_2 compared to darkness. This increase was also observed with CO_2 concentrations giving 40-80 % of the saturated rate of photosynthetic O_2 production (results not shown).

When photosynthesis was inhibited by lack of CO_2 - resulting in no carboxylation but oxygenation of RuBP, or by stopping the Calvin cycle by glycolaldehyde (Sicher 1984) - resulting in neither carboxylation nor oxygenation of RuBP, the ATP/ADP ratio in the total extract was high (Table V). In both cases the ATP/ADP ratio in the chloroplast stroma was high whereas the mitochondrial ATP/ADP ratio was relatively low. When photosynthesis was inhibited by glycolaldehyde the cytosolic ATP/ADP ratio was similar to the value obtained in darkness. In contrast when photosynthesis was inhibited by lack of substrate the cytosolic ATP/ADP ratio was very high.

Table V. The ATP/ADP ratios in chloroplasts, mitochondria and cytosol of barley protoplasts in the light when photosynthesis was inhibited by lack of CO_2 or glycolaldehyde (20 mM). The values are the mean of two experiments.

	Total extract	Chloroplasts	Mitochondria	Cytosol
no CO_2	4.1	3.6	1.4	10.6
glycolaldehyde	4.1	5.2	1.2	3.4

DISCUSSION

From the distribution of (ATP+ADP) between the three filtrates (Table II) the intracellular distribution of adenylates can be estimated to be: 68% in the chloroplasts, 26% in the cytosol and 6% in the mitochondria (taking into account mitochondria filtered off on the 8 µm filter and chloroplast breakage during filtration). For wheat protoplasts Stitt et al. (1982) reported the distribution: chloroplasts 47%, cytosol 44%, mitochondria 9%. In the present study AMP was not determined, but according to the results by Stitt et al. (1982) AMP make up only about 10% of the total adenylate pool and is found in the chloroplasts and the cytosol. Thus AMP would not significantly change the distribution calculated above and the difference compared to the results by Stitt et al. (1982) might be due to different growing conditions, or species differences.

As the mitochondria contain a small part of the adenylates in the cell it is evident that the results obtained for mitochondria are sensitive to experimental errors.

The method used in this study: "rapid fractionation of protoplasts by membrane filtration" is the quickest method available to determine intracellular metabolite concentrations. A problem is of course to interpret the results obtained to the *in vivo* situation. A big problem in this respect is to estimate the contribution from broken protoplasts in the preparation. In the present study no correction was made for this. This should mainly affect the values obtained for the cytosol fraction. Thus, the values reported should not be directly interpreted as *in vivo* values. If it is assumed that the contribution from broken protoplasts is similar in the different incubations the difference between incubations should reflect true differences. It is therefore reasonable to conclude that the cytosolic ATP/ADP ratio is higher in the light at limiting CO_2 than in darkness (Table IV). Under normal atmospheric conditions CO_2 is limiting for photosynthesis in C_3 plants. Thus, the cytosolic ATP/ADP ratio is normally higher in the light compared to darkness in contrast to the results obtained at saturating CO_2 (Stitt et al. 1982, and Table IV). This observation is important for the understanding of how the metabolism in the cytosol is affected by light.

A major difference between the experiments in Table V is that with "no CO_2" oxygenation of RuBP will yield phosphoglycolate which is converted to glycine in the photorespiratory glycolate pathway. Glycolaldehyde blocks the regeneration of RuBP in the Calvin cycle (Sicher 1984) and thus no RuBP is available and as a consequence no substrate for photorespiration can be produced. The concentrations of PGA and triosephosphate are probably relatively low when photosynthesis is inhibited. Therefore the rate of ATP transport via the triosephosphate-PGA shuttle is probably low in both situations. Upon conversion of glycine to serine in the mitochondria NAD is reduced to NADH (Gardeström and Edwards 1985). NADH can be reoxidized via the electron transport chain with subsequent ATP production. This ATP is then transported into the cytoplasm in exchange for ADP. A possible explanation for the difference in the cytosolic ATP/ADP ratio between the experiments in Table V is therefore ATP production as a consequence of photorespiration.

It has been suggested that the NADH-produced from glycine oxidation should be shuttled out and used for the reduction of hydroxypyruvate to glycerate in the glycolate pathway (Woo and Osmond 1976). Alternatively this reducing power could be supplied by the chloroplasts via the malate/oaa translocator recently described by Hatch et al. (1985). It is interesting to note that glycine is a preferred substrate for respiration in leaf mitochondria (Day et al. 1983, Bergman and Ericson 1983).

An effect of photorespiration can thus be export of ATP from the chloroplast stroma into the cytosol. This might be of importance under certain metabolic conditions. It is possible that the increased cytosolic ATP/ADP ratio observed at limiting CO_2 (Table IV) is due to such ATP export since photorespiration is suppressed under saturating CO_2 but not at limiting CO_2.

ACKNOWLEDGEMENTS

I am greatful to Gunilla Malmberg for excellent technical assistance, to Monica Nilsson for typing the manuscript, to Gerald E. Edwards and Bo Wigge for comments and suggestions and to Gunnar Öquist for encouraging support. This project was supported by the Swedish National Science Research Council.

REFERENCES

Bergman, A., and Ericson, I., 1983, Effects of pH, NADH, succinate, and malate on the oxidation of glycine in spinach leaf mitochondria, Physiol. Plant., 59:421.

Bruinsma, J., 1986, A comment on the spectrophotometric determination of chlorophyll, Biochim. Biophys. Acta, 52:576.

Day, D. A., and Wiskich, J. T., 1984, Transport processes of isolated mitochondria, Physiol. Veg., 22:241.

Dry, S. B., Day, D. A., and Wiskich, J. T., 1983, Preferential oxidation of glycine by the respiratory chain of pea leaf mitochondria, FEBS Lett. 158:154.

Flügge, V. I. and Heldt, H. W., 1984, The phosphate - triose phosphate - phosphoglycerate translocator of the chloroplasts, Trends Biochem. Sci., 9:530.

Gardeström, P., and Edwards, G. E., 1983, Isolation of mitochondria from leaf tissue of Panicum milineceum, a NAD-malic enzyme type C_4 plant, Plant Physiol., 71:24.

Gardeström, P., and Edwards, G. E., 1985, Leaf mitochondria (C_3+C_4+CAM), in: Encyclopedia of Plant Physiology, New Series, Vol. 18:314. R. Douce and D. A. Day, eds, Springer Verlag, Berlin.

Hampp, R., Goller, M., and Ziegler, H., 1982, Adenylate levels, energy charge, and phosphorylation potential during dark-light and light-dark transition in chloroplasts, mitochondria and cytosol of mesophyll protoplasts from Avena sativa L., Plant Physiol., 69:448.

Hatch, M. D., Dröscher, L., Flügge, U. I., and Heldt, H. W., 1984,A specific translocator for oxaloacetate in chloroplasts, FEBS Lett., 178:15.

Hatch, M. D., 1978, A simple spectrophotometric assay for fumarase hydratase in crude tissue extract, Analyt. Biochem. 85:271.

Lilley, R. McC., Stitt, M., Mader, G., and Heldt, H. W., 1982, Rapid fractionation of wheat leaf protoplasts using membrane filtration, Plant Physiol., 70:965.

Nishimura, M., Douce, R., and Akazawa, T., 1985, A simple method for estimating intactness of spinach leaf protoplasts by glycolate oxidase assay, Plant Physiol., 78:343.

Pradet, A., and Raymond, P., 1983, Adenine nucleotide ratios and adenylate energy charge in energy metabolism, Ann. Rev. Plant Physiol., 34:199.

Santarius, K. A., and Heber, U., 1965, Changes in the intracellular levels of ATP, ADP, AMP and Pi and regulatory function of the adenylate system in leaf cells during photosynthesis, Biochim. Biophys. Acta, 102:39.

Sicher, R. C., 1984, Glycolaldehyde inhibition of photosynthetic carbon assimilation by isolated chloroplasts and protoplasts, in: Advances in Photosynthesis Research, Vol III, 6:413. C. Sybesma, ed., Martinus Nijhoff, The Hague.

Stitt, M., Lilley, R. McC., and Heldt, H. W., 1982, Adenine nucleotide levels in the cytosol, chloroplasts, and mitochondria of wheat leaf protoplasts, Plant Physiol., 70:971.

Winter, K., Foster, J. G., Edwards, G. E., and Holtum, J. A. M., 1982, Intracellular location of enzymes of carbon metabolism in Mesembryanthemum crystallinum exhibiting C_3 photosynthetic characteristics or performing crassulacean acid metabolism, Plant Physiol., 69:300.

Woo, K. C., and Osmond, C. B., 1976, Glycine decarboxylation in mitochondria isolated from spinach leaves, Aust. J. Plant Physiol., 3:771.

ON THE FUNCTION OF MALATE-OXALOACETATE SHUTTLES IN A PLANT CELL

H. Ebbighausen, M.D. Hatch, R.McC. Lilley, S. Krömer, M. Stitt and H.W. Heldt

Institut für Biochemie der Pflanze
Untere Karspüle 2
Universität Göttingen
3400 Göttingen, F. R. G.

INTRODUCTION

The photorespiratory cycle, as summarized in Fig. 1, involves three metabolic compartments, the chloroplastic the peroxisomal and the mitochondrial compartment (14,26). Phosphoglycolate, the primary product of photorespiration, is hydrolyzed in the stroma, the resultant glycolate transported into the peroxisomal compartment for oxidation to glyoxylate and transamination to glycine. This glycine is transferred to the mitochondria for oxidation. Here the conversion of two moles glycine to one serine results in the formation of one mole NADH. The serine traverses back to the peroxisomes for transamination to hydroxypyruvate, followed by reduction to glycerate at the expense of one mole of NADH. Glycerate is then transported into the stroma, and phosphorylation to phosphoglycerate closes the loop (for ref. see 12). Since the NADH required for hydroxypyruvate reduction is equimolar to the NADH generated from glycine oxidation, it has been suggested that redox equivalents are transferred from the mitochondrial matrix to the peroxisomal compartment by malate-oxaloacetate shuttle, this transfer being mediated by the NADH-malate dehydogenase which is present in the mitochondria, peroxisomes and mitochondria (27,28,29). The question remained whether such a malate-oxaloacetate shuttle was of physiological significance. In order to give an answer to this, we studied the transport properties of isolated mitochondria, peroxisomes and also chloroplasts. These studies clearly demonstrate that a green plant cell has indeed the capacity to catalyze malate oxaloacetate shuttles at physiological rates not only for transfer of redox equivalents from the mitochondria to the peroxisomes but also for the export of redox equivalents from the chloroplasts. The latter reaction may be of importance for providing redox equivalents generated from photosynthesis as substrate for mitochondrial oxidative phosphorylation. Results will be presented, which indicate that also in the light, mitochondrial oxidative phosphorylation plays an important role in supplying a green plant cell with ATP.

RESULTS AND DISCUSSION

The effect of oxaloacetate on glycine oxidation

The notion of a malate-oxaloacetate shuttle between the mitochondrial matrix and the cytosol was supported by experiments with plant mitochondria, in which the respiration of NADH linked substrates such as glycine was found to be

OAA : oxaloacetate

This work has been supported by the Deutsche Forschungsgemeinschaft.

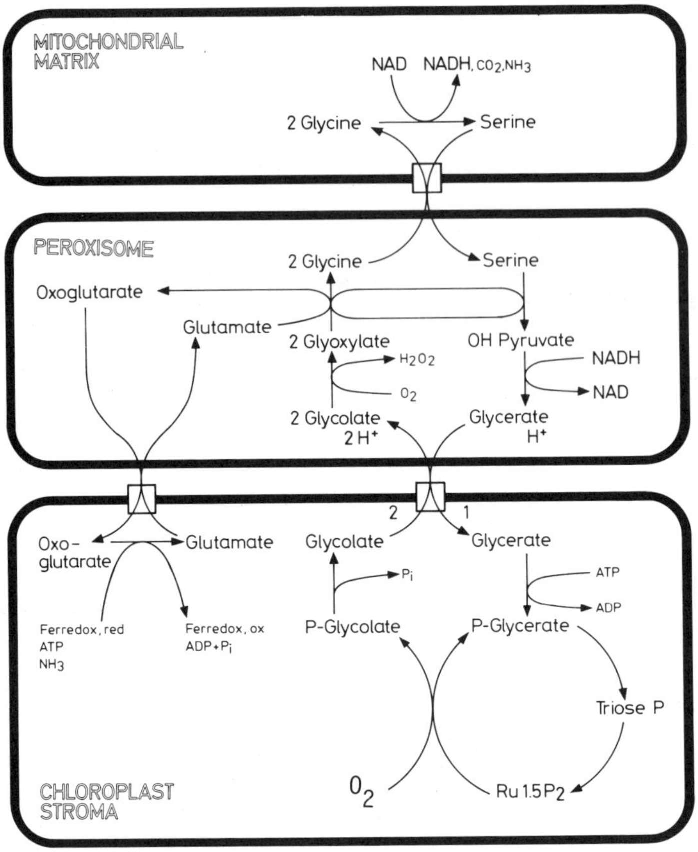

Fig. 1 Photorespiratory cycle

inhibited, due to the oxidation of the mitochondrial NADH by oxaloacetate (2,13,27). This is demonstrated in the experiment of Fig. 2. Mitochondria isolated with an improved method from spinach leaves (5), were allowed to oxidize glycine, and by a newly developed CO2/O2 electrode (16) O2 uptake and CO2 release were measured. In the presence of ADP (state 3) the release of CO2 parallels the consumption of O2. The subsequent addition of OAA inhibited O2 uptake drastically but stimulated CO2 release by about 50%, indicating an increased glycine oxidation rate. The lower rate of glycine oxidation in the absence of OAA does not appear to reflect a limitation by the capacity of the respiratory chain, since the respiration in the presence of glycine can be increased when malate is being added as additional substrate, and similar results were also found with mitochondria from pea leaves. The increase of glycine oxidation on the addition of OAA indicated in Fig. 2 may be due to a redox control of glycine decarboxylation. As shown by Douce (4) in this symposium, the glycine decarboxylase multienzyme complex is inhibited by NADH. The increase of glycine decarboxylation may be therefore the result of a lowering of the stromal NADH level upon the addition of OAA. These experiments demonstrate that the rate of mitochondrial glycine decarboxylation could considerably exceed the respiratory capacity of the mitochondria with glycine as substrate, if the NADH generated in the mitochondria were withdrawn by an efficient malate-oxaloacetate shuttle. The question therefore arises, whether such a shuttle is operating _in vivo_.

Properties of mitochondrial oxaloacetate transport

Because the equilibrium of malate dehydrogenase reaction is far on the side of malate (Keq 2×10^{-5}), the OAA concentrations in the cytosol of a leaf cell

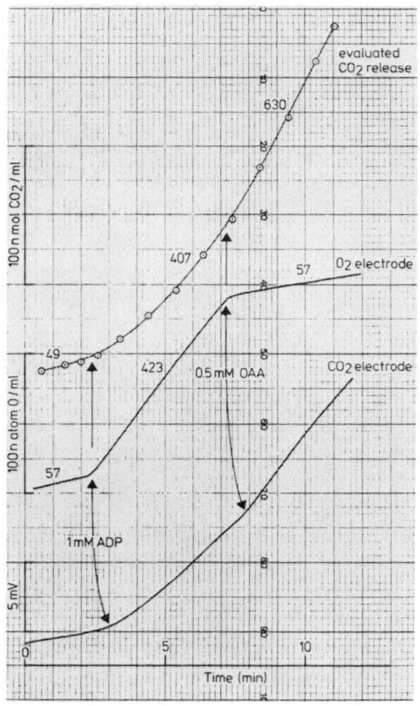

Fig. 2 Original traces of a simultaneous measurement of CO2 release and of oxygen consumption by a suspension of isolated spinach leaf mitochondria oxidizing glycine (20 mM). The concentration of mitochondria was 0.16 mg prot/ml. The upper curve has been evaluated from the CO2 electrode trace (lower curve) by use of a calibration curve. For details see ref. (16).

are to be expected in the µmolar range, and that of malate about 2-3 orders of magnitude higher. An efficient malate-oxaloacetate shuttle across separating membranes therefore requires that there is a translocator for oxaloacetate with a very high affinity to its substrate, and which is not inhibited by a more than hundred fold excess of malate. The uptake of OAA into isolated plant mitochondria fulfils these requirements. As shown in Table 1, the very rapid uptake of OAA into mitochondria from spinach and pea leaves is half saturated with µmolar concentrations of OAA, whereas the inhibitory constants of other dicarboxylates on this transport are three orders of magnitude higher. Similar results were also obtained by Oliver and Walker (18). Phthalonate, which had been found earlier to inhibit OAA uptake into plant mitochondria (2,17,18,19), in our experiments turned out as a very powerful competitive inhibitor of OAA transport with a Ki being even lower than the Km for OAA. It can be summarized from these data that in these mitochondria the OAA translocator is well suited to enable an efficient malate-oxaloacetate shuttle under physiological conditions. In this respect plant mitochondria appear to be different from mitochondria from animal tissues, such as rat live. and heart, where such a specific OAA translocator could not be demonstrated (see ref. (5)).

Malate driven hydroxypyruvate reduction in peroxisomes

The question arises whether it is possible to drive the peroxisomal hydroxypyruvate reduction with malate as the source of reduction equivalents. To answer this, a suspension of intact peroxisomes was incubated with glycolate and serine to generate hydroxypyruvate, and with malate (Fig. 3). As shown in Table 2, a considerable glycerate formation is observed, which is strictly dependent on the presence of all three substrates, and, in addition, on the presence of small concentrations of NAD. The formation of glycerate is coupled to malate oxidation ; glycerate and malate were formed in a 1:1 stoichiometry (Fig. 4). About the same stoichiometry is also found with peroxisomes which had been

Table 1. OXALOACETATE UPTAKE INTO LEAF MITOCHONDRIA. Temp. 20°. Data from ref (5).

		Spinach	Pea	
Oxaloacetate	Km	0.006	0.007	mM
	Vmax	550	570	nmol/mg prot per min
Inhibitors:				
Phthalonate	Ki	0.005	0.003	mM
Malonate	"	5	3	
2-Oxoglutarate	"	2	1	
Citrate	"	11	9	

lysed by treatment with detergent, but in this case a lag period is observed. Apparently, following the loss of integrity of these peroxisomes, the reaction partners are too much diluted at the beginning. As shown in Table 3, the apparent Km values for malate driven glycerate formation by peroxisomes are similar to the apparent Km values of the isolated enzymes (7,20,21) and are within the range of the cellular concentrations. Further work is required to elucidate how these substrates enter the peroxisomal compartment, either by translocator mediated transport across the peroxisomal membrane, or by diffusion through unspecific pores.

Comparison of activities of malate-oxaloacetate shuttles

It appears from the preceeding that a malate-oxaloacetate shuttle between the mitochondrial and peroxisomal compartment is principally possible under *in vivo* conditions. In order to assess the physiological significance of this shuttle, its rates have been compared in relation to photosynthesis in a

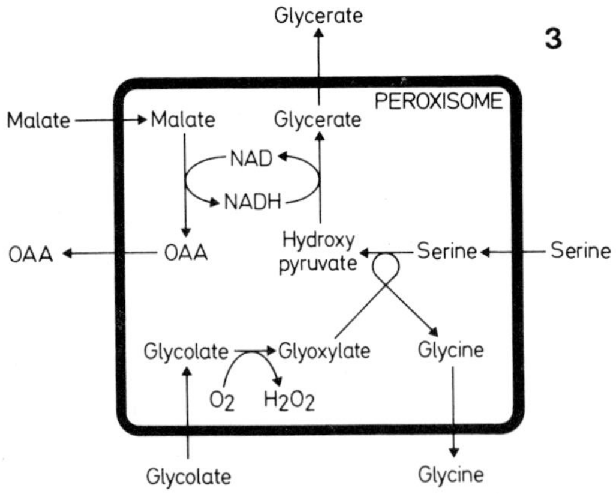

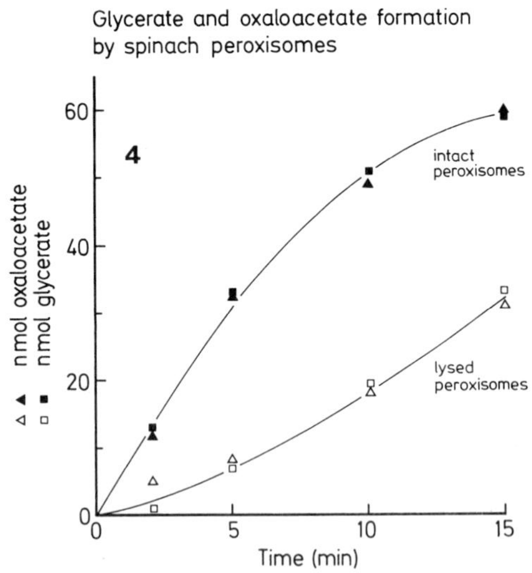

Fig. 3. Reaction scheme for malate dependent glycerate formation by isolated peroxisomes.

Fig. 4. Time course for the formation of glycerate and oxaloacetate by isolated peroxisomes supplied with glycolate, serine, malate and NAD according to the experiment of Table 2. The amount of peroxisomes in the experiment was equivalent to a hydroxypyruvate reductase activity of 57 nmol/min. The lysed peroxisomes were treated with 0.1% Triton X-100.

Table 2. GLYCERATE FORMATION BY INTACT SPINACH PEROXISOMES The peroxisomes were obtained essentially according to ref. (15) by differential centrifugation of a spinach leaf homogenate (5 min 1000 x g, 2 times 10 min 10 000 x g) without further purification. Ebbighausen, in prep. The complete incubation contained 20 mM malate, 15 mM serine, 2 mM glycolate and 1 mM NAD in a medium containing 400 mM sucrose, 20 mM Hepes-KOH, pH 7.5, 1 mM EDTA and 0.1% BSA and spinach peroxisomes equivalent to a hydroxypyruvate reductase activity of 144 mol/min in a total volume of 2 ml. Incubation time 3 min. Temp. 20°.

	Glycerate formation (nmol/assay)
1) complete	38.8
2) minus malate	2.2
3) minus serine	1.5
4) minus glycolate	2.1
5) minus NAD	2.2

Table 3. KINETIC CONSTANTS OF MALATE DRIVEN GLYCERATE FORMATION BY INTACT SPINACH LEAF PEROXISOMES Temp. 20°

Substrate	Km (mM)	Vmax (µmol/mg Chl per h)
Serine	4.3	67.6
Malate	3.3	56.6
Glycolate	0.2	62.3
NAD	0.1	45.1

whole leaf. For this, by using citrate synthase and hydroxypyruvate reductase as marker enzymes for mitochondria and peroxisomes, respectively, the activities obtained with isolated mitochondria and peroxisomes were related to the chlorophyll content in a spinach leaf (Table 3,4). With a net photosynthesis of about 200 µmol CO2/mg Chl per h, as measured with our spinach under growth conditions, and an assumed ratio of carboxylation to oxygenation of 5 (6) the CO2 release and NADH formation from glycine oxidation can be calculated to occur at a rate of 27.5 µmoles/Chl per h. Compared to this, the maximal rates of glycine oxidation and of OAA transport measured with isolated mitochondria, and the rate of malate dependent glycerate formation by isolated peroxisomes are about twice as high. These findings clearly show that a plant leaf cell has the means to shuttle the NADH formed during glycine oxidation to the peroxisomes for serving the reduction of hydroxypyruvate. Since the equilibrium of hydroxypyruvate reduction is far on the site of glycerate (Keq 3.10^{-5}) this reaction is expected to act as a drain, sequestering the redox equivalents from the mitochondria (Fig. 5A). Thus, despite the high rates of glycine oxidation by the mitochondria there would be no NADH available for oxidation by the respiratory chain, unless there is another source of redox equivalents available to the peroxisomes. As will be shown in the following, chloroplasts appear to be such an alternative source for the provision of redox equivalents.

Export of the redox equivalents from the chloroplasts

Intact isolated chloroplasts from C3 and C4 plants are known to photoreduce added oxaloacetate (1,3,8,11). Recently, in chloroplasts from Maize mesophyll

Table 4. COMPARISON OF PHOTOSYNTHETIC AND PHOTORESPIRATORY ACTIVITIES IN SPINACH LEAVES. Temp. 20°

Spinach Leaf: (in air)	net photosynthesis	200	µmol CO_2/mg Chl per h
	ribulosebisphosphate oxigenation (6)	55	µmol O_2/mg Chl per h
	corresponding CO_2 release from glycine	27.5	µmol/mg Chl per h
Isolated Organelles:			
mitochondria	glycine oxidation	46	µmol NADH/mg Chl per h
	oxaloacetate transport	51	µmol/mg Chl per h
peroxisomes	malate dependent glycerate formation	59	" " " "
chloroplasts	oxaloacetate transport	51	" " " "

cells and from Spinach leaves a specific OAA translocator was characterized (9), with a high affinity to OAA and a low inhibition by malate (Table 5), which, however, differs from the mitochondrial OAA translocator by not being specifically inhibited by phthalonate. From its properties the chloroplastic OAA translocator is suited to catalyze a rapid malate-oxaloacetate shuttle. Since the chloroplast malate dehydrogenase is NADP specific, and the redox potential of the NADPH/NADP couple in the chloroplast stroma is much more negative than the redox potential of the cytosolic NADH/NAD couple (10), an uncontrolled shuttle would drain redox equivalents from the chloroplasts. Evidence has been presented that the chloroplast malate dehydrogenase is under redox control, in being active only when the NADPH/NADP ratio is high (23). The enzyme therefore appears to have the function of a valve, in allowing the efflux of redox equivalents by the malate-oxaloacetate shuttle from the chloroplasts only when the stromal NADPH/NADP is high. In this way, excessive reducing power from photosynthesis may be made available for peroxisomal reduction (Fig. 5A). Alternatively, chloroplasts could deliver redox equivalents to the cytosol by triosephosphate-phosphoglycerate shuttle (Fig. 5B). Such a delivery of redox equivalents from the chloroplasts would allow the redox equivalent of NADH generated by mitochondrial glycine oxidation to remain available for mitochondrial electron transport. Thus, on balance, excessive redox equivalents, formed in the chloroplasts by photosynthetic electron transport, could be supplied as fuel for mitochondrial oxidative phosphorylation in the light.

Table 5. OXALOACETATE CONVERSION BY SPINACH CHLOROPLASTS IN THE LIGHT IN THE ABSENCE OF NH_4Cl.
Data from ref. (9).

Oxaloacetate	Km	0.009	mM
	Vmax	52	µmol/mg Chl per h
Inhibitors:			
Malate	Ki	1.4	mM
2-Oxoglutarate	Ki	0.42	mM

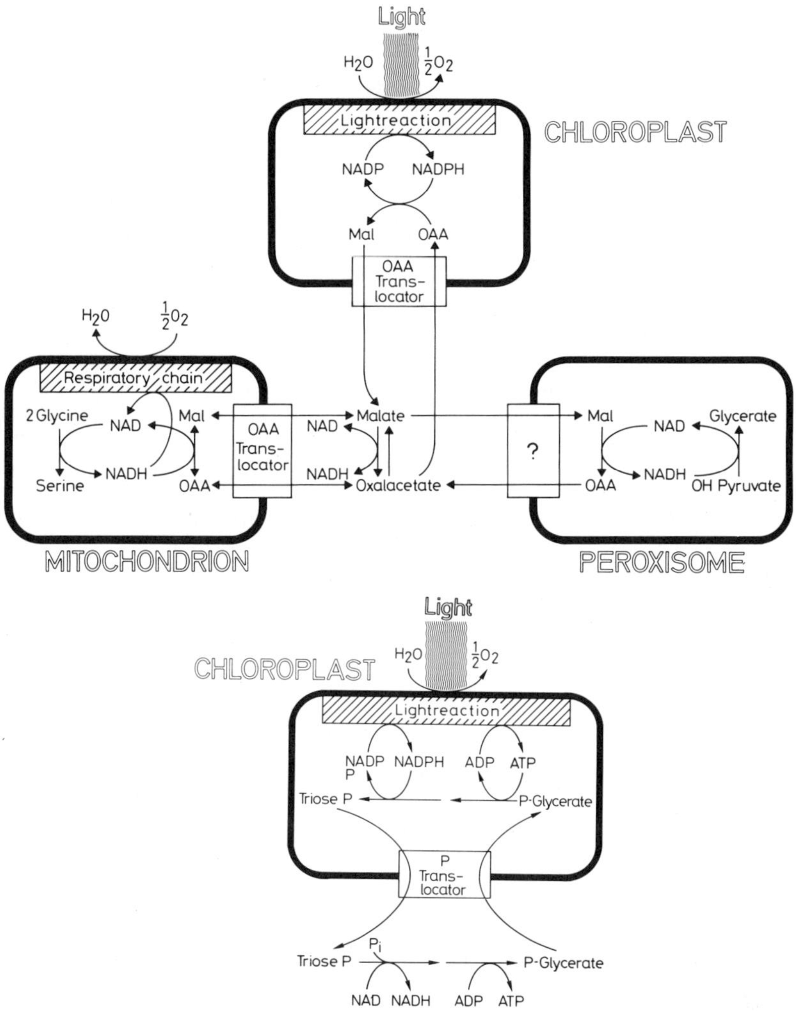

Fig. 5. Transfer of redox equivalents by malate-oxaloacetate shuttle or by triosephosphate-phosphoglycerate shuttle.

On the function of mitochondrial ATP synthesis for the metabolism of a plant cell in the light

The role of mitochondrial ATP synthesis in light metabolism of a green plant cell is not clear. It has been often assumed in the past, that during illumination the cytosol is supplied with ATP from the chloroplasts e.g. via the triosephosphate-phosphoglycerate shuttle, and that mitochondrial oxidative phosphorylation would be primarily important for the generation of ATP in the dark period.

The following consideration may demonstrate that the interplay of photosynthetic and mitochondrial electron transport could be an efficient way to generate ATP from light energy. In cyclic photophosphorylation, according to a widely accepted stoichiometry, 2 quanta of light yield 0.66 molecules of ATP, whereas in noncyclic electron transport, 1.3 ATP and 1 NADPH are generated per 4 quanta. If the redox equivalents of this NADPH were made available for mitochondrial electron transport, as discussed before, with an ATP/O of about 2.5, another 2.5 molecules of ATP could be formed. Such a cooperation of chloroplast noncyclic electron transport with mitochondrial oxidative phosphorylation could therefore yield about 3.8 ATP per 4 quanta of light, which would be a three times higher quantum efficiency than in cyclic photophosphorylation.

One might therefore expect that a plant may utilize such an efficient device, especially for supplying ATP to the cytosol. This notion is suported by the results of an experiment shown in Table 6. In this experiment, oligomycin is used as a specific inhibitor to suppress mitochondrial oxidative phosphorylation (for ref. see (24)). When oligomycin, in a concentration which completely inhibited oxidative phosphorylation by isolated pea leaf mitochondria, was added to intact barley leaf protoplasts, CO_2 dependent O_2 evolution was inhibited by 38%. When these protoplasts, in which photosynthesis was inhibited by oligomycin, were disrupted by forcing them through a 5 µm nylon net, photosynthesis was found to be stimulated, back to the rate found without oligomycin in either intact or disrupted protoplasts. Obviously, oligomycin did not inhibit the photosynthesis of the chloroplasts released from the ruptured protoplasts. In other experiments with spinach chloroplasts, not shown here, 0.1 µg/ml oligomycin did not affect photosynthesis. These results clearly demonstrate that the observed inhibition of photosynthesis by oligomycin in intact protoplasts is not due to a direct effect of oligomycin on chloroplasts photosynthesis as such. The inhibition by oligomycin probably affects a reaction involved in the utilization of the photosynthate outside the chloropasts. It has been previously shown that in a leaf sucrose synthesis can be limiting the overall process of photosynthesis (22). It is therefore feasible that the inhibitory effect of oligomycin is caused by a decrease of sucrose synthesis due to ATP shortage in the cytosol. Although many details are still remaining to be elucidated, our results suggest that the inhibition of mitochondrial ATP synthesis in the light has a severe effect on the metabolic performance

Table 6. EVIDENCE FOR THE PARTICIPATION OF MITOCHONDRIAL ATP SYNTHESIS IN THE OVERALL PROCESS OF PHOTOSYNTHESIS IN BARLEY LEAF PROTOPLASTS, AS REVEALED FROM THE INHIBITORY EFFECT OF OLIGOMYCIN, A SPECIFIC INHIBITOR OF MITOCHONDRIAL OXIDATIVE PHOSPHORYLATION. Mean values from 4 different protoplast preparations $\pm$ standard deviation.

	Photosynthesis µmol O_2 evolved/mg Chl per h	
	control	Oligomycin added (0.1 µg/ml)
Protoplasts, intact	139 ± 9	87 ± 5
Protoplasts, disrupted	132 ± 11	127 ± 17

of a plant leaf cell. It may be therefore concluded that mitochondrial oxidative phosphorylation plays an important contribution in suppying the cytosol of a plant leaf cell with ATP also in the light.

With a photosynthesis rate of 200 µmol CO_2/mg Chl per h and all photosynthate being converted to sucrose, the cytosolic ATP demand for sucrose synthesis alone would be 17 µmoles/mg Chl per h, with a ATP/O of 2.5 equivalent to a respiration rate of about 7 µatom O/mg Chl per h. On top of this comes the ATP demand for other reactions, e.g. active transport and biosynthesis processes. In leaf discs from spinach, as used in our experiments, average rates of dark respiration were 11 µatom O/mg Chl per h (25). If one assumes that this respiration may be necessary for the supply of the cytosol in the light, a total respiration rate of about 20 μatom O per mg chl per h might be required for meeting the ATP demand of the cytosol. These considerations suggest that in a plant cell, a considerable part of the NADH generated in the mitochondria during glycine oxidation may be oxidized by the respiratory chain in order to supply the cytosol with ATP.

Conclusion

The data and the discussions of the present paper show that green plant cells have the capacity for an efficient distribution of redox equivalents by an interplay of the various subcellular compartments. It can be concluded that in the light mitochondria are able to generate ATP from the oxidation of substrate hydrogen, delivered from photosynthetic electron transport. In this way, mitochondrial respiration proceeding independently from the operation of the Krebs cycle, may play an important role in the energy metabolism of an illuminated plant cell.

REFERENCES

1. J. W. Anderson and C. M. House, Polarographic study of oxaloacetate reduction by isolated pea chloroplasts,Plant Physiol. 65, 1058-1063 (1979)
2. D. A. Day and J. T. Wiskich, Effect of phthalonic acid on respiration and metabolite transport in higher plant mitochondria, Arch. Biochem. Biophys. 211, 100-107 (1981)
3. D. A. Day, C. L. D. Jenkins and M. D. Hatch, Isolation and properties of functional mesophyll protoplasts and chloroplasts from Zea mays, Aust. J. Plant Physiol. 8, 21-29 (1981)
4. R. Douce, Organisation of the respiratory chain, this Symposium
5. H. Ebbighausen, J. Chen and H. W. Heldt, Oxaloacetate translocator in plant mitochondria, Biochim. Biophys. Acta 810, 184-199 (1985)
6. C. D. Farquhar and S. von Caemmerer, Modelling of photosynthetic response to environmental conditions, in: "Encyclopedia of Plant Physiol.", New Series, O. L. Lange, P. S. Nobel, C. B. Osmond and H. Ziegler, eds., Springer, Berlin, Vol. 12, pp. 549-587 (1982)
7. N. A: Frigerio and H. A. Harburg, Preparation and some properties of crystalline glycolic acid oxidase, J. Biol. Chem. 231, 246-249 (1958)
8. M. D. Hatch and C. B. Osmond, Compartmentation and transport in C4 photosynthesis, in: "Encyclopedia of Plant Physiol.", New Series, C. R. Stocking and U. Heber, eds., Springer, Berlin, Vol. 3, pp. 144-154 (1976)
9. M. D. Hatch, L. Dröscher and H. W. Heldt, A specific translocator for oxaloacetate in chloroplasts. FEBS Letts. 178, 15-19 (1984)
10. U. Heber and K. A. Santarius, Biochim. Biophys. Acta 109, 390-408 (1965)
11. U. Heber and G. H. Krause, Transfer of carbon, phosphate energy, and reducing equivalents across the chloroplast envelope, in: "Photosynthesis and Photorespiration", M. D. Hatch et al. eds., Wiley-Interscience, New York, pp. 218-225 (1971)
12. H. W. Heldt and U. I. Flügge, Subcellular transport of metabolite, in: "The Biochemistry of Plants", D. D. Davies, ed., Academic Press, New York, Vol. 2, Supplement in print
13. E.-P. Journet, M. Neuburger and R. Douce, Role of glutamate-oxaloacetate transaminase and malate dehydrogenase in the regeneration of NAD^+ for glycine oxidation by spinach leaf mitochondria. Plant Physiol. 67, 467-469 (1981)
14. A. J. Keys, I. F. Bird, M. J. Cornelius, P. J. Lea, R. M. Wallgrove and B. J. Miflin, Photorespiratory nitrogen cycle. Nature 275, 741-743 (1978)
15. Z. Liang, C. Yu and A. H. C. Huang, Isolation of spinach leaf peroxisomes in 0.25 M sucrose solution by Percoll density gradient centrifugation, Plant Physiol. 70, 1210-1212 (1982)
16. R. McC. Lilley, H. Ebbighausen and H. W. Heldt, The simultaneous determination of carbon dioxide release and oxygen uptake in suspensions of

plant leaf mitochondria oxidizing glycine, Plant Physiol. submitted
17. A. L. Moore, C. Jackson, J. Dench, P. Morris and D. O. Hall The relationship of glycine decarboxylase to the phosphorylation potential, Plant Physiol. Supplement 63, 613 (1979)
18. D. J. Oliver and G. H. Walker, Characterization of the transport of oxaloacetate by pea leaf mitochondria, Plant Physiol. 76, 409-413 (1984)
19. M. O. Proudlove and A. L. Moore, Metabolite fluxes across the inner membrane of plant mitochondria - inhibition by phthalonic acid, Planta 160, 407-414 (1984)
20. D. W. Rehfeld and N. E. Tolbert, Aminotransferases in peroxisomes from spinach leaves, J. Biol. Chem. 247, 4803-4811 (1972)
21. V. Rocha and I. P Ting, Malate dehydrogenases of leaf tissue from Spinacea oleracea: Properties of three isoenzymes, Arch. Biochem. Biophys. 147, 114-122 (1971)
22. T. D. Sharkey, M. Stitt, D. Heineke, R. Gerhardt, K. Raschke and H. W. Heldt, Limitation of photosynthesis by carbon metabolism. II. O_2-insensitive CO_2 uptake results from limitation of triose phosphate utilization, Plant Physiol. in print (1986)
23. R. Scheibe, D. Wagenpfeil and J Fischer, J., NADP-malate dehydrogenase activity during photosynthesis in illuminated spinach chloroplasts, J. Plant Physiol. 124, 103-110 (1986)
24. E. C. Slater and H. F. Ter Welle, Applications of oligomycin and related inhibitors in bioenergetics, in "Inhibitors, Tools in Cell Research", Th. Bücher and H. Sies, eds., Springer, Berlin, pp. 258-278 (1969)
25. M. Stitt, W. Wirtz, R. Gerhardt, H. W. Heldt, C. Spencer, D. Walker and C. Foyer, A comparative study of metabolite levels in plant leaf material in the dark, Planta 166, 354-364 (1985)
26. N. E. Tolbert, Photorespiration, in "The Biochemistry of Plants", D. D. Davies, ed., Academic Press, New York, Vol. 2, pp. 487-525 (1980)
27. K. C. Woo and C. B. Osmond, Glycine decarboxylation in mitochondria isolated from spinach leaves, Aust. J. Plant Physiol. 3, 771-785 (1976)
28. R. K. Yamazaki and N. E. Tolbert, Malate dehydrogenase in leaf peroxisomes, Biochim. Biophys. Acta 178, 11-20 (1969)
29. W. C. Zschoche and J. P. Ting, Malate dehydrogenase of pisum sativum Plant Physiol. 51, 1076-1081 (1973)

MECHANISM AND CONTROL OF MITOCHONDRIAL GLYCINE OXIDATION

David J. Oliver and Joan L. Walker

Department of Bacteriology and Biochemistry
University of Idaho
Moscow, Idaho 83843 U.S.A.

The glycine oxidation system of plant mitochondria provides an excellent model for a range of problems in biochemistry and molecular biology. Glycine oxidation is the major decarboxylative step in photorespiration and is, therefore, an important reaction in the photosynthetic carbon oxidation cycle. The reaction occurs within the mitochondria, thereby providing a photosynthetic reaction within this organelle. Glycine oxidation produces NADH and competes with citric acid cycle enzymes for available NAD, thus is important in the regulation of "dark" respiration. The enzyme that catalyzes the decarboxylation of glycine is composed of four independent proteins that react sequentially with enzyme-bound reaction intermediates and provides an interesting model for protein - protein interactions.

As with several other photosynthetic enzymes, the synthesis of glycine decarboxylase is light-dependent, presumably under phytochrome control. The four component proteins are coded in the nucleus, translated on 80S cytosolic ribosomes and imported into the mitochondria. This enzyme complex, therefore, provides a potential model for molecular biologists interested in light induction, coordinated synthesis of multicomponent complexes, and polypeptide targeting to organelles.

Given the myriad of important biological questions that could be approached with studies of glycine oxidation in plant mitochondria, it is surprising that so little is known about the process. To a large degree this has resulted from the perceived difficulties in working with the enzymes involved. Until quite recently it was assumed that the enzyme complex was tightly associated with the inner mitochondrial membrane (10) and, therefore, in possession of all the unpleasant characteristics of membrane bound proteins. This membrane association was largely deduced indirectly from the inability to solubilize the enzyme in an active form. In fact, it now appears that the enzyme is not tightly bound to the membrane and that it can be solubilized and purified in a straightforward manner (13). With the advent of techniques for purifying and assaying the enzyme, this whole system now becomes available to detailed molecular studies. The purpose of this paper is to summarize our current understanding of how this enzyme complex works in the hopes of attracting increased attention to this problem.

MECHANISM OF THE GLYCINE DECARBOXYLASE COMPLEX

In order to understand the experimental problems previously encountered, one needs to appreciate the mechanism of the reaction (Figure 1). Glycine is first bound _via_ a Schiff base with the pyridoxal 5-phosphate cofactor of the 98 kD P-protein (contains _P_yridoxal 5-phosphate). Simultaneous with the loss of the carboxyl carbon of glycine as CO_2, the remaining methylamine moiety is passed to the bound lipoamide cofactor of the 15 kD H-protein (carries _H_ydrogen). At this point the carbon skeleton has been oxidized and the lipoamide reduced. The H-protein now dissociates from the P-protein and binds to the 45 kD T-protein (binds _T_etrahydrofolate). The T-protein catalyzes the transfer of the methylene group of the lipoamide-bound methylamine intermediate to form 5', 10'-methylenetetrahydrofolate with the simultaneous release of the amino nitrogen as ammonia. That finishes the carbon and nitrogen reactions, but for continued enzyme function, the reduced dihydrolipoamide of the H-protein needs be reoxidized. The H-protein, therefore, dissociates from the T-protein and binds to the 59 kD L-protein (_L_ipoamide dehydrogenase). The L-protein catalyzes the reoxidation of the lipoamide cofactor of the H-protein with the reduction of NAD (7).

$$H_2N\text{-}CH_2\text{-}COOH \;+\; PLP\text{-}CH{=}O \;\text{-->}\; PLP\text{-}CH{=}N\text{-}CH_2\text{-}COOH \qquad (1)$$

$$PLP\text{-}CH{=}N\text{-}CH_2\text{-}COOH + Lip\begin{matrix}\diagup S\\ \vert\\ \diagdown S\end{matrix} \;\text{-->}\; PLP\text{-}CH{=}O + Lip\begin{matrix}\diagup S\text{-}CH_2\text{-}NH_2\\ \diagdown SH\end{matrix} + CO_2 \qquad (2)$$

$$Lip\begin{matrix}\diagup S\text{-}CH_2\text{-}NH_2\\ \diagdown SH\end{matrix} \;+\; THF \;\text{-->}\; Lip\begin{matrix}\diagup SH\\ \diagdown SH\end{matrix} \;+\; \text{methylene THF} \;+\; NH_4^+ \qquad (3)$$

$$Lip\begin{matrix}\diagup SH\\ \diagdown SH\end{matrix} \;+\; FAD \;\text{-->}\; Lip\begin{matrix}\diagup S\\ \vert\\ \diagdown S\end{matrix} \;+\; FADH_2 \qquad (4)$$

$$FADH_2 \;+ NAD^+ \;\text{-->}\; FAD \;+ NADH \;+\; H^+ \qquad (5)$$

Figure 1. Mechanism of the Glycine Decarboxylase Multienzyme Complex. The following abbreviations are used: PLP-CH=O, pyridoxal 5-phosphate bound to the P-protein; Lip, lipoamide cofactor of the H-protein; THF, tetrahydrofolate; FAD, flavin adenine dinucleotide of the L-protein. Reactions (1) and (2) are catalyzed by the P-protein. The T-protein catalyzes reaction (3). Reactions (4) and (5) are catalyzed by L-protein. Note the central role of the H-protein which acts as a cosubstrate in reactions (2), (3) and (4).

It is important to note the unique role of the H-protein. In its three forms, lipoamide, methylamine lipoamide, and dihydrolipoamide, it behaves as a substrate in the bisubstrate reactions catalyzed by the P, T, and L proteins, respectively. As a cosubstrate it shows Michaelis-Menten saturation kinetics for each reaction (5,6,15). The Km values for the H-protein as a substrate for the P-protein were 2.3 uM for the pea leaf enzyme (15) and 3.4 uM for the chicken liver enzyme (5). The H-protein exists at a concentration of less than 1 mM in the mitochondrial matrix, but when the mitochondria are broken to release the enzyme, it becomes diluted several hundred-times. The enzyme preparation that results is so strongly limited by the H-protein concentration that the resulting activity is nearly impossible to measure.

An additional difficulty with the study of this enzyme complex is that glycine decarboxylase is not a stable supramolecular complex like pyruvate dehydrogenase. Following dilution of the mitochondrial matrix, H-protein binds to the other proteins thus making it possible to isolate P_2-H_2 complexes and L_1-H_1 complexes by chromatography and immunoprecipitation (15). While we have reported the isolation of a high molecular complex that contained all the subunits of glycine decarboxylase and the associated enzyme serine hydroxymethyltransferase, we have been unable to stabilize this complex *in vitro* sufficiently to allow detailed analysis and purification (13). It is possible that these low affinity interactions between the large subunits become important at the high protein concentrations found in the matrix, but if these interactions exist they are not strong enough to keep the complex stable following dilution. Elsewhere in this volume Neuberger et al. report their ability to stabilize this multienzyme complex *in vitro*.

ISOLATION AND PURIFICATION OF GLYCINE DECARBOXYLASE

Glycine decarboxylase activity and/or the associated proteins have been solubilized from leaf mitochondria by acetone extraction of mitochondrial lipids (12), by sonication of mitochondrial pellets (3), and by repeated freeze thawing (Neuburger et al. - this volume). Because of the ease of handling large amounts of mitochondria, we have continued to isolate the enzyme from acetone powders of pea leaf mitochondria. The individual subunits of the enzyme are separated initially from the acetone-extracted water-soluble proteins by ammonium sulfate fractionation. Following this initial separation, the individual subunit proteins can be purified by conventional techniques as summarized in Figure 2. The purification is greatly facilitated because the H-protein

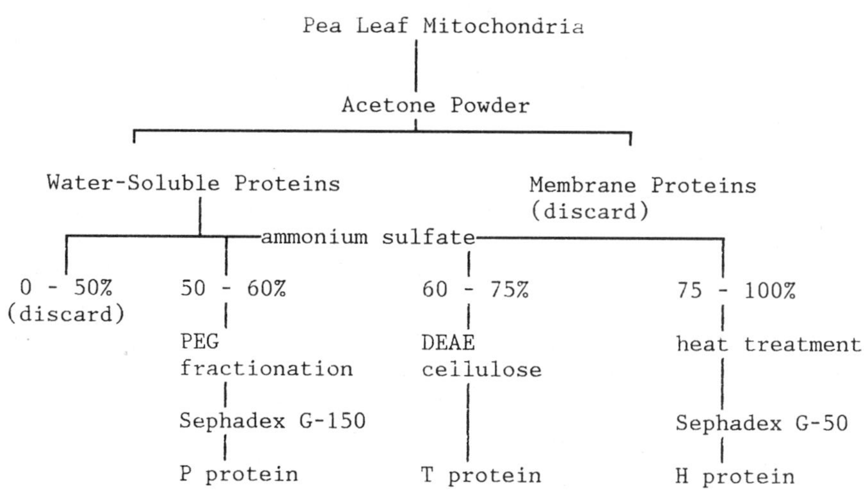

Figure 2. Flow chart for the purification of the P, T, and H proteins of the glycine decarboxylase complex.

is stable for 5 min at 90°C and the T-protein is a basic protein and is positively charged at pH 7.0. Both characteristics allow nearly complete purification in one or two steps. Following these procedures, the three separated enzymes are over 90% pure as indicated by SDS-PAGE (15).

The L-protein has not been purified because we currently have no techniques to differentiate it from the other potential isoforms of lipoamide dehydrogenase found in mitochondria. The protein was identified, however, through the use of monoclonal antibodies (15). The L-protein was immunologically indistinguishable from the other lipoamide dehydrogenases (including that of pyruvate dehydrogenase) found in pea leaf mitochondria possibility indicating that a common lipoamide dehydrogenase is shared between all mitochondrial multienzyme complexes. Furthermore, the commercial yeast enzyme was shown to substitute in the glycine decarboxylase reaction.

In terms of subunit size, specific activity, and reactions catalyzed, the glycine decarboxylase complex from pea leaves appears nearly identical to that isolated from animal tissues. The one obvious difference is the molecular weight of the T-protein which is 45 kD from plants but 33 kD from rat liver and 38 kD from avian liver (6,9). Despite the altered molecular masses, the proteins from different sources are similar in the reactions catalyzed and the observation that they are all basic proteins (6).

Following the separation and purification of the individual component enzymes of the complex, activity can be reconstituted by recombining the P, T, H, and L preparations. This complex will catalyze the release of the carboxyl carbon and amino group from glycine as well as forming 5', 10' methylene tetrahydrofolate and NADH in a ratio of 1:1:1:1. The preparation will also catalyze the exchange of CO_2 into the carbonyl carbon of glycine. This glycine-bicarbonate exchange reaction requires only the P and H proteins (15).

METABOLIC REGULATION OF GLYCINE DECARBOXYLASE

Because of the pivotal position of glycine decarboxylase in photosynthetic as well as mitochondrial metabolism, we were interested in determining if the activity of this complex is under metabolic control. The active glycine decarboxylase was reconstituted and a wide range of potential metabolic modifiers were tested for their effect on its activity. Included were adenylates, nucleotides, organic acids of the citric acid and photosynthetic carbon oxidation cycles, amino acids and enzyme cofactors (Table I). Only three compounds gave strong inhibition of the enzyme. These were glyoxylate, a peroxisomal intermediate in the glycolate pathway (14), pyridoxal 5-phosphate, an essential coenzyme for the reaction (14), and serine, the endproduct of mitochondrial glycine metabolism (11).

Glyoxylate contains a reactive aldehyde group that has been shown to react with essential lysine residues in a number of enzymes including ribulose bisphosphate carboxylase (2). Our data suggests that this is the mechanism of the inhibition observed (14). Given the high concentrations of glyoxylate needed to inhibit the enzyme (7 to 9 mM for 50% inhibition) and the peroxisomal localization of this compound, it is unlikely that this inhibition is physiologically important.

The inhibition by pyridoxal 5-phosphate was specific for an available aldehyde and the 5-phosphate on the cofactor as shown by the lack of inhibition caused by pyridoxine 5-phosphate, pyridoxamine 5-

Table I: Effect of Common Metabolites on $^{14}CO_2$ Release from [1-^{14}C] Glycine by Isolated Glycine Decarboxylase

Compounds With Little or No Effect		
Amino Acids	Cofactors	Organic Acids
alanine	acetyl CoA	aconitate
aminoacetonitrile	ADP	citrate
2-aminobutyrate	AMP	fumarate
hydroxyproline	ATP	glycolate
isoleucine	CoASH	isocitrate
leucine	NAD	ketoglutarate
proline	NADP	malate
threonine	pyridoxal	oxaloacetate
valine	pyridoxamine-P	pyruvate
	pyridoxine-P	succinate
	succinyl CoA	
	thiamine pyrophosphate	

Compounds that Inhibit and Concentration to Give 50% Inhibition	
Glyoxylate	8 mM
Pyridoxal 5-P	0.15 mM
Serine	3 mM

phosphate and pyridoxal (14). The inhibition caused by pyridoxal 5-phosphate likely results from Schiff base formation with the same lysyl residue that reacts with glyoxylate.

The inhibition caused by serine is more likely to be of physiological consequence. Serine is competitive with glycine in the glycine decarboxylation reaction and has a Ki of 3 mM compared to the Km for glycine of 2.4 mM (Figure 3). The conversion of glycine to serine within the mitochondrial matrix should result in a low glycine/serine ratio at the active site of glycine decarboxylase. In order to prevent a strong inhibition of glycine metabolism in the mitochondria, it is essential that some mechanism exist to either concentrate glycine in or remove serine from the matrix. This role would be filled by a glycine-serine exchanger.

Interestingly, serine is not an effective inhibitor of glycine oxidation by isolated mitochondria. While 10 mM serine can inhibit the oxidation of 10 mM glycine by the isolated enzyme over 90%, glycine oxidation by mitochondria is inhibited less than 20%. The amount of inhibition with mitochondria can be increased by disrupting the electrochemical gradient across the inner mitochondrial by the addition of ADP, uncouplers, or oxaloacetate. This suggests that the inner mitochondrial membrane provides an effective barrier to serine diffusion and that serine is actively excluded from the mitochondrial matrix by the electrochemical gradient. The inhibition of glycine metabolism by serine may serve to decrease the oxidation of glycine under nonphotorespiratory conditions, either in the dark or in nongreen tissues.

Current data suggests that inhibition of glycine decarboxylase by serine under photorespiratory conditions is prevented by a glycine-serine exchange mechanism. This transport protein exchanges cytosolic glycine for L-serine in the mitochondrial matrix. The ratio of two glycine entering the mitochondria per one serine leaving as required for the photorespiratory carbon oxidation cycle necessitates an additional transport function. A glycine-OH^- exchanger has been postulated to fill this role (see Oliver - this volume).

All four subunits of the glycine decarboxylase complex are encoded on nuclear genes (as opposed to the mitochondrial genome) and translated on 80S cytosolic ribosomes (16). This has been shown by 1) the lack of formation of these proteins during translation experiments with isolated pea leaf mitochondria, 2) the sensitivity of the synthesis of these proteins to cycloheximide but not chloramphenicol in pea shoots, and 3) the isolation of polyadenylated mRNA for the P and T proteins (16). Mitochondrial mRNAs are apparently not polyadenylated (8).

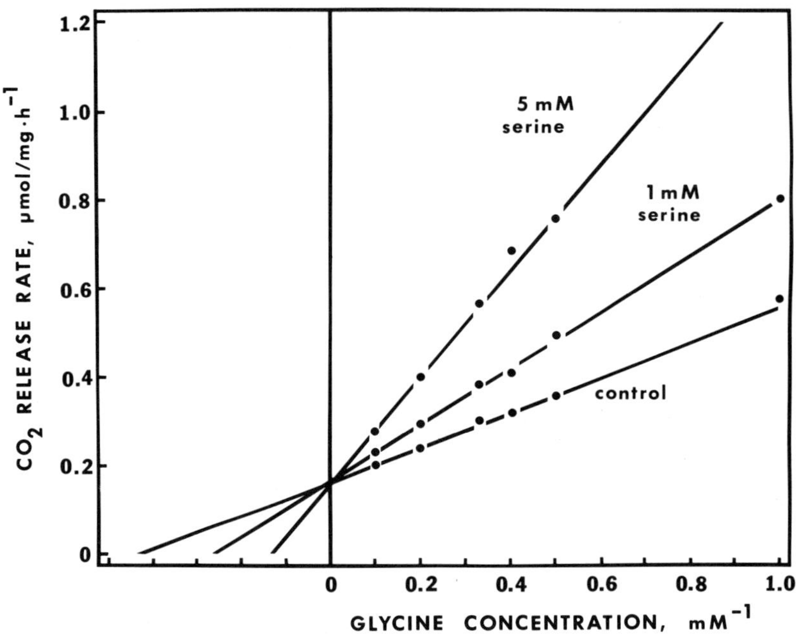

Figure 3. Kinetic analysis of the effects of serine on the rate of glycine decarboxylation by isolated, purified, and reconstituted glycine decarboxylase from pea leaf mitochondria.

Mitochondria isolated from etiolated pea leaves have very low rates of glycine oxidation. Following illumination of the plants the rates increase about 10- to 20-fold (1,3,16). This increase resulted from similar increases in activities of the four component enzymes (Table II). The most dramatic changes were observed in the levels of the P and T proteins, both of which increased at least 10-fold. The increases in the H and L proteins were smaller. Lipoamide dehydrogenases (L-protein) are used by several multienzyme complexes and the smaller proportional increase results from the relatively high level of this enzyme found in etiolated tissues. The H-protein has a unique role in the reaction in that it shuttles reaction intermediates between the active sites of the other three enzymes. In order to maintain even limited rates of glycine decarboxylase in the dark, this protein need to be present at moderate concentrations (substrate levels as opposed to catalytic levels). Thus, the H-protein concentration does not need to change as dramatically following greening of the tissue.

Table II. Changes in the Activity and Concentration of Glycine Decarboxylase and Component Enzymes from Mitochondria Isolated from Etiolated and Green Pea Leaves Measured by Different Methods

Preparation	Change Light/Dark	Method
mitochondria	13.1	activity
acetone powder	12.6	activity
P protein	12.0	activity
	14.5	immunodetection
H protein	4.2	activity
	5.0	SDS-PAGE
T protein	9.6	activity
	10.8	immunodetection
L protein	3.6	activity

The increases in the activities of the component enzymes do not result from the activation of preexisting enzymes but rather from the *de novo* synthesis of the four proteins as demonstrated by the appearance of new bands on SDS-PAGE (3) or by Western blotting and immunodetection (16). Preliminary experiments with the *in vitro* translation of mRNAs isolated from etiolated and green pea leaves indicates that the messengers for the P protein exist at much higher concentrations in light-treated plants (Kim and Oliver - unpublished). This suggests that the photocontrol of the expression of this protein is at the level of transcription. Similar results have been obtained for a range of other photosynthetic enzymes (4).

CONCLUSIONS

Glycine decarboxylase is an ideal model system for studying a range of problems in biochemistry and molecular biology. The isolation, purification, and initial characterization of the component enzymes now makes this system more available to plant scientists. Important questions that remain to be answered include:

1. What is the degree of association of the individual enzymes with the inner mitochondrial membrane?
2. Do the four component enzymes form a true complex within the matrix?
3. What are the differences and similarities of the sites on the P, T, and L protein which bind the H protein?
4. Is there a physical association between the glycine decarboxylase enzymes and serine hydroxymethyltransferase?
5. How is the feedback inhibition by serine prevented?
6. Are there differences between the lipoamide dehydrogenases (L protein) used by glycine decarboxylase and the other enzyme complexes within the mitochondrial matrix?

7. Is the photocontrol of glycine decarboxylase mediated by phytochrome?

8. Are the component enzymes of the complex synthesized as precursors?

9. How are the newly synthesized proteins directed into the mitochondria?

10. How is the differential expression of the component enzymes accomplished?

The answers to these questions will lead to important insights into the molecular mechanism by which plants regulate and balance various aspects of their metabolism.

ACKNOWLEDGEMENTS

This work was funded by a grant from the Metabolic Biology Division of the National Science Foundation and the Idaho Agricultural Experiment Station.

REFERENCES

1. Arron, G.P. and G.E. Edwards (1980) Plant Sci. Lett. 18: 229-235

2. Cook, C.M., M. Spellman, N.E. Tolbert, C.D. Stringer, and F.C. Hartman (1985) Arch. Biochem. Biophys. 240: 402-412

3. Day, D.A., M. Neuburger, and R. Douce (1985) Aust. J. Plant Physiol. 12: 219-228

4. Fluhr, R., C. Kuhlemeier, F. Nagy, and N.H. Chua (1986) Science 232: 1106-1112

5. Fujiwara, K. and Y. Motokawa (1983) J. Biol. Chem. 258: 8156-8162

6. Fujiwara, K., K. Okamura-Ikeda, and Y. Motokawa (1984) J. Biol.Chem. 259: 10664-10668

7. Kikuchi, G. (1973) Mol. Cel. Biochem. 1: 169-187

8. Leaver, C.J. and M.W. Gray (1982) Annu. Rev. Plant Physiol. 33: 373-402

9. Motokawa, Y. and G. Kikuchi (1974) Arch. Biochem. Biophys. 164: 624-633

10. Moore, A.L., C. Jackson, B. Halliwell, J.E. Dench and D.O. Hall (1977) Biochem. Biophys. Res. Commun. 78: 483-491

11. Oliver, D.J. and G. Sarojini (1986) Proceedings of the Seventh International Congress on Photosynthesis in press

12. Sarojini, G. and D.J. Oliver (1983) Plant Physiol. 72: 194-199

13. Sarojini, G. and D.J. Oliver (1984) Advances in Photosynthesis Research (C. Sybesma, ed.) Martinus Nijhoff/Dr. Junk Publishers, The Hague, Vol. 3: 553-556

14. Sarojini, G. and D.J. Oliver (1986) Proceedings of the Seventh International Congress on Photosynthesis in press

15. Walker, J.L. and D.J. Oliver (1986) J. Biol. Chem. 261: 2214-2221

16. Walker, J.L. and D.J. Oliver (1986) Arch. Biochem. Biophys. 248: 626-638

HIGHER PLANT MITOCHONDRIAL PYRUVATE DEHYDROGENASE COMPLEXES

Jan A. Miernyk [1 2], Barbara J. Rapp[2], Nancy R. David[2], and Douglas D. Randall[2]

[1]Seed Biosynthesis Research Unit
Northern Regional Research Center USDA, ARS
Peoria, IL 61604 USA

[2]Biochemistry Department
University of Missouri
Columbia, MO 65211 USA

INTRODUCTION

The pyruvate dehydrogenase complex (PDC) is a very large, multi-enzyme structure which catalyzes the oxidative decarboxylation of pyruvate, formation of a hydroxyethyl-TPP intermediate, subsequent reductive acetylation of a lipoic acid cofactor, and ultimately acetylation of Coenzyme-A. NAD then serves as the final electron acceptor for the internal redox coupling which is a portion of the reaction cycle. Divalent cations and FAD are additional required cofactors. Multiple copies of each of the enzymatic components of the PDC are present including, pyruvate dehydrogenase (EC 1.2.4.1)(E_1,PDH), lipoyl transacetylase (EC 2.3.1.12)(E_2,LTA), and the flavoprotein lipoyl dehydrogenase (EC 1.6.4.3)((E_3,LAD). Mitochondrial PDCs contain two types of regulatory enzymes, PDH-kinase (EC 2.7.1.99) and phospho-PDH-phosphatase (EC 3.1.3.43). Recently, an additional component protein "X" has been identified. It has been suggested that this protein may play some role in anchoring PDH-kinase to the core of the complex (37). The overall reaction and the relationship among the enzymes which make up the complex are presented in Figure 1.

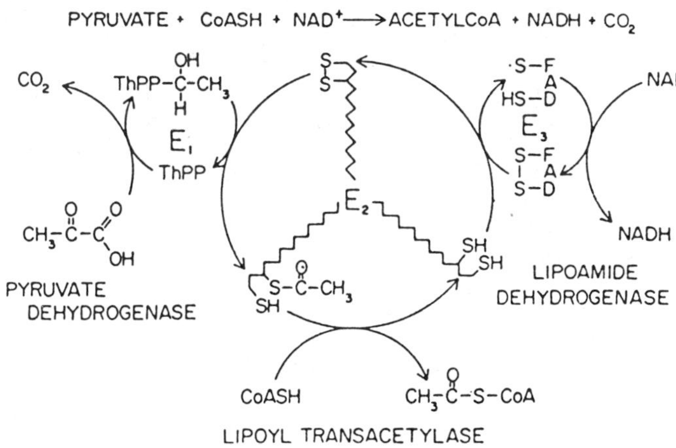

Figure 1. The reaction sequence of the pyruvate dehydrogenase complex.

SUBUNIT STRUCTURE

Bacterial PDH is a homodimer with a native M_r of 192,000 and subunit M_r of 96,000 (22). Mammalian PDH is an $\alpha_2\beta_2$-heterotetramer with a native M_r of 154,000 and subunit M_r values of 41,000 and 36,000 (22,37). It appears that plant mitochondrial PDH is also an $\alpha_2\beta_2$-heterotetramer. Native M_r values have not yet been determined for any plant PDH, but subunit M_rs are 43,000 and 37,000.

Bacterial LTA is a monomer with a native and subunit M_r values of 70,000 (22). Mammalian LTA is a monomer with native and subunit M_r values of 52,000 (1,37). Lipoyl transacetylase has not yet been purified from any plant source and nothing is known about its structure or subunit mass. There is a subunit band with an M_r of 56,000 on Western blots, which gives a positive signal when probed with antibodies to whole complex, and which may be plant LTA.

Lipoyl dehydrogenase is the only individual component of plant mitochondrial PDCs which has been purified and characterized to any significant extent. Matthews and Reed partially purified LAD from whole spinach leaves (8), Yanagawa and Egami purified the enzyme to homogeneity from etiolated shoot mitochondria of *Asparagus officinalis* L. (36), and we have purified and characterized LAD from mitochondria and chloroplasts of green pea shoots (manuscript in preparation). Monoclonal antibodies have been prepared against LAD from pea leaf mitochondria and used in the characterization of the glycine decarboxylase complex (35). Polyclonal antibodies have been prepared against lipoyl dehydrogenase from pea shoot mitochondria, porcine heart mitochondria and *E. coli*. When used to probe Western blots these antibodies revealed substantial homology among the LADs from all three sources (J.A. Miernyk, unpublished).

The number of species of LAD has been a point of some controversy. Lipoyl dehydrogenase is a component of the PDC, the 2-oxoglutarate dehydrogenase complex (OGDC), and the glycine decarboxylase complex (GDC) of bacteria; the PDC, the branched chain 2-oxo-acid dehydrogenase complex (BCOADC) and OGDC of mammalian mitochondria; the PDC, OGDC and GDC of plant mitochondria; and the PDC of plant plastids. It may also exist in the free, uncomplexed form (7). Guest and associates have unequivocally demonstrated that *E. coli* has a single structural gene encoding LAD (6,30), the transcription and translation of which provides components for both the PDC and OGDC. Biochemical and genetic studies have yielded similar results for the LADs of PDC, OGDC, and BCOADC of yeast and mammalian mitochondria (5,6,9). It has been reported that LAD can be removed from several of the multi-enzyme complexes, including cauliflower mitochondrial OGDC (13), and then full complex activity can be reconstituted using LAD from a heterologous source. In contrast, the LAD of bacterial GDCs is a protein distinct from that associated with the other multi-enzyme complexes (29). When preparations from plant mitochondria were analyzed by Western blotting using either polyclonal, or monoclonal antibodies (35), only a single subunit gives a positive signal. Additionally, when preparations from pea mitochondria were analyzed by chromatofocusing or ion-filtration chromatography, only a single symmetrical peak of activity was obtained (J.A. Miernyk, unpublished). It appears then, with the exception of the bacterial glycine carboxylase complex, that there is a single species of LAD which serves as a component of each of the multi-enzyme complexes. If this is true, a highly responsive transcriptional regulation would seem to be necessary as each of the complexes is expressed at different levels during different physiological states. Clearly this is one area of plant mitochondria which deserves considerable additional study.

Lipoyl dehydrogenases from all sources are homodimers. Native M_r values are 98,000 to 112,000 for bacterial (22,29), 112,000 for mammalian (6,22) and 92,000 to 102,000 for plant (8,35,36) LADs. Subunit M_r values are 49,000 (GDC) and 55,000 (PDC, OGDC); 55,000-56,000; and 58,000-59,000, respectively.

Bacterial PDCs are not regulated by reversible phosphorylation and do not have either PDH-kinase or phosphatase. Mammalian PDH-kinase is an αβ-heterodimer with a native M_r of 95,000 and subunit M_r values of 48,000 and 45,000 (31,37). The larger polypeptide appears to be the catalytic subunit while it has been suggested that the smaller may be a regulatory subunit (37). The PDH-kinase has not been purified from any plant source and nothing is known concerning its subunit structure. We have recently approached this question using photo-affinity labeling. When pea mitochondrial PDC was incubated with 8-azido-ATP, photolyzed, and then immunoprecipitated, a single subunit with an M_r of 53,000 was labeled. We feel that this is the catalytic subunit of the kinase (manuscript in preparation.

Mammalian phospho-PDH-phosphatase is an αβ-heterodimer with native M_r of 150,000 and subunit M_r values of 50,000 and 97,000 (32). Phosphatase activity is associated with the smaller subunit, while the large subunit is a flavoprotein of potential regulatory significance. We have recently reported some of the catalytic capabilities of pea mitochondrial phospho-PDH phosphatase (12), but this component has not yet been purified from any plant source. Western blots of pea mitochondrial PDC, when probed with antibodies to whole complex, show a positive signal at M_r 98,000 which may be a subunit of the phosphatase (2). This band is very faint on Western blots of spinach mitochondrial PDC. Phospho-PDH-phosphatase is the least tightly associated component of the PDC. When the pea mitochondrial complex was subjected to several cycles of pelleting in the ultracentrifuge and then resuspension, this high molecular weight component was lost, as was any detectable phosphatase activity (J.A. Miernyk, unpublished).

STRUCTURE OF THE COMPLEX

The *E. coli* PDC has an M_r of 4,600,000 (22). The complex is composed of a core of 24 LTA subunits arranged in an octahedral structure. Associated with this core are 12 PDH and 6 LAD components. The mammalian mitochondrial PDC has an M_r of 7,000,000 (22). The complex is composed of a core of 60 LTA subunits arranged in an icosahedral structure. Associated with this core are 30 PDH and 10 LAD components, and approximately 5 each of PDH-kinase and phospho-PDH-phosphatase (22). The PDC has been purified from a plant source only a single time (14,27), and very little is known about the structure of the complete complex. By rate-zonal sedimentation on linear glycerol gradients, it was demonstrated that the M_r value of the complex from broccoli floral bud mitochondria was intermediate between those of the *E. coli* and mammalian complexes (14). In Figure 2 we present a Western blot of partially purified spinach mitochondrial PDC.

KINETIC MECHANISM

The reaction catalyzed by the PDC is the sum of the reactions of three distinct enzymes acting in concert. For kinetics analysis, the overall sequence has been considered as a single irreversible reaction. Based upon the results of product inhibition and substrate interaction studies, Cleland (3) has described a hexa-uni ping-pong mechanism for the PDC (Figure 3).

This reaction mechanism predicts that acetyl-CoA will be a linear competitive inhibitor with respect to CoA, and NADH will be a linear competitive inhibitor with respect to NAD. All other substrate-product combinations should exhibit uncompetitive inhibition. Most data from kinetic analyses of various plant mitochondrial PDCs have been in agreement with the proposed mechanism (20,25,28). An exception was reported by Thompson et al. (33) who found that for pea mitochondrial PDC inhibition between acetyl-CoA and CoA was noncompetitive. We recently reexamined the kinetic properties of pea mitochondrial PDC, and observed the predicted competitive inhibition pattern (11). The disparate results reported by Thompson et al. and ourselves could have resulted from differences in the concentrations of the other substrates and cofactors in the reaction mixtures, and possibly substrate-dependent ligand-induced steric interactions among components of the complex.

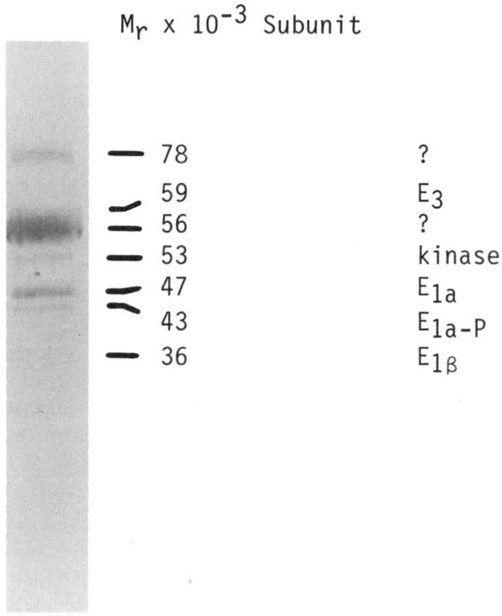

Figure 2. The subunit composition of the spinach mitochondrial pyruvate dehydrogenase complex. Components were separated by electrophoresis on a 12.5% acrylamide SDS-gel, transferred to a nitrocellulose membrane, then probed with antibodies to whole broccoli mitochondrial complex.

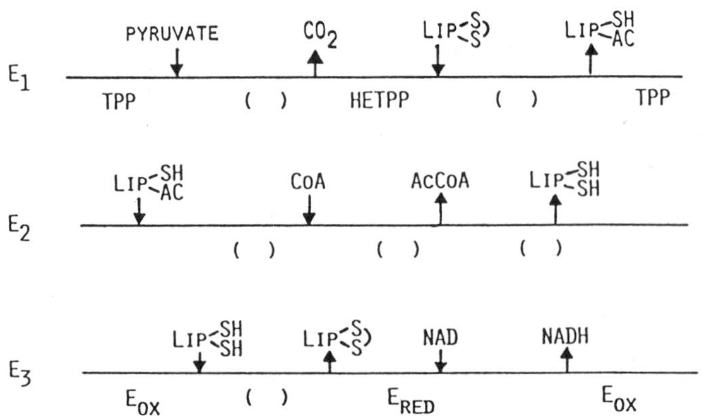

Figure 3. The reaction mechanism of the pyruvate dehydrogenase complex.

REGULATION

The PDC occupies a key position in intermediary metabolism, at the juncture between carbohydrate catabolism and the energy transducing and anabolic reactions of the mitochondria. This position makes the PDC an ideal candidate for regulation, and in all cases where this reaction has been examined it is displaced far from equilibrium. We are currently aware of several types of both fine and coarse control mechanisms. It is likely that the *in situ* activity state of the PDC reflects in part each of these mechanisms.

All PDCs are sensitive to product inhibition by acetyl-CoA and NADH (22,37). The inhibition constants calculated for plant mitochondrial PDCs are in the low micromolar range for both acetyl-CoA and NADH, although it is thought that NADH may be of greater regulatory importance. The K_i for NADH is lower than the K_m for NAD, and increasing the mole percent of NADH in reaction mixtures results in a logarithmic increase in inhibition (4,27,34).

Metabolite inhibition is a major contributor to the overall regulation of bacterial PDC activity (22). While this aspect has not been extensively studied with plant mitochondrial PDCs, we have recently reported that ADP is inhibitory at levels which occur within the mitochondrial matrix (11). This inhibition is linear competitive with respect of CoA. It is then possible that metabolite effects could contribute to the overall control of PDC activity.

Thiamine pyrophosphate is a cofactor required for both PDC activity and for the oxidation of pyruvate by isolated, intact mitochondria. Rubin *et al*. (28) observed a lag in the initial activity of broccoli mitochondrial PDC which was related to the concentration of Mg-TPP in the reaction mixture. They demonstrated that TPP must bind to PDH prior to the binding of pyruvate, and that TPP could rapidly dissociate from PDH resulting in a loss of activity. This rapid loss of activity could offer a mechanism for the fine control of PDC activity. The rapid dissociation of Mg-TPP from broccoli mitochondrial PDC is not, however, characteristic of all plant mitochondrial PDCs (e.g 11).

The most spectacular regulatory feature of mammalian mitochondrial PDCs is the reversible inactivation by phosphorylation (18,23). Pyruvate dehydrogenase kinase, an intrinsic component of the PDC, catalyzes the multi-site phosphorylation of the $E_{1\alpha}$ - subunit of PDH. Nearly complete inactivation of the complex corresponds to phosphorylation of the serine residue at site one. Additional serine residues at sites two and three are subsequently phosphorylated. The phosphorylation of sites two and three may play some role in controlling the reactivation of the complex by phospho-PDH-phosphatase.

The mitochondrial PDC was the first plant enzyme that was identified as regulated by covalent modification (15,16). We now know that plant mitochondrial PDCs also contain an intrinsic PDH-kinase, and that incubation with gamma-^{32}P-ATP rsults in inactivation of the complex and concommitant incorporation of ^{32}P into the $E_{1\alpha}$-subunit of PDH (10,17,19,21,26,34) Figure 4. While many features of this reversible phosphorylation are similar between mammalian and plant mitochondrial PDCs, there are also distinctions. Probably the most outstanding of these is the amino acid residue phosphorylated, threonine in all instances thus far examined (manuscript in preparation). This and other differences probably reflect the structural divergence between plant and animal PDHs.

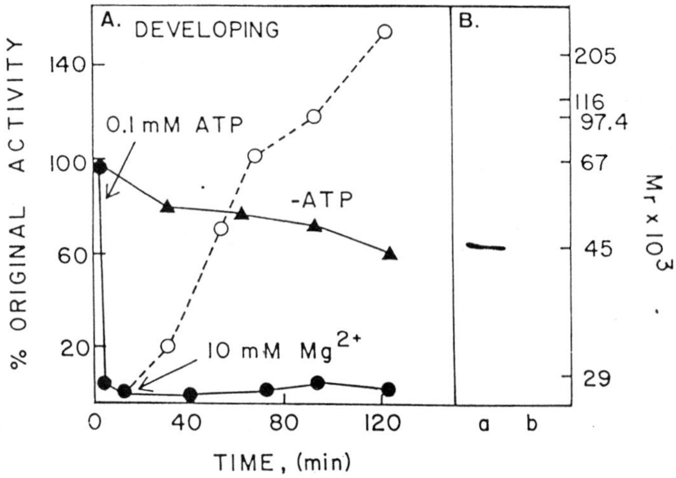

Figure 4. A. The ATP-dependent inactivation and magnesium-dependent reactivation of the developing castor oil seed mitochondrial pyruvate dehydrogenase complex. B. Autoradiography of SDS-polyacrylamide gel electrophoresis gel following [γ^{32}-P]-ATP treatment of the PDC in A. The PDC is immune precipitated before dissociation and electrophoresis. (a) 10 min after addition of [^{32}P]-ATP. (b) after 100 min in presence 10 mM $MgCl_2$following [^{32}P]-ATP.

The steady-state phosphorylation status of mitochondrial PDC is representative of the regulatory cascade involving both PDH-kinase and phospho-PDH-phosphatase (10). Many mitochondrial metabolites, including NADH, acetyl-CoA, ADP, Krebs cycle intermediates, and monovalent cations can inhibit the activity of PDH-kinase in vitro. It seems likely that some of these are also of in vivo impact. Divalent cations appear to be the most important controllers of phospho-PDH-phosphatase activity, with magnesium and manganese activating and calcium inhibiting. The inhibition by calcium is in contrast to the calcium stimulation seen with mammalian phosphatases (23,37). Inhibition occurs at micromolar levels of calcium, in the presence of millimolar magnesium, and we have recently suggested that the effect could be mediated by calmodulin (12). This proposal is in need of further study.

We recently verified that plant mitochondrial PDCs are regulated by phosphorylation in vivo. There are distinct differences in the phosphorylation state of PDC when mitochondria are isolated from developing versus germinating castor oil seeds (21). Also, there are organ specific differences in phosphorylation in light grown pea seedlings (10).

Results from developmental studies suggest that there are mechanisms responsible for the long-term coarse control of mitochondrial PDC activity (e.g. 20). We have no information with regard to transcriptional or translational controls of the synthesis of mitochondrial PDC, and very little information with regard to post-translational controls of synthesis or degradation. In many instances phosphorylation of regulatory enzyme plays two roles. It has an immediate effect upon catalytic activity and also may serve as a signal for proteolytic removal. Preliminary results suggest that this second mechanism is not a major feature in the control of pea mitochondrial PDC activity (10).

SUMMARY

The importance of understanding the PDC to an understanding of the metabolism of microbes and mammals has been adequately perceived, and there have been many studies on the PDC from these organisms. Unfortunately, this has not been true in the plant sciences. Mitochondrial PDC has been examined in only a handful of plants and tissues; broccoli and cauliflower florets, pea and spinach leaves, and castor oil seed endosperm. Very little is known about the protein chemistry of the complex itself or any of its components, and there is virtually no knowledge of the synthesis and assembly of the complex. In this chapter we have attempted to briefly outline what research progress has been made in the study of plant mitochondrial PDCs, and to point out areas where there is need for additional study. Hopefully when the proceedings of the next mitochondrial congress are published, the answers to many of todays questions will be well in hand.

ACKNOWLEDGEMENT

This work was supported by NSF Grant DMB-8506473.

REFERENCES

1. Bleile, D.M., Hackert, M.L., Pettit, F.H., and Reed, L.J., 1981, Subunit structure of dihydrolipoyl transacetylase component of pyruvate dehydrogenase complex from bovine heart, J. Biol. Chem. 256:514-519.
2. Camp, P.J., and Randall, D.D., 1985, Purification and characterization of the pea chloroplast pyruvate dehydrogenase complex. A source of acetyl-CoA and NADH for fatty acid biosynthesis, Plant Physiol. 77:571-577.
3. Cleland, W.W., 1973, Derivation of rate equations for multisite ping-pong mechanisms with ping-pong reactions at one or more sites, J. Biol. Chem. 248:8353-8355.
4. Crompton, M., and Laties, G.G., 1971, The regulatory function of potato pyruvate dehydrogenase, Arch. Biochm. Bioplys. 143:143-150.
5. Dickinson, J.R., Roy, D.J., and Dawes, I.W., 1986, A mutation affecting lipoamide dehydrogenase, pyruvate dehydrogenase and 2-oxoglutarate dehydrogenase activities in *Saccharomyces cerevisiae*, Mol. Gen. Genet. 204:102-107.
6. Guest, J.R., 1978, Aspects of the molecular biology of lipoamide dehydrogenase, Adv. Neurol. 21:219-244.
7. Lusty, C.J., and Singer, T.P., 1964, Lipoyl dehydrogenase. Free and complexed forms in mammalian mitochondria, J. Biol. Chem. 239:3733-3742.
8. Matthews, J., and Reed, L.J., 1963, Purification and properties of a dihydrolipoic dehydrogenase from *Spinacia oleracea*, J. Biol. Chem. 238:1869-1876.
9. Matuda, S., Shirahama, T., Saheki, T., Miura, S., and Mori, M., 1983, Purification and immunochemical studies of pyruvate dehydrogenase complex from rat heart, and cell-free synthesis of lipoamide dehydrogenase, a component of the complex, Biochem. Biophys. Acta 741:86-93.
10. Miernyk, J.A., Camp, P.J. and Randall, D.D., 1985, Regulation of plant pyruvate dehydrogenase complexes, Curr. Topics Plant Biochem. Physiol. 4:175-190.
11. Miernyk, J.A., and Randall, D.D., 1986, Some kinetic and regulatory properties of the pea mitochondrial pyruvate dehydrogenase complex, Plant Physiol. In press.

12. Miernyk, J.A., and Randall, D.D., 1986, Some properties of pea mitochondrial phospho-pyruvate dehydrogenase-phosphatase, Plant Physiol. In press.
13. Poulsen, L.L., and Wedding, R.T., 1970, Purification and properties of the α-ketoglutarate dehydrogenase complex of cauliflower mitochondria, J. Biol. Chem. 245:5709-5717.
14. Randall, D.D., 1982, Pyruvate dehydrogenase complex from broccoli and cauliflower mitochondria, Meth. Enzymol. 89:408-414.
15. Randall, D.D., and Rubin, P.M., 1977, Plant pyruvate dehydrogenase complex, II. ATP-dependent inactivation and phosphorylation, Plant Physiol. 59:1-3.
16. Randall, D.D., Rubin, P.M. and Fenko, M., 1977, Plant pyruvate dehydrogense complex purification, characterization and regulation by metabolities and phosphorylation, Biochim. Biophys. Acta 485:336-349.
17. Randall, D.D., Williams, M. and Rapp, B.J., 1981, Phosphorylation-dephosphorylation of pyruvate dehydrogenase complex from pea leaf mitochondria, Arch. Biochem. Biophys. 207:437-444.
18. Randle, P.J., 1981, Phosphorylation-dephosphorylation cycles and the regulation of fuel selection in mammals, Curr. Topics Cell. Reg. 18:107-129.
19. Rao, K.P., and Randall, D.D., 1980, Plant pyruvate dehydrogenase complex: Inactivation and reactivation by phosphorylation and dephosphorylation, Arch. Biochem. Biophys. 200:461-466.
20. Rapp, B.J., and Randall, D.D., 1980, Pyruvate dehydrogenase complex from germinating castor bean endosperm, Plant Physiol. 65:314-318.
21. Rapp, B.J., Miernyk, J.A., and Randall, D.D., 1986, Pyruvate dehydrogenase complexes from Ricinus communis endosperm, J. Plant Physiol. In press.
22. Reed, L.J., 1974, Multienzyme complexes, Acc. Chem. Res. 7:40-46.
23. Reed, L.J., 1981, Regulation of mammalian pyruvate dehydrogenase complex by a phosphorylation-dephosphorylation cycle, Curr. Topics Cell. Reg. 18:95-106.
24. Reid, E.E., Lyttle, C.R, Canvin, D.T., and Dennis, D.T., 1975, Pyruvate dehydrogenase complex activity in proplastids and mitochondria of developing castor bean endosperm, Biochem, Biophys. Res., Commun. 62:42-47.
25. Reid, E.E., Thompson, P., Lyttle, C.R., and Dennis, D.T., 1977, Pyruvate dehydrogenase complex from higher plant mitochondria and proplastids, Plant Physiol. 59:842-848.
26. Rubin, P.M., and Randall, D.D., 1977, Regulation of plant pyruvate dehydrogenase complex by phosphorylation, Plant Physiol. 60:34-39.
27. Rubin, P.M., and Randall, D.D., 1977, Purification and characterization of pyruvate dehydrogenase complex from broccoli floral buds, Arch. Biochem. Biophys. 178:342-349.
28. Rubin, P.M., Zahler, W.L., and Randall, D.D., 1978, Plant pyruvate dehydrogenase complex: Analysis of the kinetic properties and metabolite regulation, Arch. Biochem. Biophys. 188:70-77.
29. Sokatch, J.R., and Burns, G., 1984, Oxidation of glycine by Pseudomonas pudita requires a specific lipoamide dehydrogenase, Arch. Biochem. Biophys. 228:660-666.
30. Stephens, P.E., Lewis, H.M., Darlison, M.G., and Guest, J.R., 1983, Nucleotide sequence of the lipoamide dehydrogenase gene of Escherichia coli K12, Eur. J. Biochem. 135:519-527.
31. Stepp, L.R., Pettit, F.H., Yaeman, S.J., and Reed, L.J., 1983, Purification and properties of pyruvate dehydrogenase kinase from bovine kidney, J. Biol. Chem. 258:9454-9458.

32. Teague, W.M., Pettit, F.H., Wu, T-L., Silberman, S.R., and Reed, L.J., 1982, Purification and properties of pyruvate dehydrogenase phosphatase from bovine heart and kidney, Biochemistry 21:5585-5592.

33. Thompson, P., Reid, E.E., Lyttle, C.R., and Dennis, D.T., 1977, Pyruvate dehydrogenase complex from higher plant mitochondria and proplastids: Kinetics, Plant Plysiol. 59:849-853.

34. Thompson, P., Reid, E.E., Lyttle, C.R., and Dennis, D.T., 1977, Pyruvate dehydrogenase complex from higher plant mitochondria and proplastids: Regulation, Plant Physiol. 59:854-858.

35. Walker, J.L., and Oliver, D.J., 1986, Glycine decarboxylase multi-enzyme complex. Purification and partial characterization from pea leaf mitochondria, J. Biol. Chem. 261:2214-2221.

36. Yanagawa, H., and Egami, F., 1976, Asparagusate dehydrogenases and lipoyl dehydrogenase from asparagus mitochondria. Physical, chemical, and enzymatic properties, J. Biol. Chem. 251:3637-3644.

37. Yeaman, S.J., 1986, The mammalian 2-oxoacid dehydrogenases: a complex family, Trends Biochem. Sci. 11:293-296.

In vitro STUDIES ON THE INTERACTIONS IN MALATE METABOLISM BETWEEN MITOCHONDRIA AND CHLOROPLASTS FROM CAM-PERFORMING *Kalanchoe blossfeldiana*

Véronique Adam and Pierre Rustin

Laboratoire de Biologie Végétale IV (CNRS, UA 1180)
Université Pierre et Marie Curie, Paris, France

INTRODUCTION

Crassulacean acid metabolism (CAM) is characterized by a nocturnal fixation of atmospheric CO_2 into malate through the conjugate activities of phosphoenolpyruvate carboxylase and malate dehydrogenase. This leads to an accumulation of malate in the vacuole. In the early morning hours, malate is decarboxylated by the cytosolic $NADP^+$-dependent and the mitochondrial NAD^+-dependent malic enzymes . During this last period, the simultaneous activation of the photosynthetic process and of the mitochondrial malate decarboxylation constitutes a potential source of interaction and cooperation between the two organelles. In particular, it has been established that the CO_2 released during malate decarboxylation can be recycled by the photosynthetic process .

The aim of this work was to study *in vitro* during mitochondrial malate decarboxylation the interactions between mitochondria and chloroplasts extracted from the leaves of the CAM-performing plant, *Kalanchoe blossfeldiana*.

MATERIALS AND METHODS

Cuttings of *Kalanchoe blossfeldiana* L. cv. Tom Thumb were grown at the Phytotron in Gif-sur-Yvette under long days of 16 h (fluorescent and incandescent lamps, 100 W m^{-2}, 27°C/17°C day/night temperature). After two months, the plants were transferred to short days (9 h) during 30 to 40 days to present full CAM activity .

Mitochondria and chloroplasts were prepared as previously described . Rapidity of the whole extraction procedure and the use of a high ratio of extraction medium to plant material was found an absolute requirement in order to overcome the deleterious effects of the high concentrations of tanins and malate present in the leaf vacuoles .

O_2 evolution and consumption were measured polarographically with a Clark oxygen electrode at 27°C. The medium for the CO_2-dependent O_2 evolution by chloroplasts contained 350 mM sorbitol, 50 mM HEPES (pH 7,5), 2 mM EDTA, 1 mM $MgCl_2$, 1 mM $MnCl_2$, 1 g l^{-1} BSA and catalase (10 µkat). Illumination was provided by a cold light source (Schott KL 1500) at saturating light (4,5 mE $s^{-1}m^{-2}$). The medium for malate oxidation by mitochondria was 300 mM mannitol, 10 mM phosphate buffer (pH 7,3), 10 mM KCl, 5 mM $MgCl_2$, 50 mM malate, 2.3 mM thiamine pyrophosphate (TPP), 0.57 mM

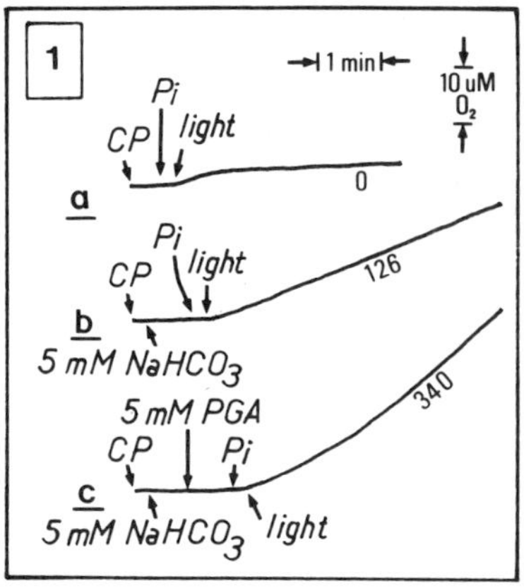

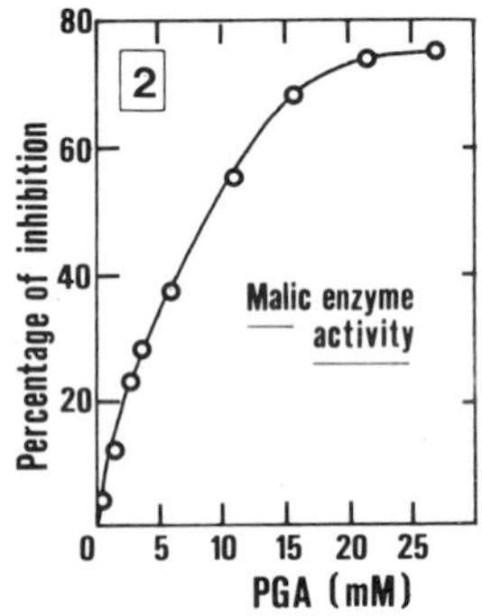

Fig. 1. CO_2-dependent O_2 evolution by K. blossfeldiana chloroplasts. **a**, control; **b**, 5 mM $NaHCO_3$; **c**, 5 mM $NaHCO_3$ and 5 mM PGA. The experimental conditions are described in Materials and Methods. Numbers along the traces are nmol O_2 min^{-1} mg^{-1} Chl.

Fig. 2. Effect of PGA on the NAD^+-dependent malic enzyme activity.

coenzyme A, 1.8 mM NAD^+, 0.28 mM ADP, 20 mM EGTA and 1 g l^{-1}BSA. Pyruvate and oxaloacetate were measured enzymatically .

RESULTS AND DISCUSSION

Chloroplast preparations obtained were 95% intact on the basis of their ability to reduce ferricyanure. Such a high degree of intactness was required to measure the O_2 evolution by illuminated chloroplasts without interferences with photooxidative processes occurring with illuminated broken chloroplasts.

Activity of chloroplasts

Figure 1 (trace a) shows that chloroplasts are unable to evolve O_2 in the absence of added $NaHCO_3$ as a source of CO_2. Adding $NaHCO_3$ stimulates O_2 uptake markedly (Fig. 1, trace b). However, the highest rates of CO_2-dependent O_2 evolution are obtained in the presence of added phosphoglycerate (PGA) (Fig. 1, trace c). This O_2 evolution depends on bicarbonate availability on the one hand, and on the furniture of intermediates of the Calvin cycle on the other hand. Beside the intactness of the chloroplast preparation, these two factors control the rate of CO_2-dependent O_2 evolution.

Activity of mitochondria

Mitochondria actively oxidize malate in the presence of ADP (state 3) (Tab. I). The production of pyruvate and oxaloacetate establishes the participation of both malate dehydrogenase and malic enzyme to this oxidation. In the presence of ATP (state 4) (Tab. I), the production of pyruvate accounts for all the O_2 uptake. Malic enzyme is then the only mitochondrial enzyme involved in malate oxidation and produces 2 mol of pyruvate and 2 mol of CO_2 per mol of oxygen consumed.

Table I. Oxygen uptake and enzymatic measurement of the products of malate oxidation by K. blossfeldiana mitohondria at pH 7.3 in the presence of arsenite (4 mM). Enzymatic determination of the products were made after 1 min incubation of the mitochondria (1.4 mg ml^{-1}) in the presence of the indicated reagents.

Conditions	O_2 uptake (natom min^{-1} mg^{-1}protein)	Substrate oxidation (nmol min^{-1} mg^{-1}protein)	
		Pyruvate	Oxaloacetate
20 mM malate + 200 µM ADP	97	60	36
20 mM malate + 200 µM ATP	33	37	0
20 mM malate + 200 µM ATP + 20 mM PGA	4	4	0

In vitro cooperation between organelles

Interaction between mitochondria, chloroplasts and their respective substrates, PGA, $NaHCO_3$, malate and ADP were investigated.

It is shown that mM concentrations of PGA affect the oxidation of malate by isolated mitochondria (Tab. I). The oxygen uptake under state 4 conditions appears to be significantly inhibited. It is also shown (Fig. 2) that the activity of solubilized malic enzyme is strongly inhibited by PGA. As malic enzyme plays a prominent role in malate oxidation under state 4 conditions (Tab. I), the inhibitory effect of PGA can be ascribed to its inhibitory effect on malic enzyme. It is noteworthy that the inhibition of malic enzyme by PGA can be released by adding manganese (not shown).

In order to avoid such an inhibition of mitochondrial activity by PGA,

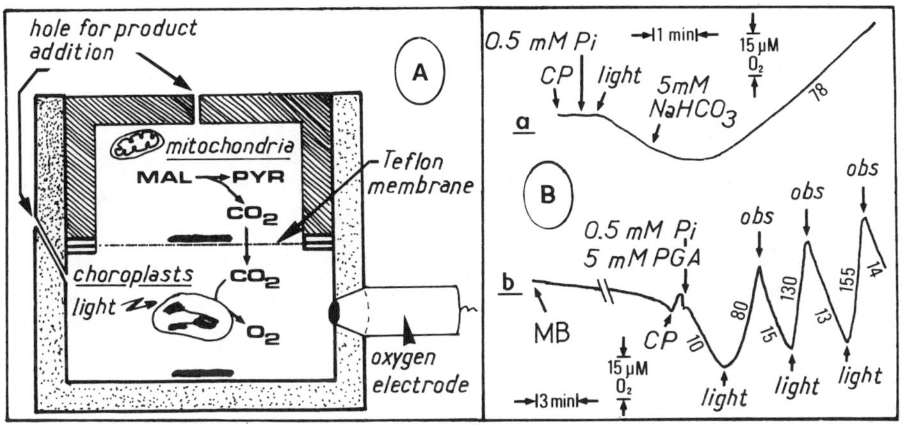

Fig. 3. A, Cell with two compartments separated by a gas-permeable membrane (Teflon). The lower compartment is fitted with a Clark O_2 electrode in order to measure O_2 concentration. B, In vitro reconstitution of the cooperation between chloroplasts and mitochondria occurring during the decarboxylation of malate. a, O_2 evolution by chloroplasts placed in the lower compartment. $NaHCO_3$ was added in the upper compartment. b, O_2 evolution by chloroplasts placed in the lower compartment using the CO_2 produced through malic enzyme activity of the mitochondria placed in the upper compartment 10 min before the addition of chloroplasts. The experimental conditions were described in Materials and Methods.

the reassociation experiments were carried out using a two-compartment cell (Fig. 3 A), the two compartments being separated by a gas-permeable teflon membrane.

When chloroplasts were added to the lower compartment fitted with a Clark oxygen electrode, O_2 evolution could be measured only after addition of $NaHCO_3$ to the upper compartment (in the absence of mitochondria) (Fig. 3 B, trace a).

The addition of $NaHCO_3$ in the upper compartment can be replaced by mitochondria decarboxylating malate through malic enzyme and releasing CO_2 (Fig. 3 B, trace b). Under these conditions, it is possible to measure an O_2 evolution by the chloroplasts in the lower compartment. In the first period, before light is on, O_2 uptake by the mitochondria placed in the upper compartment can be measured in the lower compartment. Then, when light is on, the O_2 evolution by chloroplasts, using the CO_2 released from malate by the mitochondria, can be measured. However, in this experiment a direct relationship between the measured rates of photosynthesis and respiration is presently difficult to establish since (a) mitochondria were allowed to produce CO_2 before light was on (in order to avoid the deleterious effect of light on chloroplasts in the absence of CO_2), and (b)a re-equilibration, not yet quantified, of the O_2 concentrations between the two compartments does occur.

CONCLUSION

Under suitable conditions, isolated chloroplasts and mitochondria, extracted from leaf tissues from the CAM-performing plant *K. blossfeldiana*, can perform *in vitro* the central step of the diurnal crassulacean acid metabolim.

Such a model of organelle cooperation will be further used to investigate the factors (substrates, cofactors, adenylates, experimental conditions) affecting the nature and the degree of the interactions between chloroplasts and mitochondria in CAM-plant cells.

REFERENCES

1. M. Kluge, A. Fischer and J.C. Buchanan-Bollig, Metabolic control of CAM, *in*: "Crassulacean acid metabolism", I.P. Ting and M. Gibbs, ed., Waverly Press, Baltimore (1982).
2. P. Rustin and C. Queiroz-Claret, Changes in oxidative properties of *Kalanchoe blossfeldiana* leaf mitochondria during development of crassulacean metabolism. *Planta*, 164: 415 (1985).
3. V. Adam and P. Rustin, Isolation purification and photosynthetic activity of chloroplasts from *Kalanchoe blossfeldiana*, a CAM--performing plant, *Physiol. Vég.*, 24: 663 (1986).
4. R.W. Von Korff, Purity and stability of pyruvate and ketoplutarate. *Methods Enzymol.*, 26: 159 (1969).

REGULATION OF GLYCOLLATE METABOLISM IN MITOCHONDRIA ISOLATED FROM *Euglena gracilis* KLEBS STRAIN Z PRINGSHEIM

Michael J. Merrett

Plant and Microbial Metabolism Research Group
School of Biological Sciences
University College, Swansea, U.K. SA2 8PP

In higher plants the enzymes of the glycollate pathway, with the exception of the glycine to serine conversion, are located in the peroxisome whereas in *Euglena* two key enzymes of glycollate metabolism, glycollate dehydrogenase and glutamate-glyoxylate aminotransferase are present in both peroxisome type particles and mitochondria (1,2). Glycollate oxidation in isolated mitochondria is linked to phosphorylation the P:0 ratio being consistent with glycollate electrons entering the respiratory chain at the flavoprotein level (2).

By using trypsin to digest the pellicle, hypotonic shock to disrupt the spheroplasts and centrifugation on iso-osmotic Percoll-sorbitol gradients highly purified mitochondria can be obtained from phototrophic, organotrophic and mixotrophic cells. The protein: chlorophyll ratio in mitochondria from green cells always being greater than 10^3:1 (3). Glycollate dehydrogenase and glyoxylate reductase were present in mitochondria from all cell types showing that these are constitutive enzymes of *Euglena* mitochondria (Table I).

Table I Enzyme activities in mitochondria isolated from cells of *Euglena* grown under different conditions.

Enzyme activities (μmole/h/mg^{-1} protein)	Growth conditions		
	Photo-trophic	Mixo-trophic	Organo-trophic
Succinate dehydrogenase	0.6	1.2	1.3
Citrate synthase	1.9	2.9	3.2
Malate dehydrogenase	51.0	156.0	169.0
NADPH-glyoxylate reductase	3.6	9.7	16.0
Glycollate dehydrogenase	0.4	0.4	0.4
Lactate dehydrogenase	0.6	1.1	0.5

Mitochondria isolated on a Percoll gradient show good respiratory control rates. Rates of glycine oxidation were very low in mitochondria isolated from all cell types, the ratio of malate to glycine oxidation always being greater than 4:1 (Table 2).

Euglena mitochondria contain glycine decarboxylase and serine hydroxymethyltransferase but low rates of glycine oxidation by isolated mitochondria (Table II) make it unlikely that exogenous glycine provides the substrate for serine formation in the mitochondrion. Glycollate dehydrogenase and glutamate-glyoxylate aminotransferase in the mitochondrion will allow the formation of glycine from glycollate. The rate of glycollate dependent oxygen uptake by isolated mitochondria is more than doubled on the addition of glutamate (3).

The fate of glycollate carbon in the mitochondrion is determined by several factors. The oxidation of glycollate provides ATP and glyoxylate. Glyoxylate reductase, purified from _Euglena_ mitochondria (4) is specific for NADPH and has a Km for glyoxylate of 45 μM. Glutamate-glyoxylate aminotransferase, purified from _Euglena_ mitochondria (5) has a Km of 250μM for glyoxylate and 2.00 mM for glutamate. When NADPH is available glyoxylate can be reduced to glycollate whereas in the presence of glutamate glyoxylate can be converted to glycine and serine. (Fig I). It is probable that the availability of the amino-donor is the key factor in determining the fate of glycollate since in the presence of NH_4^+ over 50% of the carbon-14 from exogenous (1-^{14}C-glycollate) supplied to _Euglena_ is converted to amino-acids and protein whereas in the absence of NH_4^+ the bulk of the radioactivity is lost as $^{14}CO_2$ (6). Cells growing under normal conditions produce 0.4 μmoles glycollate/h/mg protein^{-1}. (7) so the activity of mitochondrial glycollate dehydrogenase 0.4 μmoles/h/mg protein^{-1} is adequate to metabolise the total glycollate produced under these conditions.

The transfer of cells to photorespiratory conditions results in the derepression of peroxisomal glycollate dehydrogenase which constitutes the major part of glycollate dehydrogenase in air-grown cells (8). In _Euglena_ serine-glyoxylate aminotransferase the enzyme catalysing the conversion of L-serine to hydroxypyruvate is located exclusively in the peroxisomes (1) as is hydroxypyruvate reductase (1) catalysing the reduction of hydroxypyruvate to glycerate. _Euglena_ peroxisomes can therefore effect the conversion of glycollate to glycine and serine to glycerate but unlike the situation in green leaves glycine formed in the peroxisome will not participate in mitochondrial metabolism. Since the mitochondrial pathway of glycollate metabolism is constitutive in _Euglena_ this pathway may be concerned with biosynthesis whereas glycollate metabolism outside the mitochondrion may be concerned with photorespiratory metabolism. Such a separation of biosynthetic and photorespiratory metabolism may be necessary in a unicellular organism such as _Euglena_ to allow cell-division and growth to continue under photorespiratory conditions when nitrogen is not limiting.

Table 2. Oxidative phosphorylation in mitochondria isolated from cells of _Euglena_ grown under different conditions

Growth conditions	Phototrophic		Mixotrophic		Organotrophic	
Substrate	Malate (10mM0	Glycine (20mM)	Malate (10mM)	Glycine (20mM)	Malate (10mM)	Glycine (20mM)
Oxygen uptake (state 3) μatom min^{-1} mg^{-1} protein	0.012	0.006				
Oxygen uptake (state 4) μatom min^{-1} mg^{-1} protein	0.015		0.012		0.009	
R.C.R.	1.5		2.3		3.0	
P:O ratio	2.5		2.8		2.7	
Ratio $\frac{\text{malate}}{\text{glycine}}$ oxidation		4.1		4.1		6.1

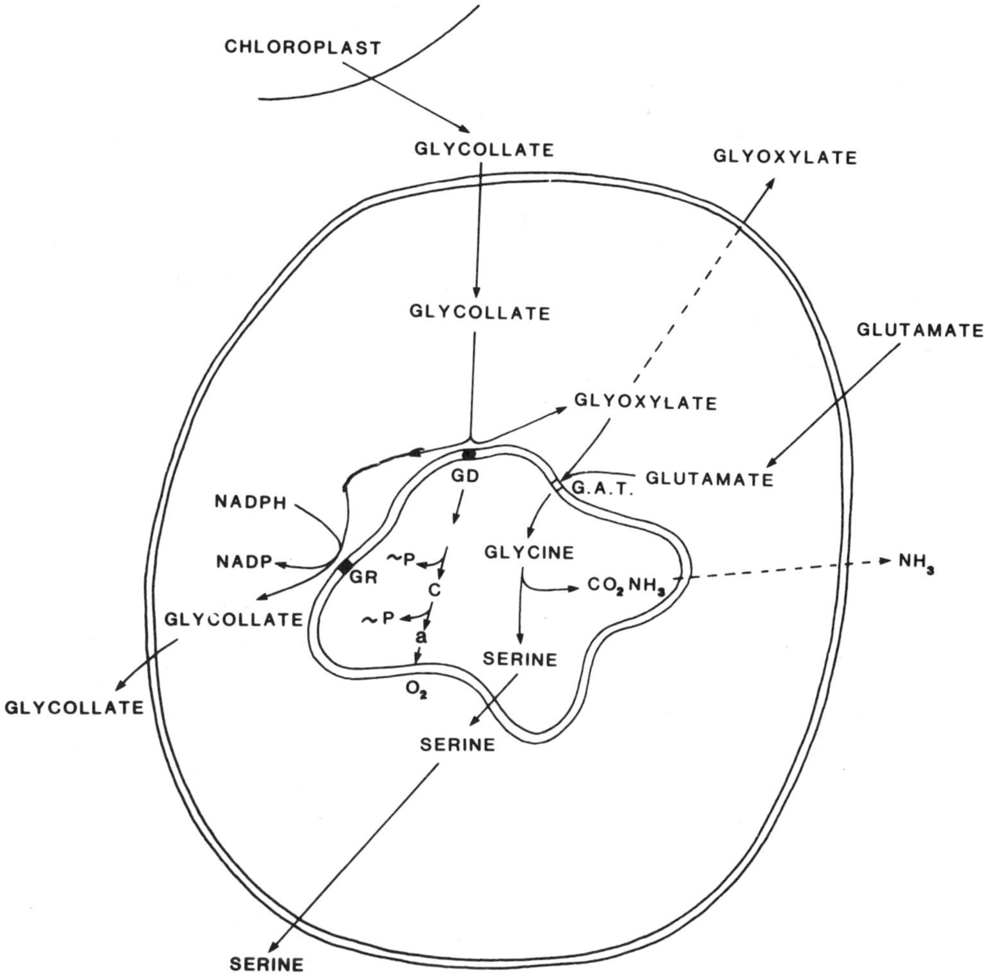

Fig. 1 Glycollate metabolism in *Euglena* mitochondria. GD, glycollate dehydrogenase. GR, glyoxylate reductase. G.A.T. glutamate-glyoxylate aminotransferase.

REFERENCES

1. N.Collins and M.J.Merrett. The localization of glycollate-pathway enzymes in *Euglena*. Biochem.J. 148:321 (1975).
2. N.Collins, R.H.Brown and M.J. Merrett. Oxidative phosphorylation during glycollate metabolism in mitochondria from phototrophic *Euglena gracilis*. Biochem.J. 150:373 (1975).
3. M. Fayyaz-Chadhary and M.J. Merrett. Glycollate pathway enzymes in mitochondria from phototrophic organotrophic and mixotrophic cells of *Euglena*. Planta 162:518 (1984).
4. A. Yokota, S. Haga and S. Kitaoka. Purification and some properties of glyoxylate reductase ($NADP^+$) and its functional location in mitochondria in Euglena gracilis z. Biochem J. 221:211 (1985).
5. A. Yokota, S. Suehiro and S. Kitaoka. Purification and some properties of mitochondrial glutamate: glyoxylate aminotransferase and the mechanism of its involvement in the glycolate pathway in *Euglena gracilis* z. Arch.Biochem.Biophys. 242:507 (1985).
6. A. Yokota, Y. Nakano and S. Kitaoka. Metabolism of glycolate in mitochondria of *Euglena gracilis*. Agric.Biol.Chem. 42:121 (1978).
7. G.A. Codd and M.J. Merrett. The regulation of glycolate metabolism in division synchronized cultures of *Euglena*. Plant Physiol 47:640 (1971).
8. A. Yokota, Y. Nakano and S. Kitaoka. Different effects of some growing conditions on glycolate dehydrogenase in mitochondria and microbodies in *Euglena gracilis*. Agric.Biol.Chem. 42:115 (1978)

A COMPARISON OF MITOCHONDRIA FROM SOYBEAN NODULES, ROOTS AND COTYLEDONS

David A. Day, G. Dean Price and Peter M. Gresshoff

Botany Department, Australian National University
Canberra, A.C.T., 201, Australia.

INTRODUCTION

Nitrogen fixation in the nodules of legume roots involves complex metabolic interactions between the plant host cytoplasm and the Rhizobium bacteroids within that cytoplasm. The plant provides carbon substrate to the bacteroid and in return receives NH_3 which is assimilated by the host. These reactions are energy demanding and it is probable that much of the required ATP is provided by the mitochondria of infected and uninfected cells. Mitochondria may also be involved in production of organic acids for bacteroid substrate supply and production of oxtoglutarate as an amine acceptor in the cytosol. The function of mitochondria in infected tissue is complicated by the very low O_2 tension within these tissues, which may be as low as 10-30 nM in soybeans (Sheehy et al 1985). This micro-aerobic environment is necessary for the function of nitrogenase, which is O_2 labile, and O_2 transport within infected cells is facilitated by leghaemoglobin. Mitochondrial electron transport may be restricted by such low O_2 concentrations; conversely, mitochondria may play an ancillary role in O_2 protection.

MATERIALS AND METHODS

Thirty to 50 g of nodules were harvested, washed in distilled water and stored on ice until utilized. Storage for up to 24h did not affect the activity of isolated mitochondria. Nodules were homogenized with a Phillips kitchen grinder (Model HR1194) in 300 ml of ice-cold grinding medium which contained 0.4 M sorbitol, 50 mM TES buffer, 2 mM EDTA, 10 mM KH_2PO_4, 30 mM ascorbate, 2% (w/v) PVP-40 and 1% (w/v) BSA. The final pH was adjusted to 7.6 with KOH. The homogenate was filtered through 4 layers of miracloth and centrifuged at 4,000g for 5 min in a Sorvall RC-5B (SS34 rotor; Dupont Instruments) centrifuge. The pellet, containing bacteroids and cell debris, was resuspended in wash medium (0.4 M sorbitol, 10 mM TES buffer pH 7.2 and 0.1% BSA) and layered over 30 ml of wash medium containing 70% Percoll (v/v), in a Sorvall SS-34 tube, and centrifuged at 40,000 g for 30 min. The purified bacteroids were located in a broad band near the bottom of the tube. The 4,000 g supernatant (above) was carefully decanted and recentrifuged at 10,000 g for 15 min. The pellet, which contained mitochondria, plastids, bacteroids and membrane fragments, was resuspended in about 10 ml of wash medium, layered over 30 ml of wash medium containing 45% (v/v) Percoll, in a Sorvall SS-34 tube, and centrifuged at 40,000 g for 30 min. The mitochondria were located in a tight brown band near the top of the tube, together with plastids, peroxisomes and membrane fragments (determined by electron microscopy). This band was removed by suction, diluted at least 5-fold with wash medium and concentrated by centrifuging at 15,000 g for 10 min. The loose pellet was resuspended in about 5 ml of wash medium and applied to the top of 30 ml of wash medium containing 28% (v/v) Percoll and a linear gradient of 0-10% (w/v) PVP-25 (see Day et al 1985), in a Sorvall SS-34 tube, and centrifuged at 40,000 g for 30 min. The mitochondria were found in a pale brown band near the bottom of the tube (Fig.1) and were washed and concentrated by centrifugation.

Mitochondria were purified from 5 day old roots and etiolated cotyledons by the method of Neuburger et al (1982), after homogenization as described above for nodules.

All enzymes were assayed according to published procedures. O_2 uptake was measured as described by Day et al (1985).

RESULTS

The method used yielded mitochondria largely free from contamination by other membrane components, despite the very large number of other membranes in the infected tissue of soybean nodules. The crude mitochondrial fraction was heavily contaminated by membrane fragments, plastids, peroxisomes and some bacteroids, as judged by electron microscopy (not shown). Enzyme assays (Table 1) confirmed this contamination. The fraction from the bottom of the final 28% Percoll gradient was essentially composed of mitochondria. Enzyme analysis (Table 1) showed that intact bacteroids had been completely removed, as judged by the absence of β-hydroxybutyrate dehydrogenase (a specific marker for the bacteroid cytoplasm). Catalase (a peroxisomal marker) was reduced by more than 50%, but the remaining activity suggests some peroxisomal contamination. Lipoxygenase activity was dramatically decreased in the final mitochondrial fraction, while cytochrome oxidase activity showed that a 3.7 -fold purification (on a protein basis) had been achieved by the final gradient (Table 1).

Nodule mitochondria readily oxidized all substrates tested, with good respiratory control, confirming previous studies (Mueke and Wiskich 1969; Rawsthorne and La Rue 1985). The oxidation of two NAD-linked substrates

TABLE 1. Enzyme activity of mitochondrial and bacteroid fractions.

Enzyme	Crude mitochondria (10,000g pellet)	Purified mitochondria	Purified bacteroids
β-hydroxybutyrate[a] dehydrogenase	11.5	n.d.[c]	65.7
catalase[b]	2.571	0.938	1.210
lipoxygenase[a]	327	40	n.d.
cyt.c oxidase[a]	171	625	165
NADH oxidase[a]	--	278	n.d.

[a] $nmol.min^{-1}. mg^{-1}$ protein; [b] $umol. min^{-1}. mg^{-1}$ protein; [c] n.d., not detectable.

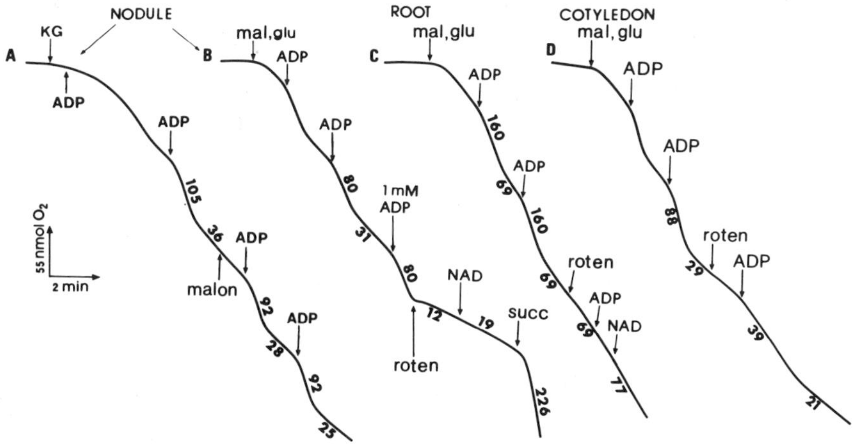

Fig 1. NAD-linked substrate oxidation. Additions: 10mM substrate, 0.2mM ADP, 15 M rotenone, 0.3mM NAD. Protein was 0.9 mg (A, B) and 0.4 mg (C, D).

is illustrated in Fig 1. oxoglutarate was rapidly oxidized, even in the presence of malonate which inhibits oxidation of the succinate produced (Fig 1A). Typical respiratory control ratios were 2.8-4.5 and ADP/O ratios 3.1-3.3 with OG as substrate; with malate these parameters were 2.5-4.0 and 2.2-2.5 respectively. Neither malate nor oxoglutarate oxidation responded to addition of NAD^+, suggesting that nodule mitochondria possess a full complement of this variable co-factor. Other substrates tested included exogenous NADH and succinate (Table 2). NADH in particular was rapidly oxidized.

Malate oxidation of nodule mitochondria was much more sensitive to inhibition by rotenone than that of root and cotyledon mitochondria, and this difference could not be attributed to different levels of matrix NAD^+ (Fig 1; exogenous NAD^+ did not stimulate). In root and nodule mitochondria, solubilized malic enzyme activity was approximately equal; that in cotyledon mitochondria was two-fold greater. Soybean mitochondria evidently differ from those from cowpea nodules which lack malic enzyme (Rawsthorn et al, these proceedings).

Rates of KCN-resistant O_2 uptake were dramatically slower, with all substrates tested, in nodule mitochondria than in their root and cotyledon counterparts (Table 2). This result was obtained with all preparations and a range of KCN concentrations (a typical experiment is shown in Table 2).

TABLE 2 Alternative oxidase activity of soybean mitochondria. KCN was presented at a final concentration of 0.5 mM. Resistant O_2 uptake was inhibited by SHAM. Rates shown are averages of at least two experiments.

Substrate	Oxygen consumption					
	Nodules		Root		Cotyledons	
	State 3	+KCN	State 3	+KCN	State 3	+KCN
	nmol O_2 . min^{-1} . mg^{-1} protein					
Malate + glu	113	15	150	68	95	48
Succinate	227	28	320	150	112	66
NADH	276	0	265	96	149	43

DISCUSSION

The isolation procedure outlined here provided intact and functional nodule mitochondria and bacteroids, free from contamination by each other. Although there was some contamination of the mitochondrial fraction by peroxisomes (Table 1) this did not interfere with the assay of mitochondrial function; any cross-contamination between bacteroids and mitochondria, on the other hand, would make independent assay of their respective oxidases very difficult. Cross-contamination between mitochondria and bacteroids can be estimated using NADH oxidase and β-hyroxybutyrate dehydrogenase respectively as markers.

Ammonia fixation in the soybean nodule cytoplasm requires, in addition to ATP, NADH and oxoglutarate. Yet both of these compounds are rapidly oxidized by isolated mitochondria. The oxoglutarate could be provided by transamination reactions in the cytosol but net synthesis must take place in the tricarboxylic acid cycle; in either case, oxidation of oxoglutarate must be restricted *in vivo*. The same is also true of NADH oxidation. Clearly, further studies are needed to identify possible control mechanisms.

The plant mitochondrial respiratory chain is branched, with bypasses of all the three energy transducing (proton-translocating) sites associated with the conventional chain. The non-phosphorylating branches can be distinguished by their insensitivity to the inhibitors rotenone and KCN. The high sensitivity of nodule mitochondria to these inhibitors shows that the non-phosphorylating pathways are virtually absent from the nodule (and may be completely absent from the infected tissue since our preparations used the whole nodule, with some organelles no doubt originating from the uninfected cells). Root and cotyledon mitochondria, on the other hand, were highly cyanide and rotenone insensitive.

The low capacity of non-energy conserving respiratory pathways of nodule mitochondria fits well with the postulated high energy demand of nodule cells during ammonia fixation. These results are also consistent with the analysis of root and nodule respiration performed by *deVisser and Lambers* (1983), who showed that nodule respiration was more efficient. Since the O_2 tension within infected cells of nodules is much less than that in root cells, it is possible that synthesis of the alternative oxidase is subject to regulation by O_2 concentration.

REFERENCES

Day, D.A., Neuburger, M., Douce, R. (1985) Aust. J. Plant Physiol. 12, 219-228.

deVisser, R., Lambers, H. (1983) Physiol. Plant. 58, 533-543.

Mueke, P.S., Wiskich, J.T. (1969) Nature 221, 674-675.

Neuburger, M., Journet, E-P., Bligny, R., Carde, J-P, Douce, R. (1982) Arch. Biochem. Biophys. 217, 312-323.

Rawsthorne, S., LaRue, T.A. (1985) In: Nitrogen Fixation Research Progress, pp 351, Evans, H.J., Bottomley, P.J., Newton, W.E. (eds), Martinus Nijhoff Publishers, Boston.

Sheehy, J.E., Minchin, F.R., Witty, J.F. (1985) Ann. Bot. 55, 549-562.

RESPIRATORY PARTICIPATION BY MITOCHONDRIA DURING EARLY PLASTIDOGENESIS IN BARLEY SEEDLINGS

John H. Owen and Alan R. Wellburn

Department of Biological Sciences, University of Lancaster
Bailrigg, Lancaster LA1 4YQ, U.K.

INTRODUCTION

Many aspects of plastid development have been extensively studied using the greening of etiolated angiosperm seedlings as a model system. Little work has been done, however, concerning the possible role of respiratory involvement by mitochondria during early plastid development in light-grown angiosperm leaves or the nature of the developmental controls which might synchronise such an involvement (see Wellburn, 1982).

Monocotyledonous plants differ from dicotyledons by possessing a basal intercalary meristem above which a linear gradient of cell development occurs with older, more differentiated cells towards the leaf tip (Leech, 1984). Recently Owen, Laybourn-Parry and Wellburn (1986_a, 1986_b) have prepared homogenous populations of protoplasts from primary leaves of seven day old, light-grown barley (Hordeum vulgare L. cv. Patty) seedlings corresponding to distinct stages of cellular and plastid development.

We have examined mitochondrial activity of protoplasts from different developmental regions of barley seedlings using both oxygen electrodes and a more sensitive technique of oxygen exchange measurement, Cartesian-diver microrespirometry (Birkenhead, Laybourn-Parry and Willmer, 1985). In addition, in order to evaluate respiratory controls during meristematic development, the influences of light of different wavelengths, plant growth regulators and inhibitors of organellar or cytosolic protein synthesis on respiratory activity of meristematic protoplasts have been investigated.

MITOCHONDRIAL RESPIRATORY BURSTS PRECEDE PLASTID DEVELOPMENT

Figure 1 shows that oxygen consumption in the dark of protoplast preparations decreases down the leaf blade towards the intercalary meristem, until a pronounced increase in dark respiration is observed in protoplasts from 0.0-0.5 cm above the meristem. Subdivision of this region reveals that protoplasts from the 0.0-0.2 cm region closest to the meristem display the highest dark respiratory activity of any leaf region.

These changes in protoplast dark respiratory activity clearly reflect changes in cellular metabolic activity which is related to developmental status within the linear array of cells along the primary leaf blade of the

Table 1. Effects of light of different wavelengths on respiration of meristematic protoplasts. Mean values of oxygen uptake are given in f.mol O_2 protoplast^{-1} h^{-1} (±SD) where n was at least 3.

Light treatment	Respiratory activity	% change from control
Dark control	128 ± 10	-
10min red light	215 ± 24	+68%
10min far-red light	116 ± 26	-9%
10min blue light	317 ± 25	+148%

seedling (Wellburn et al., 1986). Interestingly, the highest respiratory activity occurred below the amyloplast region in the true proplastid region and is probably due to enhanced mitochondrial activity prior to or concurrent with early plastid development, which is supported by previous ultrastructural observations of increased mitochondrial frequency close to the intercalary meristem (Wellburn, Robinson and Wellburn, 1982).

Proplastids have a heterotrophic requirement for nutritional materials and large molecules such as enzymes and structural proteins (Mohr, 1984). Energy requirements for early plastid development can only be satisfied by adequate supplies of ATP from mitochondria (or high rates of glycolysis) because photophosphorylation is not yet established (Wellburn, 1982). But what are the developmental controls co-ordinating mitochondrial participation in providing the early energy supplies for developing proplastids?

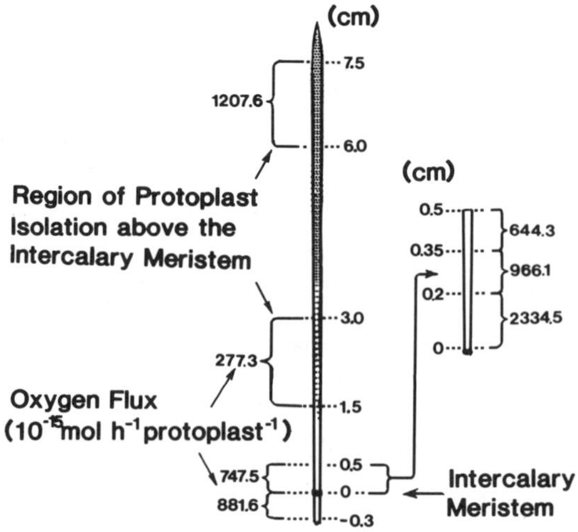

Figure 1. Oxygen consumption in the dark of protoplasts from different regions of light-grown barley seedlings as measured by Cartesian-diver microrespirometry.

Table 2. Effects of plant growth regulators on respiration of meristematic protoplasts. Mean values of oxygen uptake are given in f.mol O_2 protoplast^{-1} h^{-1} (±SD) where n was at least 3.

Treatment	Respiratory activity	% change from control
Control	756 ± 104	-
10^{-6}M GA_3	995 ± 90	+32%
10^{-6}M KN	796 ± 146	+ 5%
10^{-6}M IAA	702 ± 84	- 7%
10^{-6}M ABA	473 ± 140	-37%

DEVELOPMENTAL CONTROLS INFLUENCING MITOCHONDRIAL ACTIVITY

Table 1 shows that blue and red light treatments markedly increase dark respiratory activities of meristematic protoplasts, but far-red light causes an insignificant decrease in dark respiration. These and previous results (Owen et al., 1986$_b$) demonstrate that light, probably associated with nuclear, cytosolic and plasmalemma components of the cell, has a partial but not insignificant effect on mitochondrial activity close to the intercalary meristem of barley seedlings. There is also a greater, light-independent component of enhanced mitochondrial activity in this meristematic region which forms part of basic pattern formation.

Table 2 shows that while the plant growth regulators kinetin and IAA have no effect on mitochondrial activity in meristematic protoplasts, GA_3 and ABA have opposing effects and markedly increase or decrease, respectively, protoplast dark respiration.

Nucleic acid and protein synthesis require much ATP which results in reduced cytosolic ATP to ADP ratios which, in turn, increase rates of mitochondrial oxidative phosphorylation. GA_3, ABA, phytochrome and the blue light receptor apparently all affect nucleic acid and protein biosynthesis

Table 3. Effects of chloramphenicol and cycloheximide on respiration of meristematic protoplasts. Mean values of oxygen uptake are given in f.mol O_2 protoplast^{-1} h^{-1} (± SD) where n was at least 3.

Treatment	Respiratory activity	% change from control
Control	187 ± 24	-
0.15μM chloramphenicol	192 ± 13	-3%
0.15μM cycloheximide	397 ± 20	+112%
1.50μM chloramphenicol	217 ± 18	+ 16%
1.50μM cycloheximide	408 ± 40	+118%

and hence may regulate the rate of mitochondrial oxidative phosphorylation through control of cytosolic ATP to ADP ratios. GA_3 and ABA may be acting as alternative second messengers to phytochrome and/or the blue light receptor, but may also act in a light-independent role in basic pattern formation to regulate mitochondrial co-operation with early proplastid development.

Table 3 shows that while the eukaryotic organellar protein synthesis inhibitor chloramphenicol does not significantly alter the dark respiratory activity of meristematic protoplasts, the cytosolic protein synthesis inhibitor cycloheximide markedly increases protoplast dark respiratory rates. Plastid and mitochondrial genomes code for relatively few proteins and must interact with the nuclear genome to develop their full protein complement. It is likely that a component of cytosolic protein synthesis is important for the initiation of enhanced mitochondrial activity in meristems. If cytosolic protein synthesis is inhibited the increased cytosolic ATP to ADP ratio which would result as ATP demand drops would reduce mitochondrial respiratory chain phosphorylation. However in meristematic protoplasts cycloheximide may prevent cytosolic translation of negative control regulatory proteins which normally restrict mitochondrial ATP formation, some of which are removed by light signals. This would explain the marked increases in dark respiration of meristematic protoplasts observed after cycloheximide treatments.

ACKNOWLEDGEMENTS

JHO thanks the SERC for the award of a research studentship.

REFERENCES

Birkenhead, K., Laybourn-Parry, J.E.M., and Willmer, C., 1985, Measurement of guard cell respiration rates using a Cartesian-diver technique. Planta., 163:214.

Leech, R.M., 1984, Chloroplast development in angiosperm seedlings: current knowledge and future prospects, in: "Chloroplast Biogenesis," N.R.Baker and J.Barber, eds. Elsevier, Amsterdam.

Mohr, H., 1984, Phytochrome and chloroplast development, in: "Chloroplast Biogenesis," N.R.Baker and J.Barber, eds., Elsevier, Amsterdam.

Owen, J.H., Laybourn-Parry, J.E.M., and Wellburn, A.R. 1986a. Leaf respiration during early plastidogenesis in light-grown barley seedlings. Physiol. Plant., (in press).

Owen, J.H., Laybourn-Parry, J.E.M., and Wellburn, A.R., 1986b, Influences of light quality, growth regulators and inhibitors of protein synthesis on respiration of protoplasts from meristematic cells in barley seedlings. Ann. Bot., (in press).

Wellburn, A.R., 1982, Bioenergetic and ultrastructural changes associated with chloroplast development. Int. Rev. Cyt., 80:133.

Wellburn, A.R., Gounaris, I., Owen, J.H., Laybourn-Parry, J.E.M., and Wellburn, F.A.M., 1986, Development of bioenergetic function in light-grown seedlings, in: "Regulation of chloroplast differentiation," G.Akoyunoglou and H.Senger, eds., Alan R.Liss, New York.

Wellburn, A.R., Robinson, D.C., and Wellburn, F.A.M. 1982, Chloroplast development in low light-grown barley seedlings. Planta., 154: 259.

CO-ORDINATED ORGANELLE BIOGENESIS DURING PHOTORESPIRATORY DEVELOPMENT IN WHEAT LEAVES

Alyson K. Tobin, [1]Nazira Sumar, [1]Mina Patel, Michael O. Proudlove, [1]George R. Stewart and Anthony L. Moore

Department of Biochemistry, University of Sussex, Falmer, Brighton, Sussex, BN1 9QG, U.K. and [1]Department of Botany and Microbiology, University College, London WC1E 6BT, U.K.

INTRODUCTION

All cells of a wheat primary leaf are derived from meristematic cell division at the leaf base. Consequently, cell maturity increases towards the leaf tip, as does the size, number and photosynthetic activity of the chloroplasts[1-3]. Although developmental changes in chloroplast morphology and biochemistry have received much attention, the extent to which these influence the metabolism of other leaf cell organelles has been largely overlooked.

The photorespiratory pathway involves the cycling of carbon and nitrogen compounds between four separate subcellular compartments: chloroplasts, peroxisomes, mitochondria and cytosol[4]. The development of this pathway within a growing leaf must therefore require controlled organelle biogenesis to avoid disruption of the cycle, accumulation of toxic substances (e.g. ammonia) or depletion of the Calvin cycle. We report here a study of the changes in photorespiratory ammonia production and associated enzyme activities within leaf sections at different stages of maturity.

PROCEDURE

Materials and Methods

Measurement of photorespiratory NH_3 production: Wheat seeds (Triticum aestivum cv Maris Huntsman) were imbibed for 18h at 20°C, planted in John Innes No.3 compost and grown in a 16h light (300 mol.m^{-2}.s^{-1} PAR; 20°C)/ 8h dark (10°C) cycle for 8 days. The primary leaf was cut into 0.5cm transverse sections at 1cm intervals from the meristem (mean leaf length = 10cm). For each treatment fifty, 0.5cm leaf sections were floated on 2mM L-methionine sulphoximine (MSO; glutamine synthetase (GS) inhibitor), water, 10mM pyrid-2-yl hydroxy-methane sulphonate (HPMS; glycollate oxidase inhibitor) or 10mM $KHCO_3$ (all solutions adjusted to pH 7.0). After 30min dark preincubation at 20°C, all the sections were illuminated (200 mol.m^{-2}.s^{-1} PAR) for 30min then frozen in liquid N_2. Leaf sections were extracted in chloroform: methanol: water (5:12:13 (v/v)), fractionated through Dowex 50W, Na^+-form[5] and assayed for ammonia according to McCullough[6].

Determination of enzyme activities: Freshly harvested transverse sections of wheat primary leaves were ground in liquid N_2 with a pestle and mortar, extracted in appropriate buffers (see references[2] for details) and centrifuged for 20min at 10000 x g. The supernatants were assayed for glutamine synthetase[3], nitrate reductase[7], glycollate oxidase[8], NADH-glutamate dehydrogenase[9] and cytochrome oxidase[10]. GS isoenzymes were fractionated by FPLC[3]. In vivo nitrate reductase[11] was determined using fresh, unfrozen leaf sections.

Determination of mesophyll cell number: Leaf sections (0.5 cm transverse) were incubated in 5% chromium trioxide for 7 days at $4^{o}C$ and mesophyll cells counted in a 0.2mm deep haemocytometer[3].

RESULTS AND DISCUSSION

Figure 1a shows that ammonia accumulates when wheat leaf sections are treated with the GS inhibitor, MSO. Extraction and assay of treated leaf sections showed that the 30min dark preincubation with MSO resulted in complete inhibition of GS in all sections (data not shown). The rate of ammonia accumulation increases ca. 15-fold from the base to the tip of 8 day old wheat primary leaves. This accumulation is light-dependent (data not shown) and is inhibited by elevated CO_2 concentrations (Figure 1b) and by treatment with 10mM HPMS (glycollate oxidase inhibitor) and therefore appears to be of photorespiratory origin. Addition of 10mM glycine only stimulated NH_3 production when HPMS was also present, suggesting that glycine supply was not limiting its deamination under these conditions. This method of measuring photorespiration is known to underestimate rates for various reasons (see, for example, Walker et al[12]). Indeed, our maximum rate of ammonia accumulation (23mole x 10^{-15}.min^{-1}.$cell^{-1}$ in mature sections; equivalent to 12 mol.h^{-1}.g fwt^{-1}) is rather lower than rates measured by gas exchange methods (80 mol.h^{-1}.g fwt^{-1} ref.[13]). It is, however, comparable to other MSO-dependent rates[14]. Although absolute rates are low, the results indicate that photorespiratory ammonia production increases with leaf development and reaches a maximum when chloroplast division is complete.

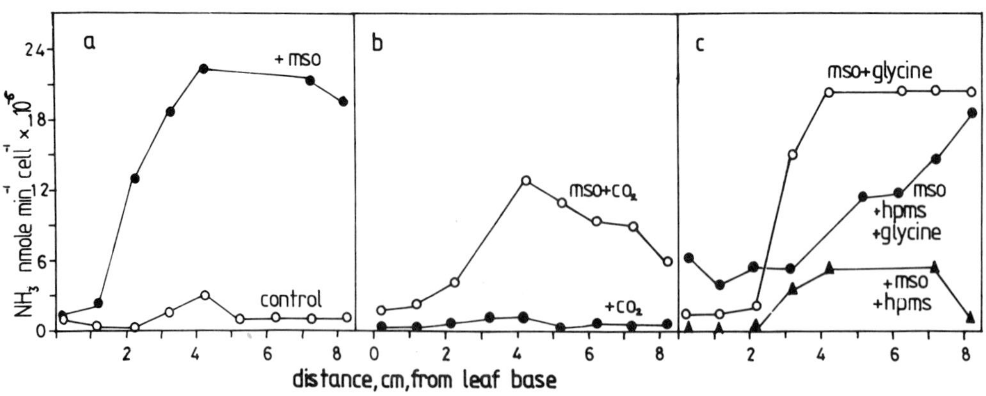

Fig. 1. Ammonia accumulation in serial transverse sections of 8-day old wheat primary leaves. Note, the x-axis refers to the distance from the leaf base, i.e. sections increase in age towards the leaf tip.

There was no significant change in cellular nitrate concentration during incubation of the leaf sections (data not shown), indicating that nitrate reduction did not contribute significantly to ammonia accumulation under these conditions. This is supported by the relatively low activity of nitrate reductase (NR) in these sections (Figure 2). It is interesting to note that, whereas NR activity in extracts (in vitro) increases with leaf cell maturity, the in vivo NR activity remains constant beyond 3cm from the meristem, suggesting a physiological limitation on nitrate reduction in mature leaf cells. Furthermore, the pattern of development of NR activity (in vivo) closely follows that of cytosolic GS (GS1; Figure 2) and may indicate a role for the latter enzyme in primary ammonia assimilation.

Chloroplast GS activity (GS2; Figure 3) develops in parallel with peroxisomal glycollate oxidase, indicating a close control of synthesis of these photorespiratory enzymes at all stages of cell maturity. The pattern of development of these photorespiratory enzymes, similar to that of RubisCO[2], differs markedly from that of non-photorespiratory enzymes. For example, the activities of mitochondrial glutamate dehydrogenase and cytochrome oxidase remain relatively constant during leaf development (Figure 3). This suggests that there is a controlled, co-ordinated synthesis of specific enzymes involved in photosynthesis and photorespiration, rather than a general increase in all metabolic pathways, during wheat leaf and chloroplast development. The parallel development of GS2 and glycollate oxidase and the low activity of GS1, relative to photorespiratory activity, lend further support to the chloroplast being the principal site for the reassimilation of photorespiratory ammonia.

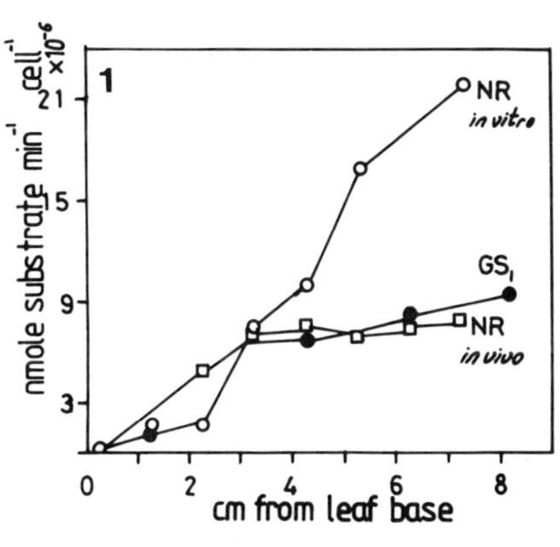

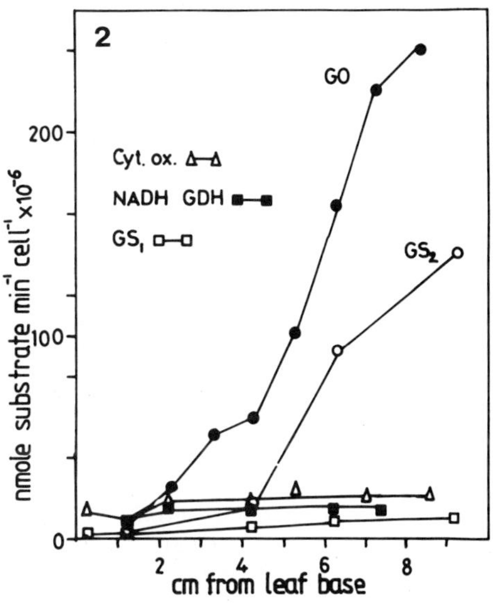

Fig. 2. Changes in activities of nitrate reductase (NR: in vivo and in vitro) and cytosolic glutamine synthetase (GS1) during leaf development.

Fig. 3. Changes in activities of chloroplast glutamine synthetase (GS2), glycollate oxidase (GO), NADH-dependent glutamate dehydrogenase and cytochrome oxidase during leaf development.

ACKNOWLEDGEMENTS

Financial support was provided by the Royal Society (I.C.I. Alfred Spinks Fellowship to AKT) and by AFRC.

REFERENCES

1. S. A. Boffey, R. J. Ellis, G. Sellden and R. M. Leech, Chloroplast division and DNA synthesis in light-grown wheat leaves, Plant Physiol. 64:502 (1979).
2. C. Dean and R. M. Leech, Genome expression during normal leaf development. I. Cellular and chloroplast numbers and DNA, RNA and protein levels in tissues of different ages within a seven-day old wheat leaf, Plant Physiol. 69:904 (1982).
3. A. K. Tobin, S. M. Ridley and G. R. Stewart, Changes in the activities of chloroplast and cytosolic isoenzymes of glutamine synthetase during normal leaf growth and plastid development in wheat, Planta 163:544 (1985).
4. A. J. Keys, I. F. Bird, M. F. Cornelius, P. J. Lea, R. M. Wallsgrove and B. J. Miflin, Photorespiratory nitrogen cycle, Nature 275:741 (1978).
5. P. A. Fentem, P. J. Lea and G. R. Stewart, Ammonia assimilation in the roots of nitrate- and ammonia- grown Hordeum vulgare (cv Golden Promise), Plant Physiol. 71:496 (1983).
6. H. McCullough, The determination of ammonia in whole blood by a direct colorimetric method, Clinica Chimica Acta 17:297 (1967).
7. M. J. Emes and M. W. Fowler, The intracellular location of the enzymes of nitrate assimilation in the apices of seedling pea roots, Planta 194:249 (1979)
8. W. Behrends, U. Rausch, H-G. Loffler, H. Kindl, Purification of glycollate oxidase from greening cucumber cotyledons, Planta 156:556 (1982).
9. E. M. Lees, and D. T. Dennis, Glutamate dehydrogenase in developing endosperm, chloroplasts, and roots of castor bean, Plant Physiol. 68:827 (1981).
10. N. E. Tolbert, A. Oeser, T. Kisaki, R. H. Hageman and R. K. Yamazaki, Peroxisomes from spinach leaves containing enzymes related to glycolate metabolism, J. Biol. Chem. 243:5179 (1968).
11. T. E. Ferrari and J. E. Varner, Control of nitrate reductase activity in barley aleurone layers, Proc. Nat. Acad. Sci. USA 65:736 (1970).
12. K. A. Walker, A. J. Keys and C. V. Givan, Effect of L-methionine sulphoximine on the products of photosynthesis in wheat (Triticum aestivum) leaves, J. Exp. Bot. 35:1800 (1984).
13. K. S. Kumarasinghe, A. J. Keys and C. P. Whittingham, The flux of carbon through the glycolate pathway during photosynthesis by wheat leaves, J. Exp. Bot. 28:1247 (1977).
14. F. M. Martin, M. Winspear, J. D. Macfarlane and A. Oaks, Effect of methionine sulphoximine on the accumulation of ammonia in C_3 and C_4 leaves. The relationship between NH_3 assimilation and photorespiratory activity, Plant Physiol. 71:177 (1983).

GLYCINE UPTAKE BY PEA LEAF MITOCHONDRIA: A PROPOSED MODEL FOR THE MECHANISM OF GLYCINE - SERINE EXCHANGE

David J. Oliver

Department of Bacteriology and Biochemistry
University of Idaho
Moscow, Idaho 83843 USA

Mersalyl was used as an inhibitor-stop reagent, to analyze the mechanism of glycine transport by its carrier in pea leaf mitochondria. Preloaded glycine exchanged for exogenous glycine and L-serine but not D-serine. In addition, the pH response of glycine metabolism suggested that glycine can exchange for hydroxyl ions. A mechanism is proposed where glycine can exchange for both serine, the end product of mitochondrial glycine metabolism, and hydroxyl ions. This mechanism would allow for the 2:1 glycine:serine ratio observed during photorespiratory glycine metabolism in leaf mitochondria.

INTRODUCTION

Mitochondrial glycine metabolism is recognized as a key reaction in photorespiration. The enzymes responsible for this reaction, the glycine decarboxylase multienzyme complex, are in the mitochondrial matrix. Two glycine molecules cross the inner mitochondrial membrane and are metabolized to serine, which leaves the mitochondria to complete the cycle. Glycine and serine are zwitterions at neutral pH values and should permeate lipid bilayers very slowly. As a consequence of the very limited permeability of biological membranes to neutral amino acids, many biological systems have glycine transporters.

Despite the evidence that glycine and serine transporters are common in a number of biological systems, including rat liver (1,2) and reticulocyte (5) mitochondria, there is no consensus that the much larger flux of glycine and serine across the inner membrane of plant leaf mitochondria is carrier-mediated. Glycine uptake measured with leaf mitochondria does not show saturation kinetics and does not occur at rates sufficient to account for measured rates of mitochondrial glycine oxidation (3,4). Yu et al. (7) measured glycine uptake but concluded this was only a minor component at physiological glycine concentrations.

Glycine oxidation is sensitive to the nonpenetrating sulfhydryl reagents, mersalyl and p-chloromercuribenzoate, and the amino acid analog aminoacetonitrile (6). The site of this inhibition occurs before the glycine decarboxylase reaction. This supplies strong indirect evidence for the existence of a glycine transporter. In this paper additional evidence is presented to support the functioning of a neutral amino acid transporter in plant leaf mitochondria, and a potential mechanism for the transport of glycine and serine is proposed.

MATERIALS AND METHODS

Glycine uptake by Percoll-purified arsenite-treated pea leaf mitochondria was measured using the silicone oil technique (4,6,7). The uptake was stopped by the addition of mersalyl and the mitochondria separated by centrifugation through silicone oil.

RESULTS AND DISCUSSION

The results of a glycine uptake experiment using mersalyl as the inhibitor-stop reagent are shown in Fig. 1. The glycine concentration was 25 uM and any noncarrier-mediated diffusion was factored out by subtracting the amount of radiocarbon in the pellet from a parallel experiment where mersalyl was added 10 sec before glycine. There was a rapid rate of glycine transport that reached a saturation point within 2 to 5 sec after the exposure to glycine. During the next 25 sec, no additional influx of glycine was measured. The rate of glycine uptake in the first 2 sec was equivalent to a rate of 12 nmol/mg protein/min. This was sufficient to account for the rate of metabolism of 25 uM glycine. Higher glycine concentrations could not be used with mersalyl (6).

Mitochondria were preloaded with [1-^{14}C] glycine, washed, and allowed to exchange the internal labelled glycine for exogenous unlabelled amino acids. The exchange reactions were stopped with mersalyl after 5 sec and the remaining internal glycine measured after the mitochondria were pelleted through silicone oil. Exogenous unlabelled glycine exchanged for internal [1-^{14}C] glycine (Fig. 2). As the internal labelled glycine was exchanged out of the organelles, less label was detected in the pellet. L-serine also exchanged for internal preloaded glycine and decreased the amount of radiocarbon that appeared in the pellet (Fig 3). The specificity of the carrier reaction was shown when D-serine did not substitute for L-serine in exchanging glycine from the mitochondrial matrix. These data indicate that glycine can exchange for either glycine or L-serine.

Glycine metabolism by mitochondria results in the unusual stoichiometry of two glycine molecules entering the mitochondrial matrix in exchange for one serine leaving. The glycine-serine exchanger described has a 1:1 glycine:serine exchange ratio (data not presented). For sustained glycine metabolism an additional transport feture is needed. Indirect evidence for this exchange mechanism came from studies on the effect of pH of glycine oxidation (Fig. 4). In this experiment

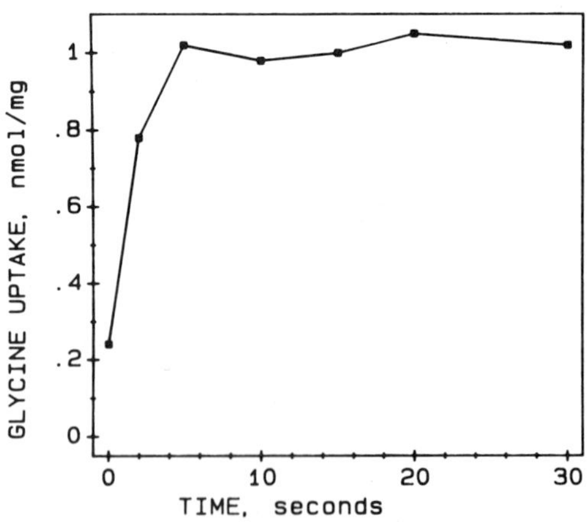

Fig 1. Glycine uptake by pea leaf mitochondria measured by the mersalyl-stop technique.

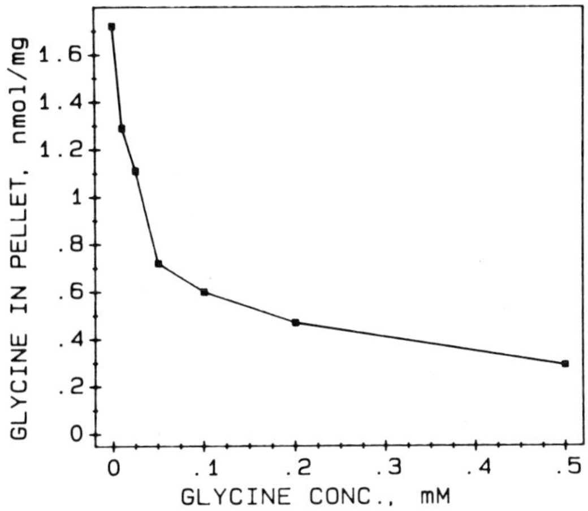

Fig 2. The exchange of preloaded [1-^{14}C] glycine for external unlabelled glycine by pea leaf mitochondria.

oxaloacetate was used as the terminal electron acceptor and FCCP was added to equilibrate the matrix and media pH. The apparent Km for glycine during glycine oxidation was high at low pH values (18 mM at pH 6.0) and decreased as the pH increased (3 mM at pH 8.4). As the pH of the matrix and the medium became more alkaline, the affinity of the mitochondria for glycine increased (Km decreased). This change in activity could not be explained by changes in the activity of glycine decarboxylase or malate dehydrogenase (data not presented). The change in mitochondrial activity, therefore, most likely resulted from changes in the glycine transport reaction. The transporter would have a high binding affinity for glycine at high pH values and a lower affinity under low pH conditions. This pH-dependent shift in the affinity of the carrier for glycine could provide the mechanism by which glycine is exchanged for hydroxyl ions by the transporter (see below).

CONCLUSIONS

The rates of glycine uptake by pea leaf mitochondria are sufficient to account for the measured rates of glycine metabolism by these organelles. Because the rate of transport is very rapid, it is not possible to use conventional oil techniques without inhibitor-stop

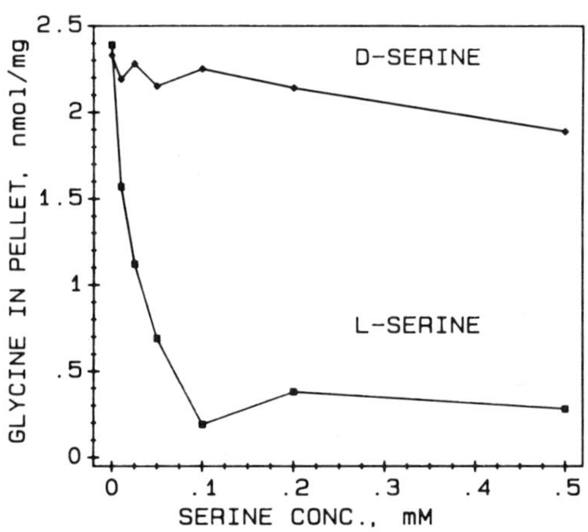

Fig. 3. The exchange of preloaded [1-^{14}C] glycine for external unlabelled D- and L-serine by pea leaf mitochondria.

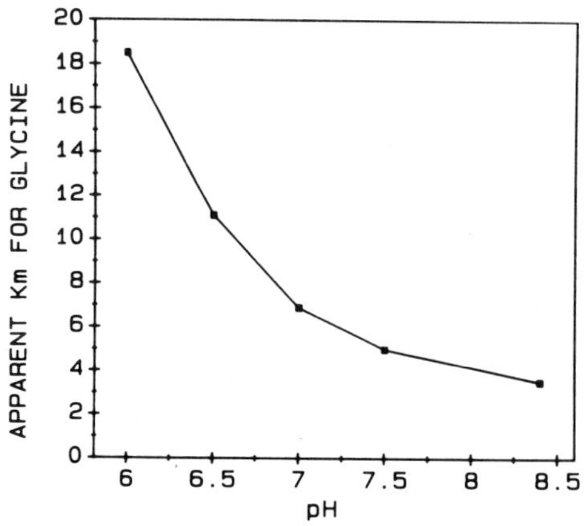

Fig. 4. The effect of pH on the apparent Km (mM) for glycine oxidation by pea leaf mitochondria.

reagents and obtain meaningful kinetics. A detailed kinetic analysis of the uptake reaction will require either the discovery of a better transport inhibitor than mersalyl or the application of fast kinetics to the problem.

Any acceptable explanation of glycine and serine transport by leaf mitochondria must allow for the two glycine to one serine ratio observed during glycine metabolism. The data presented suggest that glycine can exchange directly with with serine and hydroxyl ions. This would allow a flexible stoichiometry for the exchange reaction that could readily accommodate the 2:1 ratio needed for the glycine metabolism reaction.

The mechanism for glycine hydroxyl ion exchange was deduced from the pH response of glycine metabolism by mitochondria. The exchanger is envisioned as binding a hydroxyl ion in the alkaline matrix of the mitochondria. This causes an increase in the carriers affinity (decreased Km) for glycine at the external face of the membrane. This is followed by a transfer reaction where the amino acid binding site moves to the matrix face and the hydroxyl ion binding site moves to the external membrane surface. As the hydroxyl ion dissociates into the more acidic cytosol, the resulting decrease in the affinity of the carrier for glycine (higher Km when this site is in the less alkaline environment) causes the glycine to be released into the matrix. The shift in pH optima that resulted when the glycine concentration was changed (data not presented) suggests that, at low glycine concentrations, the rate limiting step was the binding of hydroxyl ions to the carrier. At high glycine concentrations, the limiting reaction was the dissociation of the bound hydroxyl ion at the outer mitochondrial surface.

ACKNOWLEDGEMENTS

This work was supported by the Metabolic Biology Division of the National Science Foundation and the Idaho Agricultural Experiment Station.

REFERENCES

1. Benavides, J., M.L. Garcia, J. Lopez-Lahoya, M. Urgarte, F. Valdivieso 1980 Biochim. Biophys. Acta 598: 588-594
2. Cybulski, R.L. and R.R. Fisher 1977 Biochem. 16: 5116-5120
3. Day, D.A. and J.T. Wiskich 1980 FEBS Lett. 112: 191-194
4. Proudlove, M.O. and A.L. Moore 1982 FEBS Lett. 147: 26-30
5. Rapport, S., M. Muller, R. Dumdey, J. Rathmann 1980 Eur. J. Biochem. 108: 399-455
6. Walker, G.H., D.J. Oliver, and G. Sarojini 1982 Biochem. Biophys. Res. Commun. 107: 856-861
7. Yu, C., D.L. Claybrook, and A.H.C. Huang 1983 Arch. Biochem. Biophys. 227: 180-187

SOME PROPERTIES OF PLANT MITOCHONDRIAL PYRUVATE DEHYDROGENASE KINASES

Jan A. Miernyk[1,2] and Douglas D. Randall[2]

[1]Seed Biosynthesis Research Unit
Northern Regional Research Center, USDA, ARS
Peoria, IL 61604 USA

[2]Biochemistry Department
University of Missouri
Columbia, MO 65211 USA

INTRODUCTION

Mammalian mitochondrial pyruvate dehydrogenase complexes (PDCs) are regulated in part by reversible phosphorylation (1-3). Pyruvate dehydrogenase kinase (EC 2.7.1.99) is an integral component of these complexes. This kinase catalyzes the multi-site phosphorylation of the $E_1\alpha$-subunit of the pyruvate dehydrogenase (EC 1.2.4.1, PDH) component of the complex, resulting in complete inactivation (4). Dephosphorylation, catalyzed by phospho-pyruvate dehydrogenase phosphatase, reactivates the complex. Regulation of mammalian PDH-kinase plays an important role in the overall regulation of PDC activity (5,6).

It has been clearly demonstrated that plant mitochondrial PDCs are also regulated by reversible phosphorylation (7-11). However, very little is known about the regulatory components from plant sources. Herein we present a partial characterization of PDH-kinase from plant mitochondrial PDCs.

MATERIALS AND METHODS

General. Most of the methods and materials used in this laboratory for growth of plants, isolation of mitochondria, partial purification, phosphorylation, and assay of PDC activity have been previously described (7-11).

Assay of PDH-kinase activity. Three different assays were used during the characterization of mitochondrial PDH-kinase: a) the nucleotide-triphosphate dependent inactivation of PDC; b) incorporation of ^{32}P from gamma-^{32}P-ATP into material insoluble in cold 10% trichloroacetic acid (PDH is virtually the only mitochondrial matrix protein phosphorylated under the conditions of our assay); c) the method of Braun *et al*. (12) for assay of protein-kinase activity using peptide substrates. In comparable experiments, methods 1 and 2 gave comparable results. Method 3 was used only in studies of substrate specificity.

RESULTS AND DISCUSSION

Unless otherwise stated, the data presented were obtained with PDH-kinase from mitochondria of either green pea seedlings or the endosperm from developing *Ricinus communis* seeds. Activity of PDH-kinase was optimal at pH 7.5, with Mg-ATP as the phosphoryl donor. Free ATP was not a substrate. While Mg-ATP was superior, most nucleoside-triphosphates could serve as substrates for the kinase, indicating a low degree of phosphoryl donor specificity.

The use of synthetic peptides modeled upon the structure of phosphorylation site one of mammalian PDH (4) has allowed the assessment of phosphoryl acceptor specificity. In contrast to the results with phosphoryl donor, there was a very high degree of phosphoryl acceptor specificity. Unlike mammalian PDHs, which are phosphorylated exclusively on serine residues, all plant PDHs we have thus far examined are phosphorylated on threonine residues. The most effective peptide substrate for plant PDH-kinases to date has been an analog of the eight amino acid sequence surrounding mammalian phosphorylation site one; tyr-his-gly-his-*ser*-met-ser-asn, where the serine at position five was replaced with a threonine. It is interesting to note that while this single substitution resulted in the conversion of the peptide from a non-substrate to an effective substrate for plant PDH-kinases, it conversely resulted in the conversion of a good substrate into a non-substrate for the mammalian kinase (13).

Adenosine-diphosphate, one product of the kinase reaction, was an effective inhibitor of both mammalian and plant PDH-kinases, as was pyruvate, the substrate for PDH (TABLE 1). Acetyl-CoA and NADH, products of overall PDC activity, stimulated mammalian kinase activity while inhibiting plant kinases. Neither an explanation nor a rationale for this contrast is presently available. Citrate was also an effective inhibitor of plant PDH-kinase activity *in vitro*. This inhibition was not due to the chelation of required magnesium. Citrate levels could serve as a good indicator of overall Krebs cycle activity and of acetyl-CoA demand, and thus might be an effective modulator of kinase activity. Both mammalian and plant PDH-kinases were inhibited *in vitro* by sodium, but not other monovalent cations.

The relative molecular mass of mammalian PDH-kinase has been reported as 48,000 (14). While the low abundance of PDH-kinase has thus far precluded its isolation from plants, we have attempted to estimate the relative molecular mass by incubation with 8-azido ATP, followed by photolysis, immunoprecipitation and analysis by SDS-PAGE. Using this strategy we arrived at a value of 53,300, significantly higher than for the mammalian kinase. We have previously reported that the subunit sizes of other components of plant PDCs differ from those of the mammalian and bacterial complexes (11,15). This is perhaps not surprising if one considers the evolutionary distances between these organisms.

Thus while there are many similarities in the regulation of mitochondrial PDCs by reversible phosphorylation and in the characteristics of the PDH-kinases of mammals and plants, there are also significant differences. It seems likely that these distinctions will increase as does our knowledge of this intriguing system in plant mitochondria.

TABLE 1. SOME CHARACTERISTICS OF PYRUVATE DEHYDROGENASE KINASES

	SOURCE		CITATION
	PLANT	MAMMALIAN	
pH optimum	7.5	7.0	(16)
K_m - MgATP	2.5 μM	10 μM	(17)
Specificity, phosphoryl donor	LOW	?	--
K_m - peptide acceptor	250 μM	20 μM	(18)
Specificity, phosphoryl acceptor	HIGH	HIGH	(13)
Effectors: ADP	-	-	(16)
pyruvate	-	-	(16)
acetyl-CoA	-	+	(6)
NADH	-	+	(6)
citrate	-	?	--
Na^+	-	-	(16)
Relative molecular mass	53,300	48,000	(14)

Acknowledgement: This research was supported by NSF Grant DMB-8506473 and the Missouri Agriculture Experiment Station.

REFERENCES

1. Reed, L.J., 1974, Multienzyme Complexes, Acc. Chem. Res. 7:40-46.

2. Reed, L.J., 1981, Regulation of Mammalian Pyruvate Dehydrogenase Complex by Phosphorylation-Dephosphorylation Cycle, Curr. Top. Cell. Reg. 18:95-106.

3. Randle, P.J., 1981, Phosphorylation-Dephosphorylation Cycles and the Regulation of Fuel Selection in Mammals, Curr. Top. Cell. Reg. 18:107-129.

4. Yeaman, S.J., Hutcheson, E.T., Roche, T.E., Pettit, F.H., Brown, J.R., Reed, L.J., Watson, D.C., and Dixon, G.H., 1978, Sites of Phosphorylation on Pyruvate Dehydrogenase from Bovine Kidney and Heart, Biochemistry 17:2364-2370.

5. Pettit, F.H., Pelley, J.W., and Reed, L.J., 1975, Regulation of Pyruvate Dehydrogenase Kinase and Phosphatase by Acetyl CoA/CoA and NADH/NAD Ratios, Biochem. Biophys. Res. Commun. 65:575-582.

6. Cate, R.L., and Roche, T.E., 1978, A Unifying Mechanism for Stimulation of Mammalian PDH Kinase by NADH, Dihydrolipoamide, Acetyl CoA or Pyruvate, J. Biol. Chem. 253:496-503.

7. Rubin, P.M., and Randall, D.D., 1977, Regulation of Plant Pyruvate Dehydrogenase Complex by Phosphorylation, Plant Physiol. 60:34-39.

8. Randall, D.D., Rubin, P.M., and Fenko, M., 1977, Plant Pyruvate Dehydrogenase Complex Purification, Characterization and Regulation by Metabolites and Phosphorylation, Biochim. Biophys. Acta. 485:336-349.

9. Rao, K.P., and Randall, D.D., 1980, Plant Pyruvate Dehydrogenase Complex: Inactivation and Reactivation by Phosphorylation and Dephosphorylation, Arch. Biochem. Biophys. 200:461-466.

10. Randall, D.D., Williams, M., and Rapp, B.J., 1981, Phosphorylation-Dephosphorylation of Pyruvate Dehydrogenase Complex from Pea Leaf Mitochondria, Arch. Biochem. Biophys. 207:437-444.

11. Miernyk, J.A., Camp, P.J., and Randall, D.D., 1985, Regulation of Plant Pyruvate Pyruvate Dehydrogenase Complexes, Curr. Top. Plant Biochem. Physiol. 4:175-190.

12. Braun, S., Abdel Ghany, M., and Racker, E., 1983, A Rapid Assay for Protein Kinases Phosphorylating Small Peptides and Other Substrates, Analyt. Biochem. 135:369-378.

13. Mullinax, T.R., Stepp, L.R., Brown, J.R., and Reed, L.J., 1985, Synthetic Peptide Substrates for Mammalian Pyruvate Dehydrogenase Kinase and Pyruvate Dehydrogenase Phosphatase, Arch. Biochem. Biophys. 243:655-659.

14. Stepp, L.R., Pettit, F.H., Yeaman, S.J., and Reed, L.J., 1983, Purification and Properties of Pyruvate Dehydrogenase Kinase from Bovine Kidney, J. Biol. Chem. 258:9454-9458.

15. Camp, P.J., and Randall, D.D., 1985, Purification and Characterization of the Pea Chloroplast Pyruvate Dehydrogenase Complex. A Source of Acetyl-CoA and NADH for Fatty Acid Biosynthesis, Plant Physiol. 77:571-577.

16. Hucho, F., Randall, D.D., Roche, T.E., Burgett, M.W., Pelley, J.W., and Reed, L.J., 1972, α-Keto Acid Dehydrogenase Complexes. XVII. Kinetic and Regulatory Properties of Pyruvate Dehydrogenase Kinase and Pyruvate Dehydrogenase Phosphatase from Bovine Kidney and Heart, Arch. Biochem. Biophys. 151:328-340.

17. Kerbey, A.L., Radcliffe, P.M., Randle, P.J., and Sugden, P.H., 1979, Regulation of Kinase Reactions in Pig Heart Pyruvate Dehydrogenase Complex, Biochem. J. 181:427-433.

18. Davis, P.F., Pettit, F.H., and Reed, L.J., 1977, Peptides Derived from Pyruvate Dehydrogenase as Substrates for Pyruvate Dehydrogenase Kinase and Phosphatase, Biochem. Biophys. Res. Commun. 75:541-549.

THE UPTAKE AND THE OXIDATION OF PYRUVIC ACID BY CASTOR BEAN MITOCHONDRIA

R. Brian Beechey, Michael A. Brailsford and Andrew G. Thompson

Department of Biochemistry and Agricultural Biochemistry
The University College of Wales
Aberystwyth, Dyfed SY23 3DD

INTRODUCTION

The transport of pyruvate from the cytosol to the mitochondria is an essential feature of the energy metabolism of mammalian and the non-photosynthesising plant cell.

The functional aspects of the pyruvate translocator have been well characterised in mammalian mitochondria[1-7]. Pyruvate has been shown to enter rat liver and heart mitochondria on a saturable electroneutral carrier which is inhibited by a series of analogues of α-cyanocinnamic acid. In contrast, little is known about the pyruvate translocator of plant mitochondria.

Here we present data on; 1) the characteristics of the uptake and the oxidation of pyruvate by mitochondria isolated from the endosperm of *Ricinus communis*, the castor bean, and 2) the extraction and partial purification (using detergent solubilisation followed by hydroxyapatite chromatography) of the pyruvate transporter from castor bean mitochondria.

RESULTS AND DISCUSSION

Oxidative Properties of Castor Bean Mitochondria

Mitochondria were isolated from 5 day germinated castor bean endosperms by Percoll density gradient centrifugation as described previously[8]. Mitochondrial respiration was measured polarographically using an oxygen electrode in a closed cell at 30°C.

Typical rates of mitochondrial respiration, measured using an oxygen electrode and expressed as (nmol of oxygen consumed/mg of protein/min) are; malate (16), succinate (78), external NADH (67) and pyruvate (38). Pyruvate is oxidised only in the presence of "sparker" concentrations of malate (1.5mM) and the cofactors, coenzyme A (0.05mM) and thiamin pyrophosphate (0.1mM), which presumably exert their effects via the pyruvate dehydrogenase complex.

Antimycin A and cyanide, which are inhibitors of the electron transport pathway that utilises cytochromes a and a_3 as the terminal oxidase, were potent inhibitors of the oxidation of succinate and external

NADH. These compounds gave greater than 85% inhibition at concentrations of 1μM. In contrast, the oxidation of pyruvate in the presence of malate and the cofactors, CoA and thiamin pyrophosphate could only be 50% inhibited by either antimycin A or cyanide. Salicylhydroxamic acid (1mM), the inhibitor of the respiratory pathway through the cyanide-insensitive alternative terminal oxidase[9], similarly would not completely inhibit the pyruvate-dependent oxygen uptake. A combination of 0.3mM-salicylhydroxamic acid and either 1.0μM-antimycin A or 1.0μM-cyanide totally inhibited the oxidation of pyruvate. The addition of salicylhydroxamic acid (0.3mM) did not give more than 10% inhibition of the respiration rate with either succinate or external NADH.

These data indicate that castor bean mitochondria have both a functional cyanide-sensitive electron transport pathway, and a salicylhydroxamic acid-sensitive cyanide-insensitive terminal oxidase. These are organised in such a manner that the electrons from the succinate and external NADH dehydrogenases are channelled via the cyanide-sensitive pathway, whilst the electrons from the oxidation of pyruvate appear to be partitioned between these two pathways.

The partitioning of the electrons between the two systems is not a function of the rate of electron transport. Malonate (1mM) was used to decrease the oxygen consumption associated with the oxidation of succinate to that normally observed with pyruvate. The potency and the extents of inhibition shown by the various inhibitors was unchanged.

The inaccessibility to some substrates of the cyanide-insensitive alternative terminal oxidase has also been reported in cassava[10], sweet potato[11], and spinach leaf[12] mitochondria. In common with these findings our data is difficult to rationalise on the basis of a homogeneous pool of quinone which mediates the flow of electrons between all the dehydrogenases and the two terminal oxidases. These results therefore require one to hypothesise some sort of functional compartmentation of the ubiquinone pool or a significant degree of association between specific dehydrogenases and terminal oxidases.

The inhibition of pyruvate oxidation by a series of α-cyanocinnic acid analogues was studied. All of the α-cyanocinnamates were non-competitive inhibitors, and the titres of these compounds required to induce 50% inhibition of oxygen uptake (I_{50}) are shown in Table 1. The values reported are similar to those found previously in rat liver mitochondria. The compounds listed in Table 1 do not affect the oxidation of succinate, added NADH or malate, indicating that they do not inhibit the activity of the electron transport chain and the associated dehydrogenases.

Table 1. The inhibition of pyruvate oxidation in castor bean mitochondria by α-cyanocinnamate and its analogues.

INHIBITOR	I_{50} (nmol/mg of protein)
UK5099	0.9
α-cyanocinnamate	10.5
α-cyano-4-hydroxycinnamate	14.7

The Accumulation of Pyruvate by Castor Bean Mitochondria

The accumulation of [2-^{14}C]pyruvate by castor bean mitochondria was studied using a silicone oil centrifugation procedure[8]. The uptake of pyruvate into intact mitochondria could be rapidly inhibited in a non-competitive manner by UK5099 at a titre of 1.0nmol/mg of protein. We have determined the kinetics of this uptake using a inhibitor-stop technique with UK5099. The uptake of pyruvate into castor bean mitochondria was saturable, with a Km for pyruvate of 0.10mM and a V_{max} of 0.95nmol/min/mg of mitochondrial protein.

It is clear however, that the observed rates of pyruvate accumulation are insufficient to account for the pyruvate oxidation rates shown earlier. Furthermore, the levels of pyruvate accumulation are unaffected by the addition of 1μM-antimycin A to the incubation medium.

We have further studied the apparent loss of radiolabel from castor bean mitochondria by determining the tricarboxylic acid composition of the supernatants from the uptake studies, using two dimensional cellulose t.l.c.[8,13]. This was done in the presence and absence of either UK5099 or antimycin A (Table 2).

In the absence of UK5099 and antimycin A there is a gradual loss of radiolabel associated with pyruvate, and a concomitant increase in radiolabel associated with malate and citrate. No significant levels of radiolabel were found in other TCA cycle intermediates or CO_2 (which suggests that no cycling of the [^{14}C]malate occurs). In the incubations where UK5099 and antimycin A had been added prior to the addition of pyruvate most of the radiolabel remains associated with pyruvate.

We conclude that in the absence of antimycin A, pyruvate does not accumulate appreciably, as it is metabolised to malate and citrate which are removed from the mitochondrial matrix. In the presence of antimycin A or UK5099 most of the radiolabel remained in the supernatant as pyruvate. However, the addition of antimycin A to the uptake assays does not signficantly increase the levels of [2-^{14}C]pyruvate accumulated in the castor bean mitochondria. It seems therefore that whilst pyruvate can be accumulated in the presence of antimycin A, the continued uptake of pyruvate by castor bean mitochondria appears to be linked to the

Table 2. The metabolism of [2-^{14}C]pyruvate by castor bean mitochondria.

Incubation time(s)	Additions	% of ^{14}C labelled compounds in supernatants		
		Pyruvate	Citrate	Malate
30	-	69	18	13
180	-	43	36	21
300	-	12	59	29
300	10μM-UK5099	94	4	2
300	1μM-antimycin A	95	3	2

metabolism of the pyruvate and the subsequent movement of the products from the mitochondrial matrix.

The Extraction and Purification of the Pyruvate Transporter from Castor Bean Mitochondria

Detergent extraction followed by hydroxyapatite chromatography has been used to extract and purify to varying degrees the carriers for ADP/ATP[14,15] phosphate[16-18] tricarboxylates[19], dicarboxylates[20], α-oxoglutarate[21,22] glutamate/aspartate[23] and pyruvate[22] from mammalian mitochondria. We have used a similar two step technique to extract and partially purify the pyruvate transporter from castor bean mitochondria[24].

The first step involves the solubilisation of the mitochondrial membranes with 3% (v/v) Triton X114 in the presence of cardiolipin for 20min at $0^{o}C$. This is followed by purification of the X-114 extract using hydroxyapatite (HTP) column chromatography. The activity of the solubilised pyruvate transporter was determined by reincorporation of the extracted protein into pyruvate loaded liposomes, which were then assayed for the exchange of intraliposomal pyruvate for externally added [2-^{14}C]pyruvate.

It was found that the total exchange activity and the protein solubilised increased as the Triton X114 was increased from 1-3%. Further increases in the Triton concentration solubilised more protein but the total reconstituted pyruvate-exchange activity did not change.

A comparison of the SDS-polyacrylamide gel electrophoresis patterns of the proteins extracted from the mitochondrial membranes with Triton X-114 (Fig 1, track 2) and the HTP-eluate (Fig 1, track 3) shows the effectiveness of HTP-chromatography as a purification step. More than 90% of the Triton X-114 solubilised proteins bind to the HTP-column, and associated with this purification is a 151 fold increase in the specific transport activities measured (Triton X114 extract = 0.92nmol/min/mg of protein; HTP-eluate = 139nmol/min/mg of protein). It can also be seen that the HTP-eluate contains 6 major protein bands of M_r 74kDa, 66kDa, 34kDa, 32kDa, 30kDa and 12kDa. It is not known at the present time which of these bands corresponds to the pyruvate carrier in castor bean mitochondria.

The time course for the α-cyanocinnamatesensitive exchange of intra-proteoliposomal pyruvate for external [2-^{14}C]pyruvate was linear for approximately 1min and approached a plateau after 5min at 320nmol/mg of protein. The α-cyanocinnamate-sensitive pyruvate exchange comprises between 55-65% of the total pyruvate accumulation. α-cyanocinnamate-insensitive accumulation and/or binding (carryover) of radiolabelled

Table 3. Effects of α-cyanocinnamates on the pyruvate exchange activity of proteoliposomes reconstituted with HTP-eluate.

INHIBITOR	I_{50} (μmol/mg or protein)
α-cyanocinnamate	0.03
UK5099	0.70
α-cyano-4-hydroxycinnamate	27.00

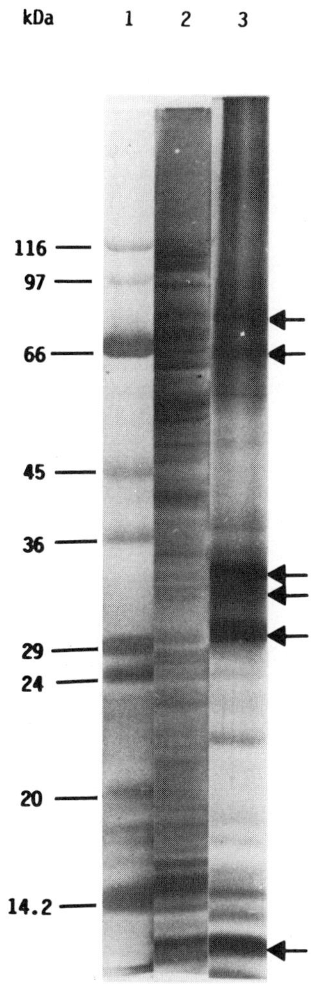

Fig. 1. SDS-Electrophoretic Pattern of the Triton X-114 Extract and the HTP-Eluate from Castor Bean Mitochondria

SDS-polyacrylamide gel electrophoresis was performed in a 10-16% linear gradient and the gel silver stained. Track 1 = marker proteins (20μg); Track 2 = Triton X-114 extract (100μg); Track 3 = HTP eluate (10μg).

pyruvate was determined in control proteoliposomes which were preincubated with α-cyanocinnamate prior to the start of the assay. In identical experiments using liposomes without any incorporated protein α-cyanocinnamate-sensitive pyruvate exchange was not observed.

In common with pyruvate transport in castor bean mitochondria the pyruvate exchange activity in proteoliposomes reconstituted with HTP-eluate is sensitive to various analogues of α-cyanocinnamate (Table 3). The relative potencies of these inhibitors in proteoliposomes (α-cyanocinnamate>UK5099>α-cyano-4-hydroxycinnamate) is somewhat different from those found in the intact mitochondria where UK5099>α-cyanocinnamate>αcyano-4-hydroxycinnamate.

From these data it can be seen that the reconstituted transport activity exhibits properties and inhibitor sensitivities which are characteristic of the pyruvate carrier in intact castor bean mitochondria. Experiments are presently underway to purify the translocator to homogeneity and to further study its activity.

REFERENCES

1. S. Papa, A. Francavilla, G. Paradies and B. Meduri (1971) FEBS Lett. 12, 285-287.
2. G. Paradies and S. Papa (1975) FEBS Lett. 52, 149-152.
3. G. Paradies and S. Papa (1977) Biochim. Biophys. Acta 462, 333-346
4. A. P. Halestrap and R. M. Denton (1974) Biochem. J. 138, 313-316.
5. A. P. Halestrap (1975) Biochem. J. 148, 85-96.
6. A. P. Halestrap (1978) Biochem. J. 172 377-387.
7. A. P. Halestrap, R. D. Scott & A. P. Thomas (1980) Int. J. Biochem. 11, 97-105.
8. M. A. Brailsford, A. G. Thompson, N. Kaderbhai and R. B. Beechey (1986) Biochem. J. 239, 355-361.
9. G. R. Schonbaum, W. D. Bonner, B. T. Storey and J. T. Bahr (1971) Plant Physiol. 44, 115-128.
10. P. F. Tomlinson and D. E. Moreland (1975) Plant Physiol. 55, 365-369.
11. R. Douce, A. L. Moore and M. Neuberger (1977) Plant Physiol. 60, 626-628.
12. S. Huq and J. M. Palmer (1978) in Plant Mitochondria (G. Ducet and C. Lance, eds.) pp. 225-232, Elsevier/North-Holland Biomedical Press, Amsterdam.
13. W. F. Myers and K. Y. Huang (1969) Methods Enzymol. 13, 431-434.
14. M. Klingenberg, P. Riccio and H. Aquila (1978) Biochim. Biophys.Acta 503, 193-210.
15. R. Kramer and M. Klingenberg (1979) Biochemistry 18, 4208-4215
16. Wehrle, J. P. and P. L. Pederson (1983) Arch. Biochem. Biophys. 223, 477-483.
17. H. V. J. Kolbe, D. Costello, A. Wong, R. C. Lu and H. Wohlrab (1984) J. Biol. Chem. 259, 9115-9120.
18. H. Wohlrab, A. Collins and D. Costello (1984) Biochemistry 23, 1057-1064.
19. I. Stipani and F. Palmieri (1983) FEBS Lett. 161, 269-274.
20. R. S. Kaplan and P. L. Pederson (1985) J. Biol. Chem. 260, 1023-1029
21. F. Bisaccia, C. Indiveri and F. Palmieri (1985) Biochim. Biophys. Acta 810, 362-369.
22. M. J. Nalecz, K. A. Nalecz, C. Broger, R. Bolli, L. Wojtczak and A. Azzi (1986) FEBS Lett. 196, 331-336.
23. R. Kramer (1984) FEBS Lett. 176, 351-354.
24. M.A. Brailsford, A.G. Thompson, N. Kaderbha, and R.B. Beechey (1986) Biochem. Biophys. Res. Comm. <u>in press</u>.

PYRUVATE TRANSPORT BY THERMOGENIC TISSUE MITOCHONDRIA

Michael O. Proudlove, [1]R. Brian Beechey and Anthony L. Moore

Department of Biochemistry, University of Sussex, Falmer, Brighton BN1 9QG, U.K. and [1]Department of Biochemistry and Agricultural Biochemistry, University College of Wales, Penglais, Aberystwyth, Dyfed SY23 3DD, U.K.

INTRODUCTION

In the thermogenic spadices of *Arum maculatum* and *Sauromatum guttatum* the breakdown of stored starch leads to the production of large quantities of cytosolic pyruvate. Further metabolism of this keto acid, by pyruvate dehydrogenase in the matrix, requires the movement of pyruvate across the inner mitochondrial membrane. The rate of influx or oxidation may then constitute a rate limiting step in TCA cycle activity and the subsequent alternative oxidase-dependent heat generation, necessary for the volatalization of insect attracting compounds required for pollination of these species[1]. The compounds α-cyano-4-hydroxycinnamate (CHCA) and α-cyano-β-(1-phenylindol-3-yl) acrylate (UK5099) have been found to inhibit pyruvate oxidation in rat liver and heart mitochondria[2,3] and the former to non-competitively inhibit pyruvate oxidation in mitochondria isolated from etiolated maize hypocotyls[4], reportedly an effect on a pyruvate-specific carrier. We therefore decided to investigate the effect of these cinnamate derivatives on the kinetics of both pyruvate oxidation and transport in mitochondria isolated from thermogenic tissues. Their rapid rates of pyruvate-dependent oxygen consumption allows a more accurate estimate of cinnamate inhibition to be made in the presence of higher concentrations of these compounds.

MATERIALS AND METHODS

Corms of *Sauromatum guttatum* were purchased from Bees Ltd., Chester, U.K., stored at 4°C until required and spadices germinated by transferring to a glasshouse at 20-25°C. Spadices of *Arum maculatum* were collected from local woodlands and used the same day. Aldrich (Gillingham, U.K.) supplied the CHCA and UK5099 and silicone oils (AR200 and AR20) were gifts from Pfitzer (U.K.) Ltd., Sandwich, U.K. and Wacker-Chemie, Munich, W. Germany, respectively. Radiolabelled, [1-^{14}C]-, pyruvate was purchased from Amersham International plc, Amersham, U.K. and all other chemicals from Sigma Chemical Co. Ltd., Poole, U.K. or BDH Chemicals Ltd., Chadwell Heath, U.K.. Mitochondria were prepared from each tissue as previously described[5,6], final pellets being resuspended in medium A (0.3M mannitol; 10mM KCl; 5mM $MgCl_2$; 10mM phosphate; 10mM HEPES, pH 7.2) at a concentration of 60mg/ml. Oxygen consumption was measured with a Hansatech (King's Lynn, U.K.) and pyruvate uptake by silicone oil centrifugal filtration[6], using antimycin A

(0.2nmoles/mg protein) and n-propyl gallate (100μM) to prevent pyruvate oxidation and 1mM ATP to generate a Δ p. Protein was estimated by the method of Lowry et al.[7].

RESULTS

Mitochondria from thermogenic spadices oxidize the glycolytic end-product pyruvate at higher rates than observed for most other plant tissues and have no requirement for exogenous cofactors, necessary for dehydrogenase activity in some plant mitochondria[8] (Fig 1A). With certain preparations from A. maculatum, however, pyruvate was only oxidized slowly, the rate being markedly enhanced by the addition of 5mM aspartate (Fig 1B). Reversal of this stimulation by 5mM aminooxyacetate suggests that the increase in O_2 consumption is due to pyruvate transaminating with aspartate, to produce oxaloacetate and alanine, and the former may subsequently condense with acetyl CoA, a product of pyruvate decarboxylation, to produce citrate. The removal of matrix acetyl CoA will reduce feedback inhibition of pyruvate dehydrogenase[9] and probably results in a mixture of pyruvate- and citrate-dependent O_2 consumption, citrate also being readily oxidized by these mitochondria[6] (Fig 1C). Direct addition of low [oxaloacetate] had an inhibitory, rather than stimulatory, effect on pyruvate oxidation. This is because this keto acid, <u>via</u> a reversal of malate dehydrogenase activity, will lower the matrix level of NADH and hence the flux through the respiratory chain.

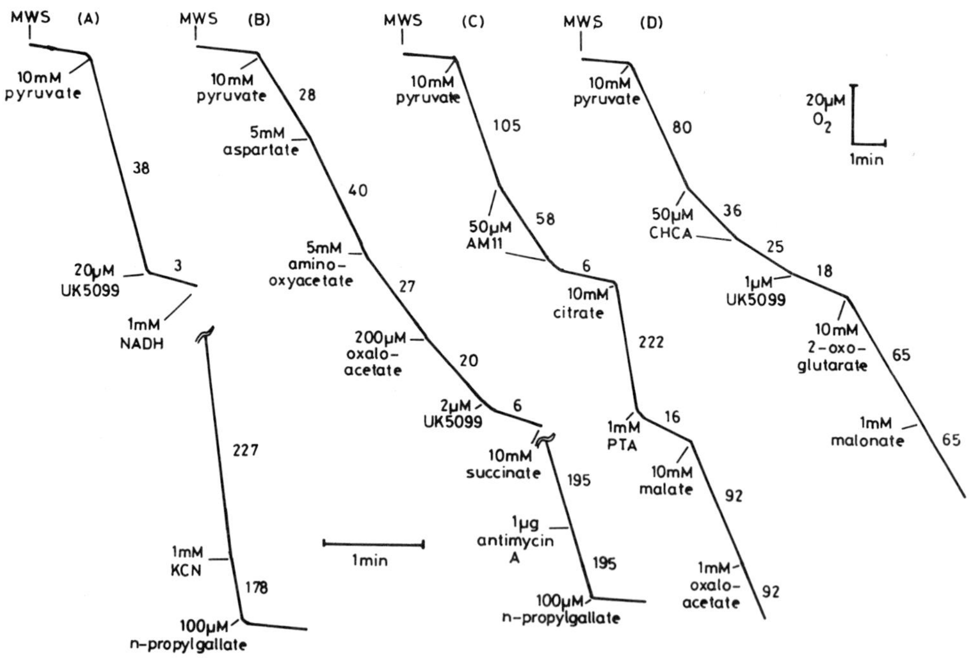

Fig 1. Substrate oxidation by <u>Sauromatum guttatum</u> (A) and <u>Arum maculatum</u> (B-D) mitochondria. Additions were made as indicated and numbers by traces refer to the rates of oxygen consumption, in nmoles/min/mg protein. Breaks in traces A & B represent increases in chart speed, to equate with the time bar in the lower left corner.

The long chain acid maleimide, AM11[10] was found to inhibit pyruvate oxidation[11] (Fig 1C). Shorter chain AM also inhibited, the degree of inhibition being proportional to chain length, there being no inhibitory effect below AM7 (results not shown). Limited pyruvate transport experiments, with *S. guttatum* mitochondria and a range of [AM11], proved inconclusive. For, although there was a measurable inhibition of pyruvate uptake, it was not as marked as seen for oxidation studies. This suggests that the maleimide was reacting with thiol groups which were associated with both pyruvate transport and pyruvate oxidation. It further suggests that, as was found for the glycine decarboxylase complex[12], pyruvate dehydrogenase is located, at least partially, in the inner mitochondrial membrane.

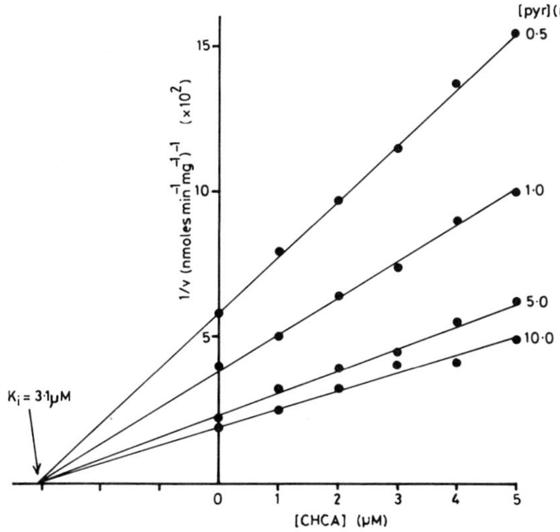

Fig 2. Dixon plot of pyruvate oxidation by *Arum maculatum* mitochondria. Mitochondria were pre-incubated with the required [CHCA] for 2min prior to the addition of different [pyruvate].

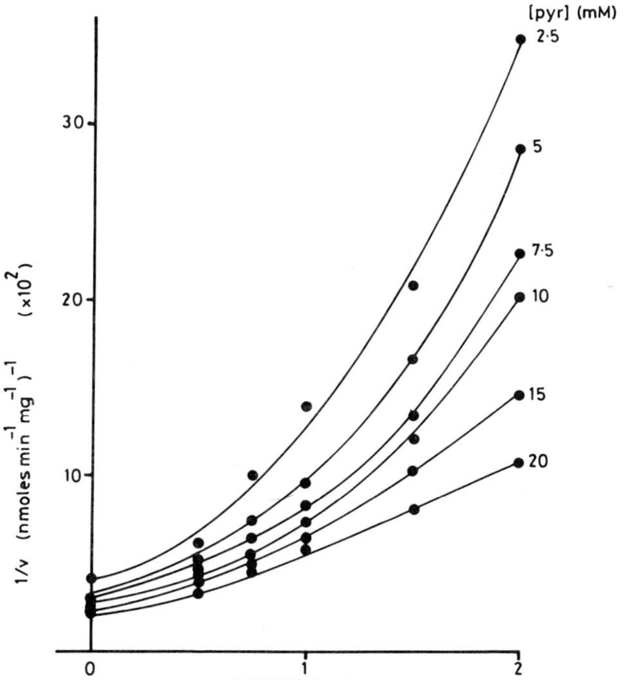

Fig 3. Dixon plot of UK5099 inhibition of pyruvate oxidation by *Sauromatum guttatum* mitochondria. Different [UK5099] were pre-incubated with mitochondria for 2min prior to the addition of specified [pyruvate].

The cinnamate derivatives, CHCA and UK5099, also inhibited pyruvate oxidation, at [micromolar], the latter having the greater effect. Their selective inhibition of pyruvate is emphasized by the restoration of O_2 consumption by complex I, complex II and external dehydrogenase substrates[2] (Fig 1). Dixon plots for the effect of the two derivatives on pyruvate oxidation show that, as found for other systems[3,4], CHCA inhibits non-competitively but UK5099 exhibits more complex kinetics (Figs 2_3 and 3). The estimated Ki for CHCA (3μM) is similar to that reported[3] for rat heart mitochondria (2μM) but is much lower that that calculated[4] for pyruvate oxidation in etiolated maize mitochondria (approximately 600μM). This dissimilarity in values for plant mitochondria may reflect the differences in the roles these organelles play in their respective tissues.

When transport was measured directly, by centrifugal filtration, it was found that values for Vm and Km varied from those calculated for oxidation studies (Table 1). These results suggest that movement of pyruvate

Table 1. The kinetic constants of pyruvate oxidation and transport by thermogenic tissue mitochondria. Experimental details were as outlined in Methods and values were derived by recalculating sections of the data, from Figs 2-5, by Eadie-Hoftsee plots.

Tissue	Vm (nmoles uptake/min/mg)		Km (mM pyruvate)	
	O_2	pyruvate	oxidation	uptake
<u>A. maculatum</u>	52.6	83.4	1.35	0.85
<u>S. guttatum</u>	48.2	82.1	2.1	1.2

across the inner mitochondrial membrane does not limit the further metabolism of this keto acid by pyruvate dehydrogenase. The effect of CHCA and UK5099 on the uptake of pyruvate into the mitochondrial matrix are presented in Figs 4 and 5. Non-competitive, Michaelis-Menton kinetics were seen for the effect of CHCA on pyruvate influx but, again, UK5099 inhibition resulted in a non-linear response. The estimated Ki, for the inhibition of transport by CHCA, was calculated to be 1.75μM (Dixon plot, not shown).

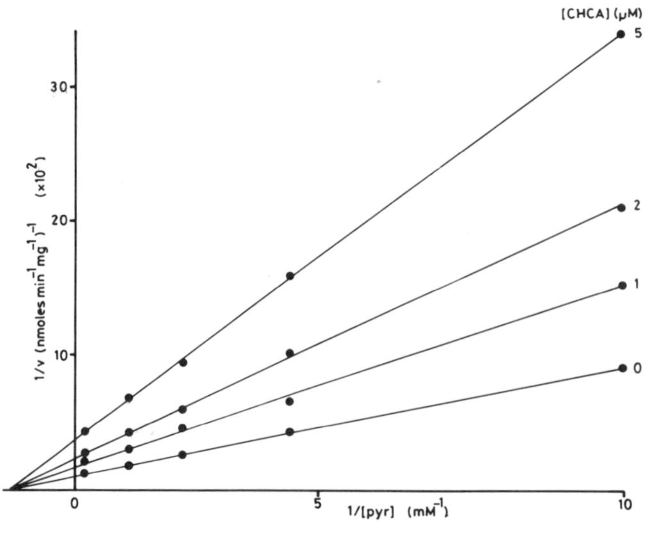

Fig 4. Lineweaver-Burk plot of CHCA inhibition of pyruvate uptake by <u>Arum maculatum</u> mitochondria. The transport of different [pyruvate] into the matrix was followed after pre-incubating mitochondria in the appropriate [CHCA] for 2min. Reactions were terminated after 20s by the addition of 1mM CHCA, followed by rapid centrifugal filtration, as outlined in Methods.

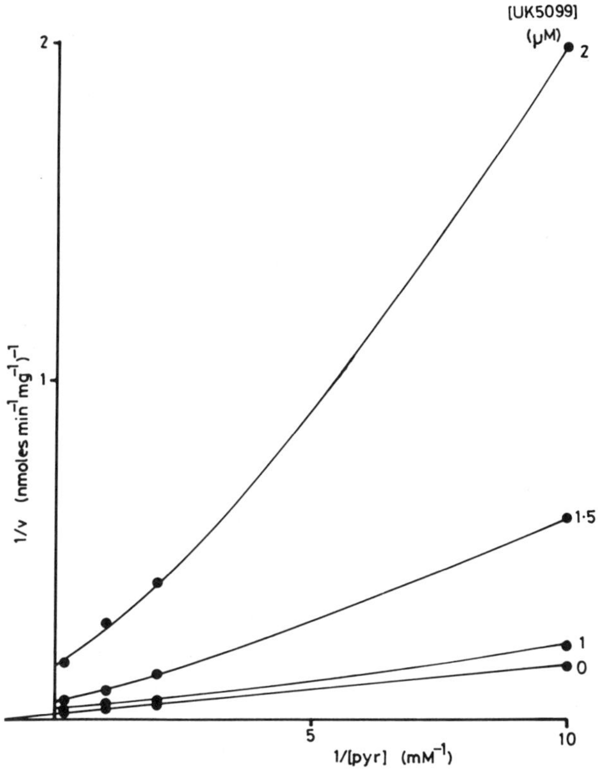

Fig 5. Lineweaver-Burk plot of UK5099 inhibition of pyruvate uptake by *Arum maculatum* mitochondria. Conditions were as described in Fig 4, using varying [UK5099], and the inhibitor-stop was 100μM UK5099.

In rat heart mitochondria, Shearman and Halestrap[3] explain this type of response to UK5099 as being due to this compound having a Ki much lower than the concentration of carrier molecules. In some ways, a simpler view would be that UK5099 dissociates only very slowly from the enzyme-inhibitor complex and this will constitute a rate limiting step in the transport, and subsequent oxidation, of pyruvate. At high [substrate]: low [inhibitor], pyruvate is able to bind the carrier more readily and kinetics tend to Michaelis-Menton (see Figs 3 and 5). When the [inhibitor] is increased and [pyruvate] decreased the kinetics deviate from linearity, UK5099 having a far greater effect than would have been expected. Under these conditions, more of the carrier molecules exist as the enzyme-inhibitor complex and pyruvate binding and transport will be a function of the dissociation of this complex and the successful formation of the enzyme-substrate complex.

CONCLUSIONS

1. Pyruvate is transported across the inner membrane of thermogenic tissue mitochondria by a specific carrier.
2. The carrier is inhibited by the substituted cinnamate derivatives CHCA and UK5099.
3. CHCA inhibits non-competitively, for UK5099 the kinetics are more involved, depending on the dissociation complex of the enzyme-inhibitor complex.
4. Both oxidation and transport of pyruvate are sensitive to the long chain maleimide, AM11.
5. Transport of pyruvate does not limit pyruvate dehydrogenase activity.

ACKNOWLEDGEMENT

This work was supported by a grant from the AFRC to ALM.

REFERENCES

1. B. J. D. Meeuse, Physiological and Biochemical Aspects of Thermogenic Respiration in the Aroid Appendix, in: "The Physiology and Biochemistry of Plant Respiration," J. M. Palmer ed., University Press, Cambridge (1984).
2. A. P. Halestrap, The mitochondrial pyruvate carrier; kinetics and specificity for substrates and inhibitors, Biochem. J. 148:85 (1975).
3. M. S. Shearman and A. P. Halestrap, The concentration of the mitochondrial pyruvate carrier in rat liver and heart determined with α-cyano-β-(1-phenylindol-3-yl)acrylate, Biochem. J. 223:673 (1984).
4. D. A. Day and J. B. Hanson, Pyruvate and malate transport and oxidation in corn mitochondria, Plant Physiol. 59:630 (1977).
5. A. L. Moore and M. O. Proudlove, Mitochondria and Sub-Mitochondrial Particles, in: "Isolation of Membranes and Organelles from Plant Cells," J. L. Hall and A. L. Moore eds., Academic Press, London (1983).
6. M. O. Proudlove and A. L. Moore, Metabolite fluxes across the inner membrane of plant mitochondria - inhibition by phthalonic acid, Planta 160:407 (1984).
7. O. H. Lowry, N. J. Rosebrough, A. L. Farr and R. J. Randall, Protein measurement with the Folin phenol reagent, J. Biol. Chem. 193:265 (1951).
8. M. Neuburger and R. Douce, Transport of Various Coenzymes in Percoll Purified Plant Mitochondria, in: "Plant Mitochondria: Structural, Functional and Physiological Aspects," A. L. Moore and R. B. Beechey eds., Plenum Press, New York (1986).
9. I. M. Moller and J. M. Palmer, Regulation of the Tricarboxylic Acid Cycle and Organic Acid Metabolism, in: "The Physiology and Biochemistry of Plant Respiration," J. M. Palmer ed., University Press, Cambridge (1984)
10. D. G. Griffiths, M. D. Partis, R. N. Sharp and R. B. Beechey, N-poly-methylene-carboxymaleimides - a new class of probes for membrane sulphydryl groups, FEBS Lett. 134:261 (1981).
11. D. G. Griffiths, M. D. Partis, A. L. Moore, M. O. Proudlove and R. B. Beechey, N-polymethylene-carboxymaleimides: probes for sulphydryl groups associated with peripheral and integral membrane proteins, Biochem. Soc. Trans. 11:753 (1983).
12. A. L. Moore, M. O. Proudlove, M. D. Partis and R. B. Beechey, The effect of N-polymethylene-carboxymaleimides on glycine movement into pea leaf mitochondria, Adv. Phos. Res. 3:909 (1984).

AN ENZYME PROFILE OF ISOLATED PLANT MITOCHONDRIA

K.L. Soole, I.B. Dry and J.T. Wiskich

Botany Department
Adelaide University
Adelaide, Australia

The most common technique used to determine the location of enzymes associated with mitochondria involves measuring the latency of enzyme activity in mitochondria treated with different concentrations of detergent[3]. The interpretation of such data is not always straightforward. In this study we demonstrate how proteolysis may be used to selectively remove enzymes from the various mitochondrial compartments and hence be used as a tool to determine enzyme location. The results presented demonstrate that interpretation of enzyme location based on latency experiments may not always be valid.

METHODS

Preparation

Turnip mitochondria were prepared as previously described[2] and purified on a discontinuous Percoll density gradient composed of 5 ml 50% (v/v); 10 ml 33% (v/v); 12 ml 28% (v/v); 7 ml 15% (v/v) Percoll containing 0.3M sucrose, 10 mM phosphate buffer pH 7.4 and 0.1% BSA and centrifuged at 43,000 g for 40 min. The mitochondrial fraction was collected and washed twice with 0.4 M sucrose and 10 mM TES pH 7.4. Mitochondria were resuspended to a protein concentration of 1.0 - 1.2 mg protein/ml.

Proteolysis

The mitochondrial suspension (1.0 mg protein/ml) was incubated with specified concentrations of Pronase, a mixture of non-specific bacterial proteases (Sigma, Protease IX), for 15 min in a shaking waterbath at 30°C. Proteolysis was stopped by the addition of 1 mM PMSF (phenylmethylsulfonyl fluoride - Sigma) and the suspension was placed immediately on ice. Aliquots were taken for enzyme assays. All enzymes investigated were when solubilized, found to be completely sensitive to Pronase digestion.

Enzyme Assays

Hexokinase (HK) was assayed as in Tanner et al.[6] but in 10 mM TES buffer at pH 7.2. Other assays were adenylate kinase (AK)[1], isocitrate dehydrogenase (ISCDH) and aspartate aminotransferase (ASAT)[4]. Phthalonate (0.4 mM) was included in the ASAT assay. Alanine aminotransferase (ALAT) was assayed using a similar method to ASAT, except the reaction was started with 10 mM alanine and 10 mM 2-oxoglutarate and malate dehydrogenase (MDH) was replaced by lactate dehydrogenase. MDH activity was measured by monitoring NADH oxidation at 340 nm in a total of 3 ml containing 0.25 M sucrose, 10 mM KH_2PO_4 buffer, 10 mM TES, 5 mM $MgCl_2$ at pH 7.2 and 0.01 - 0.02 mg protein. The reaction was initiated with 1 mM OAA. Protein concentrations were estimated by the method of Lowry[5].

RESULTS AND DISCUSSION

On the basis of classical latency experiments (Table 1), it is clear that ISCDH is entirely located in the matrix and that HK and AK are present external to the inner membrane. When considering MDH, ASAT and ALAT the location of the enzymes is not as clear. The bulk of the activity of these enzymes is in the matrix but there appears to be some activity external to the inner membrane (Table 1). The location of these enzymes was investigated using Pronase digestion. Figure 1 represents enzyme assays in the presence of a detergent (Triton X-100) after intact mitochondria had been treated with varying concentrations of Pronase. It can be seen that high levels of Pronase were required to decrease ISCDH activity whereas HK activity was completely removed at very low concentrations. These results are consistent with an outer membrane location for HK[6] and a matrix location for ISCDH. A concentration of Pronase marginally higher than that required to remove HK activity completely removed AK activity (Fig. 1) confirming the presence of this enzyme in the intermembrane space[3]. The hydrolysis profiles of MDH, ASAT and ALAT under these conditions were similar to that of ISCDH indicating a matrix location for these enzymes. These results are consistent with the latency experiments.

When considering the external location of MDH, ASAT and ALAT the assays were performed with intact mitochondria after they were treated with Pronase (Fig. 2). This enabled specific consideration of the apparent external activity of these enzymes. MDH and ASAT activity decreased at a very low Pronase concentration concomitantly with HK suggesting that some MDH and ASAT is associated with the outer membrane. Proteolysis did not completely remove all the MDH and ASAT activity. Further investigation (data not shown)

Table 1. Enzyme activities of Turnip mitochondria after treatment with detergent or Pronase

	Activity with Detergent	Intact Mito-chondria	% External	Pronase Sensitive	% External
MDH	43821	5783	13	3627	8
ASAT	387	108	18	37	10
ALAT	400	120	30	0	0
HK	105	105	100	105	100
AK	1057	1057	100	1057	100
ISCDH	232	0	0	0	0

Enzyme rates nmoles NADH/min/mg protein. Pronase sensitive activity was determined on intact mitochondria after treatment with 15 μg Pronase/mg protein for 15 min at 30°C.

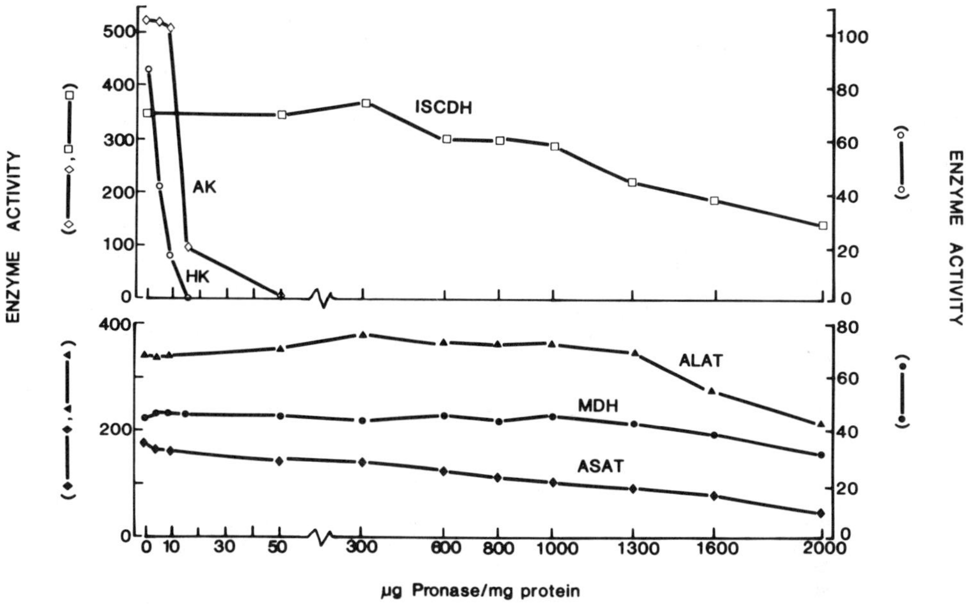

Fig. 1. Enzyme assays of Pronase-treated mitochondria performed in the presence of 0.04% Triton X-100. Rates are expressed in nmoles NADH/min/mg protein except for MDH where the units are µmoles NADH/min/mg protein.

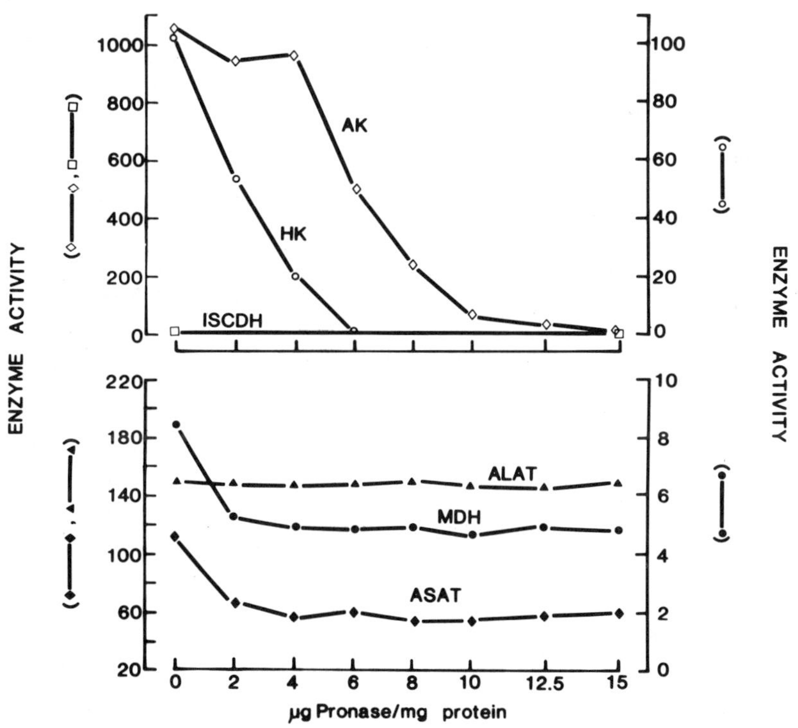

Fig. 2. Enzyme assays in the absence of Triton X-100 of Pronase-treated mitochondria. Rates are expressed in nmoles NADH/min/mg protein except for MDH where the units are µmoles NADH/min/mg protein.

revealed that some of this "residual" activity could be accounted for by peroxisomal contamination. In the case of ASAT a further contribution may be from matrix-produced OAA leaking out of the mitochondria and contributing to the assay. Phthalonate, an OAA uptake inhibitor[4], was included in the assay to minimize the efflux of matrix OAA, but some efflux is still occurring. There was no decrease in ALAT activity at this Pronase concentration indicating that there is no ALAT associated with the outer membrane (Fig. 2). The ALAT assay measures pyruvate production and because there is no change in activity after Pronase treatment and no ALAT activity could be detected in turnip peroxisomes the "external" activity represents the efflux of pyruvate produced by matrix enzyme. If the treatments with and without detergent (Table 1) are considered, the efflux of pyruvate produced in the matrix gives the misleading impression of some ALAT being located external to the inner membrane.

In conclusion, it was possible to identify the location of some enzymes within the various mitochondrial compartments after treatment with increasing concentrations of Pronase and recording the hydrolysis profiles of HK, AK, and ISCDH as marker enzymes for each compartment; that is, the outer membrane the intermembrane space and the matrix space respectively. Comparison of the hydrolysis profiles of MDH, ASAT and ALAT with the marker enzymes placed them in the matrix space. Some MDH and ASAT was found to be associated with the outer membrane but ALAT was completely contained in the matrix space.

Pronase treatment appears to give a more accurate estimation of the amount of enzyme present in the outer compartments of mitochondria than estimations derived from the comparison of enzyme rates with and without detergent as shown here with ASAT and ALAT (Table 1). Enzymic hydrolysis is a useful technique for the selective and progressive removal of proteins from the outside to the inside of the mitochondrion and hence is a useful tool in determining enzyme location. The results presented here may represent the true distribution of mitochondrial enzymes _in vivo_. However the association of MDH and ASAT with the outer membrane may occur during mitochondrial isolation.

ACKNOWLEDGEMENTS

Financial support for this research was provided by an Australian Commonwealth Postgraduate Award to K.L. Soole and by the Australian Research Grants Committee. The authors would like to thank Carol Robinson for typing the document.

REFERENCES

1. H.U. Bergmeyer, "Methods of Enzymatic Analysis", Academic Press, New York (1974).
2. D.A. Day and J.T. Wiskich, _Arch. Biochem. Biophys_. 171:117 (1975).
3. D.A. Day, G.P. Arron, and G.G. Laties, _J. Exp. Bot_. 30(116):539 (1979).
4. D.A. Day and J.T. Wiskich, _Arch. Biochem. Biophys_. 271(1):100 (1981).
5. O.H. Lowry, N.J. Rosebrough, A.L. Farr and R.J. Randall, _J. Biol. Chem_. 193:265 (1951).
6. G.J. Tanner, L. Copeland and J.F. Turner, _Plant Physiol_. 72:659 (1983).

ANOTHER TYPE OF ALTERNATIVE ELECTRON TRANSFER PATHWAY IN THE YEAST *Candida parapsilosis*

Martine Guerin, Nadine Camougrand, Alain Cheyrou and Michèle-France Henry

Institut de Biochimie Cellulaire et Neurochimie du CNRS
1 rue Camille Saint-Saëns
33077 Bordeaux cedex, France

INTRODUCTION

C.parapsilosis SP1 (CBS 7154) is devoid of alcohol dehydrogenase I and therefore dependent on oxidative metabolism for growth (1). In spite of this constraint, it is able to grow on high glucose concentrations and on a glycerol medium supplemented with antimycin A or drugs acting at the mitoribosomal protein synthesis level, owing to the peculiar organisation of its respiratory chain (1, 2). Indeed, besides the normal respiratory pathway, *C.parapsilosis* develops a secondary mitochondrial electron transfer chain, which specifically reoxidizes cytoplasmic NADH failing to be oxidized at the fermentative level. This secondary pathway is insensitive to catabolic repression, resistant to antimycin A, therefore not branched at the ubiquinone level, but is inhibited by amytal, SHAM and high cyanide concentrations (1, 3).

Determination of the energetic yields of *C.parapsilosis* as a function of the energy source, in the presence or in the absence of antimycin A, allowed to demonstrate that this pathway was branched at phosphorylation site III (Camougrand et al., submitted to publication).

This paper aims to describe the different segments and oxidases of the respiratory chains of *C.parapsilosis*.

I. THE NAD(P)H DEHYDROGENASE-TRANSHYDROGENASE COMPLEX

It was previously demonstrated that in this yeast, glucose was mainly catabolized through the hexose monophosphate pathway, leading to a high NADPH production (Caubet et al., submitted to publication). NADPH is effectively reoxidized by mitochondria, though to a lesser rate than NADH. Determination of NADPH cytochrome c oxidoreductase activity on a crude mitochondrial suspension indicated that two kinetic systems were in presence, one with a high affinity for NADPH (Km = 10 $\pm$ 17 µM, Vmax = 11 $\pm$ 5 µmol NADPH-oxidized.min^{-1}.mg^{-1}) which is amytal-insensitive, and a second one with a low affinity (Km = 817 $\pm$ 200 µM, Vmax = 24 $\pm$ 2 µmol) NADPH-oxidized.min^{-1}.mg^{-1}) which is amytal-sensitive. Purification of mitochondria on a Percoll gradient led to the loss of the high affinity system insensitive to amytal and antimycin A. One can assume that this system resulted from a microsomal contamination of the crude mitochondrial preparation. It must be pointed out that purified mitochondria always retained the amytal-sensitive NADPH cytochrome c oxidoreductase activity.

Some organisms are able to reoxidize NADH and NADPH through the respiratory chain, but the optimum pH varies with the origin of the mitochondria (4). In C.parapsilosis, two pH optima were obtained equal to 6.5 and 5.25 for the reoxidation of NADH and NADPH respectively.

The oxidation of NADH and NADPH as only partly insensitive to antimycin A and cyanide (40-70%). However, the oxidation of these substrates differed in sensitivity to inhibition by mersalyl, amytal and SHAM (Table I)

TABLE I : EFFECT OF DIFFERENT INHIBITORS ON MITOCHONDRIAL RESPIRATION

Inhibitors	Oxygen uptake : percentage of inhibition NADH	NADPH
Amytal	20	42
SHAM	20	66
Mersalyl	30	10

It was demonstrated that in plant mitochondria NADPH dehydrogenase and NADH dehydrogenase are different enzymes which can be distinguished by their sensitivity to amytal, NADPH dehydrogenase only being sensitive to this inhibitor (5). It however appeared that in C.parapsilosis both dehydrogenases NAD(P)H were amytal sensitive, but the amytal pathway was more important with NADPH as substrate than with NADH. As for NADH, the combined effects of either amytal and antimycin A, or amytal and cyanide were additive leading to a 100% inhibition of oxygen uptake.

By submitting mitochondria to a digitonin-osmotic shock treatment to remove the outer membrane, it was demonstrated that both NAD(P)H dehydrogenase activities, amytal-sensitive, were recovered with the inner membrane fraction.

Is there a transhydrogenase ?

In view of the differencies existing between the sensitivities of oxygen uptake with NADH or NADPH towards amytal and antimycin A, the problem was to know whether the appearance of the sensitivity of NADH reoxidation to amytal was not in fact due to a partial reoxidation of NADH by NADPH dehydrogenase through the functioning of a transhydrogenase : $NADP^+ + NADH + nH^+ \rightleftharpoons NADPH + NAD^+ + nH^+$

On the one hand, it appeared that NADH oxidation was inhibited by successive $NADP^+$ additions whereas NADPH oxidation was NAD^+-insensitive. On the other hand, the arginine binding reagent, butanedione, a specific inhibitor of transhydrogenase (6) had not effect on NADPH reoxidation, but inhibited by 25% the reoxidation of NADH. In the presence of butanedione, the inhibition by amytal of NADH oxidation was largely prevented, leaving however an oxygen uptake which remained partly sensitive to the inhibitor. Butanedione, however, had no effect on NADPH reoxidation already inhibited by amytal.

From these data, we hypothesized that inner mitochondrial membrane of C.parapsilosis contained an enzymatic complex composed of two specific dehydrogenases specific to NADH and NADPH and of a transhydrogenase.

II. BRANCHING POINT OF THE SECONDARY PATHWAY

The fact that C.parapsilosis easily grows on a glycerol medium supplemented with antimycin A and therefore enhances the secondary route of electron flow, proved to be a useful tool to study the alternative pathway in this yeast.

When cytochrome spectra of mitochondria isolated from cells grown in the presence of antimycin A were recorded at room temperature as a function of cyanide concentration, cytochrome b was immediately reduced, whereas the reduction levels of cytochrome aa_3 and c were dependent on the cyanide concentration. Cytochorme aa_3 reduction was biphasic : a first maximum was reached at 0.1 mM KCN and a seocnd one at 10 mM KCN. Complete cytochrome c reduction is reached at only 10 mM KCN (7).

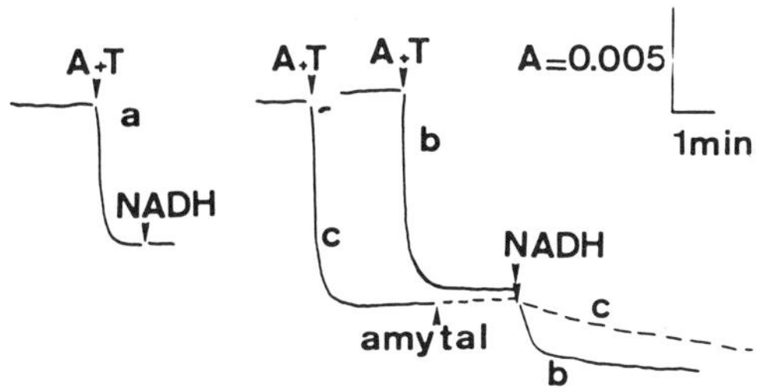

FIG.1 : KINETIC MEASUREMENTS OF CYTOCHROME c REDUCTION. Addition : 5 mM NADH, 5 mM ascorbate-TMPD (A+T), 2 mM amytal. (a) mitochondria from S.cerevisiae; (b)(c) antimycin A mitochondria from C.parapsilosis

Further kinetic experiments were performed with S.cerevisiae and C.parapsilosis mitochondria. It appeared that in S.cerevisiae used as a control, cytochrome c could be fully reduced by the couple ascorbate-TMPD (Fig. 1a). After addition of NADH, no extrareduction occurred. On the opposite, in C.parapsilosis only part of cytochrome c was reduced after ascorbate-TMPD addition. A subsequent addition of NADH led to a further cytochrome c reduction (Fig. 1b), which was sensitive to amytal (Fig. 1c). Thus, electron flow from NADH proceeded through a specific pool of cytochrome c, unaccessible to ascrobate. Afterwards, cytochrome c delivered electrons either to cytochrome aa_3 or to an alternative oxidase (7).

III. THE ALTERNATIVE OXIDASE

First, spectrophotometric measuremnts were performed on cells grown in the presence of antimycin A. Under these culture conditions, cellular respiration was 85% inhibited by SHAM alone. Cytochrome spectra were recorded on cell suspensions, glucose-reduced versus H_2O_2-oxidized. Under these conditions, all cytochromes b, c and aa_3 appeared. However, when the spectrum was recorded on glucose-reduced cells in the presence of 10 mM cyanide versus untreated cells, only two peaks appeared, cytochrome b at 562 nm and a cytochrome absorbing at 590 nm (Fig. 2A). The peak height between 575-590 nm of this latter cytochrome was very dependent of the cyanide concentration, reaching its maximum at nearly 20 mM KCN (7).

The same spectrophotometric studies were performed on mitochondria (Fig. 2B) isolated from cells grown on glycerol control) or on glycerol + antimycin A (ANTI-mitochondria). Spectra of ANTI-mitochondria, NADH-reduced + SHAM versus oxidized, were recorded at 77°K. On the control experiment (trace a) with control mitochondria, it can be seen that cytochromes b, C_1, c and aa_3 were reduced, the max of cytochrome aa_3 being shifted to 597 nm upon cyanide addition. In ANTI-mitochondria, NADH induced the reduction of the oxidase at 597 nm and the reduction of both cytochromes b (562 and 558 nm) but not the one of C_1 (trace b). Addition of SHAM promoted the reduction of part of cytochrome c but not that of C_1, increased the peak at 597 nm and above all induced the reduction of a cytochrome absorbing at 590 nm (trace c) (7).

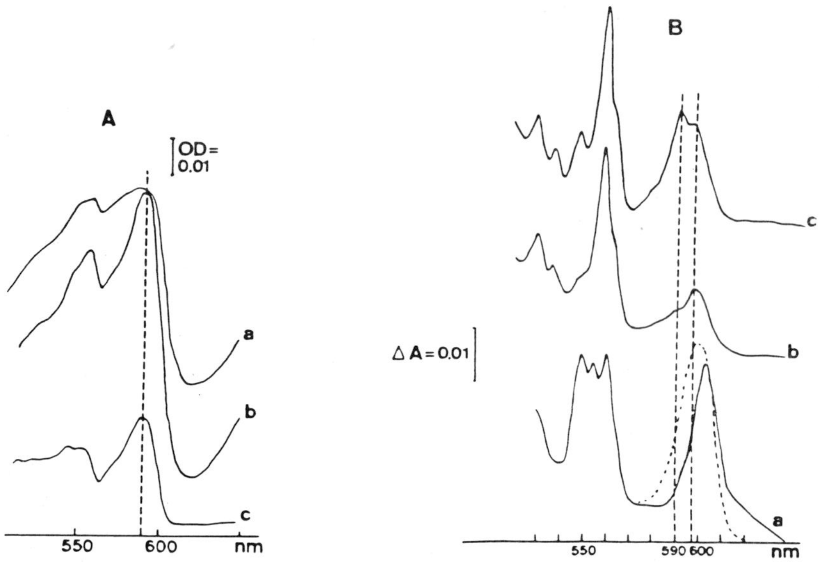

FIG. 2 A : CYTOCHROME SPECTRA OF GLUCOSE-REDUCED + KCN VERSUS UNTREATED CELLS at 20°C. (a) 0.1 mM KCN. (b) 10 mM KCN. (c) 10 mM KCN-reduced versus 0.1 mM KCN-reduced. B : CYTOCHROME SPECTRA OF MITOCHONDRIA AT -196°C. (a) control, NADH-reduced versus oxidized; ----- + 0.1 mM KCN. (b) antimycin A mitochondria NADH-reduced versus oxidized. (c) identical to (b) + preincubation with 1mM SHAM.

In conclusion from all these data, it appeared that *C.parapsilosis* develops a secondary mitochondrial electron transfer chain different from the alternative respiratory chains described in higher plants and some microorganisms, which specifically oxidizes NADH failing to be oxidized at the fermentative level and NADPH produced in excess at the hexose monophosphate pathway. At this time, three steps of this pathway were evidenced : 1. a NAD(P)H dehydrogenase-transhydrogenase complex located on the inner membrane; 2. a specific pool of cytochrome c and 3. an alternative oxidase absorbing at 440 and 590 nm.

REFERENCES

1. M. Guérin, N. Camougrand, G. Velours, and B. Guérin, New mutants resistant to glucose repression affected in the regulation of the NADH reoxidation, *Eur. J. Biochem.* 124:457 (1982).
2. N. Camougrand, G. Velours, and M. Guérin, Resistance of *C.parapsilosis* to drugs, *Biol. Cell.* 58: (1986).
3. N. Camougrand, R. Caubet, and M. Guérin, Evidence for an alternative and non-phosphorylating pathway for NADH reoxidation in a yeast strain resistant to glucose repression, *Eur. J. Biochem.* 135:367 (1983).
4. I.M. Moller, and J.M. Palmer, Properties of the oxidation of exogenous NADH and NADPH by plant mitochondria, *Biochim. Biophys. Acta* 638:225 (1981).
5. D.E. Koeppe, and R.J. Miller, Oxidation of NADPH by isolated corn mitochondria, *Plant Physiol.* 49:353 (1971).
6. A.I. Moody, and R.A. Reid, Inhibition of nucleotide transhydrogenase in rat liver mitochondrial particles by DCCD and butanedione, *Biochem. J.* 209:899 (1983).
7. M. Guérin, and N. Camougrand, The alternative oxidase of *C.parapsilosis*, *Eur. J. Biochem.* (1986).

EVIDENCE OF A POSITIVE MEMBRANE POTENTIAL AND A CALCIUM ACCUMULATION IN EUGLENA GRACILIS Z MITOCHONDRIA BY IN VIVO MICROELECTRODE STUDIES

Christian Tamponnet (1,2), Jean-Pierre Rona (3), Jean-Nöel Barbotin (1) and Régis Calvayrac (2)

(1) Lab. de Technologie Enzymatique, UA CNRS 523, Université de Technologie de Compiègne,BP 233, F-60206 Compiègne, France
(2) Lab. des Membranes Biologiques,Tour 54-53,2°étage,Université Paris VII, 2 Place Jussieu,F-75251 Paris Cedex 05, France
(3) Lab. d'Electrophysiologie des Membranes, UA CNRS 1180 Tour 54-64, 4°étage, Université Paris VII, 2 Place Jussieu F-75251 Paris Cedex 05, France

Recently, the biotechnological interest in plant cells has increased considerably and immobilization processes have emerged as an important tool for increasing their longevity (3). The immobilization of Euglena gracilis Z cells in a calcium alginate gel exhibited the conservation of their cellular structures and functions (10). To explain such a behaviour, the role of the high external Ca^{2+} concentration is suspected. So, the ionic concentrations and mitochondrial membrane potential differences have been measured.

Euglena gracilis strain Z (Pringsheim) cells were grown at 25°C in darkness in a defined culture medium (4). Stored in darkness at 4°C in different storage media SM(i) at different Ca^{2+} concentrations (iM), cells fixed with 10% (W/V) KI, are counted with a Malassez slide. The concentration of the different ions tested after acid extraction, the potential differences and membrane resistances are measured as previously described (2,9).

Table 1 demonstrates the accumulation of Ca^{2+} by Euglena cells when transferred from culture medium to the storage medium ($SM(10^{-1})$, Ca^{2+} = 100 mM). When etiolated Euglena cells exhibit a giant, unique mitochondrion during one phase of their cellular cycle: the interphase (5), the mitochondrial membrane potential difference can be measured because this mitochondrial network is located next to the plasma membrane. Indeed, a second intracellular potential Ex, positive inside (+38mV) appears whereas the plasma membrane potential difference Em is negative inside (-58mV, see fig. 1). At the same time, we measured the membrane resistance to determine in which compartment the microelectrode had penetrated. Our results indicated a double unit-membrane surrounded compartment (table 2),and only the mitochondrion is a closed double unit-membrane compartment wide enough to be penetrated by the microelectrode. Direct measurements with microelectrodes of mitochondrial membrane potential have been already made, but only with isolated giant mitochondria of Drosophila or caprizone-fed mice (8,11). Usually, measurements of mitochondrial membrane potential are obtained with the use of dyes or ionic probes (7). The comparison of the two types of measurements demonstrates a great difference: with microelectrodes the membrane potential difference is positive inside, whereas negative inside with ionic probes.

Table 1. Relation of internal ionic concentrations to external calcium concentration.

External Medium	L medium (Ca^{2+})=2mM	SM (10^{-1}) (Ca^{2+})=100mM
(Ca^{2+}) [a]	1.6±0.3	16.6±1.2
(Cl^{-}) [a]	6.1±0.5	19.2±2.1
(Na^{+}) [a]	6.6±0.6	2.9±0.3
(K^{+}) [a]	98.6±4.9	85.7±4.7

[a] The different ionic measurements are expressed as fmol. per cell.

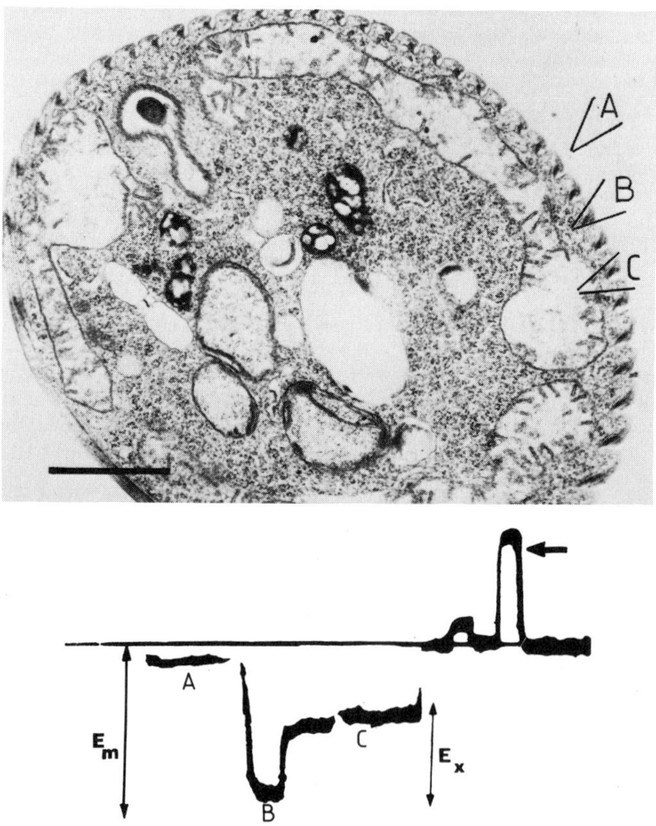

Fig. 1.(a) The transmembranes electrical profile exhibits a new potential difference Ex appearing after the plasmalemma's one Em, when the microelectrode gradually penetrates the cell. The arrow indicates a standart potential of + 50 mV.(b) Micrograph of an Euglena cell exhibiting an unique, giant mitochondrion. The relative position of the microelectrode (A,B and C) corresponds to the potential differences registered. The bar indicates 5 μm.

Table 2. Measurements of encountered membranes resistances and potential differences.

L medium	Resistance [a]	Potential [b]
Plasma Membrane	48±12	- 58± 9
Plasma Membrane + Endocellular Membrane(s)	130±21	- 20± 4
Endocellular Membrane(s)	72±29	+ 38±11

[a]Resistances are expressed as MΩ.
[b]Potentials are expressed as mV.

At this point, we propose that the difference registered between microelectrodes and ionic probes can be explained by an artefact due to the ionic probes as has been demonstrated in isolated plant vacuoles (1).
In an effort to explain the role played by the high external Ca^{2+} concentration, the mitochondrial membrane-potential difference has been measured as a function of the external Ca^{2+} concentration (fig. 3), and a direct correlation has been demonstrated.

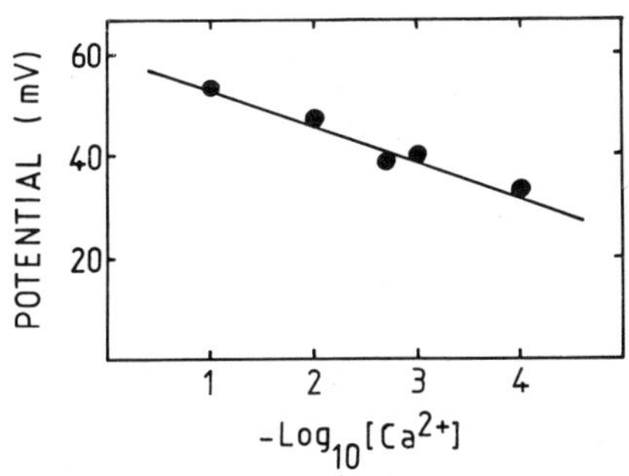

Fig. 2. Relation of the mitochondrial membrane potential difference to the external calcium concentration. The potential differences are expressed as mV, and calcium concentrations as $mol.l^{-1}$.

Table 3. Evolution of the electrochemical potential of calcium through the mitochondrial membranes as a function of the external calcium concentration.

Mitochondrial Membrane	(Ca^{2+}) = 2 mM r=0.01[a]	r=0.1	(Ca^{2+}) = 100 mM r=0.01	r=0.1
E eq.(mV)[b]	- 30	- 60	- 60	- 90
E x. (mV)[c]	+ 38		+ 53	
$\Delta\bar{\mu}$[d] (kJ mol.$^{-1}$)	+ 13.1	+ 18.9	+ 21.8	+ 34.2
$\Delta\|\Delta\bar{\mu}\|$[e] (kJ mol.$^{-1}$)	8.7		8.7	

[a] r is the ratio: free Ca^{2+} / bound Ca^{2+}
[b] E eq. = calcium equilibrium potential (mitochondrial matrix related to cytoplasm)
[c] E x. = measured potential difference (mitochondrial matrix related to cytoplasm)
[d] $\Delta\bar{\mu}$. = electrochemical potential of Ca^{2+} (mitochondrial matrix)
[e] $\Delta|\Delta\bar{\mu}|$. = increase of electrochemical potential of Ca^{2+} when cells are transferred from 2 mM-Ca^{2+} to 100 mM-Ca^{2+} medium. The values correspond to r = 0.1 and r = 0.01.

The average size of Euglena cells (Vt=3pl), organites (Vo=1.2pl) and cytoplasm (Vc=1.8pl) have been calculated by statistical analysis of micrographs. The cytoplasmic Ca^{2+} concentration, the global quantity of Ca^{2+} in cells are known. As the ration free Ca^{2+}/ bound Ca^{2+} is between 0.1 and 0.01 for microalgae (6), using the Nernst equation and the electrochemical potential of Ca^{2+} through the mitochondrial membranes, we demonstrated the existence of a mitochondrial Ca^{2+} pump which is activated when cells are transferred from a low-Ca^{2+} concentrated medium to a high-Ca^{2+} concentrated one (table 3).

By extrapolation of these results to other organites such as the chloroplasts, we then postulate the existence of a Ca^{2+} accumulation by internal membranes to explain the conservation of cellular strucutres and functions by immobilized Euglena cells.

REFERENCES

(1)-Barbier-Brygoo,H., Gibrat,R., Renaudin,J.P., Brown,S., Pradier,J.M., Grignon,C., Guern,J. (1985) Biochim. Biophys. Acta 819:215-224
(2)-Brenneke,R., Lindemann,B. (1971) J. Life Sci. 1:52-58
(3)-Brodelius,P., Deus,B., Mosbach,K., Zeus,M.H. (1979) FEBS Lett. 103:93-97
(4)-Calvayrac,R., Douce,R. (1970) FEBS Lett. 96:8-17
(5)-Calvayrac,R., Van Lente,F., Butow,R.A. (1971) Science 173:252-254
(6)-Campbell,A.K. Intracellular Calcium (Gutfreund ed.) J.Wiley and Sons, New York, 1983
(7)-Ducet,G. (1984) Physiol. Vég. 22:675-686
(8)-Maloff,B.L., Scordilis,S.P., Tedeschi,H. (1977) Science 195:898-900
(9)-Rona,J.P., Van de Sype,G., Cornel,D., Grignon,C., Heller,R. (1980) Bioelectrochem. Bioenerg. 7:377-391
(10)-Tamponnet,C., Costantino,F., Barbotin,J.N., Calvayrac,R. (1985) Physiol. Plant. 63:277-283
(11)-Tupper,J.T.,Tedeschi,H. (1969) Proc. Natl. Acad. Sci.USA 63:370-377

STRUCTURE AND BIOGENESIS OF THE PLANT MITOCHONDRIAL INNER MEMBRANE

Malcolm J. Hawkesford and Christopher J. Leaver

Department of Botany, University of Edinburgh
Edinburgh EH9 3JH, Scotland

INTRODUCTION

Growth and differentiation in plants depends upon the integrated metabolism of, and provision of energy by, both mitochondria and chloroplasts. While in recent years considerable progress has been made in our understanding of the factors which regulate chloroplast biogenesis and function, comparable studies on mitochondria have received far less attention (for a review see 1). This is surprising given the many developmental transitions throughout the plant life cycle that are associated with, or are dependent upon, marked changes in mitochondrial number, structure and metabolism. A priority is therefore to gain an understanding of the role of the mitochondrion in meeting the changing energy requirements of the cell, and hence in determining plant yield and viability. In addition to the elucidation of metabolic pathways, and determination of important regulatory or limiting steps, there is a need to understand the factors regulating mitochondrial biogenesis. This will involve an understanding of how genetic, developmental and environmental factors regulate biogenesis and how this, in turn, influences plant development.

The capacity for oxidative phosphorylation in a tissue can be considered to be dependent upon total mitochondrial number, inner membrane area, and the composition, density and optimal arrangement of the energy conserving units. The efficiency and capacity of any metabolic pathway depends upon both the total number of functional units present and their metabolic state. Although short term regulation of enzyme activity, proton and electron flow etc. is important in maintaining optimal efficiency, the major developmental changes that occur in a plants lifetime will dictate fundamental changes that will be dependent on biosynthesis of the polypeptide components involved. A specific example is the change in the importance of the uncoupled pathway in some tissues or under some conditions (e.g. aroid appendices, wounded tissue, possibly C_4 bundle sheath cells and during photorespiration).

The inner mitochondrial membrane has a crucial function in regulating the flow of all substrates and metabolites into and out of the mitochondrion, and also in determining the energy balance of the cell as the site of oxidative phosphorylation. Except by analogy with fungal and animal mitochondria relatively little is known of the structure of the plant inner mitochondrial membrane. There are however a number of unique features: (a) a variety of dehydrogenases are present in addition to the complex I NADH dehydrogenase; possibly a rotenone insensitive by-pass of this complex and a dehydrogenase oxidising external NADH, neither of which are coupled to transmembrane proton transfer, (b) the existence of a cyanide insensitive alternative quinol oxidase, which is also uncoupled from proton transfer, (c) the synthesis of unique variant mitochondrially encoded polypeptides which are components of the inner mitochondrial membrane and appear to be associated with a developmentally regulated mitochondrial dysfunction. This is recognised phenotypically by the failure to produce functional pollen and is termed cytoplasmic male sterility (CMS).[1]

In order to further characterise these unique features, and increase our understanding of how mitochondrial biogenesis and function is coupled to changing energy and metabolic demands during plant development, it is essential to have a sound knowledge of the structure and composition of the plant inner mitochondrial membrane.

Of particular importance to the study of mitochondrial biogenesis is the ability to identify and assign a functional role to specific polypeptides and to determine the genomic location and structure of the genes which encode them. The purified polypeptides and genes can then be used to raise monospecific antibodies and generate specific DNA probes, which can be used to investigate developmental regulation and coordinated expression at the level of transcription and translation.

This paper presents a brief review of the current position with regard to the isolation and characterisation of genes for mitochondrial proteins and reports on our initial progress in the purification and characterisation of polypeptide components of the inner mitochondrial membrane of higher plants.

THE PLANT MITOCHONDRIAL GENOME: STRUCTURE AND INFORMATION CONTENT

Mitochondria contain at least 300-400 proteins which are encoded by both nuclear and mitochondrial genomes.[1] The mitochondrial genome contains two functionally different classes of genes. The first encodes the ribosomal and transfer RNA components of a mitochondrion specific, protein-synthesising system which is responsible for the translation of the messenger RNA (mRNA) transcripts of the second set of genes. The majority of these genes code for a small number (ca. 10-15) of hydrophobic polypeptides which are components of the electron transport chain and F_1F_0 ATPase complex of the inner membrane.

Less than 10% of mitochondrial protein mass is synthesised in the organelle. The remaining proteins, including those of the outer membrane, the inter-membrane space, the matrix space (including the enzymes of the TCA cycle and those required for replication and transcription of mtDNA and translation of mtRNA), and the remaining components of the inner membrane, are encoded in the nuclear genome. The nuclear encoded proteins destined for the mitochondrion are normally synthesised as higher molecular weight precursors on cytosolic ribosomes. Amino terminal extensions are involved in targetting the polypeptide to the mitochondrion where import is dependent upon an electrochemical potential across the inner membrane. Import is usually accompanied by the proteolytic cleavage of the precursor by specific chelator-sensitive proteases in the matrix space. Thus mitochondrial biogenesis requires the controlled and integrated expression of both the nuclear and mitochondrial genetic systems which in turn must be related to the overall energy demands of the cell.

MITOCHONDRIAL GENOME STRUCTURE

In most organisms mitochondrial DNAs (mtDNA) are present as covalently closed, circular molecules, which occur in multiple copies per mitochondrion and are autonomously replicated and transcribed in the organelle. The mitochondrial genome of higher plants is much larger and more variable in size than its animal (16-17Kb) and yeast (ca. 70Kb) counterparts. It varies in size from ca. 218Kb in various *Brassica* sp, 430Kb in *Triticum* sp, 530Kb in *Zea mays* and up to 2400Kb in muskmelon. It is now generally accepted that the genome can be represented as a covalently closed, circular molecule which contains a number of short repeated sequences. It is thought that recombinations between some homologous sequences can generate a number of sub-genomic circular molecules, leading to the formation of a multipartite genome. The number and relative frequency of recombination between the different pairs of repeated DNA sequences and the efficiency of different replication origins will determine the genome organisation *in vivo*.[2]

Recent evidence suggests that multiple aberrant recombination events could modify the genome environment of certain genes and in some extreme cases generate unique chimaeric (mosaic) genes.[3,4] These genes are apparently transcribed and translated into variant mitochondrial polypeptides which may in some cases lead to mitochondrial dysfunction resulting in the cytoplasmic male-sterile phenotype.

MITOCHONDRIAL GENES

The large and variable size of the plant mitochondrial genome, even within a single family such as the Cucurbits cannot be explained as due to the presence of repetitive DNA, which constitutes less than 10% of the genome.[5] It is also unlikely that plant mtDNA contains proportionally more genes, as the range of genes identified so far in higher plants, is with one or two exceptions, similar to those described in other organisms (Table 1).

In higher plants three mitochondrial specific ribosomal RNA genes (26S, 18S and 5S) and a number of transfer RNA genes have been identified, together with several genes encoding integral components of the energy transducing complexes of the inner mitochondrial membrane.[1,6] These include genes encoding the apocytochrome *b* polypeptide (COB)[7] of the bc_1 complex, the three largest hydrophobic polypeptides of the cytochrome *c* oxidase complex (COX I, II and III),[8,9] the alpha-subunit of the F_1-ATPase complex (ATPA)[10] and subunits 6[11] and 9[12] of the F_0-ATPase (ATP6 and 9). A recent report has also identified a gene in *Cucumis* showing high homology to the URF1 gene from the animal mitochondrial genome which has been shown to encode one of the hydrophobic subunits of the complex I NADH dehydrogenase.[13] It is likely that, in common with animal mitochondria, higher plant mtDNA will be shown to encode additional polypeptide components of complex I.

Perhaps the most surprising findings to date are that the plant mitochondrial genome encodes a 5S rRNA[1] and the alpha subunit of the F_1-ATPase.[10] In all other organisms examined to date the mitochondrial ribosomes does not contain a 5S rRNA component and the alpha-subunit together with the other four subunits of the F_1-ATPase are encoded in the nuclear genome.

A comparison of the sequence of specific genes isolated from *Z. mays* with the homologous genes in other organisms reveals homology at the predicted amino acid level (Table 2), particularly in regions of the polypeptide which have functional importance, for example at the haem binding sites.

Table 1. Mitochondrial genes and their products in plant, yeast and human mitochondria.

	Zea mays	Saccharomyces cerevisiae	Homo sapiens
Mitochondrial Genome Size (Kb)	570	80	16.6
Protein coding genes			
Complex I NADH dehydrogenase			
URF1 product	+	–	+
URF2 product	?	–	+
URF3 product	?	–	+
URF4 product	?	–	+
URF4L product	?	–	+
URF5 product	?	–	+
Ubiquinol cytochrome c reductase			
Apocytochrome b	+	+	+
Cytochrome c oxidase			
Subunit I	+	+	+
Subunit II	+	+	+
Subunit III	+	+	+
F_1 - F_0 ATPase			
Alpha subunit	+	–	–
Subunit 6	+	+	+
Subunit 8 (E.coli, 17Kd protein)	?	(aap1)	(urf a6L)
Subunit 9 (DCCD binding protein)	+	+	–
Small ribosomal subunit protein	+	+	–
RNAs			
Large- rRNA	26S	21S	16S
Small- rRNA	18S	15S	12S
5S - rRNA	+	–	–
Transfer RNAs	ca.30	25	22

The sequence data has also revealed several major differences in gene structure and codon usage, compared to other organisms. In contrast to a number of fungal systems where the COXI and COB genes contain multiple, noncoding intervening sequences (introns), the homologous genes in those plants examined to date do not. However in contrast to other organisms a single, centrally located intron is found in the COXII[9] gene of several monocotyledonous plants, as well as in artichoke and sunflower, although not in a number of other dicotyledons (Pisum, Oenothera).

Table 2. Mitochondrial Genes from Zea mays.

GENE	COB	COX I	COX II	ATP A	atp6	atp 9
DNA CODING SEQUENCE:						
INTRONS	0	0	1	0	0	0
LENGTH (Bp)	1164	1584	780	1524	873	222
HOMOLOGY:						
Human	47%	58%	52%	-	-	-
Yeast	52%	58%	56%	62%(Nuc)	45%	56%
PREDICTED POLYPEPTIDE:						
AA Residues	388	528	260	508	291	74
Mol. Weight	42,900	58,219	30,800	55,117	31,721	7,368
HOMOLOGY:						
Human	47%	67%	40%	-	-	56%
Yeast	48%	60%	47%	65%	33%	46%

In higher plants the UGA codon acts as a terminator codon as specified by the universal code, whereas in other mitochondrial systems UGA specifies tryptophan. The one codon assignment which is apparently unique to higher plants is CGG, which specifies arginine in the universal code, and functions as a tryptophan codon in plants.[9]

The availability of specific gene probes has allowed them to be mapped on the circular linkage map of mtDNA from Z. mays constructed by Lonsdale and his colleagues.[6]

MITOCHONDRIAL GENE EXPRESSION

Analysis of the transcriptional patterns of the Z. mays mitochondrial protein-coding genes reveals in most cases a complex situation suggesting extensive processing of the primary transcripts.[7,8,9,10,11,12] In contrast the number of polypeptides which can be identified as mitochondrial translation products is relatively small. Isolated intact, coupled mitochondria in the presence of an oxidisable substrate (or alternatively an ATP generating-system) will incorporate radioactively labelled amino acids into protein. The labelled mitochondrial proteins can be resolved by either one (Figure 1A) or two-dimensional IEF/SDS gel electrophoresis (Figure 1B) and visualised by autoradiography of the dried gel. Between ca. 15 and 30 polypeptides ranging in estimated molecular weight from 8000-58000 can be detected by this approach and with the exception of the two highest molecular weight species the major translation products are associated with the inner membrane. While the absolute number of translation products synthesised by mitochondria of different species remains approximately constant, there are marked and reproducible quantitative differences depending upon the developmental state of the tissue from which the mitochondria are isolated. A notable exception to this generalisation is the synthesis of unique variant translation products by mitochondria isolated from the different cytoplasmic male sterile lines of Z. mays.[1] For example in Figure 1A a comparison of the translation products of mitochondria isolated from five-day old, dark grown seedlings of the normal (N-fertile) and male sterile T (Texas) cytoplasm of Z. mays reveals the synthesis of an additional membrane bound polypeptide with an estimated molecular weight of 13000 daltons.

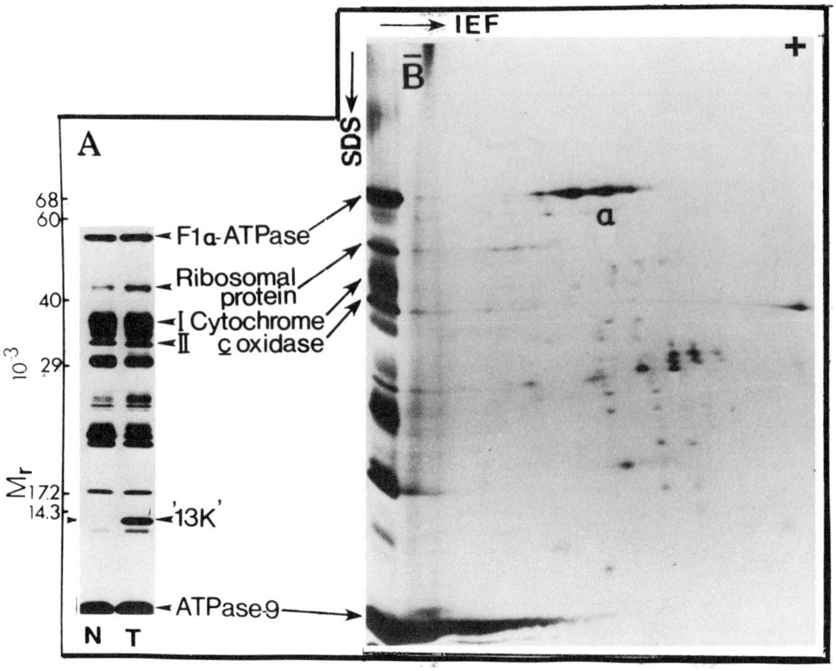

Figure 1. (A) one-dimensional, SDS-PAGE and (B) two-dimensional, IEF followed by SDS-PAGE, of ^{35}S-methionine labelled polypeptides synthesised by mitochondria isolated from male fertile (N) (A and B) and male sterile (T) cytoplasm *Z. mays* (A).

A combination of immunoprecipitation and protein immunoblot analysis with antibodies raised against purified mitochondrial polypeptides from yeast, animal and plant origins, coupled with the partial purifications of specific enzyme complexes has led to the identification of a number of mitochondrial translation products. These include the alpha-subunit[14] and subunit 9 (DCCD-binding protein)[15] of the F_1F_0-ATPase, cytochrome *c* oxidase subunits I and II,[1] apocytochrome *b* of the bc_1 complex and a ribosomal protein (see Figure 1A). The higher resolution afforded by two dimensional electrophoresis suggests that in excess of 30 polypeptides are synthesised in mitochondria, although it is possible that certain labelled polypeptides are products of premature termination, charge modification or proteolytic cleavage.

While the remaining translation products still remain to be identified and assigned a functional role (including those characteristically associated with the CMS phenotype) it is obvious that plant mitochondria synthesise, and by extrapolation encode, additional polypeptides not synthesised by animal and fungal mitochondria. However the additional DNA required to encode these gene products only represents a relatively minor portion of even the smallest plant mitochondrial genome described to date and we still have no explanation for the large and variable size of the mitochondrial genome in higher plants.

ISOLATION AND ANALYSIS OF INNER MEMBRANE ENZYME COMPLEXES

The inner mitochondrial membrane consists of a number of oligomeric complexes composed of, in total, hundreds of component polypeptides. These complexes include those responsible for energy transduction, transport of substrates and metabolites into and out of the mitochondrion, as well as enzymes involved in biosynthesis and degradation.

An essential prerequisite to establishing structural-functional relationships in an oligomeric enzyme complex is the isolation and identification of the component polypeptides. This will require the combined application of multiple extraction and separation steps, including differential solubilisation followed by chromatographic and electrophoretic fractionation.

Few plant inner mitochondrial membrane enzyme complexes and their component subunits have been purified to homogeneity, and the complexity of most of the enzymes, together with the lack of suitable antibody probes from other organisms dictates that at least partial purifications must be undertaken prior to investigation of their biogenesis. An exception is the F_1 portion of the F_1-F_0 ATPase which has been independently purified by a number of laboratories.[14,16] There are also reports of the preliminary purifications of succinate dehydrogenase[17] and cytochrome c oxidase[18] from sweet potato although the relationship of the five structural subunits to other preparations has not been established.

We have undertaken the purification of a number of these complexes from Z. mays with the aim of obtaining monospecific antibody probes to their component subunits. Figure 2 is a summary of the strategies we have applied to separate intact oligomeric membrane protein complexes. We have

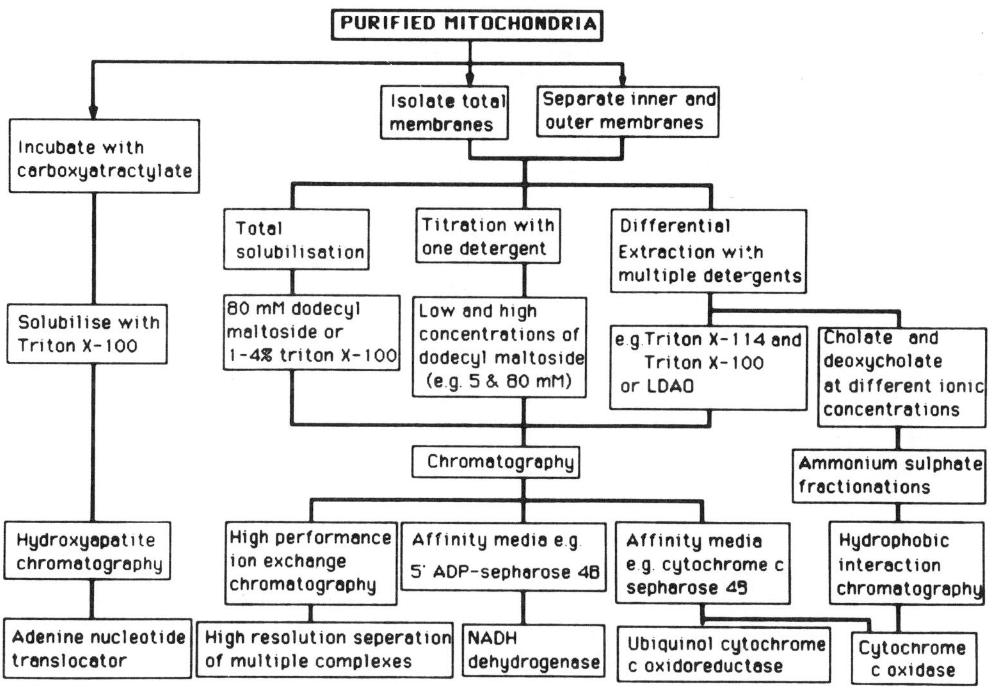

Figure 2. General strategies for the isolation of complexes of the mitochondrial inner membrane.

successfully purified the adenine nucleotide translocator from Zea mays using classically established methods[19] (see Figures 2 and 3) and raised antibodies to this polypeptide. The purified monomeric subunit has an estimated molecular weight of 30 Kd which is in close agreement with the values of 30-33 Kd reported for the Neurospora and mammalian enzymes. The Z. mays nuclear gene encoding this polypeptide has also been isolated and sequenced, and the predicted size of the polypeptide is 40,519 Kd.[20] One of the first purifications of the F_1 portion of the F_1F_0 ATPase was achieved by extraction of mitochondrial membranes with a chloroform/methanol mixture, followed by differential ammonium sulphate precipitation and sucrose density gradient centrifugation (Figure 4).[14] A typical five-subunit complex is seen in the final preparation of which the largest has been shown to be synthesised within the mitochondrion.

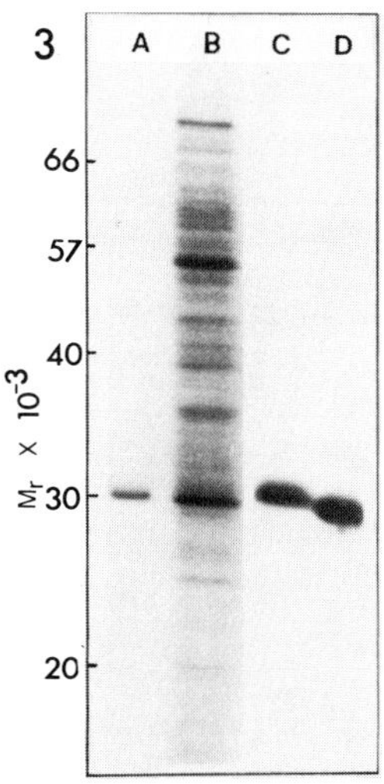

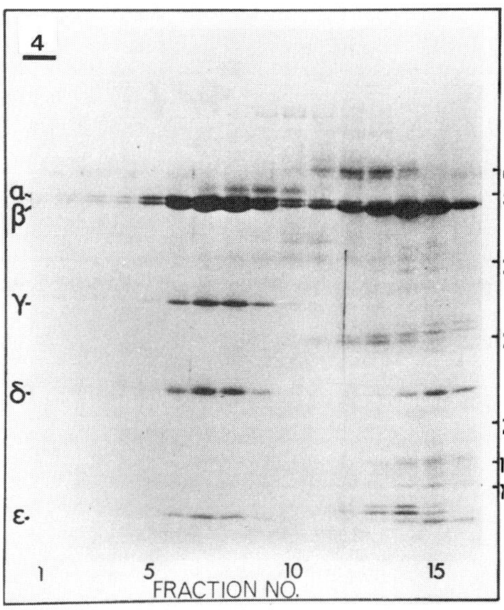

Figure 3. SDS-PAGE (A,B) and immunodecorated electroblot probed with an antibody prepared against the purified Z. mays adenine nucleotide translocator (C,D). A and C, purified adenine nucleotide translocator, B and D, total mitochondrial protein.

Figure 4. Sucrose gradient purification of the Z. mays mitochondrial F_1 ATPase complex: 0.2 ml of an ammonium sulphate fraction containing ~2 mg of protein was layered onto a 4.5 ml linear gradient of 5-20% (w/v) sucrose and centrifuged at 110000 g_{max} for 16 h in an SW50.1 rotor. Fractions of 0.3 ml were collected and 30 µl of each was mixed with sample buffer, electrophoresed on a 10-15% (w/v) acrylamide linear gradient gel and stained with Coomassie Brilliant Blue. Fraction 1 corresponds to the bottom of the gradient. The positions of the five major subunits are indicated on the left (from reference 14).

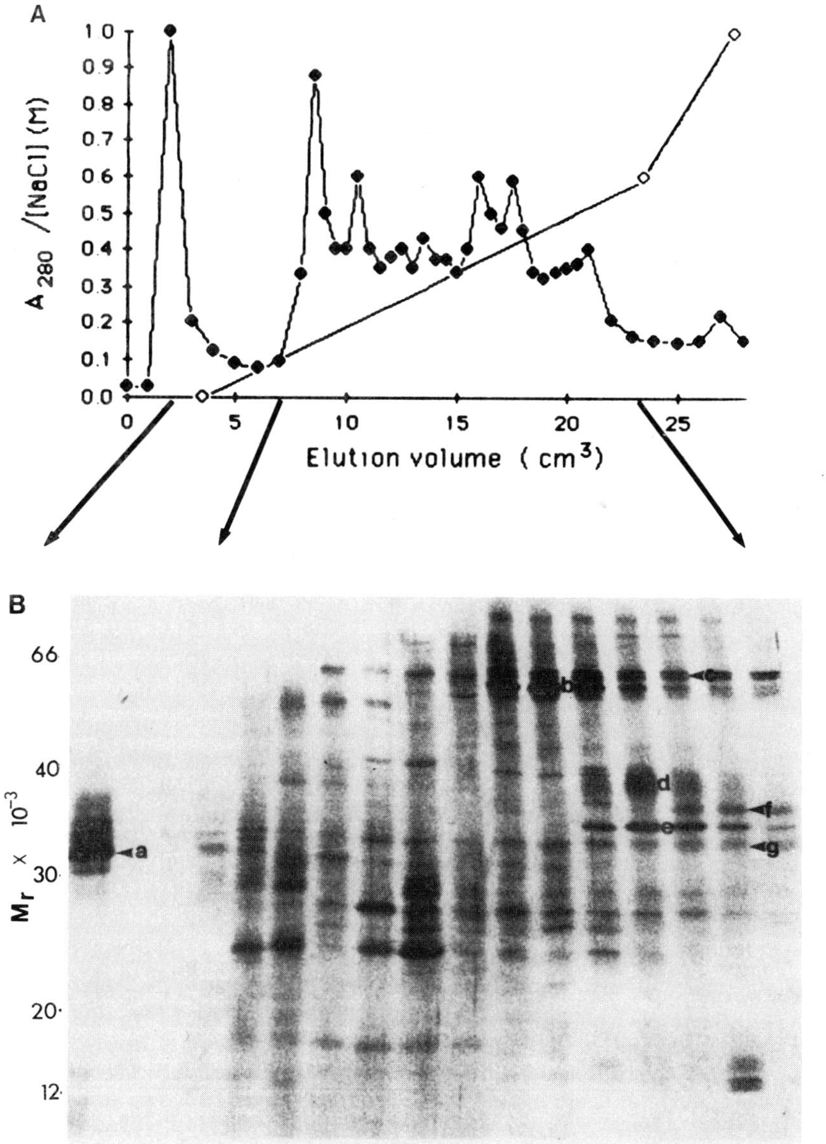

Figure 5. Separation of a dodecyl maltoside-solubilised fraction of Zea mays mitochondrial membranes by ion-exchange chromatography (mono Q). A. Protein profile and eluting salt gradient. B. Silver-stained SDS-PAGE of protein-containing fractions. A replicate gel was electroblotted onto nitrocellulose and the following polypeptides identified by immunodecoration: (a) adenine nucleotide translocator, (b) alpha subunit of the F_1 ATPase, (c) core I and II polypeptides of complex III, (d) cytochrome oxidase subunit I and (e) II, (f) and (g) cytochromes c_1 and b of complex III, respectively.

A preliminary approach to the analysis of oligomeric complexes is illustrated in Figure 5, which shows the anion exchange chromatography elution profile of a dodecyl-maltoside-solubilised total membrane extract. Fractions were further analysed by SDS-PAGE followed by protein immunoblotting using antibodies raised against specific proteins from plant, fungal and mammalian mitochondria. After electrophoretic transfer from the polyacrylamide gel to nitrocellulose and immunodecoration of the blots with cross-reacting antibodies, specific polypeptides from the major complexes were identified (Figure 5 a-g).

A more selective solubilisation was possible using differential extraction techniques either by titration with one detergent or with the use of different detergents with varying properties. This may be followed by ammonium sulphate precipitations which can achieve minor enrichments for specific complexes. Triton X-100 is one of the most effective membrane solubilising agents, however it possesses a number of disadvantages and in some instances may beneficially be replaced with alkyl-glycoside detergents such as octyl glucoside or dodecyl maltoside. Gentle extraction of less hydrophobic proteins is possible using detergents such as Triton X-114[21] or LDAO[22] (lauryl dimethylamine N-oxide).

A number of different chromatographic approaches are possible and of particular importance is the choice of detergent concentration in the elution buffer. The concentration should be above the critical micelle concentration under the specific conditions employed, and the molar ratio of detergent micelles to protein complexes should be sufficient to disperse all protein complexes. Non-ionic detergents are generally effective at dispersing aggregates whilst being sufficiently mild so as not to disrupt the complexes themselves. In addition to the alkyl-glucosides, some of the polyoxyethylene ethers such as Brij 35 (Sigma) or Thesit (Boehringer) are UV transparent and particularly useful.

Cytochrome c oxidase and the cytochrome bc_1 complex have been partially purified by solubilising membranes with dodecyl maltoside followed by chromatography on a cytochrome c affinity column and further fractionated by high performance anion exchange chromatography on mono Q (Pharmacia FPLC) (Hawkesford and Leaver, in preparation). The subunit compositions of the complexes is similar to that described for yeast, although the smaller, nuclear encoded subunits require further resolution.

SUMMARY AND FUTURE PROSPECTS

We have summarised in a diagrammatic form (Figure 6) our present state of knowledge of the structure and genetic origins of the major energy transducing components of the plant inner mitochondrial membrane. Little structural information is available for the plant enzymes and therefore the structures have been derived from published studies on animal and fungal complexes.

The structure of complex I is that of the bovine enzyme as proposed by Ragan and co-workers,[23] and not all of the suggested 25 subunits are shown individually. The mitochondrially encoded subunits (at least six in the human enzyme and probably a similar number in plant mitochondria) are hydrophobic polypeptides and are thought to form a structural sheath around the enzymic core of the complex. Antibodies raised against the bovine subunits cross-react with plant polypeptides (24, and our unpublished data) supporting the idea of an analogous complex in plant mitochondria. Little is known of the nature of the rotenone-insensitive by-pass of this complex and it is therefore not shown.

Recent reports[22,25,26] and our unpublished observations on the purification of the dehydrogenase responsible for the oxidation of external NADH have not confirmed the polypeptide composition of this enzyme or its structure within the membrane.

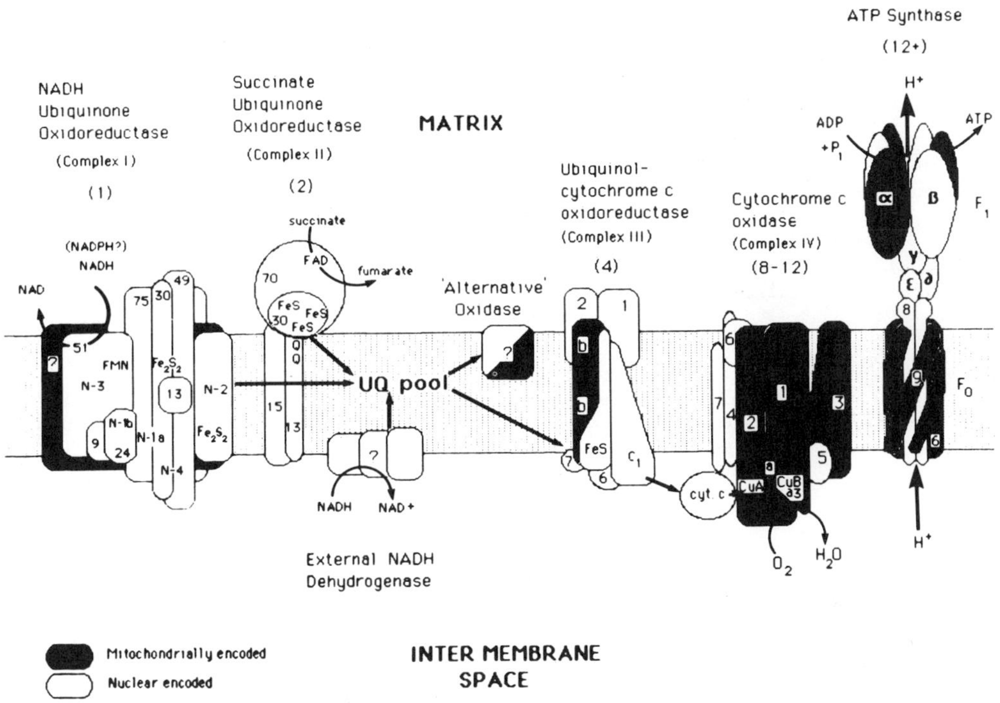

Figure 6. Summary diagram illustrating our present understanding of the contribution of the nuclear and mitochondrial genomes to the biogenesis of the energy transducing components of the inner mitochondrial membrane of higher plants.

Our partial purification of the cytochrome bc_1 complex suggests a subunit composition comparable to other isolated bc_1 complexes. The complex seems to be composed of two large, loosely associated, 'core' polypeptides, cytochromes b and c_1 polypeptides and a number of other polypeptides including the Rieske iron-sulphur protein.

The complexity of the plant mitochondrial cytochrome c oxidase compared to that isolated from other sources has not been confirmed. The mammalian enzyme reportedly consists of up to 15 subunits whilst fewer subunits have been resolved in yeast. The seven subunits depicted represents a minimal model and the structural organisation is based on an arrangement suggested by Jarausch and Kadenbach.[27]

The nature of the alternative cyanide insensitive oxidase is still controversial. Preliminary data suggest that it may contain at least a polypeptide of 30-40 Kd (these proceedings). The involvement of a product of the mitochondrial genome is only based on indirect evidence.[28]

The depicted structure of the F_1 portion of the ATPase is a generalised diagrammatic representation of the enzyme from a number of sources. The drawing of the F_0 portion is highly stylised. Several subunit 9 polypeptides (the DCCD-binding protein) constitute the proton channel, together with subunits 6 and 8 (possibly a dimer), which may have a role in the assembly of this complex and in the binding to the F_1 portion. It has been suggested that subunit 6 may be the oligomycin sensitivity conferring protein.[29] An additional soluble factor present in mammalian ATPase, designated F_6 has been omitted. The genes for subunits 6 and 9 are located in the mitochondrion, whilst the genetic origin of subunit 8 is still unclear.

Clearly our knowledge of the structure of the plant mitochondrial inner membrane is very inadequate. In future it is hoped that an understanding of physiological and metabolic processes will be related to the physical structure of the enzyme complexes involved. We believe that investigation of the genetic and environmental factors involved in regulating mitochondrial biogenesis and function will increase our appreciation of how energy metabolism interfaces with differential gene expression during plant growth and differentiation.

ACKNOWLEDGEMENTS

We thank G. Schatz and C.I. Ragan for the generous gifts of antibodies and the SERC, AFRC and the Gatsby Foundation for financial support.

REFERENCES

1. C. J. Leaver and M. W. Gray, Mitochondrial genome organization and expression in higher plants, Ann. Rev. Plant Physiol. 33:373 (1982).
2. D. M. Lonsdale, T. P. Hodge and C. M.-R. Fauron, The physical map and organisation of the mitochondrial genome from the fertile cytoplasm of maize, Nucleic Acids Res. 12:9249 (1984).
3. R. E. Dewey, C. S. Levings and D. H. Timothy, Novel recombinations in the maize mitochondrial genome produce a unique transcriptional unit in the Texas male sterile cytoplasm, Cell 44:437 (1986).
4. J. Bailey-Serres, D. K. Hanson, T. D. Fox and C. J. Leaver, Mitochondrial genome rearrangement leads to extension and relocation of the cytochrome c oxidase subunit I gene in Sorghum, Cell 47:in press (1986).
5. B. L. Ward, R. S. Anderson and A. J. Bendich, The mitochondrial genome is large and variable in a family of plants (Cucurbitaceae), Cell 25:793 (1981).
6. A. J. Dawson, T. P. Hodge, P. G. Isaac, C. J. Leaver and D. M. Lonsdale, Location of the genes for cytochrome oxidase subunits I and II, apocytochrome b, alpha-subunit of the F_1 ATPase and the ribosomal RNA genes on the mitochondrial genome of maize (Zea mays L.), Curr. Genet. 10:561 (1986).
7. A. J. Dawson, V. P. Jones and C. J. Leaver, The apocytochrome b gene in maize mitochondria does not contain introns and is preceded by a potential ribosome binding site, EMBO J. 3:2107 (1984).
8. P. G. Isaac, V. P. Jones and C. J. Leaver, The maize cytochrome c oxidase subunit I gene: sequence, expression and rearrangement in cytoplasmic male sterile plants, EMBO J. 4:1617 (1985).
9. T. D. Fox and C. J. Leaver, The Zea mays mitochondrial gene coding cytochrome oxidase subunit II has an intervening sequence and does not contain TGA codons, Cell 26:315 (1981).
10. P. G. Isaac, A. Brennecke, S. M. Dunbar and C. J. Leaver, The mitochondrial genome of fertile maize (Zea mays L.) contains two copies of the gene encoding the alpha-subunit of the F_1-ATPase, Curr. Genet. 10:321 (1985).
11. R. E. Dewey, C. S. Levings III and D. H. Timothy, Nucleotide sequence of ATPase subunit 6 gene of maize mitochondria, Plant. Physiol. 79:914 (1985).
12. R. E. Dewey, A. M. Schuster, C. S. Levings and D. H. Timothy, Nucleotide sequence of F_0-ATPase proteolipid (subunit 9) gene of maize mitochondria, Proc. Natl. Acad. Sci. USA 82:1015 (1985).

13. D. B. Stern, A. G. Band and W. F. Thompson, The watermelon mitochondrial URF-1 gene: evidence for a complex structure, Curr. Genet. 10:857 (1986).
14. E. Hack and C. J. Leaver, The alpha subunit of the maize F_1-ATPase is synthesised in the mitochondrion, EMBO J. 2:1783 (1983).
15. E. Hack and C. J. Leaver, Synthesis of a dicyclohexylcarbodiimide-binding proteolipid by cucumber (Cucumissativus L.) mitochondria, Curr. Genet. 8:537 (1984).
16. Y. Iwasaki and T. Asahi, Purification and characterisation of the soluble form of mitochondrial adenosine triphosphatase from sweet potato, Arch. Biochem. Biophys. 227:164 (1983).
17. T. Hattori and T. Asahi, The presence of two forms of succinate dehydrogenase in sweet potato root mitochondria, Plant Cell Physiol. 23:515 (1982).
18. M. Maeshima and T. Asahi, Purification and characterisation of sweet potato cytochrome c oxidase, Arch. Biochem. Biophys. 187:423 (1978).
19. M. Klingenberg, H. Aquila and P. Riccio, Isolation of functional membrane proteins related to or identical with the ADP/ATP carrier of mitochondria, Meth. Enzymol. 56:407 (1979).
20. A. Baker and C. J. Leaver, Isolation and sequencing of a cDNA encoding the ATP/ADP translocator of Zea mays L., Nucleic Acids Res. 13:5857 (1985).
21. C. Bordier, Phase separation of integral membrane proteins in triton X-114 solution, J. Biol. Chem. 256:1604 (1981).
22. N. D. Cook and R. Cammack, Purification and characterisation of the rotenone-insensitive NADH dehydrogenase of mitochondria from Arum maculatum, Eur. J. Biochem. 141:573 (1984).
23. T. Ohnishi, C. I. Ragan and Y. Hatefi, EPR studies of iron-sulphur clusters in isolated subunits and subfractions of NADH-ubiquinone oxidoreductase, J. Biol. Chem. 260:2782 (1985).
24. I. R. Cottingham, M. W. J. Cleeter, C. I. Ragan and A. L. Moore, Immunological analysis of plant mitochondrial NADH dehydrogenases, Biochem. J. 236:201 (1986).
25. I. R. Cottingham and A. L. Moore, Partial purification and properties of the external NADH dehydrogenase from cuckoo-pint (Arum maculatum) mitochondria, Biochem. J. 224:171 (1984).
26. R. R. Klein and J. J. Burke, Separation procedure and partial characterisation of two NAD(P)H dehydrogenases from cauliflower mitochondria, Plant Physiol. 76:436 (1984).
27. J. Jarausch and B. Kadenbach, Structure of the cytochrome c oxidase complex of rat liver. 2. Topological orientation of polypeptides in the membrane as studied by proteolytic digestion and immunoblotting, Eur. J. Biochem. 146:219 (1985).
28. M. E. Musgrave, I. C. Murfet and J. N. Siedow, Inheritance of cyanide-resistant respiration in two cultivars of pea (Pisum sativum L.), Plant, Cell Environ. 9:153 (1986).
29. S. Joshi, M. J. Pringle and R. Siber, Topology and function of "stalk" proteins in the bovine mitochondrial H+ ATPase, J. Biol. Chem. 261: 10653 (1984).

SYNTHESIS OF THE NUCLEAR DNA-ENCODED SUBUNITS OF HIGHER PLANT CYTOCHROME C OXIDASE AND F_1ATPase

Tadashi Asahi, Masayodhi Maeshima, Tsuyoshi Nakagawa, Kazuto Kobayashi, Yukimoto Iwasaki and Kenzo Nakamura

Laboratory of Biochemistry, Faculty of Agriculture
Nagoya University
Chikusa, Nagoya 464, Japan

INTRODUCTION

Active biogenesis of mitochondria occurs in higher plant cells without accompanying cell division in response to changes in physiological and environmental conditions. For instance, it is well known that a marked increase in respiratory activity in storage organs during seed germination is brought about by active mitochondrial biogenesis. As another example, reports from our laboratory have demonstrated that mitochondrial biogenesis takes place in sweet potato root tissue when the tissue is sliced and incubated under moist conditions at room temperature, namely wounded, or is infected with pathogens(1,2). In the former case, mitochondrial biogenesis is programmed during ontogenesis of individual plant, whereas in the latter, it is induced in response to stresses. We are interested in the question how mitochondrial biogenesis is programmed and induced in non-dividing cells of higher plants. At the present status of our knowledge, however, it seems to be very hard to make an approach toward the molecular mechanisms of the programming and induction of mitochondrial biogenesis. We should have much information about the mechanism of mitochondrial biogenesis itself in higher plant cells before we shall try to study the molecular mechanisms of the programming and induction of mitochondrial biogenesis.

Reports from our laboratory have shown that the formation of mitochondrial inner membrane (the formation of cristae-rich, active mitochondria from cristae-poor, inactive ones) in the cotyledons of imbibing pea seeds requires no protein synthesis(3,4). Dry pea cotyledons contain a soluble form of succinate dehydrogenase and free forms (forms associated without any other subunits) of soluble or membrane-bound cytochrome c oxidase subunits, and the abnormal forms of proteins are associated with the mitochondrial inner membrane and assembled into the corresponding complexes during seed imbibition(5,6). Such mitochondrial biogenesis without de novo synthesis of proteins, however, seems to occur only in imbibing seeds. In most cases, mitochondrial biogenesis in higher plant cells requires de novo synthesis of mitochondrial proteins as can be imaginated from a commonsense viewpoint. For instance, the immunologically determined amounts of mitochondrial proteins actually increase during mitochondrial biogenesis in wounded sweet potato root tissue and germinating castor bean endosperm (Figure 1). Thus we should have much understanding of the question how mitochondrial proteins are synthesized and localized in definite mitochondrial compartments in higher plant cells in order to have detailed information about the mechanism of mitochondrial biogenesis in higher plant cells. We should also have information about the mechanism of subunit assembly since most of important proteins in mitochondria are constructed from various molecular species of subunits in definite stoichiometry and topological arrangement. Such information is prerequisite to make an approach to the mechanisms of the programming and induction of mitochondrial biogenesis in higher plant cells.

As seen in Figure 1, cycloheximide, an inhibitor for protein synthesis in cytoplasms, inhibits the increases in cytochrome c oxidase subunit II and F_1ATPase α-subunits, mitochondrially made proteins, in both wounded sweet potato root tissue and germinating castor bean endosperm. Similarly, chloramphenicol, an inhibitor for protein synthesis in mitochondria, depresses the increases in cytoplasmically made mitochondrial proteins

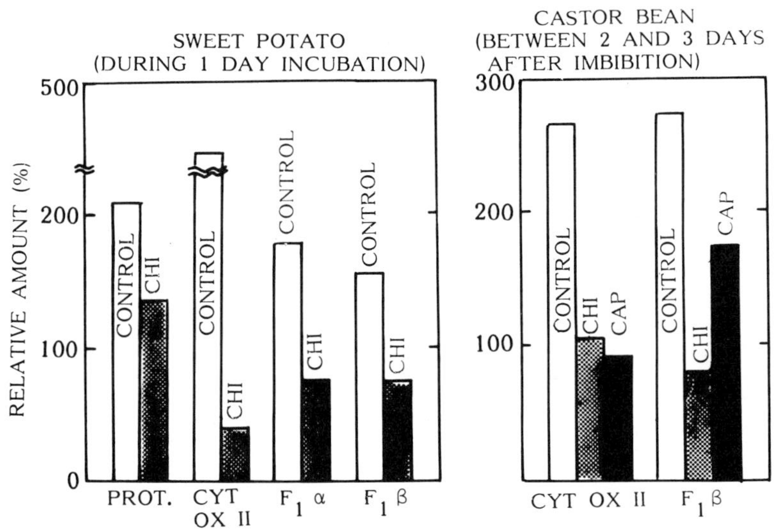

Figure 1. Increases in mitochondrial inner membrane proteins in wounded sweet potato root tissue and germinating castor bean endosperm. Cytochrome c oxidase subunit II (CYT OX II) and F_1ATPase α- (F_1α) and β- (F_1β) subunits were determined by a quantitative immunoblotting method. The amounts of the subunits and total mitochondrial proteins (PROT.) are expressed as percentage of the contents per fresh weight in sweet potato root tissue slices incubated at 30°C for 1 day or in the endosperm of castor bean germinated for 3 days relative to those in freshly prepared slices or the endosperm germinated for 2 days. Chloramphenicol (CAP) or cycloheximide (CHI) were supplied to the tissues for 1 day.

during mitochondrial biogenesis in the tissues. The facts indicate coorperation between mitochondrial and cytoplasmic protein synthesis systems, in other words, a close relationship between mitochondrial and nuclear genomic systems. We are also interested in the mechanism of the cooperation, but much understanding of the mechanisms of the synthesis, localization and assembly of mitochondrial proteins and more detailed characterization of higher plant mitochondrial genome are necessary to make an approach to this problem, too.

We have investigated the mechanisms of the synthesis and intracellular localization of higher plant cytochrome c oxidase and F_1ATPase. Here we describe the present situation of our work on this project.

OUTLINE OF PAST KNOWLEDGE ABOUT THE SYNTHESIS OF HIGHER PLANT CYTOCHROME C OXIDASE AND F_1ATPase

Higher plant cytochrome c oxidase has been reported to consist of five subunits (subunits I-V) with different molecular weights(7,8), and thus it is thought to be very simple in the subunit composition as compared with the enzymes from other eukaryotes. The larger three subunits (subunits I-III) have been thought to be encoded by mitochondrial genes, and the genes have been cloned and sequenced(9-11). The other smaller subunits (subunits IV and V) seem to be encoded by nuclear genes(12), but no conclusive evidence for this genetic control has presented so far.

Mitochondrial F_1ATPase in higher plant cells is composed of five (α-ε) or six (α-ε and δ') molecular species of subunits(13-18). Whether it contains five or six subunit species seems to depend upon plant varieties; we suspect that the monocot enzyme may contain five subunit species, whereas the dicot one, six. All F_1ATPases from other organisms including prokaryotes and similar proteins (CF_1) in chloroplasts from all plant species are constructed from five molecular species of subunits in a stoichiometry of 3:3:1:1:1, and thus it should be noted that some higher plant F_1ATPases contain an additional subunit. The α-subunit of higher plant F_1ATPase is encoded by the mitochondrial genome(13,14,19) and its gene has been cloned and sequenced(20-22). This is in contrast with the fact that all F_1ATPase subunits are encoded by the nuclear genome in all other eukaryotes. The other subunits, at least the β-, δ- and δ'-subunits, of higher plant F_1ATPase are encoded by nuclear genes(19).

As seen from the outline described above, detailed information is available for the synthesis of the mitochondrial DNA-encoded subunits, but not the nuclear DNA-encoded ones, of higher plant cytochrome c oxidase and F_1ATPase. Thus we are now directing our special attention to the synthesis and intracellular localization of the nuclear DNA-encoded subunits of these enzymes.

SYNTHESIS AND INTRACELLULAR LOCALIZATION OF THE NUCLEAR DNA-ENCODED SUBUNITS OF HIGHER PLANT CYTOCHROME C OXIDASE

Reports from our laboratory have shown that five subunits are detected when purified higher plant cytochrome c oxidase is electrophoresed on a polyacrylamide gel containing sodium dodecyl sulfate (SDS) and urea(7,8). Recently, we found, however, that additional subunits could be detected when the gel additionally contained glycerol. In the latter gel system, the subunit IV band is split into two bands (subunits IVa and IVb), and the subunit V band is into three (subunits Va, Vb and Vc with apparent molecular weights of 7,400, 6,800 and 6,200, respectively).

When purified sweet potato cytochrome c oxidase is incubated with ^{14}C-labelled N,N'-dicyclohexylcarbodiimide, both subunits IVa and IVb (but not any other subunits) are labelled, which means that the subunits correspond to the subunit III of mammalian and microorganism cytochrome c oxidases. Probably, the subunits may be encoded by mitochondrial genes and not by nuclear genes. At present time, we have no evidence against the possibility that the subunit IVb is a degradation product of the subunit IVa, namely an artifact, but we suppose that the mitochondrial genome may contain two genes for the subunit IV. The subunit III of higher plant cytochrome c oxidase may not correspond to the subunit III of the enzymes from other sources, and so whether the subunit is encoded by the mitochondrial or nuclear genome should be examined.

We have isolated the subunits Va, Vb and Vc of sweet potato cytochrome c oxidase by eluting the polypeptides from gel slices after electrophoresis of the purified enzyme on a polyacrylamide gel containing SDS and glycerol. There is evidence that these subunits are different from one another in the molecular species. First, both antibodies raised in rabbits against the purified holoenzyme and the isolated subunit V (a mixture of subunits Va, Vb and Vc) react with the subunit Vc, but not with the other two, in immunoblotting. Second, the N-terminal amino acid residue of the subunit Vc is L-alanine, whereas no amino acid is released from the subunit Va by Edman degradation, that is, the subunit Va has no free terminal amino group. Consequently, we now conclude that higher plant cytochrome c oxidase comprises seven (or eight) subunits. The subunit composition still shows a tendency to be simple compared with the composition of the enzymes from other eukaryotes; namely, the higher plant enzyme is lacking in subunits with a molecular weight range of 10,000-20,000, which are present in all the enzymes from other eukaryotes.

We are particularly interested in the synthesis of the subunit Vc among sweet potato cytochrome c oxidase subunits because the subunit seems to be synthesized tissue-specifically. In general, leaves are exposed to high temperatures by day, whereas roots are allowed to grow at moderate temperatures even in the daytime. This led us to postulate that enzymes in leaves might be relatively stable at high temperatures as compared with the corresponding enzymes in roots. This tendency has been observed with sweet potato cytochrome c oxidase associated with the mitochondrial inner membrane(23). Recently, we could confirm this proposal with pea and mung bean cytochrome c oxidases. As shown in Table 1, the enzyme in the mitochondria from the overground organs are more stable than that from the roots. The enzyme retains this character as to the thermostability even after being solubilized from the mitochondrial inner membranes and partially purified. Exogenously added phospholipid stabilizes the partially purified enzyme, but even in the presence of phospholipid, the enzyme from the overground organs more slowly losts its activity than does the enzyme from the roots. Consequently, we infer that the observed difference in the thermostability may due to a difference in the molecular species and not in the annular phospholipid species. We have been unable to find any significant difference in the subunit composition between the enzymes from the overground and underground organs of mung bean when analysed by polyacrylamide gel electrophoresis in the presence of SDS and urea. However, when the gel is subjected to immunoblotting with anti-sweet potato cytochrome c oxidase subunit V antibody, the subunit Va (the largest subunit out of the subunit V) of the enzyme from the roots, but not from the overground organs, reacts with the antibody, which indicates that at least

Table 1. Thermostability of Higher Plant Cytochrome c Oxidase

Preparation	Half-time of inactivation (min)			
	Pea		Mung bean	
	Root	Leaf	Root	Hypocotyl
Mitochondria	13(37°C) 14(42°C)	46(37°C) 29(42°C)	13(50°C)	23(50°C)
Solubilized fraction [a]	25(25°C)	53(25°C)	13(30°C)	28(30°C)
Partially purified fraction [b]				
-phosphatidylethanolamine	----	----	9(30°C)	16(30°C)
+phosphatidylethanolamine	----	----	34(30°C)	107(30°C)

[a] The submitochondrial particles were washed once with a 5% deoxycholate-2 M KCl solution, then suspended in a 1% Triton X-100-2 M KCl solution. The suspensions were centrifuged at 100,000 xg for 1 hour to obtain the supernatants.

[b] The solubilized fractions were fractionated by DEAE-cellulose column chromatography, and active fractions were collected.

the subunit Va from the overground organs is different from that from the roots in the molecular species and thus at least the subunit is synthesized tissue-specifically. Recently, smaller subunits of mammalian and slime mold cytochrome _c_ oxidases have also been shown to be tissue-specific or developmental stage-specific(24-26). It should be noted that the subunit Va of mung bean cytochrome _c_ oxidase corresponds to the subunit Vc of the sweet potato enzyme judging from their immunological properties. Thus it is very probable that the subunit Vc of sweet potato cytochrome _c_ oxidase is synthesized in a tissue-specific manner, which we are now attempting to confirm. We know from experience that it is very hard to purify mung bean cytochrome _c_ oxidase and thus to isolate its subunits, and so we are directing our attention to the synthesis of the subunit Vc of sweet potato cytochrome _c_ oxidase.

When a size-fractionated poly(A)$^+$RNA from sweet potato root tissue is translated with a wheat germ _in vitro_ translation system and the translation products are analysed by SDS-polyacrylamide gel electrophoresis and fluorography, a product with the same mobility as the subunit Vc of sweet potato cytochrome _c_ oxidase is immunoprecipitated with anti-sweet potato cytochrome _c_ oxidase subunit V antibody. The result indicates that the subunit Vc is encoded by the nuclear genome and that it is synthesized in a form indistiguishable from the mature form in the size although most of nuclear DNA-encoded mitochondrial proteins are synthesized as larger precursors with extra peptides that are thought to play important roles in post-translational import into mitochondria. Thus we propose that the mature subunit Vc has a signal for localization into the mitochondrial inner membrane in its structure. In order to deduce the structure of the signal, we determined the sequence from the N-terminal amino acid residue (L-alanine) to the 23rd amino acid residue with an amino acid sequencer. The sequenced region accounts for about half part of the total sequence, but it still contains no putative sequence for membrane-bound domains. Recently, α-helical structures with charged amino acid residues on one side and hydrophobic ones on the other side were proposed to function as a signal for recognition of mitochondria in the post-translational transport(27,28). We have observed that the amino acid residues from the N-terminal to the 11th position of the subunit Vc can form such an amphiphilic α-helical structure (Figure 2). We suspect that this α-helix may function as a signal for the post-translational transport into mitochondria.

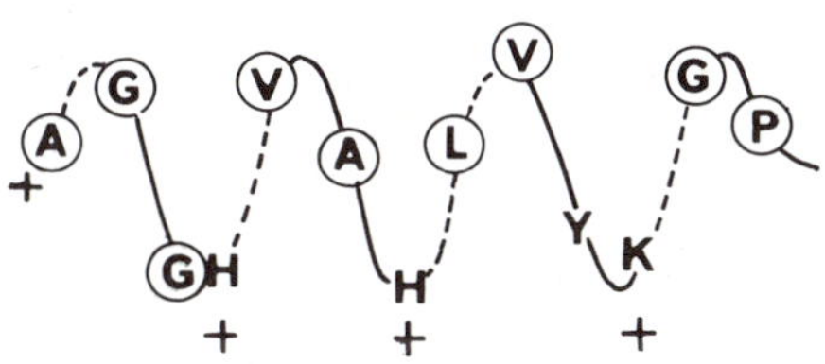

Figure 2. An amphiphilic α-helix structure deduced for the N-terminal part of the subunit Vc of sweet potato cytochrome _c_ oxidase. The amino acid residues surrounded by circles are hydrophobic.

SYNTHESIS AND INTRACELLULAR LOCALIZATION OF THE NUCLEAR DNA-ENCODED SUBUNITS OF HIGHER PLANT F_1ATPase

As described above, the α-subunit of higher plant F_1ATPase is encoded by the mitochondrial genome, and the other subunits are by the nuclear genome(13,14,19). We have observed that each subunit of the enzyme seems to be encoded by more than two genes in higher plants, at least in sweet potato(19,29).

When poly(A)$^+$RNA from sweet potato root tissue is translated with a wheat germ _in vitro_ translation system and the translation products are immunoprecipitated with anti-sweet potato F_1ATPase antibody, at least six products with molecular weights of 60,000, 58,000, 27,500, 27,000, 26,500, and 23,500 are detected after SDS-polyacrylamide gel electrophoresis (Figure 3)(19). The antibody reacts with the α-, β-, γ-(?), δ- and δ′-subunits of the enzyme in immunoblotting. The two larger products compete with sweet potato F_1ATPase β-subunit, which has been isolated by eluting the polypeptide from a gel slice after SDS-polyacrylamide gel electrophoresis of the purified enzyme, in immunological reactions, and the other four do with either the δ- or the δ′-subunit isolated as described above. The antibody used does not distinguish the δ-subunit from the δ′-subunit, although the subunits differ in structure. Thus the β-subunit with a molecular weight of 51,500 is synthesized <u>in vitro</u> in two forms of larger precursors with extra peptides with molecular weights of 8,500 and 6,500, and the δ- and δ′-subunits with molecular weights of 26,000 and 23,000, respectively, is in four forms in all. This suggests that the poly(A)$^+$RNA contains at least two kinds of mRNAs for each of the subunits.

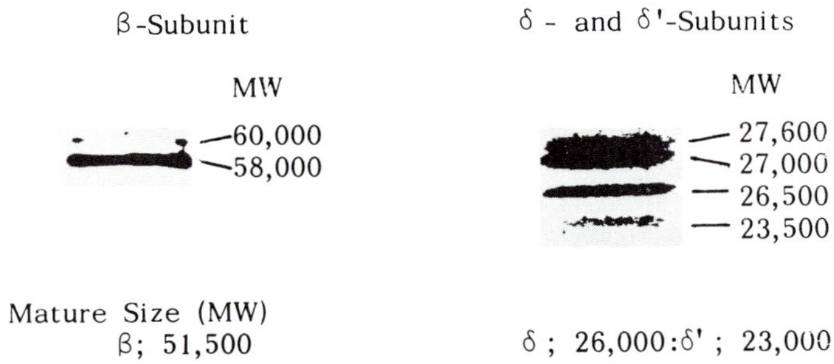

Figure 3. <u>In vitro</u> synthesis of plural larger precursors for the β-, δ- or δ′-subunit of sweet potato F_1ATPase. Poly(A)$^+$RNA from sweet potato root tissue was translated with a wheat germ <u>in vitro</u> translation system, and immunoprecipitates from the translation products with anti-sweet potato F_1ATPase antibody were analysed by SDS-polyacrylamide gel electrophoresis and fluorography. MW; molecular weight. See reference 19 for details.

We have succeeded in isolating the α- and β-subunits of sweet potato F_1ATPase on a large scale by the following method(29). First, the purified enzyme is treated with 6 M guanidine-HCl to be dissociated into the constituting subunits, then the preparation is fractionated by high performance liquid chromatography (HPLC) with a molecular sieving column. Finally, a fraction containing both α- and β-subunits is fractionated by HPLC with an ion exchange column. Either the α- or the β-subunit fraction obtained shows a single protein band after SDS-polyacrylamide gel electrophoresis. However, more than two protein bands are detected when either of the final preparations is subjected to isoelectric focussing (Figure 4). In the case of the α-subunit, one major band with an isoelectric point (pI) of 6,4, one second major band with a pI of 6.1, and two minor bands with a pI of about 6.2 and a very acidic pI are observed (in reference 29, we reported the presence of one major and several minor bands for the α-subunit, but here we more precisely describe the isoelectric focussing pattern); in this regard, it is interesting that maize mitochondrial genome contains two genes for the α-subunit(21) and so sweet potato mitochondrial genome may also contain more than two α-subunit genes which are actually expressed. In the case of the β-subunit, two major bands with pIs of 5.4 and 5.5 and one minor band with a slightly acidic pI are detected, which suggests the presence of three kinds of polypeptides for the β-subunit in sweet potato root mitochondria.

When the N-terminal part of the isolated β-subunit is sequenced by an

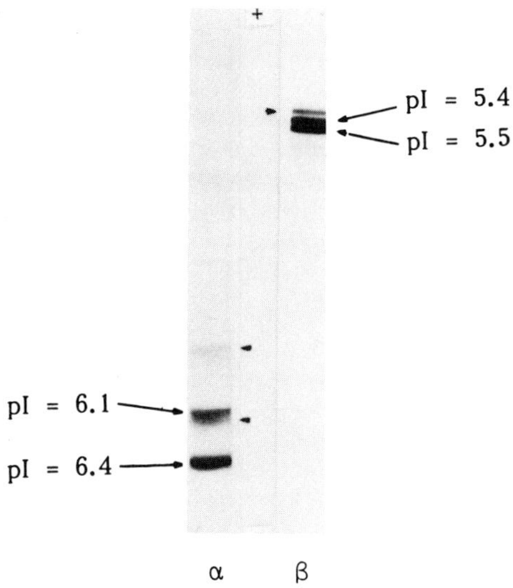

Figure 4. Isoelectric focusing of the α- and β-subunits of sweet potato F_1ATPase. The subunits were isolated by twice HPLC of the purified enzyme preparation treated with 6 M guanidine-HCl, then each subunit preparation was subjected to isoelectric focussing. See reference 27 for details. A similar photograph was presented in reference 27, but this photograph more precisely represents protein distributions than the previous one does. Arrowheads indicate minor bands.

amino acid sequencer, both L-glutamic acid and L-aspartic acid appear at the 8th position in about an equal molar ratio, although a single amino acid is detected at each of other positions in the part from the N-terminal to the 18th. This indicates that there are at least two molecular species of F_1ATPase β-subunit in sweet potato root mitochondria; probably, they differ from each other only in a few amino acid residues in the amino acid sequence.

Taken together, it is evident that sweet patato possesses more than two F_1ATPase β-subunit genes which are expressing their information at least in the root tissue. Boutry and Chua actually showed the present of two genes for the β-subunit in tobacco plant(30). Thus, higher plants are thought to have more than two genes for F_1ATPase β-subunit. We propose from the data presented in Figure 3 that higher plants may also have more than two genes for either the δ- or δ′-subunit, probably for γ- and ε-subunits, too. The question why higher plants possess plural genes for each of the subunits of F_1ATPase is now raised; that is, it is very interesting to examine whether there are differences in the regulation of expression, such as in the tissue-specificity and in the induction, among the plural genes.

We compared the amino acid sequence of the N-terminal part of the β-subunit of sweet potato F_1ATPase with that deduced from the nucleotide sequence of tobacco β-subunit gene by Boutry and Chua(30). We have observed that the former sequence has a homology with the region between the 56th and 73rd amino acid residues of the latter(29). Accordingly, we propose that the sequence from the N-terminal to the 55th for the deduced sequence of the tobacco subunit may be the extra peptide which is cleaved from the precursor to form the mature subunit during import into the mitochondria. The extra peptide can form an amphiphilic α-helix like the N-terminal part of sweet potato cytochrome _c_ oxidase subunit Vc.

CONCLUSION

Our work described above shows that the subunit compositions of higher plant cytochrome _c_ oxidase and F_1ATPase are somewhat different from those of the enzymes from other sources. Higher plant cytochrome _c_ oxidase is composed of seven (or eight) subunits but has no subunits with molecular weights of 10,000-20,000 which are present in the enzymes from other eukaryotes. F_1ATPase from higher plant mitochondria contains an additional subunit besides five subunits which are the common constituents for the soluble forms of ATP synthetase complexes from other sources.

In addition to the particularity in the structure, the higher plant enzymes have characters also in the genetic control. It is now well established that the α-subunit of higher plant F_1ATPase is encoded by the mitochondrial genome although the subunit for F_1ATPases from other eukaryotes is encoded by the nuclear genome. The results described in the present paper show that the subunits, at least some of them, are encoded by plural genes. It is now evident that higher plants, at least some species of them, possess more than two genes for the α- and β-subunits of F_1ATPase; all the genes appears to express their genes at least in some tissues. We suggest that higher plants also have plural genes for the other subunits of F_1ATPase and some (at least subunit IV and Vc) of cytochrome c oxidase subunits; in some cases, the plural genes express their information in the same tissue, but in other cases, they do in a tissue-specific manner. Now, attention should be directed to the problem whether and how the plural genes differ from one another in the regulation of gene expression.

The N-terminal sequence of the subunit Vc of sweet potato cytochrome c oxidase and the pre-sequence of the precursor of tobacco F_1ATPase β-subunit suggest that amphiphilic α-helix structures may function in the intracellular localization of mitochondrial proteins and thus there is no essential difference in the mechanism of the localization between higher plants and other eukaryotes. However, a recent report shows that the pre-sequence of *Chlamydomonas reinhardtii* ribulose 1,5-bisphosphate carboxylase/oxygenase small subunit directs the import of proteins into yeast mitochondria(31). Consequently, higher plant mitochondria are proposed to have the unique ability to reject proteins which are destined to be imported into chloroplasts, and thus such an ability should be investigated in future.

REFERENCES

1. T. Asahi, Biogenesis of cell organelles in wounded plant storage tissue cells, *in*: "Biochemistry of Wounded Plant tissues," G. Kahl, ed., Walter de Gruyter & Co., Berlin (1978).
2. T. Asahi and M. Maeshima, Biogenesis of mitochondria in higher plant cells, *in*: "The New Frontiers in Plant Biochemistry," T. Akazawa, T. Asahi and H. Imaseki, ed., Japan Scientific Societies Press, Tokyo (1983).
3. Y. Nawa and T. Asahi, Rapid development of mitochondria in pea cotyledons during the early stage of germination, *Plant Physiol.*, 48:671 (1971).
4. Y. Nawa and T. Asahi, Biochemical studies on the development of mitochondria in pea cotyledons during the early stage of germination: effects of antibiotics on the development, *Plant Physiol.*, 51:833 (1973).
5. N. Nakayama, I. Sugimoto and T. Asahi, Presence in dry pea cotyledons of soluble succinate dehydrogenase that is assembled into the mitochondrial inner membrane during seed imbibition, *Plant Physiol.*, 65:229 (1980).
6. M. Matsuoka and T. Asahi, Mechanism of the increase in cytochrome c oxidase activity in pea cotyledons during seed imbibition: the presence of free cytochrome c oxidase subunits in dry cotyledons and their probable assembly into the holoenzyme during seed hydration, *Eur. J. Biochem.*, 134:223 (1983).
7. M. Maeshima and T. Asahi, Purification and characterization of sweet potato cytochrome c oxidase, *Arch. Biochem. Biophys.*, 187:423 (1978).
8. M. Matsuoka, M. Maeshima and T. Asahi, The subunit composition of pea cytochrome c oxidase, *J. Biochem.*, 90:649 (1981).
9. T. Fox and C. J. Leaver, The zea mays mitochondrial gene coding cytochrome oxidase subunit II has an intervening sequence and does

not contain TGA codons, Cell, 26:315 (1981).
10. L. Bonen, P. H. Boer and M. H. Gray, The wheat cytochrome oxidase subunit II gene has an intron insert and three radical amino acid changes relative to maize, EMBO J., 3:2531 (1984).
11. P. G. Isaac, V. P. Jones and C. J. Leaver, The maize cytochrome c oxidase subunit I gene: sequence, expression and rearrangement in cytoplasmic male sterile plants, EMBO J., 4:1617 (1985).
12. M. Maeshima and T. Asahi, Mechanism of increase in cytochrome c oxidase activity in sweet potato root tissue during aging of slices, J. Biochem., 90:391 (1981).
13. M. Boutry, M. Briquet and A. Goffeau, The α subunit of a plant mitochondrial F_1ATPase is translated in mitochondria, J. Biol. Chem., 258:8524 (1983).
14. E. Hack and C. J. Leaver, The α-subunit of the maize F_1-ATPase is synthesized in the mitochondria, EMBO J., 2:1783 (1983).
15. Y. Iwasaki and T. Asahi, Purification and characterization of the soluble form of mitochondrial adenosine triphosphatase from sweet potato, Arch. Biochem. Biophys., 227:164 (1983).
16. A. Horak and M. Packer, Coupling factor activity of the purified pea mitochondrial F_1-ATPase, Biochim. Biophys. Acta, 810:310 (1985).
17. V. L. Spitsberg, N. E. Pfeiffer, B. Partridge, D.E. Wylier and S. M. Schuster, Isolation and antigenic characterization of corn mitochondrial F_1-ATPase, Plant Physiol., 77:339 (1985).
18. S. K. Randall, Y. Wang and H. Sze, Purification and characterization of the soluble F_1-ATPase of oat root mitochondria, Plant Physiol., 79:957 (1985).
19. Y. Iwasaki and T. Asahi, Intracellular sites of the synthesis of sweet potato mitochondrial F_1ATPase subunits, Plant Mol. Biol., 5:339 (1985).
20. C. J. Braun and C. S. Levings III, Nucleotide sequence of the α subunit gene from maize mitochondria, Plant Physiol., 79:571(1985).
21. P. G. Isaac, A. Brennike, S.M. Dunbar and C. J. Leaver, The mitochondrial genome of fertile maize (Zea mays L.) contains two copies of the gene encoding the α-subunit of the F_1ATPase, Curr. Gen., 10:321 (1985).
22. A. Morikami and K. Nakamura, Structure of pea mitochondrial F_1ATPase α-subunit gene and its pseudogene generated by recombination within the coding sequence, Submitted to Plant Mol. Biol.
23. M. Maeshima, T. Asahi and I. Uritani, Effect of temperature on the activity and stability of plant cytochrome c oxidase, Agric. Biol. Chem., 44:2351(1980).
24. L. Kuhn-Nentwig and B. Kadenbach, Isolation and properties of cytochrome c oxidase from rat liver and quantification of immunological differences between isozymes from various rat tissues with subunit-specific antisera, Eur. J. Biochem., 149:147 (1985).
25. R. Bisson and G. Schiavo, Two different forms of cytochrome c oxidase can be purified from slime mold Dictyostelium discoideum, J. Biol. Chem., 261:4373 (1986).
26. A. Stroh and B. Kadenbach, Tissue-sqecific and sqecies-specific distribution of -SH groups in cytochrome c oxidase subunits, Eur. J. Biochem., 156:199 (1986).
27. K. Kobayashi, Y. Iwasaki, T. Sasaki, K. Nakamura and T. Asahi, Putative amino-terminal presequence for β-subunit of plant mitochondrial F_1ATPase deduced from the amino-terminal sequence of the mature subunit, FEBS Lett., 203:144 (1986).
28. D. Roise, S. J. Horvath, J. M. Tomich, J. H. Richards and G. Schatz, A chemically synthesized pre-sequence of an imported mitochondrial protein can form an amphiphilic helix and perturb natural and artificial phospolipid bilayers, EMBO J., 5:1327 (1986).
29. G. von Heijine, Mitochondrial targeting sequences may form amphiphilic hilices, EMBO J., 5:1335 (1986).

30. M. Boutry and N. H. Chua, A nuclear gene encoding the beta subunit of the mitochondrial ATP synthase in *Nicotiana plumbaginifolia*, *EMBO J.*, 4:2159 (1985).
31. E. C. Hurt, N. Soltanifar, M. Goldschmidt-Clermont, J. D. Rochaix and G. Schatz, The cleavable pre-sequence of an imported chloroplast protein directs attached polypeptides into yeast mitochondria, *EMBO J.*, 5:1343 (1986).

TWO NUCLEAR GENES ENCODE THE BETA SUBUNIT OF THE MITOCHONDRIAL ATP SYNTHASE IN *Nicotiana plumbaginifolia*

Marc Boutry

Laboratoire d'Enzymologie
Université de Louvain
Place Croix du Sud, 1, Box 8
B-1348 Louvain-la-Neuve, Belgium

INTRODUCTION

Living organisms offer little variation in the final step of oxidative phosphorylation leading to the ATP production. Whether the primary energy source is light or chemicals, whether the organism is a bacteria, a yeast, a mammal or a plant, all pathways converge to build-up a transmembrane protic gradient which is used to convert ADP to ATP in the ATP synthase complex. The presence of the latter in bacterial, chloroplast and mitochondrial membranes attests a common and early origine of this enzyme during evolution. Few important structural or functional differences are found among different organisms. This indicates that the ATP synthase complex evolved rapidly as a stable and highly efficient enzyme which probably supplanted more ancient and less performing ATP-supplying enzymes.

In contrast to the well conserved protein structure, function and mechanism, dramatic variations occur in the organization of genes encoding the ATP synthase complex. The best characterized case is the *E. coli* ATP synthase. Its 8 distinct subunits are encoded by an operon which also comprises an additional open reading frame whose product has not been identified yet. These nine genes are gathered on a short DNA segment and are transcribed into a single polycistronic RNA (1). Eukaryots offer a more complicated picture: chloroplasts and mitochondria possess their own genetic information which encodes for 10 to 15% of the organelle proteins, the majority of which are dependent on the nuclear genome, synthesized on cytoribosomes and finally imported into the organelle. Consequently, the genes encoding subunits of several organelle complexes are dispersed on two genomes. For instance, six out of the nine subunits of the chloroplast ATP synthase complex are encoded by separated operons located on the chloroplast DNA (2). Interestingly, they conserve a gene arrangment similar to the one found in *E. coli*. The three other subunits of the chloroplast ATP synthase are encoded by nuclear genes (2).

Various situations are observed concerning the mitochondrial ATP synthase complex. For instance, subunit 9 (DCCD-binding proteolipid) is encoded by the mitochondrial DNA in yeast (3) and plants (4, 5) and by nuclear genes in mammals and in *Neurospora crassa* (6). Another variation is found with the alpha subunit gene which is located in the nucleus of yeast and mammals but belongs to the mitochondrial genome in plants (7, 8).

Besides diversified gene localizations, different controls of gene expression are to be expected among different organisms. In yeast, the synthesis of the ATP synthase complex, as that of other mitochondrial enzymes, is regulated by the carbon source. For instance, in the presence of high glucose concentrations, many mitochondrial functions are repressed and ATP is mainly supplied by the glycolytic pathway. As a consequence, it has been possible to isolate yeast mutants defective in the ATP synthase (9) which would be lethal in other organisms. Much less is known about a possible differential expression in higher eukaryots. However since mitochondria play roles of variable importance in different tissues or at different development stages, the expression of the ATP synthase complex could be modulated accordingly. In plants, mitochondria are a major source of ATP in non chlorophylic tissues. In green tissues, it is not clear what is the mitochondrial contribution to the cellular ATP supply in conditions where photosynthesis is fully active (see other papers in this volume).

CLONING NUCLEAR GENES FOR THE BETA SUBUNIT

To gain access to the study of differential expression of ATP synthase genes in plants, we have undertaken an analysis of the genes encoding subunits of the mitochondrial ATP synthase. We first focussed on the beta subunit for several reasons. Experiments involving chemical modification and inhibitors indicate that this subunit plays a major role in catalysis (10). From available gene sequences, it appears that beta is probably the most conserved subunit of the ATP synthases among different organisms (12). This last property let us foresee the isolation of the plant gene(s) by hybridization with the corresponding gene already isolated from yeast. Finally indirect evidences indicate that the beta subunit of the plant mitochondrial ATP synthase is probably a nuclear encoded product (7, 8). An additional incitement is the fact that previous informations are not available today concerning plant nuclear genes encoding mitochondrial proteins.

We chose to isolate the gene from *Nicotiana plumbaginifolia*, a species closely related to tobacco. Although mitochondria are not easily isolated from that species (etiolated material is hardly available), *N. plumbaginifolia* has a relatively small nuclear genome which simplifies gene screening and is amenable to genetic transformation which should allow *in vivo* characterization of sequences essential to the gene expression (see below). We constructed a genomic library of *N. plumbaginifolia* in a phage vector (EMBL4) and screened it with a yeast gene coding for the ATP synthase beta subunit. Several clones were isolated. Restriction mapping and genomic Southern blot analyses indicated that *N. plumbaginifolia* contains 2 distinct and non allelic genes for the mitochondrial ATP synthase beta subunit (11). This was unexpected since two yeast species: *Saccharomyces cerevisiae* (12) and *Schizosaccharomyces pombe* (13) contain both a single gene for the beta subunit. Another discrepancy with yeast was revealed by the sequence analysis of one (*atp2-1*) of the two plant genes: it is interrupted by 8 noncoding intervening sequences (introns) which have to be removed from the primary transcript to give a mature RNA suitable for translation (Figure 1). On the contrary, the yeast gene is uninterrupted. The high sequence homology among the beta subunits from bacteria, yeasts, mammals and plants indicates that their corresponding genes have derived from a common ancestor preexisting before the divergence between eu- and prokariotes. Introns have possibly been lost during evolution in bacteria and unicellular eukaryotes like yeast. A similar hypothesis has been suggested for genes encoding another ubiquitous protein: the globin. The function, if any, of most introns is still obscure. They may have played an important evolutionary role, for instance by promoting recombinatory events between preexisting genes and creating new genes and consequently new functions (15).

The *atp2-1* intron borders have been localized from the nucleotide sequence of cDNA clones and the coding sequence has been identified. Figure 2 reports the amino acid sequence deduced from the coding nucleotide sequence. We have compared the beta subunit sequence from the *E. coli* ATP synthase (A), the *N. tabacum* chloroplast ATP synthase (B) and the human (C), yeast (D) and *N. plumbaginifolia* (E) mitochondrial ATP synthases. A high homology is clearly observed along the entire sequence except at the N-terminal where the 3 mitochondrial polypeptides (C-E) extend to different lengths. Table 1 gives the calculated homology among the ATP synthase beta subunits from different organisms. In this comparison, we have excluded the N-terminal sequences which are involved in mitochondrial targeting (see below). The amino acid sequence homology varies between 65 to 80%. As expected, the highest homologies are obtained when the three mitochondrial polypeptides are compared.

WHY TWO GENES?

By Southern hybridization, we have observed that the different *Nicotiana* species analyzed so far as well as more distant species like petunia, tomato, rice and maize contain at least 2 genes for the mitochondrial beta subunit. We have sequenced part of a genomic clone containing the second gene (*atp2-2*) from *N. plumbaginifolia*. Both *atp2-1* and *atp2-2* genes have a similar structure (Figure 1). Introns are located at the same position. Interestingly, these noncoding sequences show less than 40% homology. This and other observations suggest that both genes diverged from a common ancestor much before the appearance of modern *Nicotiana* species.

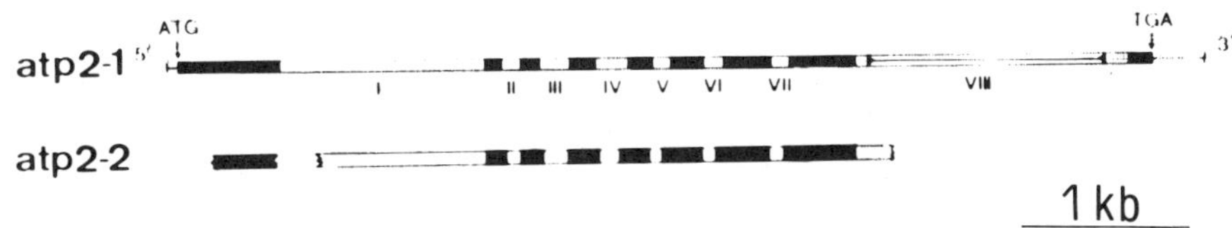

Figure 1. Structure of *atp2-1* and *atp2-2* genes.
The structure of the entire *atp2-1* gene is from reference 11. A genomic clone containing the *atp2-2* gene has been subcloned in plasmid vectors for sequencing as in (11). The regions depicted corresponds to the sequence determined so far. Exons are represented in closed boxes and introns (numbered for *atp2-1*) in open boxes.

Figure 2. (Next page) Comparison of the ATP synthase beta subunits from different organisms.
The amino acid sequences of the ATP synthase beta subunit from *E. coli* (19)(A), tobacco chloroplast (20)(B), human (21)(C), *S. cerevisiae* (12)(D) and *N. plumbaginfolia* (11)(E) mitochondria are aligned. The sequence of the second *N. plumbaginifolia* gene (*atp2-2*) has been obtained from residue 81 to the C-terminal. It is identical to the *atp2-1* amino acid sequence but four residues which are circled below the *atp2-1* sequence. Identical residues in the 5 sequences are underlined.

```
                   C MTSLWGKGTGCKLFKFRVAAAPASGALGRLTPSASLPPAQLLLRAVRR
                                               D MVLPRLYATSRAAFKAAKQS
E MASRRLLASLLRQSAQRGGGLISRSLGNSIPKSASRASSRASPKGFLLNRAVQYATSAAAPST
  1
                             A MATGKIVQVIGAVVDVEFPQDAVPRVYDALEVQNGNER-----L
 B MRINPTTSGSGVSTLEKKNPGRVVQIIGPVLDVAFPPGKNPNIYNALVVQGRDSVGQPINV
c·RSHPVRDYAAQTSPSPKAGAATGRIVAVIGAVVDVQFDEGLPP-ILNALEVQGRETR-----L
d·APLLSTSWKRCMASAAQSTPITGKVTAVIGAIVDVHFEQSELPAILNANEIKTPQGK-----L
e·PPKSGSEPSGKITDEFTGAGSIGKVCQVIGAVVDVRFDEGLPP-ILTALEVLDNQIR-----L
  64

a·VLEVQQQLGGGIVRTIAMGSSDGLRRGLDVKDLEHPIEVPVGKATLGRIMNVLGVPVDMKGEI
b·ACEVQQLLGNNRVRAIAMSATEGLTRGMEVIDTGAPISVPVGGATLGRIFNVLGEPVDNLGPV
c·VLEVAQHLGESTVRTIAMDGTEGLVRGQKVLDSGAPIKIPVGPETLGRIMNVIGEPIDERGPI
d·VLECAQMLGENTVRTIAMDGTEGLVRGEKVLDTGGPISVPVGRETLGRIINVIGEPIDERGPI
e·VLEVAQHLGENMVRTIAMDGTEGLVRGQRVLNTGSPITVPVGRATLGRIINVIGEAIDERGPI
  127                                                      (P)   (D)

a·GEEERWAIHRAAPSYEELSNSQELLETGIKVIDLMDPFAKGGKVGLFGGAGVGKTVNMMELIR
b·DTSTTSPIHRSAPAFIQLDTKLSIFETGIEVVDLLAPYRRGGKIGLFGGAGVGKTVLIMELIN
c·KTKQFAPIHAEAPEFMEMSVEQEILVTGIKVVDLLAPYAKGGKIGLFGGAGVGKTVFIMELIN
d·KSKLRKPEHADPPSFAEQSTSAEILETGIKVVDLLAPYARGGKIGLFGGAGVGKTVFIQELIN
e·TTDHFLPIHREAPAFVEQATEQQILVTGIKVVDLLAPYQRGVKIGLFGGAGVGKTVLIMELIN
 (N)

a·NIAIEHSGYSVFAGVGERTREGNDFYHEMTDSNVI---D-----KVSLVYGQMNEPPGNRLRV
b·NIAKAHGGVSVFGGVGERTREGNDLYMENKESGVINE-ENIAESKVALVYGQMNEPPGARMRV
c·NVAKAHGGYSVFAGVGERTREGNDLYHEMIESGVINLKDAT--SKVALVYGQMNQPPGARARV
d·NIAKAHGGFSVFTGVGERTREGNDLYREMKETVVINL-EG--ESKVALVFGQMNEPPGARARV
e·NVAKAHGGFSVVAGVGERTREGNDLYREMIESGVIKLGEKQSESKCALVYGQMNEPPGARARV
  257                                     (D)

a·ALTGLTMAEKFRG-EGRDVLLFVDNIYRYTLAGTEVSALLGRMPSAVGYQPTLAEEMGVLQER
b·GLTALTMAEYFRDVNEQDVLLFIDNIFRFVQAGSEVSALLGRMPSAVGYQPTLSTVMGSLQVR
c·ALTGLTVAEYFRDQEGQDVLLFIDNIFRFTQAGSEVSALLGRIPSAVGYQPTLATDMGTMQER
d·ALTGLTIAEYFRDEEGQDVLLFIDNIFRFTQAGSEVSALLGRIPSAVGYQPTLATDMGLMQER
e·GLTGLTVAEHFRDAEGQDVLLFIDNIFRFTQANSEVSALLGRIPSAVGYQPTLATDLGGLQER
  324

a·ITSTKTGSITSVQAVYVPADDLTDPSPATTFAHLDATVVLSRQIASLGIYPAVDPLDSTSRQL
b·ITSTKVGSITSIQAVYVPADDLTDPAPATTFAHLDATTVLSRGLAAKGIYPAVKPLDSTSTML
c·ITTTKKGSITSVQAIYVPADDLTDPAPATTFAHLDATTVLSRAIAELGIYPAVDPLDSTSRIM
d·ITTTKKGSVTSVQAVYVPADDLTDPSPSTSFAHLDASSVLSRGISELGIYPAVVPLDSKSRLL
e·ITTTKKGSITSVQAIYVPADDLTDPAPATTFAHLDATTVLSRQISELGIYPAVDPLDSTRSML
  391

a·DPLVVGQEHYDTARGVQSILQRYQELKDIIAILGMVELSEEDKLVVARARKIQRFLSQPFFVA
b·QPRIVGEEHYETAQRDKQTLQRIKELQDIIAILGLDELSEEDRLLVARARKIERFLSQPFFVA
c·DPNIVGSEHYDVARGVQKILQDYKSLQDIIAILGMDELSEEDKLTVSRARKIQRFLSQPFQVA
d·DAAVVGQEHYDVASKVQETLQTYKSLQDIIAILGMDELSEQDKLTVERARKIQRFLSQPFAVA
e·SPHILGEDHYNTARGVQKVLQNYKNLQDIIAILGMDELSEDDKMTVARARKIQRFLSQPFHVA
  458

a·EVFTGSPGKYVSLKDTIRGFKGIMEGEYDHLPEQAFTMVGSIEEAVEKAKKL
b·EVFTGSPGKYVGLAETIRGFQLILSGELDGLPEQAFYLVGNIDEATAKAMNLEMESNLKK
c·EVFTGHMGKLVPLKETIKGFQQILAGEYDHLPEQAFYMVGPIEEAVAKADKLAEEHSS
d·EVFTGIPGKLVRLKDTVASFKAVLEGKYDNIPEHAFYMVGGIEDVVLKAEKLAREAN
e·EVFTGAPGKYVDLKESINSFQGVLDGKYDDLSEQSFYMVGGIDEVIAKAEKIAKESAA
```

Figure 2

Table 1. Sequence homology of the ATP synthase beta subunits from different sources.

	E. coli	Tobacco chloroplast	Human mito.	Yeast mito.	*N. plumb.* mito.
E. coli	-	65 %	70 %	67 %	66 %
Tobacco chloroplast	65 %	-	70 %	65 %	69 %
Human mitochondria	70 %	70 %	-	79 %	80 %
Yeast mitochondria	67 %	65 %	79 %	-	75 %
N. plumbaginifolia mitochondria	66 %	69 %	80 %	75 %	-

Homologies have been determined from the sequences depicted in Figure 2 on the region spanning from the first to the last residues of the *E. coli* sequence. Consequently the N terminal pre-sequences of the mitochondrial beta polypeptides have been excluded from the comparison.

Coding sequences (exons) are much more conserved (about 90% homology) and the deduced amino acid sequence from *atp2-2* is almost identical to the one from *atp2-1*. On 464 amino acid residues identified so far only four have been changed (Figure 2). This indicates that a strong selective pressure has maintained an unchanged polypeptide and is consistent with the high homology observed among different organisms.

We know that, in *N. plumbaginifolia*, both genes are transcribed at a similar level in leaf tissues since we have isolated at a same frequency cDNA clones from both of them. Whether they are transcribed at a similar level in different tissues and/or at different developmental stages is unknown but the availability of hybridization probes specific to each gene will allow us to answer that question. Whether both polypeptides are assembled into a functional complex is also unknow. Since one of the four modified amino acid residues (proline-aspartate) changes the global charge of the polypeptide, two-dimensional electrophoresis analysis should allow us to identify if both polypeptides are assembled within mitochondrial membranes.

A PRE-SEQUENCE FOR MITOCHONDRIAL TARGETING

It is now clear from studies of yeast and other organisms that the majority of mitochondrial and chloroplast proteins encoded in the nucleus are synthesized on cytoribosomes as precursors containing an N-terminal extension or transient pre-sequence which targets the protein to the organelle. During or soon after entering the organelle, the pre-sequence is removed from the mature polypeptide (16,17). Pre-sequences of several chloroplast proteins and of mitochondrial proteins from yeast and animal organisms have been obtained. Their sizes are extremely variable: from 10 up to 90 amino acid residues. No consensus sequence has been found but the following common features have been noticed: many hydroxylated amino acids (serine and threonine), no or very few negatively charged residues, and a higher than average frequency of positively charged residues (mainly arginine and lysine) (16,17). Recently it has been shown that in most mitochondrial presequences amphiphilic alpha helices can be predicted with a high hydrophobic moment (18).

A first clue to the existence of a pre-sequence for the *N. plumbaginifolia* mitochondrial beta subunit comes from the comparison of the polypeptide size as deduced from the sequence (Mr = 59 866) with the size of the mature polypeptide as determined by gel electrophoresis (Mr = 51 000). Recently, the amino-terminal sequence of the mature beta subunit of the sweet potato mitochondrial ATP synthase has been reported (24). Comparing their sequence with the one deduced from the *atp2-1* gene, the authors suggest a pre-sequence of 55 residues for the *N. plumbaginifolia* beta subunit.The existence of a pre-sequence of about 8,000 was confirmed by *in vitro* uptake of the precursor by isolated mitochondria. The isolated gene was used as a template in an *in vitro* transcription-translation system in the presence of [^{35}S]-methionine. The labeled precursor had the expected size of 59 000 (Figure 3). When we incubated the precursor with purified intact mitochondria and then reisolated mitochondria through a sucrose cushion, we observed a cosegregation of the precursor with mitochondria. Moreover part of the precursor was converted to the mature size (Figure 3, lane B).

Interestingly, no homology was found when we compared the putative pre-sequences of yeast, human and plant mitochondrial beta subunits. Even their lengths vary largely from 16 amino acid residues in yeast to about 80 in plant. This indicates that, unlike the ATP synthase complex, the machinery responsible for mitochondrial targeting evolved differently in the three organisms.

PROSPECTS

The isolation and characterization of plant nuclar genes encoding the beta subunit of the mitochondrial ATP synthase has opened several possibilities for further development.

1. Why two genes?

Since *N. plumbaginifolia* and many other plant species contain two non allelic genes coding for the mitochondrial ATP synthase beta subunit,

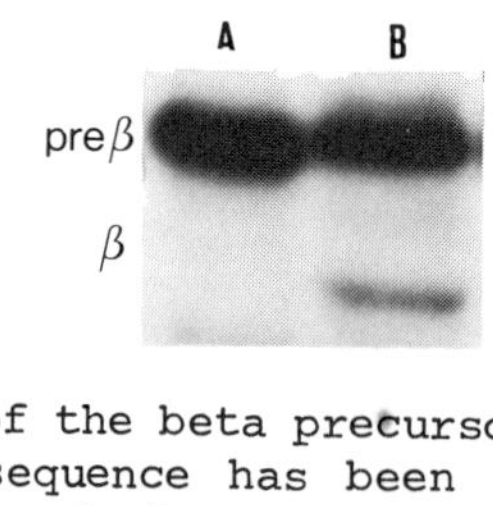

Figure 3. *In vitro* uptake of the beta precursor by mitochondria. An *atp2-1* cDNA sequence has been subcloned in the expression vector pDS5 (22) and then submited to an *in vitro* transcription-translation system as indicated (22) except that translation was carried out in a reticulocyte lysate (NEN). In lane A, 5 µl of the translation mixture were analyzed by SDS-PAGE (23) and autoradiography. In lane B, 10 µl of the translation mixture were added to purified faba bean mitochondria (23) (100 µg proteins) in a final volume of 25 µl of 0.4 M mannitol, 10 mM $MgCl_2$, 50 mM KH_2PO_4, 10 mM Na succinate, pH 7.2. After an incubation of 15 min at 25°C, mitochondria were collected through a 0.6 M sucrose cushion (500 µl in an Eppendorf tube) and analyzed as in A.

it is important to figure out why plants need two genes while yeasts can do with only one. We are developing hybridization probes specific to each of both genes in order to analyse their relative expression in different tissues.

2. A constitutive promoter

The beta subunit of the mitochondrial ATP synthase is one of the most abundant mitochondrial proteins which is expressed in every tissue and probably at all growth stages. Therefore the promoter from such a gene could be used to drive, after genetic transformation, the transcription of foreign genes. For that purpose, we have isolated the *atp2-1* promoter region and prepared several deletions in order to define the smallest fragment which still promotes normal transcription. We have fused these engineered promoter pieces to a marker gene and transferred then into tobacco by an *Agrobacterium tumefaciens* Ti-derived vector. Regenerated transformed plants will be tested for the expression of the chimeric genes.

3. An amino acid sequence for targeting to plant mitochondria

We have isolated the DNA sequence corresponding to the *N. plumbaginifolia* beta subunit pre-sequence and fused it in front of the coding sequence of a bacterial gene for the chloramphenicol acetyltransferase (which confers resistance to chloramphenicol). This chimeric gene has been placed under the control of a plant promoter (CaMV 35S promoter) and introduced into tobacco by genetic transformation. The gene is expressed in transformed plants and the chloramphenicol acetyl transferase activity is associated with mitochondria (manuscript in preparation). Experiments are underway to characterize the efficiency and the specificity of the process. Clearly this opens the possibility to target many new proteins to plant mitochondria.

ACKNOWLEDGMENTS

Most of this work has been performed in the laboratory of Plant Molecular Biology (Dr N.-H. Chua) at The Rockefeller University, New York. I gratefully acknowledge Drs N.-H. Chua (New York), A. Goffeau and M. Briquet (Louvain-la-Neuve) for their support and interest. I am "chargé de recherches" of the F.N.R.S.

REFERENCES

1. M. Futai and H. Kanazawa, Structure and function of proton-translocating adenosine triphosphatase: biochemical and molecular biological approaches, *Microbiol. Reviews* 47: 285-312.
2. J. Henning and R.G. Herrmann, Chloroplast ATP synthase of spinach contains nine nonidentical subunit species, six of which are encoded by plastid chromosomes in two operons in a phylogenetically conserved arrangement, *Mol. Gen. Genet.* 203: 117-128 (1986).
3. L.A. Hensgens, L.A. Grivell, P. Borst and J.L. Bos, Nucleotide sequence of the mitochondrial structure gene for subunit 9 of yeast ATPase complex. *Proc. Natl. Acad. Sci. USA*, 76: 1663-1667 (1979).
4. E. Hack and C.J. Leaver, Synthesis of a dicyclohexylcarbodiimide-binding proteolipid by cucumber mitochondria, *Current Genetics* 8: 537-542 (1984).
5. R.E. Dewey, A.M. Schuster, C.S. Levings III, D.H. Timothy, Nucleotide sequence of Fo-ATPase proteolipid (subunit 9) gene of maize mitochondria, *Proc. Natl. Acad. Sci. USA* 82: 1015-1019 (1985).
6. P. Van den Boogaart, J. Samallo, S. Van Dijk, E. Agsteribbe, Structural and functional analyses of the genes for subunit II of cyto-

chrome aa_3 and for a dicyclohexylcarbodiimide-binding protein in *Neurospora crassa* mitochondrial DNA, *In*: "Mitochondrial Genes", P. Slonimski, P. Borst, G. Attardi, eds, Cold Spring Harbor Laboratory, Cold Spring Harbor (1982).

7. M. Boutry, M. Briquet and A. Goffeau, The alpha subunit of a plant mitochondrial F_1-ATPase is translated in mitochondria, *J. Biol. Chem.* 258: 8524-8526 (1983).
8. E. Hack and C.J. Leaver, The α-subunit of the maize F_1-ATPase is synthesized in the mitochondrion, *EMBO J.*, 2: 1783-1789 (1983).
9. M. Boutry and A. Goffeau, Alterations of the alpha and beta subunits of the mitochondrial ATPase in yeast mutants, *Eur. J. Biochem.* 125: 471-477 (1982).
10. A.E. Senior and J.G. Wise, The proton-ATPase of bacteria and mitochondria, *J. Membr. Biol.* 73: 105-124 (1983).
11. M. Boutry and N.-H. Chua, A nuclear gene encoding the beta subunit of the mitochondrial ATP synthase in *Nicotiana plumbaginifolia*, *EMBO J.* 4: 2159-2165 (1985).
12. M. Tukeda, A. Vassarotti and M.G. Douglas, Nuclar genes coding the yeast mitochondrial adenosine triphosphatase complex, *J. Biol. Chem.* 260: 15458-15465 (1985).
13. M. Boutry, A. Vassarotti, M. Ghislain, M. Douglas and A. Goffeau, Isolation of the structural genes for the alpha and beta subunits of the mitochondrial ATPase from the fission yeast *Schizosaccharomyces pombe*, *J. Biol. Chem.* 259: 2840-2844 (1984).
14. R. Lewin, Surprise finding with insect globin genes, *Science* 226: 328.
15. W. Gilbert, Genes-in-pieces revisited, *Science* 228: 823-824 (1985).
16. R. Hay, P. Bohni and S. Gasser, How mitochondria imports proteins, *Biochim. Biophys. Acta* 779: 65-87.
17. M. Douglas, M. McCannon and A. Vassarotti, Targeting proteins into mitochondria, *Microbiol. Rev.* 50: in press (1986).
18. G. von Heijne, Mitochondrial targeting sequences may form amphiphilic helices, *EMBO J.* 5: 1335-1342 (1986).
19. M. Saraste, N.J. Gay, A. Eberle, M.J. Runswick and J.E. Walker, The atp operon: nucleotide sequence of the gene for the γ, β and ε subunits of *Escherichia coli* ATP synthase, *Nucleic Acids Res.* 9: 5287-5296.
20. K. Shinozaki, H. Deno, A. Kato and M. Sugiura, Overlap and cotranscription of the genes for the beta and epsilon subunits of tobacco chloroplast ATPase, *Gene* 24: 147-155 (1983).
21. S. Ohta and Y. Kagawa, Human F_1ATPase: molecular cloning of cDNA for the beta subunit, *J. Biochem.* 99: 135-141 (1986).
22. D. Stveber, I. Ibrahimi, D. Cutler, B. Dobberstein and H. Bujard, A novel *in vitro* transcription-translation system: accurate and efficient synthesis of single proteins from cloned DNA sequences, *EMBO J.* 3: 3143-3148.
23. M. Boutry and M. Briquet, Mitochondrial modification associated with the cytoplasmic sterility in Faba beans, *Eur. J. Biochem.* 127: 129-135 (1982).
24. K. Kobayashi, Y. Iwasaki, T. Sasaki, K. Nakamura and T. Asahi, Putative amino-terminal presequence for β-subunit of plant mitochondrial F_1-ATPase deduced from the amino-terminal sequence of the mature subunit, *FEBS Lett.* 203: 144-148.

DIVERGENCE IN mtDNA AND EFFECTS IN INTERSPECIFIC COMBINATIONS OF NUCLEAR AND MITOCHONDRIAL GENOMES IN THE YEAST GENUS SACCHAROMYCES

Maurice Claisse, François Michel, and Donald Hawthorne[+]

Centre de Génétique Moléculaire du C.N.R.S., Laboratoire associé à l'Université Pierre et Marie Curie Paris 6, 91190 Gif-sur-Yvette, France. [+]Genetics Department, University of Washington, Seattle, WA 98195, USA

INTRODUCTION

Yeast is an exceptional material for the study of the cytochromes of the respiratory chain. In the species *Saccharomyces cerevisiae* both nuclear and mitochondrial genetic maps are well defined with numerous mutants available. As this organism is a facultative aerobe it is possible to obtain mutations interrupting the respiratory chain without any lethal effect. In the First International Symposium on Plant Mitochondria held in 1978 it was reported on the modifications of mitochondrial translation products in cytochrome b deficient mutants of *S. cerevisiae*, which suggested a mosaic organisation for the cytochrome b structural gene[1]. It has since been demonstrated that as many as 6 exons and 5 introns may comprise this mitochondrial gene named *cob-box* [2, 3, 4]. About 200 mutations have been mapped on this mitochondrial DNA segment and span about 7300 base pairs[4], i.e. roughly seven times more than needed to specify the cytochrome b apoprotein sequence.

Since the UGA opal stop codon is read as a tryptophan in yeast mtDNA[5], extensive open reading frames are found in some introns. From these open reading frames, proteins can be visualized in intron mutants by autoradiographies of mitochondrial translation products[2, 4, 6]. Of special interest is the product encoded by bI4 intron which has two functions: the splicing of bI4 itself and the processing of the intron aI4 [7] from the *oxi3* gene for subunit I of cytochrome oxidase. Mutations designated *box7*, lying within bI4, are pleiotropic: they prevent the synthesis of both cytochrome b and subunit I of cytochrome oxidase[2, 4, 6, 7]. The *box7* mutations can be supplanted by the product of a nuclear suppressor, *NAM2* [8].

Two species of *Saccharomyces*, *S. cerevisiae* and *S. douglasii*, are distinguished by the infertility of the spores issuing from the diploid hybrid. The mitochondrial DNA's of these two species show differences in their introns. When the mitochondrial genomes are combined with heterologous nuclear genomes, the S. douglasii nuclear genome is necessary for the processing of the *oxi3* mRNA of the same species and allows the processing in mitochondria from *S. cerevisiae* *box7* mutants[9,10].

MATERIALS AND METHODS

Strains

The "grande" colony donors used to perform reciprocal cytoduction experiments were:

S. douglasii 4301-3A α HO HMLα HMRα ade2 ura5 [rho^{d+}]
4301-3A'-1 α HO HMLα HMRα ade2 ura5 aap1 [rho^{d+}]
S. cerevisiae 4309-3D a ho kar1 his5 trp1 leu2 can1 [rho^{c+}]

The symbols [rho^{d+}] and [rho^{c+}] mean that these strains are [rho^{+}] and that the mitochondria are the wild type mitochondria of the species, "d" for S. douglasii and "c" for S. cerevisiae.

The presence of kar1 (karyogamy defective) in one of the parental strains of a cross inhibits nuclear fusion in the zygote and fosters the issuance of buds with a single haploid nucleus as well as dicaryotic buds[11]. When one of the parents is [rho°] (i.e. mutant without mtDNA) then its nucleus can be selectively combined with the mitochondrial genome of the [rho^{+}] parent. Such cytoductants were:
4301-3A"-1-16 [rho^{c+}] : same nuclear genome as S. douglasii 4301-3A'-1 but associated with the mitochondrial genome [rho^{c+}] of S. cerevisiae 4309-6D.
4309-6D'-2 [rho^{d+}] : same nuclear genome as S. cerevisiae 4309-6D but mitochondrial genome [rho^{d+}] of S. douglasii 4301-3A-1.

In this series of haploid strains, labelling of cells by $^{35}SO_4^{--}$ in presence of cycloheximide was relatively poor. To increase the labelling, these strains were crossed with [rho°] isolates of the same species as the nuclear genome. The resulting diploids were:
DH1, S. douglasii [rho^{d+}] from 4301-3A [rho^{d+}] x 4285-1B' [rho°]
DH11, S. douglasii [rho^{c+}] from 4301-3A"-1-16 [rho^{c+}] x 4285-1B' [rho°]
CH1, S. cerevisiae [rho^{c+}] from 4309-6D [rho^{c+}] x M8-227/1/51 [rho°]
CH11, S. cerevisiae [rho^{d+}] from 4309-6D'-2 [rho^{d+}] x M8-227/1/51 [rho°]

Other S.cerevisiae strains, from the collection of the "Centre de Génétique Moléculaire, Gif-sur-Yvette", were used as references and in the box7 cytoduction experiments.
KL14-4A/I21 a his1 trp2 leu2 [rho^{+}] ω^{+} C_1^R O_1^R P_{454}^R (provided by B. Dujon).
D273-10B/A α met [rho^{+}] ω^{+}
A422/12 [rho^{-}] strain derived from D273-10B/A48 by mutagenesis with ethidium bromide and conserving several oxi3 markers[12].
CK320 a kar1 leu1 can1 [box7-V328]
V328 α ade1 op1 [box7-V328]
The diploid strain KM91 resulted from the cross
777-3A α op1 ade1 [rho^{+}] x KL14-4A/60 a OP1 his1 trp2 [rho°]

Spectrophotometric analysis

The determination of the cytochrome spectra was carried out according to Claisse et al.[13] on whole cell pastes 0.7 mm thick, frozen in liquid nitrogen after reduction of the cytochromes by dithionite. A Cary 219 cpectrophotometer was used.

Identification of mitochondrially translated polypeptides

Aerobic cultures, labelling of cells with $^{35}SO_4^{--}$ in the presence of cycloheximide, and preparation of mitochondria were carried out essentially as described by Claisse et al.[2] when glucose was the energy source, and as described by Claisse et al.[6] when galactose was the energy source, but in this latter case cells were allowed to incubate for 40 min in 1% galactose-low sulphate medium before the addition of cycloheximide (0.6 mg/ml); for either sugar $^{35}SO_4^{--}$ was added 5 min after the addition of cycloheximide. Mitochondrial proteins were analyzed by electrophoresis in the presence of sodium lauryl sulphate on gel slabs containing an exponential gradient (10-15 %) of polyacrylamide, which were fixed, dried, and autoradiographed[2].

DNA analysis

The techniques used were essentially as described by Maniatis et al.[14].

High stringent conditions for hybridizations were 8 x SSC (1 x SSC = 0.15 M NaCl, 0.015 M trisodium citrat) at 65°C ; low stringent conditions were 8 x SSC at 50°C. mtDNA was separated in cesium chloride gradient after extraction of total DNA from yeast protoplasts. Published cob-box[15] and oxi3[12] sequences from S. cerevisiae D273-10B were used to construct mtDNA probes and for nucleotide numbering of restriction sites.

RESULTS AND DISCUSSION

Cytoductants with an S. douglasii nucleus and S. cerevisiae [rho+] mitochondria are respiratory sufficient; while cytoductants with an S. cerevisiae nucleus and S. douglasii [rho+] mitochondria are respiratory deficient[9]. These latter cytoductants have a cytochrome oxidase deficiency (Fig. 1) and more precisely a deficiency in cytochrome oxidase subunit I (Fig. 2). An S. douglasii [box7-V328] cytoductant was derived from a donor S. cerevisiae (V328) bearing a nonreverting box7 defect but it was nonetheless respiratory sufficient (Table 1). Like the NAM2 suppressor[8] isolated in S. cerevisiae, a naturally occuring equivalent of NAM2 in the S. douglasii nucleus is proposed such that there is processing of both the defective box7 bI4 intron and the aI4 intron.

The reciprocal cytoductants gave some indication of the divergence of mitochondrial DNA's, particularly the mosaic genes and the intron open reading frames implicated in the mRNA processing. The next step was to pursue the molecular analysis of the mitochondrial DNA. The S. douglasii mtDNA was digested by various restriction endonucleases (Fig. 3 for example). The digestion patterns show many differences compared with S. cerevisiae. The length of the S. douglasii mtDNA is about 65 Kb whereas in S. cerevisiae it is 70 Kb in the short strain (D273-10B) and 75,8 Kb in the long strain (KL14-4A). These differences between the sizes appear to reflect fewer introns in the S. douglasii genome.

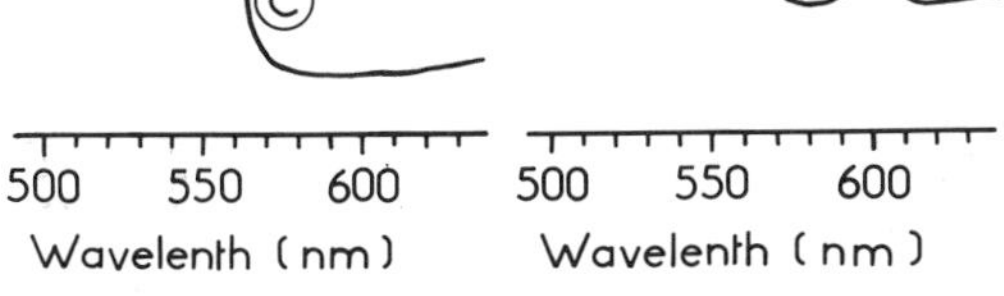

Fig. 1. Low temperature absorption spectra of reduced cytochromes.
Cells were grown at 28°C for 3 days on plates of YPGalA medium (1% yeast extract, 1% bacto-peptone, 2% galactose, 0.002% adenine) and solidified by 2.3% agar.
B: CH1, S. cerevisiae [rho^{c+}]; C: CH11, S. cerevisiae [rho^{d+}]; D: DH1, S. douglasii [rho^{d+}]; E: DH11, S. douglasii [rho^{c+}]

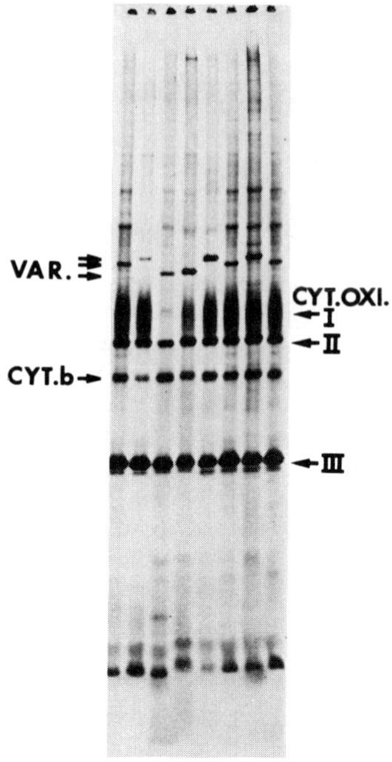

Fig. 2. Mitochondrial translation products. Glucose was the energy source. The roman numerals identify cytochrome oxidase subunits I, II, and III. VAR shows the migration of the variant I polypeptide. CYT.b indicates the cytochrome b polypeptide. A: *S. cerevisiae* KM91 [rho+] ; B, C, D, E : see Fig. 1 ; e: 4301-3A"-1-16 [rhoc+].

Table 1. Phenotypes of cytoductants compared to *S. douglasii* and *S. cerevisiae* wild types and mutant box7-V328

Strains	Growth on glycerol	Cyt. b	Cox I
S. cerevisiae [rho^{c+}]	+	+	+
S. douglasii [rho^{d+}]	+	+	+
S. douglasii [rho^{c+}]	+	+	+
S. cerevisiae [rho^{d+}]	−	+	−
S. cerevisiae [box7-V328]	−	−	−
S. douglasii [box7-V328]	+	+	+

Cox I and Cyt. b identify respectively cytochrome oxidase subunit I and cytochrome b.

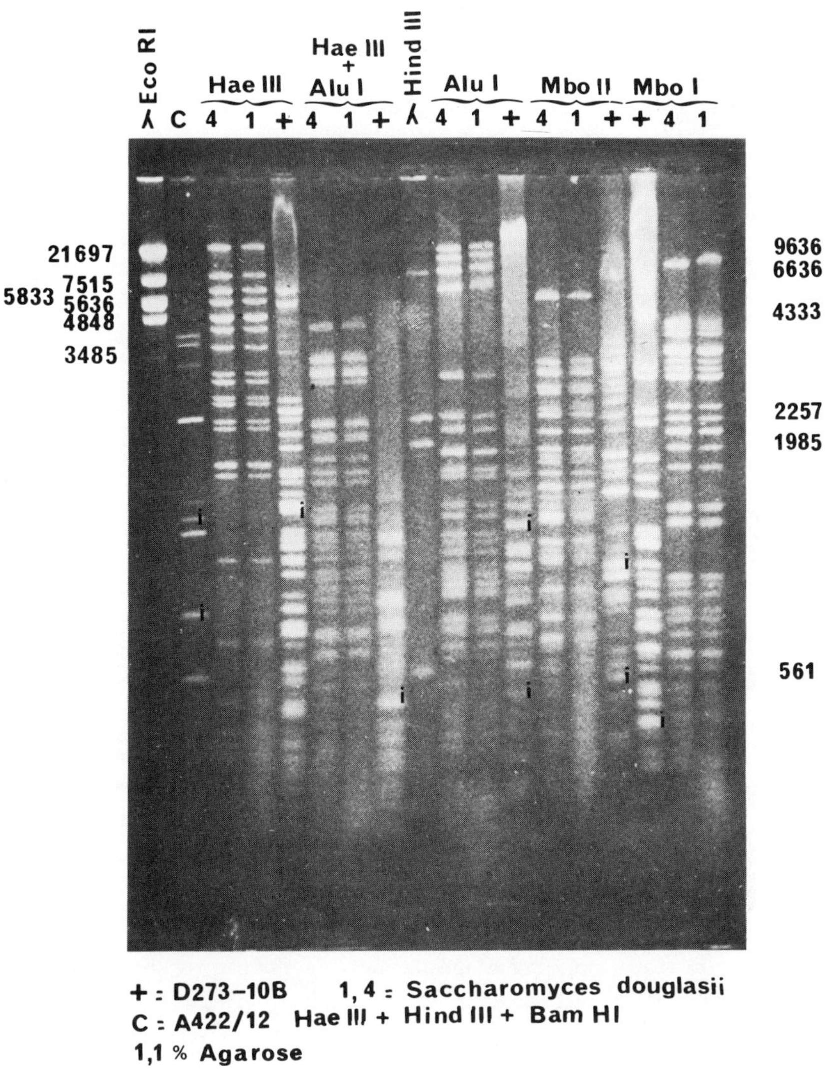

Fig. 3. Restriction endonuclease patterns of mtDNA from <u>S. cerevisiae</u> and <u>S. douglasii</u>.
Molecular weight markers: fragment mixtures from cleavage of phage λ DNA with restriction endonucleases Eco RI or Hind III; indications in base pairs.
Letters i designate fragments in mtDNA from <u>S. cerevisiae</u> hybridizing with the intron bI4 probe (Mbo I 1361 - Mbo I 1718). Agarose concentration of the gel, strains, and restriction endonucleases are as indicated in the figure.

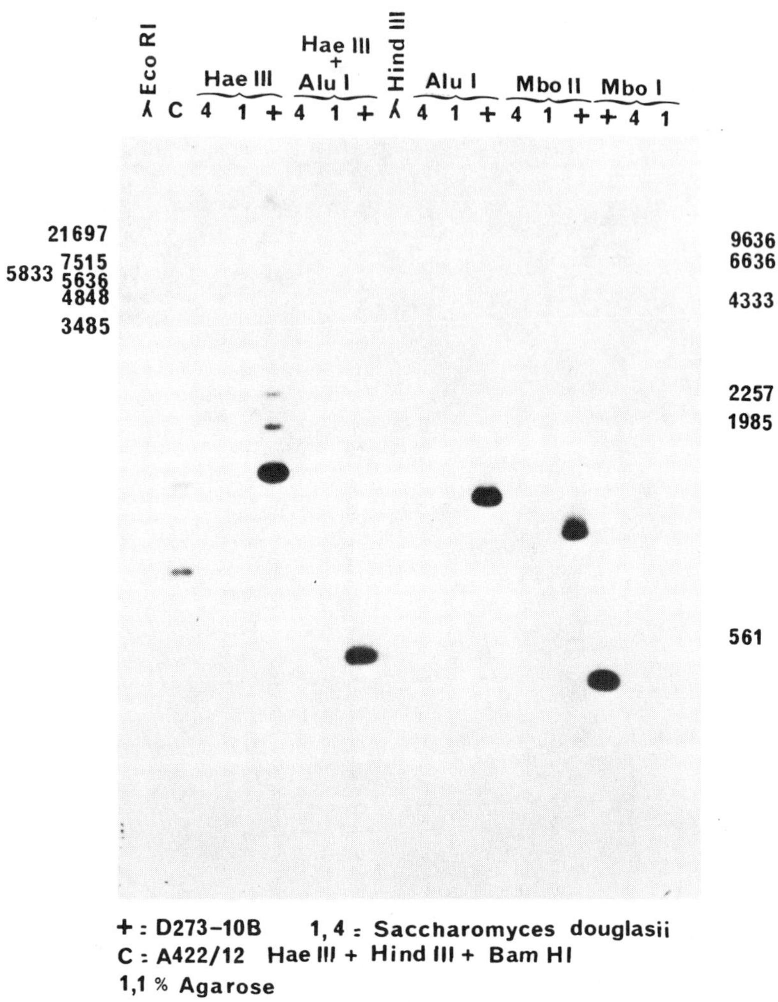

Fig. 4. Hybridization of endonuclease restriction fragments of mtDNA from S. cerevisiae and S. douglasii with the intron bI4 probe Mbo I 1361 - Mbo 1 1718.
Hybridization conditions: 8 x SSC and 50°C.
Gel, strains, markers, and restriction endonucleases are as in Fig. 3.

To detect some pertinent homologies between the mitochondrial DNA's, probes were cloned from S. cerevisiae mtDNA: on the one hand purely intron bI4 probes (Mbo I 1361 - Mbo I 1718, and Mbo I 1718 - Mbo I 2072); on the other hand a combined exon B4 - intron bI4 probe (Eco RI 654 - Mbo I 1361) and a purely exon B6 probe (Rsa I 2960 - Rsa I 3188). The exon probes for B4 and B6 hybridized to the S. douglasii mtDNA. However, the intron bI4 probes failed to hybridize to the S. douglasii restriction fragments, even under less stringent conditions that allowed cross- hybridization to aI4 fragments of S. cerevisiae (Fig. 4). An intron bI3 probe (Bam HI - Hae III; gift from A. Gargouri) did detect a homologous bI3 sequence in S. douglasii (Fig. 5). Thus the cytochrome b structural gene is also a mosaic gene in S. douglasii and we showed that bI4 is either absent or strongly modified.

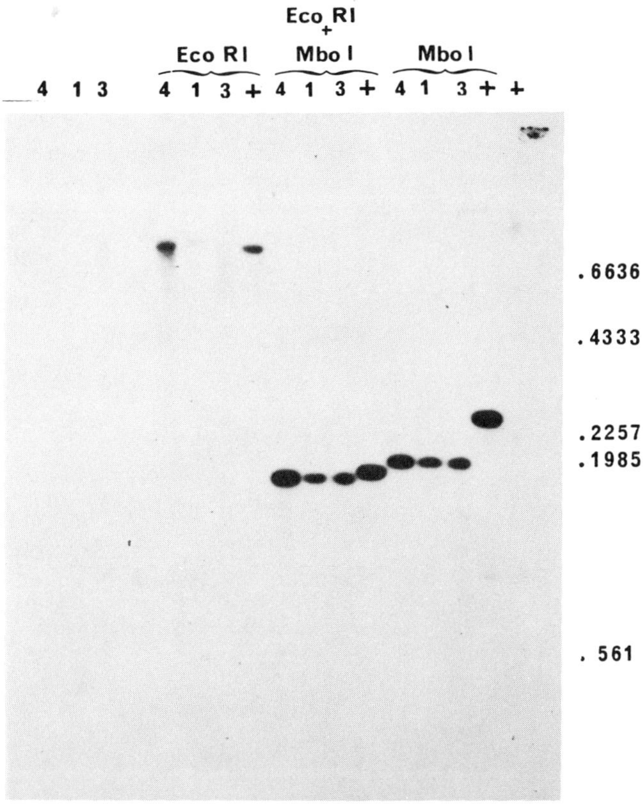

+: KL14-4A/121
1,3,4 : Saccharomyces douglasii
0,9 % Agarose

Fig. 5. Hybridization of endonuclease restriction fragments of mtDNA from S. cerevisiae and S. douglasii with the intron bI3 probe (Bam HI - Hae III; gift from A. Gargouri).
Hybridization conditions: 8 x SSC and 65°C
Restriction endonucleases, markers, agarose concentration of the transferred gel, and strains are as indicated in the figure.

We proposed by hybridization data that aI4 was also either absent or strongly modified in S. douglasii[10]. Indeed this was fully confirmed by us and others[16]. An intron aI4 probe (Hae III 7208 - Hind III 8073) that we cloned failed to detect any homologous sequence in the oxi3 gene of S. douglasii even in the low stringent condition (Fig. 6); an oxi3 exon probe (Bam HI 6625 - Hae III 7208) hybridized quite well to the S. douglasii restriction fragments (Fig. 7).

Although the S. douglasii mitochondrial genome lacks DNA sequences having homology to the oxi3 intron aI4 of S. cerevisiae we have still to postulate some nuclear mediated processing of the oxi3 messenger to account for the absence of cytochrome oxidase subunit I in the cytoductant S. cerevisiae [rho^{d+}] . In fact, a novel intron aI4-alpha, splits the DNA sequence in S. douglasii which corresponds to the oxi3 exon A4 of S. cerevisiae; and in place of the intron aI4 of S. cerevisiae there is another intron aI4-beta

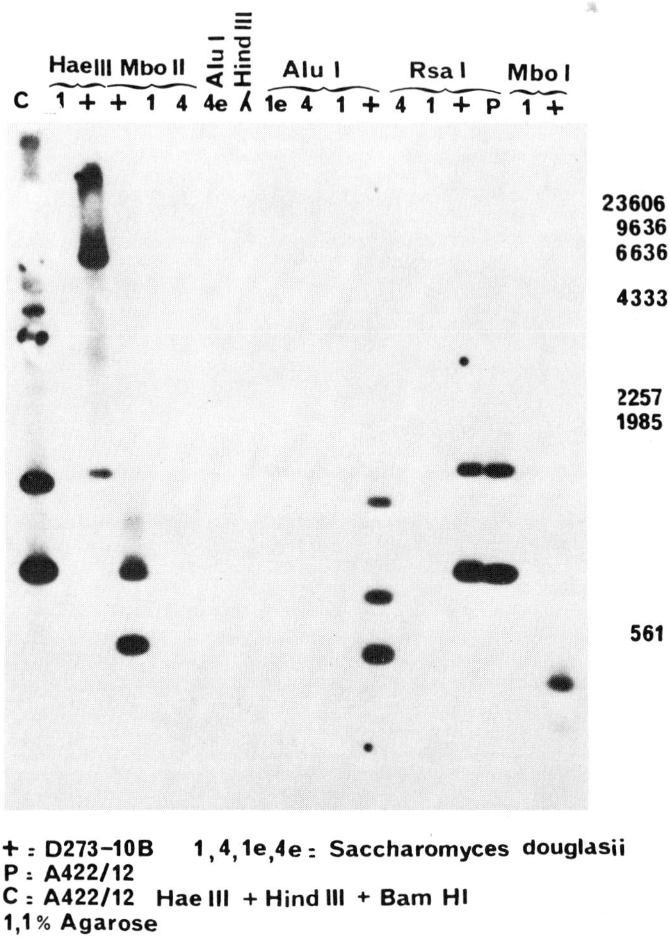

Fig. 6. Hybridization of endonuclease restriction fragments of mtDNA from S. cerevisiae and S. douglasii with the intron aI4 probe (Hae III 7208 - Hind III 8073).
Hybridization conditions: 8 x SSC and 50°C.
Restriction endonucleases, markers, agarose concentration of the transferred gel, and strains are as indicated in the figure.

which shows no homology to aI4 [16]. In the S. cerevisiae [rho^{d+}] cytoductant the oxi3 aI4-alpha intron fails to be excised. Furthermore, from genetic analysis we know that a single gene, designated NAM11, and mapped on chromosome IX of S. douglasii is needed for processing the S. douglasii mitochondrial mRNA.

Respiratory deficiency resulting from incompatibilities of interspecific combinations of wild type mitochondria and nuclei are not limited to the strains of S. cerevisiae and S. douglasii in the combination realized in the cytoductants S. cerevisiae [rho^{d+}]. Indeed, other sexually compatible Saccharomyces that we studied also yielded cytoductants deficient in mitochondrial functions. This phenomenon might have a more general significance and is consistent with criteria for speciation exemplified by the very low spore viabilities seen seen with interspecific hybrids.

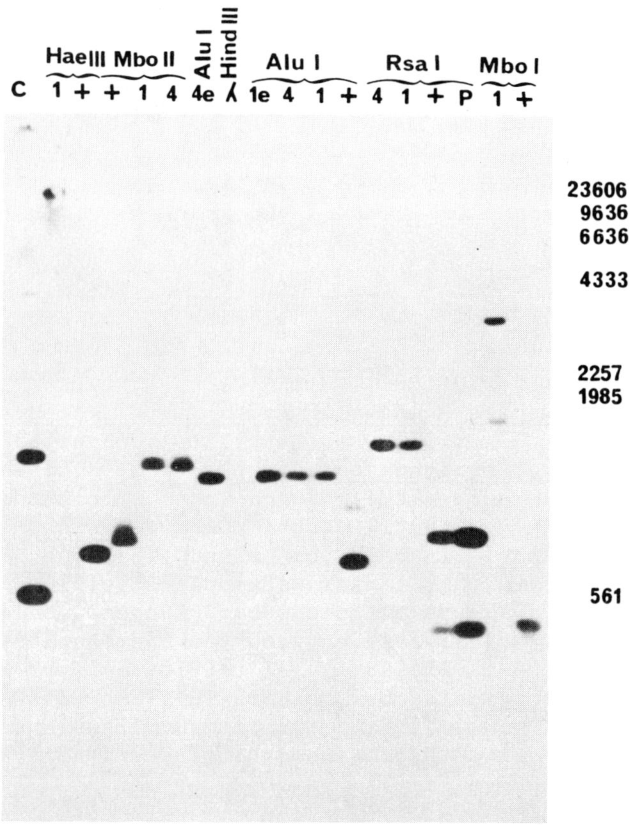

+ : D273-10B   1,4,1e,4e : Saccharomyces douglasii
P : A422/12
C : A422/12 Hae III + Hind III + Bam HI
1,1% Agarose

Fig. 7. Hybridation of endonuclease restriction fragments of mtDNA from _S. cerevisiae_ and _S. douglasii_ with an exon _oxi3_ probe (Bam HI 6625 – Hae III 7208). Hybridization conditions: 8 x SSC and 65°C. Restriction endonucleases, markers, agarose concentration of the transferred gel, and strains are as indicated on the figure.

REFERENCES

1. A. Spyridakis and M. L. Claisse, Yeast cytochrome b: structural modifications in yeast mitochondrial mutants, p. 11 _in_: "Plant Mitochondria", G. Ducet and C. Lance, eds., Elsevier/North-Holland Biomedical Press, Amsterdam (1978).
2. M. L. Claisse, A. Spyridakis, M. L. Wambier-Kluppel, P. Pajot, and P. P. Slonimski, Mosaic organisation and expression of the mitochondrial DNA region controlling cytochrome c reductase and oxidase, II, Analysis of protein translated from the _box_ region, p. 369 _in_: "Biochemistry and Genetics of Yeasts", M. Bacila, B. L. Horecker, and A. O. M. Stoppani, eds., Academic Press, New-York (1978).

3. P. P. Slonimski, M. L. Claisse, M. Foucher, C. Jacq, A. Kochko, A. Lamouroux, P. Pajot, G. Perrodin, A. Spyridakis, and M. L. Wambier-Kluppel, Mosaic organization and expression of the mitochondrial DNA region controlling cytochrome c reductase and oxidase, III, A model of structure and function, p. 391 in: "Biochemistry and Genetics of Yeasts", M. Bacila, B. L. Horecker, and A. O. M. Stoppani, eds., Academic Press, New-York (1978).
4. C. Jacq, P. Pajot, J. Lazowska, G. Dujardin, M. Claisse, O. Groudinsky, H. de la Salle, C. Grandchamp, M. Labouesse, A. Gargouri, B. Guiard, A. Spyridakis, M. Dreyfus, and P. P. Slonimski, Role of introns in the yeast cytochrome b gene: cis- and trans-acting signals, intron manipulation, expression, and intergenic communications, p. 155 in: "Mitochondrial Genes", Monograph 12, P. P. Slonimski, P. Borst, and G. Attardi, eds., Cold Spring Harbor Laboratory Press, New-York (1982).
5. T. D. Fox, Five TGA "stop" codons occur within the translated sequence of the yeast mitochondrial gene for cytochrome c oxidase subunit II, Proc. Natl. Acad. Sci. USA, 76:6534 (1979).
6. M. L. Claisse, P. P. Slonimski, J. Johnson, and H. R. Mahler, Mutations within an intron and its flanking sites: patterns of novel polypeptides generated by mutants in one segment of the cob-box region of yeast mitochondrial DNA, Molec. Gen. Genet., 177:375 (1980).
7. H. De La Salle, C. Jacq, and P. P. Slonimski, Critical sequences within mitochondrial introns: pleiotropic mRNA maturase and cis-dominant signals of the box intron controlling reductase and oxidase, Cell, 28:721 (1982).
8. O. Groudinsky, G. Dujardin, and P. P. Slonimski, Long range control circuits within mitochondria and between nucleus and mitochondria, II, Genetic and biochemical analysis of suppressors which selectively alleviate the mitochondrial intron mutations, Mol. Gen. Genet., 184:493 (1981).
9. D. Hawthorne and M. L. Claisse, Compatibility of interspecific combinations of nuclear and mitochondrial genomes in Saccharomyces, in: "XIth International Conference on Yeast Genetics and Molecular Biology", Montpellier, Abstracts, p. 221 n° M 33 (1982).
10. M. L. Claisse, F. Michel, and D. Hawthorne, Divergence of mitochondrial DNA in Saccharomyces cerevisiae and Saccharomyces douglasii and effects in compatibility of nuclear and mitochondrial genome combinations, in: XIIth International Conference on Yeast Genetics and Molecular Biology", Edinburgh, Abstracts, p. 112 n° E 54 (1984).
11. J. Conde and G. R. Fink, A mutant of Saccharomyces cerevisiae defective for nuclear fusion, Proc. Natl. Acad. Sci. USA, 73:3651 (1976).
12. S. G. Bonitz, G. Coruzzi, B. E. Thalenfeld, A. Tzagoloff, and G. Macino, Assembly of the mitochondrial membrane system, Structure and nucleotide sequence of the gene coding for subunit I of yeast cytochrome oxidase, J. Biol. Chem., 255:11927 (1980).
13. M. L. Claisse, G. A. Péré-Aubert, L. P. Clavilier , et P. P. Slonimski, Méthode d'estimation de la concentration des cytochromes dans les cellules entières de levure, Eur. J. Biochem., 16:430 (1970).
14. T. Maniatis, E. F. Fritsch, J. Sambrook, Molecular cloning, a laboratory manual, Cold Spring Harbor Laboratory (1982).
15. F. G. Nobrega and A. Tzagoloff, Assembly of the mitochondrial membrane system, DNA sequence and organization of the cytochrome b gene in Saccharomyces cerevisiae D273-10B, J. Biol. Chem., 255:9828 (1980).
16. Z. Kotylak, J. Lazowska, D. C. Hawthorne, and P. P. Slonimski, Intron encoded proteins of mitochondria: key elements of gene expression and genomic evolution, p. 1 in: "Achievements and Perpectives of Mitochondrial Research", Volume II: Biogenesis, E. Quagliariello, E. C. Slater, F. Palmieri, C. Saccone, A. M. Kroon, eds., Elsevier Science Publishers, Amsterdam (1985).

A COMPARATIVE TRANSCRIPTIONAL ANALYSIS OF THE CODING REGIONS IN THE CMS-T UNIQUE READING FRAME BETWEEN FERTILE AND T MALE STERILE CYTOPLASM

J.I. Qin, N. Walker, C. Fauron, D Gupta and A.G.Abbott

Department of Biological Sciences
Clemson University
Clemson, SC

Cytoplasmic male sterility is a phenomenon where plants fail to shed viable pollen. There is considerable evidence that altered mitochondrial gene expression is responsible for cytoplasmic male sterility (cms). In Zea mays, there are three major types of male sterile cytoplasms: cms-T, cms-C and cms-S. These three types of cytoplasms are distinguished from each other on the basis of the specific nuclear genes that restore fertility. Pollen fertility can also be reestablished by cytoplasmic reversion, which is a cytoplasmically inherited change. Cytoplasmic reversion is naturally occurring in cms-S cytoplasm and can be selected in cms-T by tissue culture toxin selection schemes.

In order to identify genetic changes that give rise to male sterility, restriction enzyme analyses of mitochondrial DNA from normal (N) and male sterile cytoplasms have been done (Pring et al., 1977; Pring and Levings, 1978). These experiments revealed considerable structural heterogenity between the two mitochondrial genomes (Figure 1). These numerous differences have made it difficult to design experiments to reveal molecular changes that give rise to cytoplasmic male sterility. In contrast comparative restriction fragment analysis of mitochondrial DNA from cms-T and fertile cytoplasmic T revertants show very few differences. In figure 1, Xho I restriction patterns of mitochondrial DNA from four different T revertant lines is compared with the restriction pattern of cms-T mitochondrial DNA. These patterns appear very similar for most of the restriction fragments, however there is one prominent difference between these restriction profiles. In the cms-T cytoplasm there is 6.6kb Xho I fragment which is absent in the four T revertant lines studied. This is true for all the T revertant lines that have been isolated with the exception of one (Umbeck and Gengenbach, 1983).

This alteration of a 6.6kb Xho I fragment is correlated with cytoplasmic reversion to fertility and toxin resistance (Gengenbach et al. 1981). Therefore it was important to know if there were any coding sequences on this region of the mitochondrial DNA; and if any gene coding sequences did exist, were there any transcriptional differences between this region and homologous regions in N and cms-T mitochondria. To be able to characterize this region cosmid clones from a T library were probed with the 6.6kb Xho I fragment. One of the clones that hybridized to this fragment, was used for subcloning an 8.5kb Bam HI fragment into pUC8, an E. coli plasmid. This clone was designated pB8.5T, and it contains in entirety, the 6.6kb Xho I fragment. Figure 2 shows a restriction map of pB8.5T.

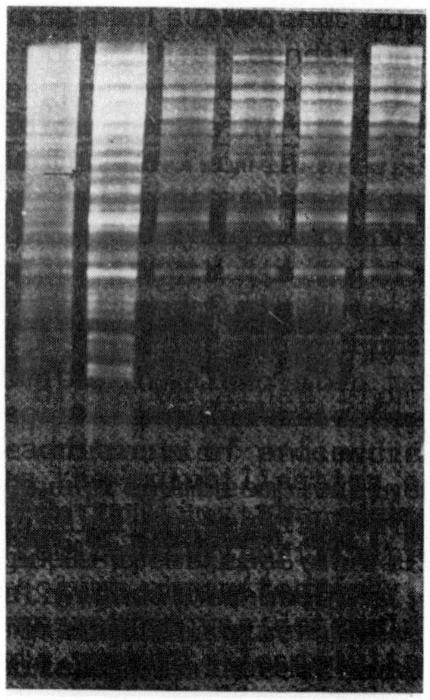

Figure 1: Xho I restriction endonuclease digestion patterns of mt. DNA from N: lane 1; cms-T: lane 2; T rev (V_3,V_6,V_{24},V_{26}): lanes 3-6; run on a 0.8% agarose gel. Arrow denotes 6.6kb XhoI I fragment.

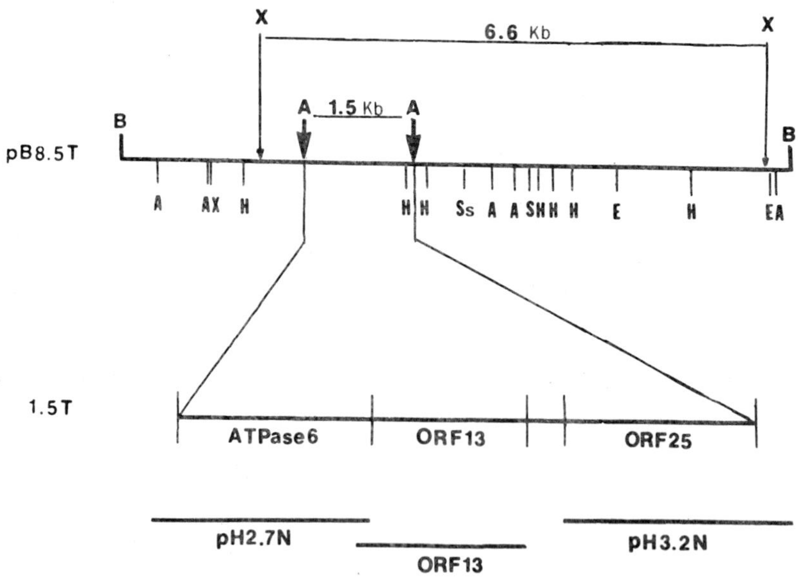

Figure 2: Restriction map of the clone pB8.5T, showing the location of the 6.6kb Xho I fragment and the 1.5kb Ava I fragment on the map. The location of the probes specific for the 5' flanking sequences of ATPase 6, ORF 13 and ORF 25 on the 1.5kb Ava I region are also shown. The restriction endonucleases used are A= Ava I, E=Eco RI, H=Hind III, S=Sal I, Ss=Sst II and X=Xho I.

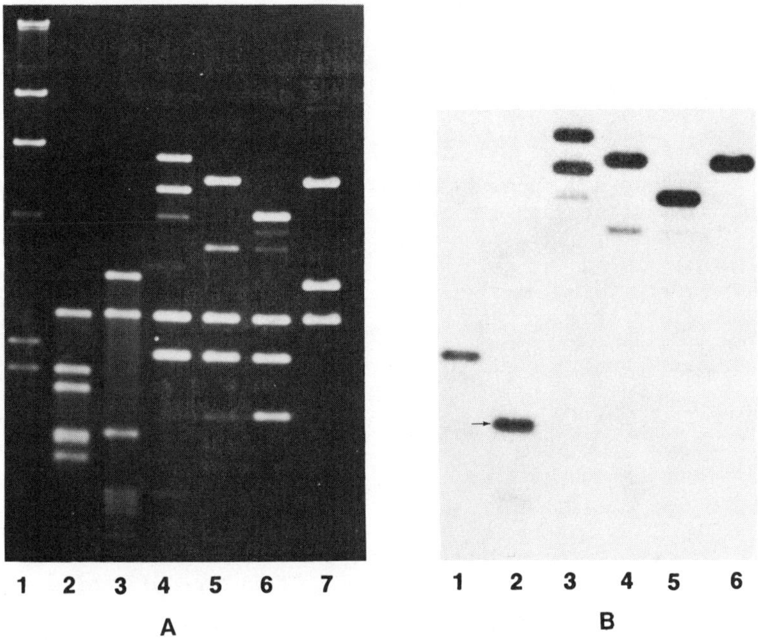

Figure 3: A) Restriction endonuclease digestions of the clone pB8.5T, run on a 0.8% agarose gel. The enzymes used were 1: Hind III; 2 : Bam HI, Hind III; 3 : Bam HI, Ava I; 4 : Bam HI, Eco RI, Xho I; 5: Bam HI, Eco RI, Sal I; 6 : Bam HI, Sst II, Eco RI; and 7 : Bam HI, Sal I.
B) Autoradiogram of a Southern blot of the gel described in Figure 3A, hybridized with ^{32}P end labelled N coleoptile mt RNA. Arrow indicates a 1.5kb Ava I fragment which has sequences homologous to the RNA.

To detect the presence of coding sequences on this altered restriction fragment, the clone pB8.5T was cut with several different restriction enzymes and run on an agarose gel (Figure 3A). A Southern blot of this gel was hybridized with radioactively labelled mitochondrial RNA from N cytoplasm (Figure 3B). Regions of the pB8.5T were found to be homologous to the mitochondrial (mt) RNA, and these coding regions could be delimited to a 1.5kb Ava I fragment (indicated by arrow in Figure 3B). On the restriction map of pB8.5T, the 1.5 kb Ava I fragment was positioned within the 6.6kb Xho I fragment (Figure 2).

The 6.6kb Xho I fragment in cms-T is altered in the N and the T revertant cytoplasms. Sequences homologous to this region are located on a 6.0kb Xho I fragment in N and T revertant genomes. Experiments were then designed to determine whether these rearrangements were positioned within the coding sequences. Therefore, the 1.5kb Ava I fragment was hybridized to a Southern blot of normal, cms-T and T revertant (V_3) mitochondrial DNAs cut with the restriction endonuclease Ava I. As seen in Figure 4, the 1.5kb Ava I fragment that is present in cms-T mitochondria is altered in both N and T revertant cytoplasms. In N mitochondria the region of homology lies on a 2.1kb Ava I fragment, while in T revertant (V_3), the homologous sequences are on a fragment of an approximate size of 2.2kb (indicated by arrows on Figure 4).

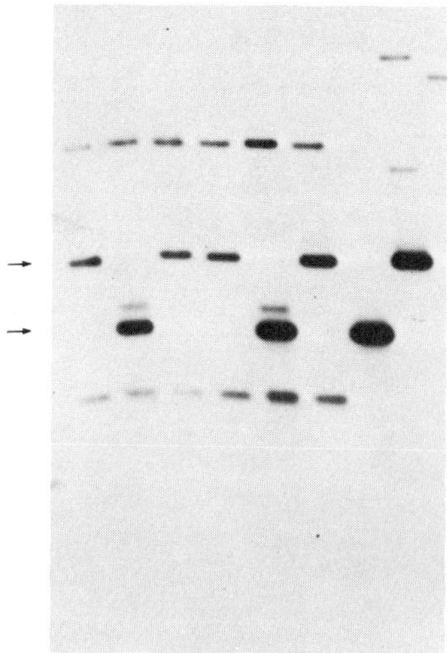

Figure 4: Autoradiogram of a Southern blot with mt DNA from N: lane 1, cms-T: lane 2 and T revertant (V_3): lane 3. Hybridized with ^{32}P labelled 1.5kb Ava I fragment. Arrow indicates the altered restriction fragment in N and T rev mitochondria.

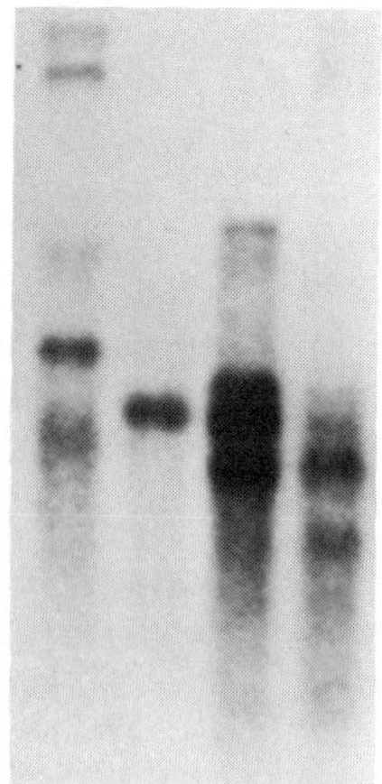

Figure 5: Autoradiogram of a Northern blot of a formaldehyde agarose gel, with mt RNA from N: lane 1; cms-T: lane 2; T revertant (V_{18}): lane 3 and T (Rf1, Rf2): lane 4. The blot was hybridized with the 1.5kb Ava I fragment of pB8.5T. Arrows indicate the major transcripts homologous to this fragment.

Thus, rearrangements within the 1.5 kb coding region gives rise to structural differences between the N, cms-T and T revertant cytoplasms. It was important to know if these structural differences result in transcriptional differences between the three cytoplasms. To address this question, the 1.5kb Ava I fragment was hybridized to a Northern blot of mitochondrial RNA from N, cms-T and the T revertant (V_{18}) (Figure 5). This unique restriction fragment of cms-T has major homology to a 2200 nucleotide long transcript in the male fertile cytoplasm. In contrast a complex transcriptional pattern is obtained in cms-T mitochondria. There is a large transcript which is 4000 bases long and several others in the range of 1900 to 1600 nucleotides. In the V_{18} mitochondria, the 1.5 kb Ava I fragment is homologous to a unique fragment, which is 1860 bases in size.

Thus, structural rearrangements in the 1.5kb Ava I fragment, which are correlated with cytoplasmic reversion to fertility and toxin resistance. result-in transcriptional differences between N, cms-T and T revertant cytoplasms.

Dewey et al. 1986, have recently characterized a unique chimeric coding frame in cms-T mitochondria, designated as TURF 2H3. Comparsion between their data and ours, positions the 1.5kb Ava I fragment within the

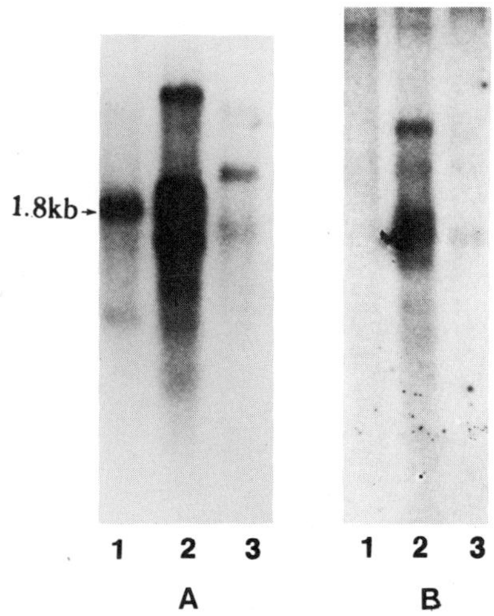

Figure 6: Autoradiograph of similar Northern blots of formaldehyde agarose gels (1.2%), with mt RNA from N: lane 1, T revertant (V_{18}): lane 2 and cms-T: lane 3.
A) The blot was hybridized with an ORF 25 specific probe. Arrow indicates the unique transcript in the T revertant mitochondria.
B) The blot was hybridized with an ORF 13 specific probe.

boundaries of the TURF 2H3. This region of the TURF 2H3, has sequences homologous to the 5' flanking sequences of the ATPase subunit 6. It carries an open reading frame designated ORF 13 and most of another open reading frame, ORF 25. The ORF 13 and ORF 25 can code for polypeptides of an approximate size of 13000 Mr and 25000 Mr respectively. In cms-T mitochondria, there is a unique protein which is 13 kdalton in size. This mitochondrial protein is not seen in N or in any of the T revertants, that have been isolated by tissue culture toxin selection (Forde et al., 1980; Leaver and Gray, 1982).

We then conducted several experiments to further characterize the specific region(s) of the 1.5kb Ava I fragment that is involved in the reestablishment of fertility and toxin resistance. Probes specific for ORF 13 and ORF 25 were hybridized to Northern blots of N, cms-T and T revertant mitochondrial RNAs. The location of these specific probes on the 1.5kb Ava I fragment, is shown on the restriction map of pB8.5T (Figure 2). As seen in Figure 5 the 1.5kb Ava I fragment hybridizes to a unique transcript in T revertant mitochondria. This novel transcript which is 1860 bases long, is homologous to the ORF 25 region of the 1.5kb Ava I fragment (Figure 6A). In contrast, when ORF 13 was hybridized to a similar Northern blot, no transcripts homologous to this region were detected in N and T revertant mitochondria (Figure 6B). Data from restriction mapping and sequencing analysis of the region homologous to the 1.5kb Ava I fragment in T revertants, suggest that recombinations in this region physically separate the ORF 13 and ORF 25 coding sequences (Fauron et al., 1986).

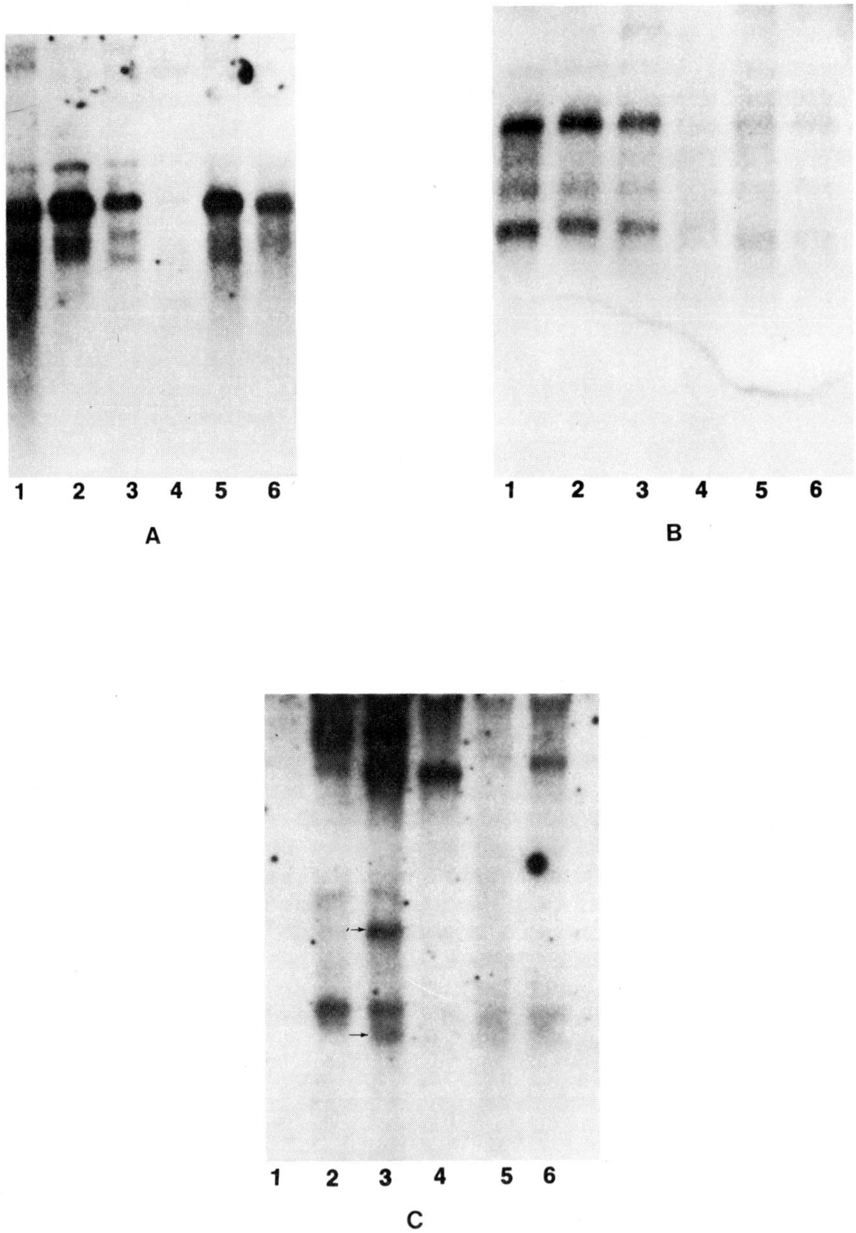

Figure 7: Autoradiograph of similar Northern blots of formaldehyde agarose gels (1.2%), with mitochondrial RNA from six different normal male fertile lines. Lane 1: B73N, lane 2: 3377 (hybrid); lane 3: A632 (carries restorer alleles for T); lane 4: OH51A (carries restorer alleles for S); lane 5: F5DD1 (carries restorers for C and T); lane 6: CE1 (inbred).

A) Hybridized with a probe specific for ORF 25

B) Hybridized with a probe specific for the 5' flanking sequences of ATPase 6.

C) Hybridized with a probe specific for ORF 13.

In the presence of restorer alleles, Rf1 and Rf2, mitochondrial transcripts homologous to the 1.5kb Ava I fragment differ from the nonrestored cms-T (Figure 5). This was originally described by Dewey et al., 1986. The exact function of these restorer alleles is not completely understood, however the present data suggests, that restorers may be processing enzymes. We decided to study in general, the effect of nuclear background on mitochondrial transcriptional patterns. For this study, six different normal male fertile lines, with varying nuclear backgrounds were used. Among these six different lines, some were inbreds, some hybrids and others had S, C or T restorer alleles. For this study we chose to look at transcripts homologous to the chimeric construct in cms-T, which we already knew to be altered in the T cytoplasm, in the presence of T restorers.

Probes specific for ATPase 6, ORF 25 and ORF 13 were hybridized to similar Northern blots of mitochondrial RNA from these six different normal male fertile lines. Figure 7A shows one such autoradiogram, where the Northern blot is hybridized to an ATPase 6 5' flanking sequences specific probe. Transcripts homologous to this region of the 1.5 kb Ava I fragment remain invariant between the different N lines. Similarly the transcriptional pattern obtained with an ORF 25 probe, shows no changes for the major homologous transcripts, between these normal male fertile lines (Figure 7B). Some of the minor transcripts homologous to ORF 25 do show variation in size. The nature of these transcripts is being studied. In contrast, an ORF 13 probe, hybridizes uniquely to one of the normal male fertile lines (A632) Figure 7C, indicated by arrow). This is the only example we have of ORF 13 homologous transcripts, in the normal male fertile cytoplasm. The other transcripts observed on the autoradiogram are due to a contamination of coding regions homologous to the 5' flanking sequence of ATpase 6.

In conclusion, recombinations in a region between ORF 13 and ORF 25, separate the two open reading frames. These recombinations create a novel ORF 25 reading frame in the T revertant mitochondria, and ORF 13 is not transcribed in that cytoplasm. Using probes specific for these regions we were also able to show that, nuclear background does not significantly influence mitochondrial transcriptional patterns. This suggests that, T restorers may act specifically on the chimeric TURF 2H3 coding unit.

REFERENCES

(1) Dewey, R.E.; Levings, C.S. III. and Timothy, D.H., Cell, 44: 439-449, Feb. 14, 1986.

(2) Fauron, C. M.-R.; Abbott, A.G.; Bretell, R.I.S., and Gesteland, R.F., Current Genetics (in press).

(3) Forde, B.G.; Oliver, R.J.C.; Leaver, C.J.; Gunn, R.E. and Kemble, R.J., 1980. Genetics, 95: 443-450.

(4) Genenbach, B.G.; Miller, R.J.; Koeppe, D.E. and Arntzen, C.J., 1973. Can. J. Bot. 51: 2119-2125.

(5) Leaver, C.J. and Gray, M.W., 1982. Ann. Rev. Plant Physiol. 33: 373-402.

(6) Pring, D.R.; Levings, C.S., III.; Hu, W.W.L. and Timothy, D.H., 1977. Proc. Natl. Acad. Sci. U.S.A. 74(7): 2904-2908.

(7) Pring, D.R. and Levings, C.S. III., 1978, Genetics, 89: 121-136.

(8) Umbeck, P.F. and Gengenbach, B.G., Crop Science, 23: 594: 588, May-June 1983.

CYTOCHROME OXIDASE FROM ETIOLATED CORN SEEDLINGS

R. Dutch, R. Ingle, J. Centeno, W. Peiffer, G. Babcock
and S. Ferguson-Miller

Michigan State University
East Lansing, Michigan 48824

INTRODUCTION

Cytochrome oxidase is a complex, multisubunit membrane protein that is the terminal member of the respiratory chain in mitochondria. The enzyme appears to act as a proton pump, but neither the mechanism of this energy conserving process nor the roles of any of the 12-13 different subunits in the mammalian oxidase are known.

Some studies have suggested the involvement of cytochrome oxidase in the mitochondrially-inherited defect that gives rise to male sterility in plants[1]. Evidence for an alteration in subunit I of cytochrome oxidase in mitochondria of male-sterile sorghum[2] is of particular interest since this largest subunit is likely to play a major role in the electron and proton transfer functions of the enzyme[3].

Studies were initiated to investigate cytochrome oxidase activity in Normal and Texas male-sterile maize mitochondria, and in the isolated form of the enzyme, to determine whether any structural or functional alterations are associated with male-sterility and to compare the characteristics of the maize oxidase with the mammalian, fungal, and bacterial forms of the enzyme.

Our results indicate no significant spectral or activity differences between the oxidases from normal and Tms maize mitochondria, but reveal that maize oxidase has unusual spectral characteristics which may help to decipher some important aspects of the mechanism of cytochrome oxidase.

RESULTS AND DISCUSSION

Respiratory parameters of normal and Texas male-sterile maize mitochondria were determined polarographically, using glutamate and malate as substrates. Average P/O ratios (normal: 2.4; Tms: 2.7) and respiratory control ratios (normal: 4.2; Tms: 4.1) are not significantly different between the two strains.

Attempts to purify maize oxidase reveal that it is highly labile, showing rapid loss of activity at room temperature especially in the presence of chromatography resins. However, a relatively pure oxidase

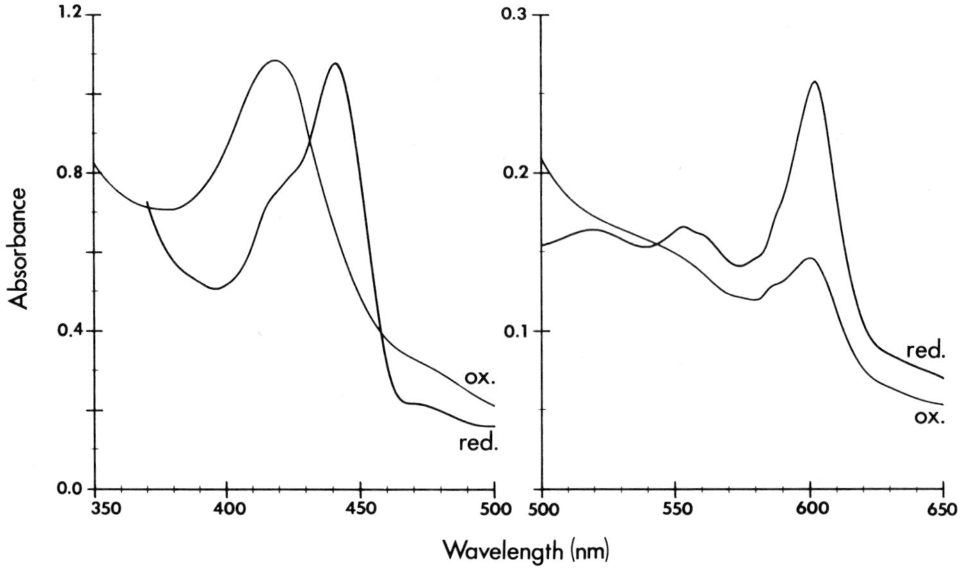

Fig. 1. VISIBLE SPECTRA OF DITHIONITE REDUCED AND AIR OXIDIZED FORMS OF PARTIALLY PURIFIED MAIZE CYTOCHROME c OXIDASE. The sample used was a concentrate of fraction I from the hydroxyapatite column shown in Figure 2. After storage at -100°C, this preparation had a turnover number of 281 sec^{-1} at 25°C in 50 mM KH_2PO_4, pH 6.5. All spectra were run in 75 mM KH_2PO_4, 45 mM Bicine, 5% sucrose, 15 mM lauryl maltoside, pH 7.9, at a final concentration of 8.3 μM $\underline{aa}_3$.

(by spectral criteria: Figure 1) that retains high activity (TN = 462 s^{-1}) can be isolated by hydroxyapatite chromatography in lauryl maltoside at 4°C (Figure 2).

Resonance Raman spectroscopy was employed to compare the purified maize cytochrome oxidase to the beef heart enzyme, and also to look for possible differences between the cytochrome oxidases in normal and male-sterile mitochondria. A significant difference between the beef and maize oxidases (both in mitochondria and in the purified state) is the marked attenuation of the 1610 cm^{-1} vibrational frequency in the maize oxidase (Figure 3). The 1610 cm^{-1} frequency is associated with the hydrogen bonded form of the formyl substituent of the heme a porphyrin ring, and its disappearance could indicate a weakening of this hydrogen bond[4]. The oxidase of the male-sterile maize mitochondria shows the same resonance Raman characteristics as the normal maize oxidase, including the attenuated 1610 cm^{-1} frequency.

Visible spectra of the purified maize oxidase (Figure 1) reveal that the maize enzyme differs from mammalian, fungal, and bacterial oxidases in the positions of the Soret and α absorption bands in the reduced forms (reduced Soret = 440 nm instead of 443 nm; reduced α band = 602 nm instead of 604 nm). These shifts can be attributed to an altered heme a environment.

CONCLUSIONS

We find no significant differences between N and Tms maize mitochondria with respect to respiratory control and ADP/O ratios. Some differences might have been expected if the proton pumping function of cytochrome oxidase in Tms maize mitochondria were altered.

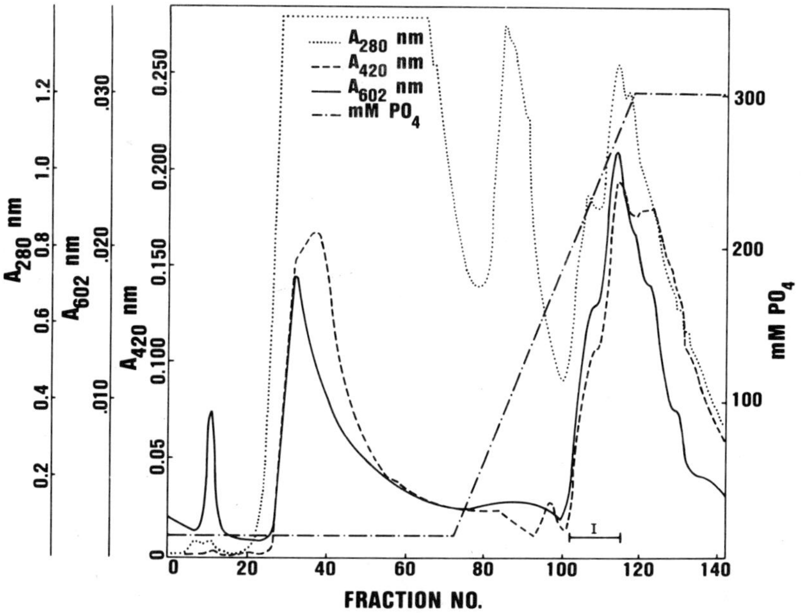

Fig. 2. ELUTION PROFILE OF HYDROXYAPATITE CHROMATOGRAPHY OF LAURYL MALTOSIDE SOLUBILIZED MAIZE MITOCHONDRIA. The sample of mitochondria (GL 466, normal) was diluted to 20 mg/ml in the column buffer (45 mM Bicine, 5% sucrose, 10 mM phosphate, 15 mM lauryl maltoside, pH 7.9) with added lauryl maltoside to 4%, 3 mM 1,2-epoxy-3-(p-nitrophenoxy) propane, 1.7 mM PMSF and 10 μM E-64 (Sigma Chem. Co.). A total of 300 mg of mitochondrial protein was loaded onto a 55 ml hydroxyapatite column, washed with 2-1/2 column volumes of buffer, and eluted with 100 ml of a 10mM-300mM phosphate gradient. Elution was completed by a 150 ml wash with 300 mM phosphate. All work was performed at 4°C.

Hydroxyapatite chromatography of maize mitochondria solubilized in lauryl maltoside results in significant purification of cytochrome oxidase with retention of activity and spectral properties. Purification of the maize seedling oxidase presents unusual difficulties because of the sensitivity of the enzyme to heat and surface denaturation.

Visible spectra of mitochondrial and purified forms of maize cytochrome oxidase reveal shifts in the Soret and α absorption bands compared to those of mammalian, fungal, and bacterial enzymes. These unique spectral properties of the plant oxidase have been noted previously[5] and appear to relate to an altered environment of heme a.

Resonance Raman analysis also indicates that the heme a environment of maize cytochrome oxidase is altered compared to the beef enzyme and suggests that this change is associated with the formyl substituent of the heme a porphyrin ring. Since this group has been proposed to play a role in the proton pumping function of cytochrome oxidase[4], further studies regarding this activity of the plant enzyme will be of particular interest.

ACKNOWLEDGEMENTS

This work was supported by NIH Grants GM26916 and BRSG (CHM) to SFM and GM25480 to G.B.

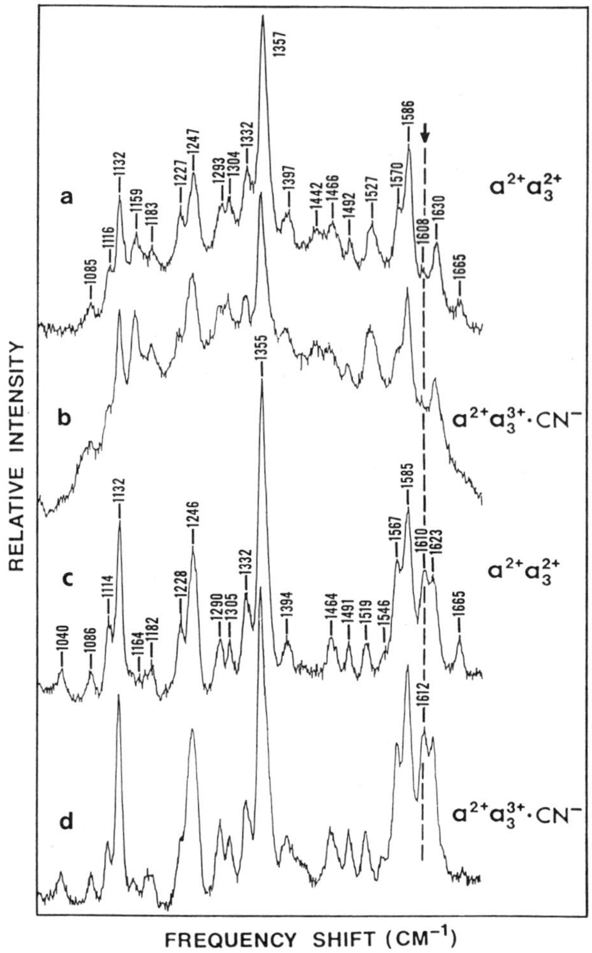

Fig. 3. RESONANCE RAMAN SPECTRA OF FULLY REDUCED AND CYANIDE BOUND CYTOCHROME OXIDASES FROM MAIZE SEEDLING AND BEEF HEART MITOCHONDRIA. The maize cytochrome oxidase was the same sample as shown in Figure 1. Beef heart cytochrome oxidase was isolated by a Hartzell-Beinert procedure. Laser excitation was at 441.6 nm (20 mW power). a. Maize cytochrome oxidase fully reduced with dithionite at 8.3 µm $\underline{aa_3}$, pH 7.9. b. Maize cytochrome oxidase in mixed valence state reduced with TMPD-ascorbate in the presence of cyanide at 8.3 µm $\underline{aa_3}$, pH 7.9. c. Beef heart cytochrome oxidase fully reduced with TMPD-ascorbate at 30 µM $\underline{aa_3}$, pH 7.4. d. Beef heart cytochrome oxidase in mixed valence state reduced with TMPD-ascorbate in the presence of cyanide at 40 µM $\underline{aa_3}$, pH 7.4.

REFERENCES

1. Watrud, L.S., Laughnan, J.R., Gabay, S.J., and Koeppe, D.E., Can.J.Biochem., 54, 2718, (1976).
2. Leaver, C.J., Hack, E., Dawson, A.J., Isaac, P.G., and Jones, V.P., Mitochondria 1983, Gruyter and Co., Berlin, p. 269, 1983.
3. Buse, G., Biewald, R., Raabe, G., and Fleischauer, J. (1984). 3rd European Bioenergetics Conf., Congress Edition, Hannover, p. 119.
4. Babcock, G.T. and Callahan, P.M., Biochemistry, 22, 2314, 1983.
5. Chance, B., Bonner, W.D., and Storey, B.T., Annual Review of Plant Physiology, 19, 295, 1968.

DICYCLOHEXYLCARBODIIMIDE PROTECTS AGAINST *HELMINTHOSPORIUM MAYDIS* RACE T TOXIN ACTION ON SUSCEPTIBLE CORN MITOCHONDRIA

Marcia J. Holden and Heven Sze

Dept. of Botany
University of Maryland
College Park, Maryland

Helminthosporium maydis is the causative agent of Southern Corn Leaf Blight. *H. maydis*, Race T, produces a pathotoxin (HmT toxin) which is responsible for the severity of the disease on cytoplasmic male-sterile T (cms T) corn. The pathotoxin has no effect on normal-fertile corn. Both cytoplasmic male sterility and sensitivity to the toxin are maternally inherited and the genes responsible for these traits are thought to be on the mitochondrial DNA[1].

A primary target for HmT toxin action is the mitochondrion. Stimulation of succinate or NADH-driven respiration (O_2 consumption) and inhibition of malate-driven respiration has been well described[2]. We have reported that HmT toxin decreased calcium uptake[3] and dissipated calcium gradients[4] in cms T, but not normal-fertile, corn mitochondria. Here we demonstrate that HmT toxin dissipated membrane potential and inhibited malate-dependent electron transport at the same concentration range effective in inhibiting calcium uptake. Dicyclohexylcarbodiimide (DCCD) at 5 - 15 μM protected against HmT toxin action on cms T mitochondria. These results suggest that all three activities may be affected through the interaction of toxin with a common "receptor" specific to cms T corn mitochondria.

Mitochondria were isolated from the roots of cms T (W64AT) and normal-fertile (W64AN) 3-4 day old corn seedlings. Calcium transport was measured by uptake of ^{45}Ca using a filtration method. Generation and dissipation of mitochondrial membrane potential was followed by monitoring absorbance changes (O.D. 511-533 nm, dual wavelength spectrophotometer) or fluorescence quenching (excitation 495 nm, emission 586 nm) of the lipophilic cation, safranine. Electron transport was monitored with dichlorophenolindophenol (DCPIP) or ferricyanide as electron acceptors[5]. Purified HmT toxin was a generous gift of Dr. J.M. Daly, University of Nebraska.

HmT toxin dissipated membrane potential in susceptible, but not resistant, corn mitochondria. The decrease in membrane potential was dependent on the toxin concentration and independent of the substrate (malate, succinate or ATP) used to generate the potential (Fig. 1). Within two minutes, 50% of the potential was dissipated by 25 ng/ml (33 nM) of HmT toxin and 50 ng/ml HmT dissipated the potential at a rate comparable to that of the protonophore, CCCP (5 μM). When mitochondria were pretreated (2 min) with increasing concentrations of dicyclohexylcarbodiimide (DCCD),

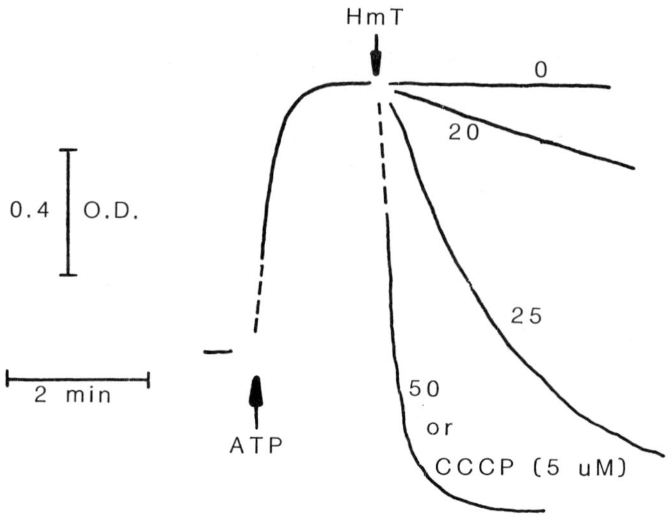

Fig. 1. HmT toxin collapsed the membrane potential generated by ATP in cms T (susceptible corn) mitochondria. Reaction mixture contained 450 μg mitochondrial protein, 0.25 M sorbitol, 2.5 mM Hepes-BTP (pH 7.4), 20 μM safranine, 1 mM $MgSO_4$, 1mM KCN, and 2 mM K_2HPO_4. The reaction was started with 1 mM ATP-BTP and absorbance changes were monitored. Numbers indicate concentration of HmT toxin in ng/ml.

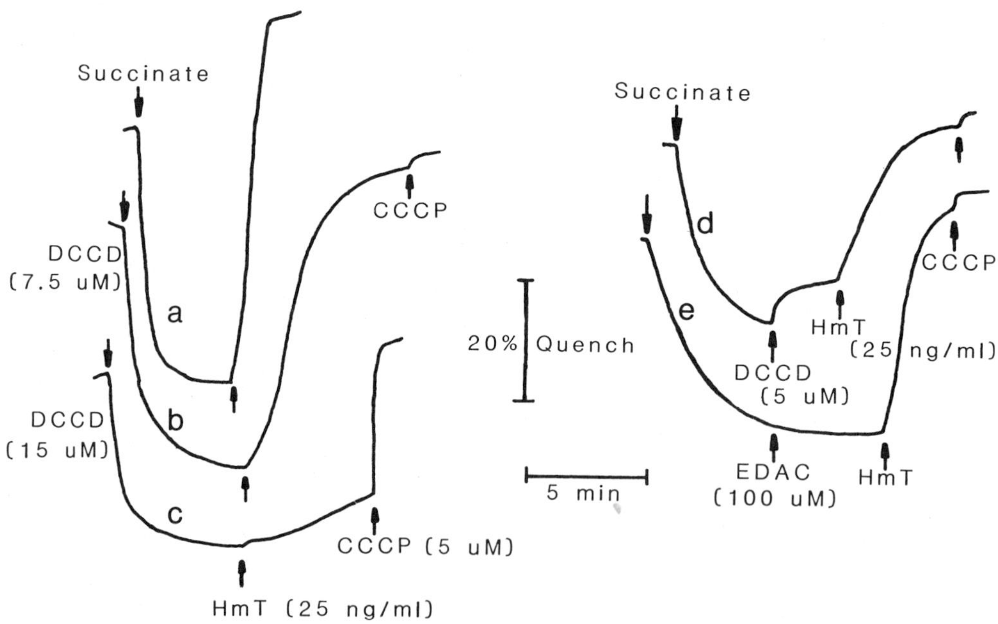

Fig. 2. DCCD prevented HmT toxin-induced dissipation of the mitochondrial membrane potential. Mitochondrial protein (250-300 μg) was incubated in a reaction mixture of 0.25 M sorbitol, 2.5 mM Hepes-BTP, 1 mM K_2HPO_4, 3-4 μM safranine and 12.5-17 μM rotenone. DCCD was included in b and c. The reaction was started with 2 mM succinate and fluorescence quenching was monitored.

HmT toxin-induced dissipation of the potential (Fig. 2) was prevented. DCCD alone (15 μM), decreased the substrate-generated membrane potential by <50%, however at this concentration, DCCD completely eliminated the collapse of the potential by HmT toxin (Fig. 2c). The water soluble carbodiimide, 1-ethyl-3-(3-dimethylaminopropyl)-carbodiimide (EDAC, 100 μM), was ineffective in preventing toxin action and has no effect on generation of membrane potential (Fig. 2e) suggesting that DCCD modified a component located in a hydrophobic environment.

Established calcium gradients in cms T mitochondria can also be dissipated by HmT toxin (13 nM) at a rate comparable to the calcium ionophore, A23187[4]. HmT toxin is likely increasing membrane permeability to calcium in addition to dissipating the membrane potential required to establish and maintain calcium gradients. Pretreatment (7 min) of mitochondria with 4 μM DCCD also prevented HmT toxin-induced loss of accumulated calcium (Fig. 3).

Malate-dependent electron transport in cms T mitochondria, measured by reduction of DCPIP or ferricyanide, was inhibited immediately by HmT toxin (33 nM), while succinate-dependent electron transport was stimulated[5]. Consistent with other activities, DCCD (8 μM) prevented the inhibition of malate-dependent electron transport by the toxin (Fig. 4). It has been postulated that inhibition of malate-dependent electron transport is a secondary effect, the result of NAD^+ loss through the NAD^+ porter[6]. Addition of NAD^+ restored malate-dependent electron transport consistent with this idea. However, NAD^+-enhanced activity was less sensitive to rotenone and may indicate activity of the rotenone-insensitive bypass. Our experiments suggest that NAD^+ loss accounts for some, but perhaps not all of the toxin inhibition of malate-dependent electron transport[5].

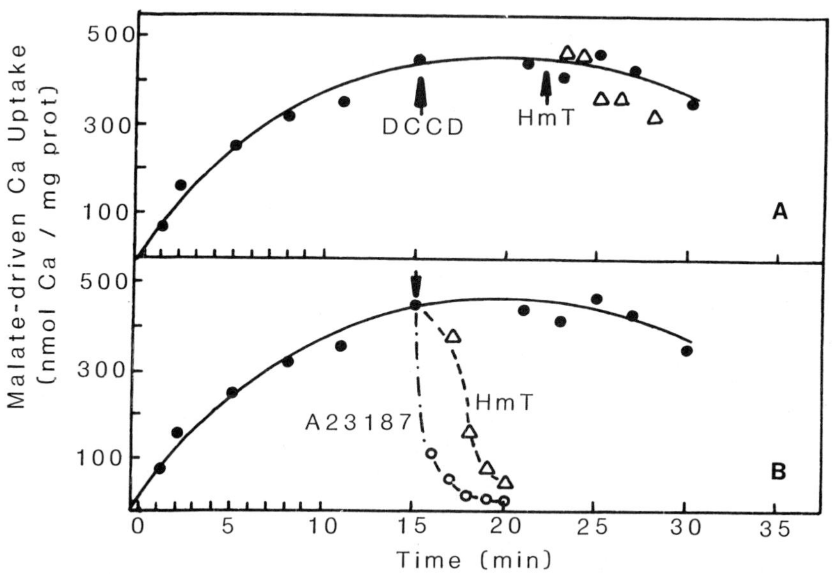

Fig. 3. DCCD prevented HmT toxin-induced dissipation of Ca^{2+} gradients in T corn mitochondria. Ca^{2+} uptake was initiated by the addition of mitochondrial protein (50-100 μg) in 0.25 M sorbitol, 2.5 mM Hepes-BTP, 5 mM K_2HPO_4, 4 mM malate, 0.2 mM $^{45}CaCl_2$ and aliquots were filtered at different time intervals. Additions were (A) 4 μM DCCD, 5 ng/ml HmT toxin and (B) 2 μg/ml A23187, 5 ng/ml HmT toxin (6.5 nM).

DCCD protection of activities inhibited by the HmT toxin in T mitochondria suggests that the toxin "receptor(s)" is proteinaceous and that the site is in a hydrophobic environment. Inhibition of malate-dependent electron transport implies the toxin may interact with Complex I. From these results, we propose a preliminary working model. The toxin binds to (a) specific subunit(s) of the NADH-UQ reductase of the T mitochondria. This modification leads to inhibition of electron transport via complex I and causes ion leakage.

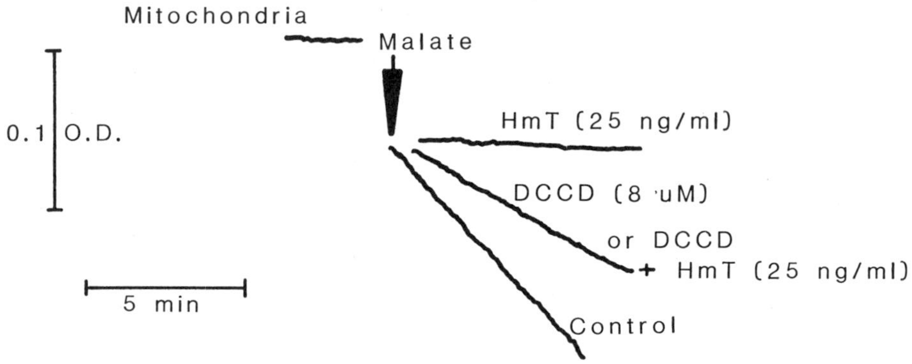

Fig. 4. DCCD prevented HmT toxin from inhibiting malate-dependent electron transport in T corn mitochondria. Electron transport was measured by following ferricyanide reduction (420 nm) in a reaction mixture of mitochondrial protein (100 μg), 0.25 M sorbitol, 10 mM Hepes-BTP (pH 7.4), 1 mM $K_3Fe(CN)_6$, 20 mM KCl, 1 mM malonate and 1 mM KCN. Mitochondria were incubated with DCCD for 5 min prior to the addition of HmT toxin. Incubation with toxin continued for an additional 5 min and the reaction was started with malate (50 mM).

ACKNOWLEDGEMENT

This work was supported in part by Department of Energy, DE-AS05-82ER13015 to H.S.

REFERENCES

1. C.J. Leaver and M.W. Gray, Mitochondrial Genome Organization and Expression in Higher Plants, Ann. Rev. Plant Physiol. 33:373-402 (1982)
2. J. M. Daly, Mechanisms of Action, in: "Toxins in Plant Disease," R.D. Durbin, ed., Academic Press, N.Y. (1981)
3. A. Kimber and H. Sze, Helminthosporium maydis T Toxin Decreased Calcium Transport into Mitochondria of Susceptible Corn, Plant Physiol. 74:804-809 (1984)
4. M.J. Holden and H. Sze, Helminthosporium maydis T Toxin Increased Membrane Permeability to Ca^{2+} in Susceptible Corn Mitochondria. Plant Physiol. 75:235-237 (1984)
5. M.J. Holden, Helminthosporium maydis, Race T (HmT) toxin effects on mitochondrial membranes from susceptible (cms T) corn. Ph.D Dissertation, University of Maryland, College Park (1985)
6. A. Berville, A. Ghazi, M. Charbonnier and J-F Bonavent, Effects of Methomyl and Helminthosporium maydis Toxin on Matrix Volume, Proton Motive Force, and NAD^+ Accumulation in Maize (Zea mays L.) Mitochondria, Plant Physiol. 76:508-517 (1984)

CASTOR BEAN MITOCHONDRIAL PROTEIN SYNTHESIS IN RESPONSE TO TEMPERATURE-INDUCED STRESS

Naheed Kaderbhai, R. Brian Beechey and Mustak Kaderbhai

Department of Biochemistry and Agricultural Biochemistry
School of Agricultural Sciences, The University College
of Wales, Aberystwyth, Dyfed SY23 3DD, Wales, UK

INTRODUCTION

Higher plants when exposed to elevated temperatures respond by rapid and increased expression of sets of proteins known as heat-shock proteins[1]. These proteins are thought to primarily protect the cellular systems from lethal thermally-exerted damage. Mitochondria have been implicated as targets of the heat-shock response on the basis of the observations that agents such as uncouplers of oxidative phosphorylation and inhibitors of respiratory chain cause induction of the heat-shock proteins[2]. The findings that isolated corn (*Zea mays*) mitochondria when exposed to elevated temperatures express *in vitro* novel polypeptides[3,4], also suggests that the organelles have a role to play in heat-shock response. Here we present some preliminary findings of the studies on the direct effects of temperature shifts on the protein synthesis by isolated, functional castor bean (*Ricinus communis*) mitochondria.

EXPERIMENTAL

Mitochondria were prepared under stringent aseptic conditions from four-day germinated castor beans as described previously[5]. Protein synthesis by freshly isolated mitochondria was performed using a modified procedure of Leaver et al[6]. The extent of protein synthesis was monitored by measuring the amount of ^{35}S-methionine incorporated into acid-insoluble products by hot trichloroacetic acid precipitation method.

RESULTS AND DISCUSSION

Freshly prepared castor bean mitochondria, provided with succinate, showed good respiration rates with respiratory controls of about 2.4 and ADP/O ratio of around 2. In the presence of exogenously added energy sources and amino acids including ^{35}S-methionine, the mitochondria showed time-dependent incorporation of the radiolabel into trichloroacetic acid-insoluble products (Fig 3). The *in vitro* protein synthesis by castor bean mitochondria exhibited many characteristics similar to those reported for mitochondria from other sources[8]; protein synthesis was strictly dependent on the presence of exogenously added energy sources supplied either as substrates for oxidation by mitochondria or through an ATP-regenerating system. It was strongly inhibited by chloramphenicol but was relatively insensitive to cycloheximide. Mitochondria supplemented with acetate as the sole exogenous substrate for energy source were not active in their protein synthetic activity. Post-translational microbial screening indicated that there were about 400 to 500 bacteria in each translational mixture. Control experiments with the isolated bacteria showed that this number of cells could not be responsible for the level of ^{35}S-methionine incorporation observed with the isolated mitochondria.

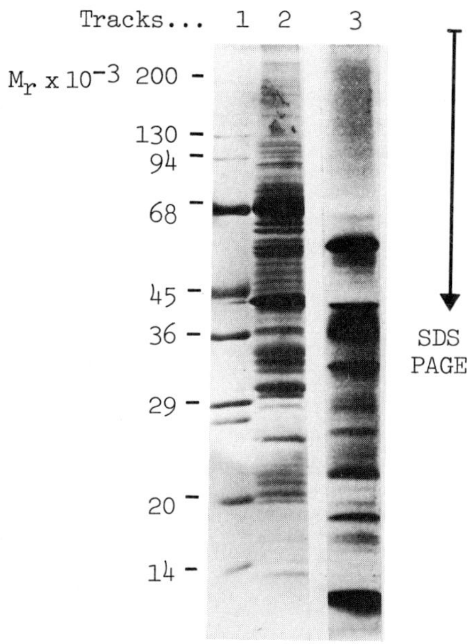

Fig 1. Proteins synthesised by mitochondria isolated from 4-days germinated castor bean endosperms. Tracks 1, marker proteins; track 2, total mitochondrial proteins detected by Coomassie Blue; track 3 fluorogram showing <u>in vitro</u> synthesised polypeptides.

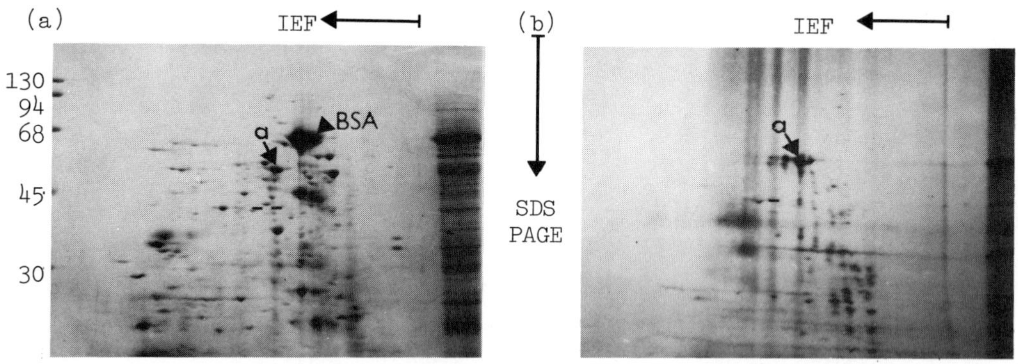

Fig 2. High resolution IEF/SDS-PAGE analysis of polypeptides synthesised by isolated castor bean mitochondria. The 2-D profiles represent a 95% protein extract of total mitochondrial protein prepared by solubilisation with 4.5M urea and 1% (w/v) TX-100: a) Coomassie Blue stained, b) fluorogram. The arrows indicate the position of a major polypeptide, tentatively identified as the α-subunit of F_1-ATPase, encoded by the mitochondrial genome.

The proteins synthesised by isolated castor bean mitochondria were separated by SDS-PAGE and analysed by fluorography (Fig 1, track 3); the profile shows about 30 to 40 distinct polypeptides ranging in M_r of 58000 to 12000. Some nine major in vitro synthsised products are clearly distinguishable of which only a few appear to correspond with their counter-part Coomassie Blue stained protein profile (Fig 1, track 2). The most intensely labelled product of about 58000 M_r has been tentatively identified as the α subunit of F_1ATPase[7]. The products when analysed by high-resolution two-dimensional IEF/SDS PAGE showed many more components (Fig 2b), in the order of 100, than those visualised in either the one dimensional SDS-PAGE (see Fig 1) or those previously reported[8]. The major protein encoded by the mitochondrial genome corresponds well with a similar polypeptide, most likely the α-subunit of F_1ATPase, in the comparable Coomassie Blue stained profile (indicated by arrows in Fig 2a and 2b). Surprisingly, this finding indicates that the vast majority of the mitochondrially-synthesised proteins are minor components of the organelle. The in vitro synthesised proteins observed here do not appear to arise from artefacts due to truncated products resulting from premature translation termination or proteolysis since the patterns observed were highly reproducible in different preparations of mitochondria. The large number of polypeptides expressed by the castor bean mitochondrial genetic system appears to be consistent with the complexity and size of mitochondrial genome reported from other higher plants[9].

Mitochondria were isolated at 4°C from castor bean endosperms that had been germinated for 4 days at 30°C. The rate and extent of ^{35}S-methionine incorporation into newly biosynthesised proteins was measured as a function of time at varying temperatures of incubation during the translation phase (Fig 3). No significant protein synthesis was observed at 0°C. Both the extent and rate of protein synthesis increased when the temperatures were shifted to 12°C and 20°C. Further increases in temperatures to 28°C and 35°C resulted in progressive decline in protein synthetic activity of the isolated mitochondria. Compared with the optimal physiological growth temperature of 30°C for castor beans, the isolated mitochondria display a sharp temperature optimum at 20°C (Fig 4).

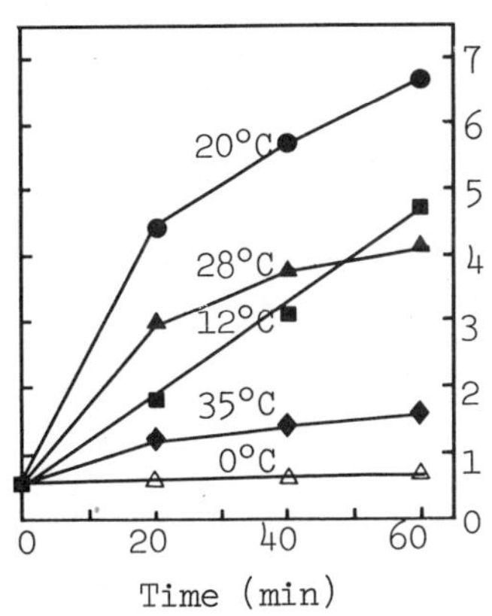

Fig 3. Time course of in vitro protein synthesis by castor bean mitochondria at various temperatures.

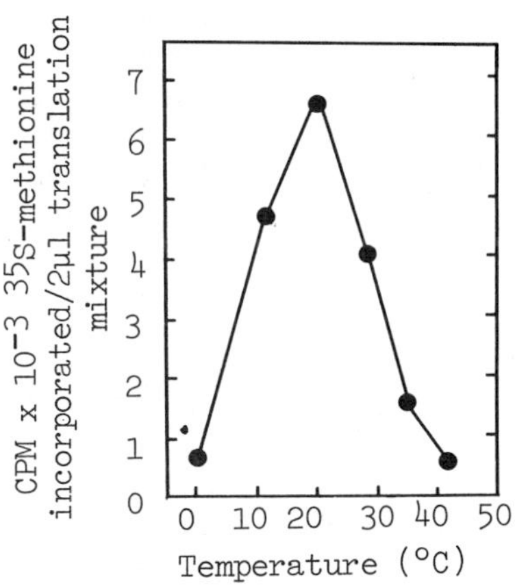

Fig 4. Temperature optima for protein synthesis by isolated castor bean mitochondria.

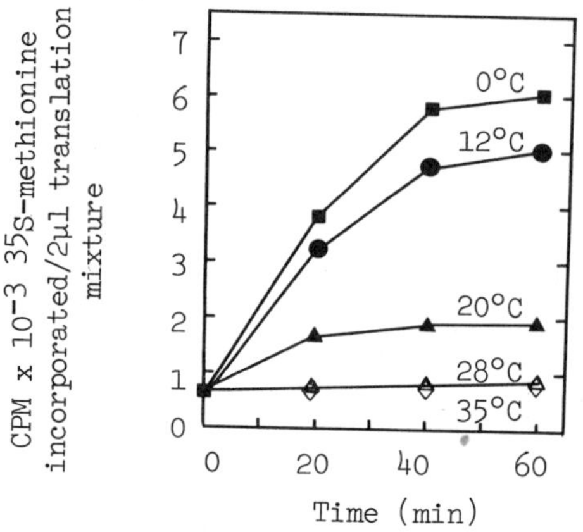

Fig 5. The effect of preincubating castor bean mitochondria for 15min at various temperatures on the *in vitro* protein synthesis at 20°C.

Mitochondria were preincubated for 15min at temperatures ranging from 0 to 35°C in resuspension buffer or translation mixture without radiolabeled methionine, and their ability to incorporate ^{35}S-methionine into newly synthesised proteins was carried out at 20°C (Fig 5). There was progressive inhibition of the extent and rate of protein synthesis with increasing preincubation temperatures. Surprisingly, the preincubation at 0°C gave higher radiolabel incorporation compared to preincubation at 20°C, the optimal temperature for the *in vitro* protein synthesis (see Fig 4). The fluorographic profile of the proteins synthesised in mitochondria heat-shocked at various temperatures remained unaltered (data not shown) but the relative incorporation of the radiolabel into newly made proteins reflected the level of biosynthesis that were observed as a result of preincubations and direct temperature shifts.

Nembiolo and White[3] recently reported that corn seedling mitochondria when incubated at 27°C and then shifted to 37°C during translation, show a marked stimulation in protein synthesis together with induction of mitochondrially-encoded 52000M_r protein. A similar finding by Sinibaldi and Turpen[4] also demonstrate that a heat-shock protein of M_r60000 is encoded by corn seedlings and *Brassica campestris* in both the C-4 plants mitochondria when the isolated organelles were heat-shocked at 41°C prior to *in vitro* translation. Our work with castor bean - a C-3 plant, mitochondrial protein synthesis show a curious response to heat-shock despite the fact that the respiratory function and the energy status of the isolated mitochondria is unaffected by the same temperature treatments.

REFERENCES

1. J. A. Kimpel and J. L. Key (1985). TIBS., 10, 35-357.
2. M. Ashburner and J. J. Bonner (1979). Cell, 17, 241-254.
3. C. M. Nebiolo and E. M. White (1985). Plant Physiol. 79, 1129-1132.
4. R. M. Sinibaldi and T. Turpen (1985). JBC, 260 (29), 15382-15385.
5. M. A. Brailsford et al., (1986). Biochem. J., 239, 355-361.
6. C. J. Leaver, E. Hack and B. G. Forde (1983). Methods in Enzymology, 97, 476-484.
7. C. J. Leaver and B. G Forde (1980). In: Genome Organisation and Expression in Plants (Edited by C.J. Leaver). Plenum Press, New York, pp 407-425.
8. E. Hack and C. J. Leaver (1983). EMBO, J., 2 (10), 1783-1789.
9. R. Sederoff (1984). Adv. Genet. 22, 1-108.

CITRATE SYNTHASE MESSENGER RNA IN RELATION TO ENZYME SYNTHESIS IN EUGLENA

Jane E. Parker

School of Biological Sciences
University College of Swansea
Swansea, SA2 8PP, U.K.

INTRODUCTION

The unicellular biflagellate Euglena gracilis is unusual amoungst eukaryotes in being able to thrive both as a phototroph in the light and as an organotroph in the dark, and rapidly adjusts its cellular machinery to accommodate a utilizable energy source most efficiently[1]. For example dark grown cells possess undifferentiated proplastids which develop into functional chloroplasts on exposure to light[2]. The associated cellular changes during chloroplast development can be conveniently studied in dark grown non-dividing cells maintained on a "resting" medium lacking exogenous carbon and nitrogen[3]. Upon illumination of resting cells there is a 12 hour lag in chloroplast development followed by a period of rapid proplastid differentiation over 72 hours[3]. During the lag phase an immediate transcient increase in the specific activities of enzymes of the Tricarboxylic Acid (TCA) cycle is observed[4,5], which may furnish the cell with the necessary reductant and substrate energy during chloroplast development. The mitochondrial enzyme citrate synthase is of particular interest since it occupies a key position, regulating the entry of carbon into the TCA cycle[6]. Cannons and Merrett[7] demonstrated that increased citrate synthase (CS) activity was proportional to an increase in enzyme protein but was not reflected in enhanced levels of translatable CS mRNA. This would suggest that control of enzyme activity operates primarily at a post-transcriptional level.

In the present study a closer examination of CS synthesis in Euglena gracilis Klebs strain z Pringsheim has been undertaken. Antibodies raised against the homogeneously pure protein have been used to assess CS mRNA in relation to enzyme synthesis by "in vitro" translation and immunoprecipitation. The analysis of mRNA in heterologous cell-free translation system is based upon the untested assumption that translatable mRNA is an accurate reflection of mRNA abundance, and it is recognised that a more stringent assessment employs specific hybridisation probes. A cDNA library has therefore been constructed from Euglena poly(A) RNA and screened for cDNA encoding CS with the aim of using it to probe CS mRNA in relation to enzyme expression at different stages of development. The cloning strategy is described.

CLONING OF CDNA ENCODING EUGLENA CITRATE SYNTHASE

Translatable CS mRNA was detected within the poly(A) containing fraction of total cellular RNA by immunoprecipitation of its nascent 46 kDa polypeptide from translation lysates. The immunoprecipitable product was not of significantly larger molecular weight than the mature CS polypeptide by the criterion of SDS/polyacrylamide gel electrophoresis. Results showed that levels of translatable CS mRNA remained constant irrespective of developmental changes in enzyme abundance, determined by immunotitration[8]. It was also noted that there were no detectable changes in the total polypeptide profiles of poly(A) RNA during drastic developmental transitions e.g. the greening and etiolating of cells. By the criterion of "*in vitro*" translation, therefore, it is proposed that a key control of protein synthesis in *Euglena* is exerted at a post-transcriptional level other than mRNA abundance.

For the purposes of cDNA cloning developmental enrichment of CS mRNA was not considered a viable proposition. Double stranded cDNA (ds cDNA) was synthesised according to the method of Gubler and Hoffman[9] from poly(A) RNA enriched in CS mRNA by fractionation on linear 10 - 30% denaturing sucrose gradients. The cDNA represented a size range of 0.8 to 2.0 kbp. This was homopolymer (dC)-tailed and inserted into the (dG)-tailed plasmid cloning vector PUC 9[10]. The recombinant plasmid was used to transform *E.coli* cells of the Rec A^+ strain JM83. "Miniprep" analysis of randomly selected colonies containing recombinant plasmid (identified by the lac Z chromogenic insertional screening procedure) showed an insert frequency of 85%. Of these 45% contained cDNA inserts of greater than 1.0 kbp. 2500 recombinant colonies were further screened for the presence of CS cDNA.

A genomic clone containing the nuclear gene encoding CS from *Saccharomyces*[11] was kindly provided by G. Schatz (Biocentre, University of Basel, Switzerland) as a possible probe for *Euglena* cDNA. However it failed to hybridise with *Euglena* total and poly(A) RNA under conditions of low stringency. An alternative screening strategy was therefore performed which employed a mixed oligonucleotide probe. The N-terminal 42 amino acid sequence of purified *Euglena* CS was determined (J. Findley, Dept. Biochemistry, University of Leeds, U.K.) and a mixture of 17'mer oligonucleotides constructed to represent all possible codon sequences of a 6 amino acid portion of low redundancy (Fig. 1). The application of such a probe in hybridisation studies depends on the different melting temperatures of cDNA with perfectly matched oligonucleotides and oligonucleotides containing mismatches at one or more base positions.

peptide		Lys	Asn	Lys	Tyr	Gly	Asp	
mRNA	5´	AA^{A}_{G}	AA^{U}_{C}	AA^{A}_{G}	UA^{U}_{C}	$GG^{C/G}_{A/U}$	GA	3´
oligo	3´	TT^{T}_{C}	TT^{A}_{G}	TT^{T}_{C}	AT^{A}_{G}	$CC^{G/C}_{T/A}$	TC	5´

Fig. 1 Region of *Euglena* citrate synthase selected for the synthesis of an oligonucleotide probe

It has been shown that duplexes with a single bp mismatch are significantly less stable than their perfectly matched counterparts[12]. The use of appropriate hybridisation and washing conditions should therefore eliminate only the formation of imperfect duplexes whilst maintaining a hybridisation signal from an oligonucleotide with 100% homology to the target cDNA. In the present study DNA from transformed E.coli colonies was immobilised on nitrocellulose membrane filters and screened with a ^{32}P-labelled mixed oligonucleotide probe at 38°C in 6 x SSC (0.9M NaCl/0.09M Na.citrate). Filters were washed in 6 x SSC at increasing temperature stringencies determined from the base composition and degree of mismatch of the probe[13]. Hybridisation was predicted at 40°C and 45°C but not at 52°C.

Using the described procedure two clones have been isolated which reproducibly hybridise at the predicted stringencies. Both contain 1.76 kbp cDNA inserts which produce identical restriction endonuclease digestion patterns, and are therefore considered to encode the same DNA sequence. The cDNA recognises a single poly(A) RNA species approximately 1.8 kb in size in Northern hybridisations. However the identified cDNA has consistently failed to select Euglena mRNA encoding CS in "Hybrid Release Translation" experiments[14,15]. No specifically hybridising translatable poly(A) RNA was detected. The reason for this is not understood since the hybrid selection manipulations did not significantly reduce poly(A) RNA translatability and CS mRNA was detectable by immunoprecipitation from the non-hybridising RNA fraction.

The cDNA clones selected using a mixed oligonucleotide probe can so far only be tentatively scored as positives. The cDNA is currently being partially sequenced to obtain unequivical identification. It is hoped that this will reveal CS mRNA coding data and be applied in hybridisation analyses of RNA and genomic DNA in order to elucidate more fully the regulation of CS expression in Euglena.

REFERENCES

1. J. A. Schiff and S. D. Schwartzbach, Photocontrol of chloroplast development, in: "The Biology of Euglena, 3" D. E. Buetow, ed., Acad. Press, New York (1982).
2. T. Osafune, S. Klein and J. A. Schiff, Events surrounding the early development of Euglena chloroplasts. 18. The structure of the developing proplastid in the first hours of illumination from serial sections of wild-type cells, J. Ultrastructure Res. 73:77 (1980).
3. A. W. Holowinsky and J. A. Schiff, Events surrounding the early development of Euglena chloroplasts. I Induction by pre-illumination, Plant Physiol. 45:339 (1970).
4. M. A. Horrum and S. D. Schwartzbach, Nutritional regulation of organelle biogenesis in Euglena: Photo- and metabolite induction of mitochondria, Planta. 149:376 (1980).
5. A. C. Cannons and M. J. Merrett, The regulation of synthesis of mitochondrial enzymes in regreening and division-synchronised Euglena cultures, Planta. 159:125 (1983).
6. S. Beeckmans, Some structural and regulatory aspects of citrate synthase, Int. J. Biochem. 16:341 (1984).
7. A. C. Cannons and M. J. Merrett, Regulation of synthesis of citrate synthase in regreening Euglena gracilis, Eur. J. Biochem. 142:597 (1984).
8. T. Yamada, A. Tanaka, S. Horikawa, S. Numa and S. Fukui, Cell-free translation and regulation of Candida tropicalis catalase messenger RNA, Eur. J. Biochem. 129:251 (1982).
9. U. Gubler and B. J. Hoffman, A simple and very efficient method for generating cDNA libraries, Gene. 25:263 (1983).

10. J. Vieira and J. Messing, The pUC plasmids, an M13mp7-derived system for insertion mutagenesis and sequencing with synthetic universal primers, Gene. 19:259 (1982).
11. M. Suissa, K. Suda and G. Schatz, Isolation of yeast genes for citrate synthase and fifteen other mitochondrial proteins by a new screening method, EMBO J. 3:1773 (1984).
12. K. Itakura, J. J. Rossi and R. B. Wallace, Synthesis and use of synthetic oligonucleotides, Ann. Rev. Biochem. 53:323 (1984).
13. J. Singer-Sam, R. L. Simmer, D. H. Keith, L. Shively, M. Teplitz, K. Itakura, S. M. Gartler and A. D. Riggs, Isolation of a cDNA clone for human X-linked 3-phosphoglycerate kinase by use of a mixture of synthetic oligodeoxyribonucleotides as a detection probe, Proc. Natl. Acad. Sci. USA. 80:802 (1983).
14. R. B. Ricciardi, J. S. Miller and B. E. Roberts, Purification and mapping of specific mRNAs by hybridisation-selection and cell-free translation, Proc. Natl. Acad. Sci. USA. 76:4927 (1979).
15. T. Maniatis, E. F. Fritsch and J. Sambrook, "Molecular Cloning, A Laboratory Manual," Cold Spring Harbor Laboratory, New York (1982).

PROTEIN SYNTHESIS, RESPIRATION AND GROWTH IN

EUGLENA GRACILIS Z IN THE PRESENCE OF ANTIMYCIN A

P. Benichou*, R. Calvayrac* and M.L. Claisse**

* Lab. Membranes Biologiques, Paris VII, t53-54, 2°et., 2, Place Jussieu, 75251 Paris Cedex 05
** Centre Génétique Moléculaire, Lab. Propre CNRS associé Paris VI, 91190 Gif-sur-Yvette

INTRODUCTION

Various organisms exhibit cyanide resistant respiration in normal or particular physiological conditions[1]. This alternative mitochondrial pathway can also be induced by some respiratory inhibitors. Antimycin A, known to block the mitochondrial respiratory chain between b and c_1 cytochromes, was added to growth medium of <u>Euglena gracilis</u> strain Z. Cells were grown in mineral media supplemented with 33 mM lactate or 17 mM glucose as sole carbon source[2]. The kinetics of adaptation to antimycin A (2μM) were first studied. Then, the growth, respiration and protein biosynthesis of cultures of Euglena in presence of this inhibitor have been characterized. These cultures were maintained for months in media containing antimycin A, which was not carried out in previous studies[3].

MATERIALS AND METHODS

<u>Biomass</u> was estimated by cellular density and dry weight. Respiration was measured by oxygen consumption, using a Clark type electrode and expressed as nmol $O_2.h^{-1}.(\mu g\ protein)^{-1}$; protein content was assayed as described by Bradford.

<u>In vivo incorporations</u> of labelled valine were carried out on cells transferred in resting medium. The conditions described in yeasts[4] were adapted to the Euglena cells.

Antimycin A (Ant.A) was dissolved in isopropanol, cyanide (KCN), azide (NaN_3) and dinitrophenol (DNP) in water, salicylhydroxamic acid (SHAM) and propylgallate (PG) in ethanol.

RESULTS

Effects of antimycin A during the logarithmic phase of growth

The addition of antimycin A during the logarithmic phase of growth was immediately followed by a growth stop for about 10 hours while respiration was markedly decreased. Then, cells started to divide again whereas O_2 consumption was enhanced (fig.1), moderately in glucose cultures but greatly in lactate medium ; respiration became resistant to cyanide concentration more

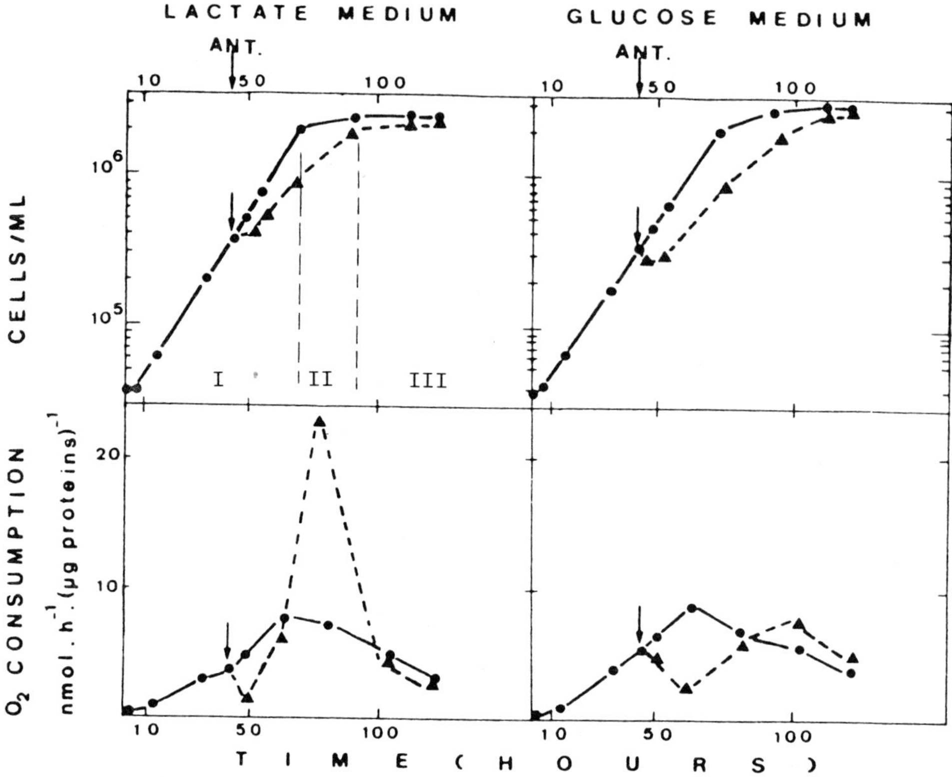

Fig. 1. Evolution versus time of cell density (a) and O_2 consumption (b). Control cells ●— After addition of antimycin A (Ant.A 2 µM)▲--

than 100 fold that necessary to inhibit control cells whereas azide sensitivity was maintained.

Growth and respiration of cells adapted to antimycin A

Following results concern Euglena that exhibited constant generation times once they had been cultured 2 months in the presence of antimycin A. When cells treated with antimycin A were reinoculated into fresh medium containing the same concentration of the inhibitor, growth could be maintained with a generation time slightly increased and a dry weight decreament compared with control cells (fig. 2 a,b). The average level of O_2 consumption was particularly high in lactate medium (2c). Furthermore, in the cells reinoculated in medium containing antimycin A, respiration remained resistant to cyanide and was more sensitive to SHAM and PG than in control cells (Table I). Comparison between I_{50} allowed to define the nature (sensitive or resistant) of cells towards inhibitors.

Incorporation of labelled valine into cellular proteins

Control cells were able to incorporate 20% of the disponible labelled valine (^3H or ^{14}C) into proteins. Similar results were obtained with cells adapted to antimycin A. The presence of exogenous carbon source (lactate or glucose) did not increase this incorporation. Thus, inhibition of the incorporation by respiratory inhibitors was studied with cells transferred in resting medium without carbon source. When variable quantities of KCN, PG, SHAM, Ant.A or NaN_3 were added, the effects of each drug were parallel on respiration and valine incorporation (Table II). DNP 0.2mM totally inhibited the incorporation and had no effect on respiration.

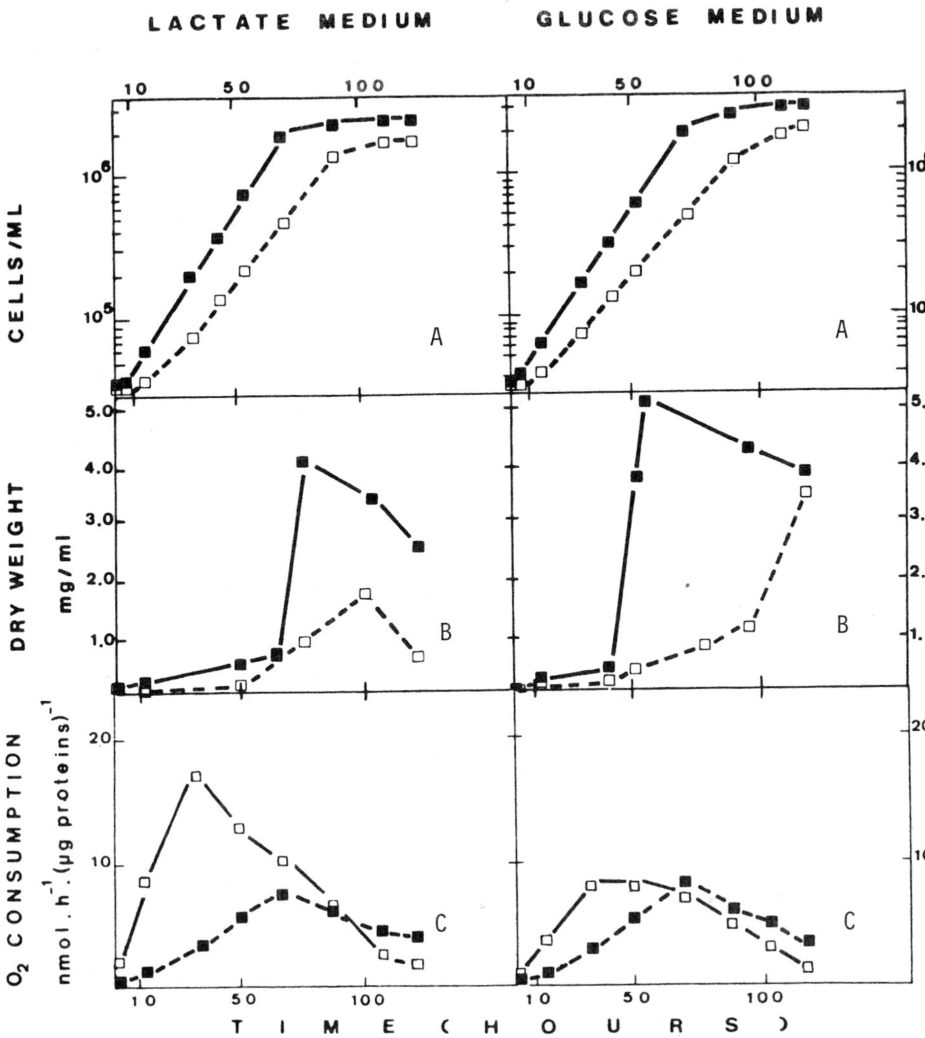

Fig. 2. Evolution versus time of cell density (a), dry weight (b) and O_2 consumption (c).
Control cells ■—— Cells adapted to antimycin A □- -

Table 1. Effects of respiratory inhibitors on O_2 consumption. Comparison between control cells and Euglena adapted to antimycin A. Cells were harvested in late exponential phase of growth.
I_{50} : Effector concentration necessary for 50% inhibition of the respiration. As cellular density was 10^6 cells/ml values of I_{50} may also be expressed in µM.

SUBSTRATE USED FOR THE CELL CULTURE		ANT.A	I_{50}			
			CN	N_3Na	SHAM	PG
	mM	µM	nmol / 10^6 cells			
LACTATE	33	0	40	40	810	2380
		2	4600	20	540	550
GLUCOSE	17	0	20	10	1748	5174
		2	1000	20	450	710

Table 2. Comparison of the action (in %) of the respiratory effectors on the incorporation and on the respiration. Cells were cultured in lactate medium and harvested in the logarithmic phase of growth.

EFFECTORS	CONTROL CELLS RESP. (%)	CONTROL CELLS INC. (%)	ADAPTED CELLS RESP. (%)	ADAPTED CELLS INC. (%)
No effector	100	100	100	100
N_3Na 1mM	0	0	0	0
Ant.A 2μM	20	60	100	110
Ant.A 4μM	0	0	10	10
KCN 0.1mM	55	45	105	105
KCN 10mM	0	0	0	0
SHAM 0.5mM	55	60	50	25
SHAM 0.8mM	50	55	20	10
PG 0.5mM	80	75	50	15
PG 2.4mM	50	5	0	0
DNP 0.2mM	100	0	100	0

CONCLUSION

Euglena adapted to antimycin A were cultured for months in the presence of this inhibitor in media containing either a fermentescible carbon source (glucose) or a non-fermentescible substrate (lactate).

In such conditions, these adapted cells exhibited a respiration resistant to cyanide — but sensitive to azide — whereas sensitivity to SHAM and PG was increased. Using these inhibitors, parallel effects were obtained on the respiration and on the incorporation of labelled valine into proteins, in control cells and in cells adapted to antimycin A. DNP, which is an uncoupler, inhibited only the incorporation.

The ratio of dry weight increament to total oxygen consumption allowed the determination of growth yield during the optimal phase of growth. In control cells grown in glucose medium, this growth yield was 1.8 higher than in cells adapted to antimycin A, while a typical value of 3 could be expected if only the first phosphorylation site was functional in the respiratory chain. Similar results were obtained with cells grown in lactate medium. However, many authors (except Wilson[5]) claimed that the alternative respiratory pathway was not energetic.

We demonstrated here that antimycin A treated Euglena develop a respiration resistant to cyanide i.e an alternative respiratory pathway and exhibit an energetic level sufficient for protein synthesis and growth. Further experiments are now in progress at the subcellular level.

REFERENCES

1. M. Henry and E. Nyns, Cyanide insensitive respiration. An alternative mitochondrial pathway.Subcell. Biochem. 4, 1-65 (1975).
2. R. Calvayrac , Relations entre les substrats, la respiration et la structure mitochondriale chez *Euglena gracilis* Z. Arch. Mikrobiol. 73, 308-314, (1970).
3. R. Calvayrac and R. Butow: Action de l'antimycine A sur la respiration et la structure des mitochondries d'*Euglena gracilis* Z. Arch. Mikrobiol. 80, 62-69 (1971)
4. P. Pajot and M. Claisse: Utilization by yeasts of D-Lactate and L-Lactate as sources of energy in the presence of antimycin A. Eur. J. Biochem.49, 275-285 (1974).
5. S. Wilson : Energy conservation by the plant mitochondrial cyanide insensitive oxidase. Some additional evidence.Biochem. J. 190, 349-360 (1980).

RESPIRATION IN INTACT TISSUES: PROBLEMS AND PERSPECTIVES

Hans Lambers and David A. Day

Botanical Laboratory, University of Utrecht, Lange Niewstraat 106, 3512 PN Utrecht, The Netherlands
and
Botany Department, Australian National University
G.P.O. Box 4, ACT 2601 Canberra, Australia

INTRODUCTION

For a full appreciation of the role and significance of mitochondrial respiration in the metabolism of the intact plant, it is imperative to complement investigations at the biochemical level with a study of the intact system. However, the methodology used for isolated mitochondria is not necessarily appropriate for intact tissues. Inhibitors which are very specific in a study of mitochondria, may have severe side effects in vivo. Also, the optimum concentration range of the applied inhibitor may be rather different for both situations, due to poor penetration, metabolism, or sequestration of the inhibitor in the living tissue. In this paper we will discuss some of the approaches and problems involved in estimating electron transport in vivo.

USE OF DIFFERENT GENOTYPES

In recent years, amazing progress has been made in selecting high yielding genotypes of Lolium perenne, using leaf dark respiration as a selection criterion[1,2]. The rate of mature leaf respiration was found to be inversely correlated with the crop growth rate. Further progress is to be expected in this area, since a similar inverse correlation can readily be found using other grass species. Unfortunately, we have very little understanding of this phenomenon. It cannot simply be explained by variation in the contribution of the alternative path in leaf respiration[3,4]. Perhaps the requirement for ATP in fast growing genotypes is lower than that in the slow growing ones, but so far this hypothesis has not yet been corroborated. Alternatively, the efficiency of ATP production may be higher in the fast growing genotypes. The recent developments in the application of ^{31}P-NMR techniques in botanical research[5,6] promises some progress in this field.

Investigations of genotypes with and without alternative path capacity, may yield results which are not only pertinent to those interested in plant respiration, but also to plant breeders. Such genotypes have recently become available for Glycine max, Helenium amarum, Plantago lanceolata, Pisum sativum and Zea mays[7]. There is ample evidence to suggest that the alternative path functions as an "energy overflow", at least under a range of conditions in some plants[8,9], but we need to gain further understanding of the significance of the alternative path for the survival of a plant. Does the capacity of this path contribute to a plant's flexibility? Some results, using plants exposed to a change in water potential in the root environment, do lead to such a suggestion[10,11], but corroboration, using genotypes with and without alternative path capacity is warranted.

The importance of respiration studies for plant breeding, combined with a lack of understanding of the exact quantitative role of repiration in the plant metabolic processes, indicates the need of further studies on respiration of intact plants. In such studies the use of inhibitors will be essential. The present paper attempts to provide information on the use of such inhibitors, problems arising from their use and suggestions to circumvent these problems.

MEASUREMENT OF ELECTRON TRANSPORT IN VIVO

The use of inhibitors

Substituted hydroxamic acids have been successfully used as inhibitors of the alternative path in isolated mitochondria[12,13]. They are also widely used in investigations of respiration in whole tissues and slices[14-20]. For inhibitor studies of isolated mitochondria the concentration of substituted hydroxamic acids, (e.g., SHAM, salicylhydroxamic acid) need not exceed 1 mM. In intact tissues concentrations as high as 25 mM are often used[17,21]. In intact tissues, SHAM can be deposited in the vacuoles, as seen from the sharpening of the vacuolar phosphate peak in ^{31}P-NMR spectra, due to complexation of SHAM with divalent cations (J.K.M. Roberts, pers. comm.). This, together with poor penetration, may explain the need for high concentrations of SHAM in in vivo studies. However, when using high concentrations of SHAM there is a risk that the cytochrome path is also inhibited. In callus-forming potato tuber discs such an inhibition was indeed found[20]. However, in roots of Pisum sativum, 25 mM only inhibited the alternative path and not the cytochrome path, at least when used in short term (less than 20 minutes) experiments. Nonetheless, great caution must be exercised when using mitochondrial inhibitors with whole tissues. For example, other oxidases may be inhibited[22], or even stimulated by SHAM[23]. KCN, an inhibitor of cytochrome oxidase, can also have other effects.

What follows is a suggested experimental strategy which in our opinion will enable a reliable estimation of electron transport activities in vivo.

(i) It is recommended that oxygen electrodes rather than manometers are used and that experiments are kept short. Use of inhibitors should be restricted to no more than 20 min.

(ii) The first step is to determine adequate inhibitor concentrations. Respiration should be titrated with KCN and a concentration chosen that maximally inhibits oxygen uptake (note, however, thatan inhibition is not always seen; in such cases, a concentration of 0.5 mM KCN is suggested and this can be checked in the presence of SHAM). Following this, respiration should be titrated with SHAM in the presence of KCN and the minimum concentration that inhibits maximally, chosen for further use.

(iii) The chosen concentration of SHAM should then be applied to the tissue in the absence of KCN. If the concentration chosen is low (less than 5 mM), a stimulation of oxygen uptake may be seen[15]. This stimulation in roots[23] is thought to be due to activation of an externally located peroxidase) which is stimulated by low SHAM concentrations and inhibited by KCN and high concentrations of SHAM. Obviously this can confound results (but note that to date this system has only been observed in roots). When stimulation of oxygen uptake by low SHAM concentrations is observed, it is necessary to go to a higher concentration. Sometimes, a stimulation of oxygen uptake by peroxidase activity at low SHAM concentrations coincides with inhibition of the alternative path. The net effect may then be no inhibition of the oxygen uptake by low SHAM concentrations. When this is the case, titration of respiration by SHAM yields non-linear -plots[24]. Also in this case, it is necessary to go to a higher SHAM concentration, e.g. 25 mM.

(iv) When very high concentrations of SHAM are used, it is important to ensure that they do not affect the cytochrome path. This can be done by employing the "reverse titration" technique, introduced by Theologis and Laties[18]. This involves measurement of respiration in the presence of varying concentrations of KCN (or antimycin) both in the absence and presence of SHAM (or another alternative path inhibitor) at a concentration sufficient to fully inhibit the alternative path and any induced peroxidase activity (see above). If the inhibition by KCN is the same under both conditions then it can be concluded that no SHAM inhibition of the cytochrome path occurred. Using this technique, De Visser and Blacquiere[15] have demonstrated that even 25 mM SHAM does not inhibit the cytochrome path in roots of Pisum sativum, in short term experiments. In addition, they demonstrated that the SHAM-stimulated, KCN-inhibited oxidase (see above) did not contribute to oxygen uptake in Pisum sativum roots, even though it was present. The presence of such a systems is easily detected by titrating respiration with SHAM, beginning at 1 mM.

Clearly, estimations of in vivo electron transport activity are not easy and are often tedious. However, if care and patience are exercised then the inhibitors can be used in conjunction with uncouplers and substrates (such as sugars and organic acids) to gain insight into the regulation of respiration (see below).

Uncouplers

The commonly applied uncouplers, both in vitro and in vivo, are CCCP and FCCP (carbonyl cyanide-m-chlorophenyl-hydrazone and p-trifluoromethoxy carbonylcyanide phenylhydrazone, respectively). Sometimes DNP (2,4-dinitrophenol) is also used, but its effects are rather dependent on the pH and very high concentrations are needed. Recent evidence, based on ^{31}P-NMR measurements, suggest that the effects of DNP may not be due to impairment of ATP synthesis, but rather to direct effect on membrane lipids[25].

Uncouplers, being protonophores, primarily uncouple the flux of electrons through the electron transport chain from the oxidative phosphorylation of ADP. Provided ADP is limiting in the first place, uncoupler will stimulate the flux of electrons through the cytochrome path. However, uncouplers will also lower the cytosolic ATP concentration and increase the ADP concentration. This may lead to enhanced glycolytic activity, accumulation of pyruvate, followed by a faster turnover of the TCA cycle[26]. Under these conditions, more NADH will now be available to the electron transport path, possibly leading to (further) engagement of the alternative path. Therefore, increased respiration after addition of uncoupler by itself is no proof for ADP limitation of the cytochrome path. Only if such a stimulation is observed in the presence of an inhibitor of the alternative path, this conclusion is justified[27]. Conversely, even when adenylates do restrict the flux of electrons through the cytochrome path, addition of uncoupler may have no effect on overall oxygen uptake, because electrons can be redirected from an engaged alternative path to the cytochrome path. Under these conditions, addition of uncoupler will prevent SHAM inhibition[14].

However, it should be noted that uncouplers can affect nonmitochondrial oxidases also. Even when both the cytochrome and the alternative pathways are saturated, stimulation of oxygen uptake can sometimes be observed, For example, in roots of Pisum sativum CCCP apparently stimulated the beforementioned peroxidase, probably due to enhanced cytosolic NAD(P)H levels[15]. Whenever the respiration in the presence of CCCP is more than the sum of that in the presence KCN (the capacity of the alternative path) and that in the presence of SHAM + CCCP (the capacity of the cytochrome path), there is reason to suspect operation of a third system. Such cases, however, are rare.

RESIDUAL RESPIRATION

In the presence of inhibitors of both the cytochrome and the alternative path, a "residual" component of respiration may persist in intact tissues. Such a residual component is invariably small in isolated mitochondria. In roots, this residual component rarely exceeds 10% of the control rate[17]. In leaves this component may be more important[17,28], though it tends to be rather small when slices of the same leaves are used[17,29].

Poor penetration into (some parts of) the tissue may offer part of the explanation for the residual component. For example, it is not known to what extent -if at all- inhibitors such as SHAM will enter the stele of intact roots. However, the rather large residual component in potato or fruit slices (sometimes 40% of the control rate[16,18]) is unlikely to be due to poor penetration. It is not known if the residual component is a part of the control rate, or induced upon addition of inhibitors. A kinetic analysis, using the affinity for oxygen of the residual component, may clarify the problem.

REGULATORY ASPECTS

Notwithstanding some problems encountered in using respiratory inhibitors and uncouplers in vivo, such compounds can provide valuable information on the regulation of respiration in intact tissues. They can also be applied in combination with each other, or with respiratory substrates, e.g., sucrose and glucose, requiring the glycolytic pathway, or glycine, which can be fed straight into the mitochondria.

Application of a combination of SHAM and CCCP and of SHAM alone, has provided the information that the cytochrome path is sometimes, but definitely not in all tissues and species, restricted by ADP[3,14,27,28]. Recent evidence from ^{31}P-NMR studies, showing much lower free ADP concentrations in the cell than previously suspected, has also led to the conclusion that it is most plausible that respiration in vivo is limited by ADP[5].

Application of glycine, in the absence and presence of SHAM, led to the conclusion that the activity of the cytochrome path in leaf slices of Lolium perenne was determined by the supply of substrates to the mitochondria. In this tissue, the control of glycolysis was so tight as to limit the flux of electrons through the cytochrome path; glycine did, but glucose alone did not, increase the flux through the cytochrome path[3]. Such a tight control of glycolysis by adenylates was also found for cultured cells of Acer pseudoplatanus, where both pyruvate and DNP did, but sugars did not, stimulate respiration[26].

Engagement of the alternative path generally occurs in the presence of a high level of respiratory substrates in the cells[29-33]. When this coincides with restriction of the cytochrome path, such as in roots of Triticum aestivum, Zea mays, Spinacea oleracea[27] and Lolium perenne[3] and in leaves of Triticum aestivum after six hours of photosynthesis[14,29], the alternative path acts as a true "energy overflow", removing respiratory substrates in an inefficient manner[8].

Engagement of the alternative path sometimes occurs when there is no restriction of the cytochrome path by adenylates, e.g., in roots of Phaseolus vulgaris[27], Pisum sativum, Plantago lanceolata and P. major[24], and in leaves of Phaseolus vulgaris[28], Carex acutiformis and C. diandra (A. Kooyman and H. Lambers, unpublished). Under those circumstances, and definitely when coinciding with a sudden increase in demand for ceilular ATP, the alternative path may function as an inefficient mechanism producing ATP ("overcharge overflow" model[21]), rather than as an "energy overflow".

There are obviously various control points in respiration. In some tissues the regulation of glycolysis may be so tight as to be the major control point, limiting the flux of electrons through the cytochrome path. In other tissues glycolysis is obviously less tightly controlled and adenylates may limit the flux of electrons through the cytochrome path, sometimes leading to engagement of the alternative path. When substrate input is very high and ATP demand great, the cytochrome path may be fully engaged and the alternative path may also contribute.

Therefore, inhibitors, uncouplers and substrates, applied alone or in various combinations, has provided information about the flexibility of the control points in respiration. They are useful tools, provided one is aware of some side effects which occasionally may occur.

THE FATE OF THE RESPIRATORY ATP

It has now been known for decades that mitochondria provide the cell with ATP. For a full appreciation of the quantitative significance of repiration in the carbon economy of a plant, it is important to establish how much respiratory energy is required for the various biosynthetic, transport and maintenance processes.

Experimental methods have been described to separate three components of respiration, providing energy for biosynthesis, transport and maintenance of solute gradients and cellular structures[34,35]. When a comparison is made of the values obtained with such experimental methods and theoretical values, based on known costs of transport and biosynthetic reactions, a large discrepancy is often found[36]. Assuming that it is valid to deduce the rate of ATP formation from the rate of oxygen uptake (with a correction for the contribution of the alternative path, if engaged), this discrepancy must reflect our lack of understanding of the costs of some major processes in the cell. This is obviously an area in which much more work is required.

PERSPECTIVES FOR CROP IMPROVEMENT

The alternative path

There is evidence that the alternative path functions as an energy overflow, removing carbohydrates when the substrate level in the cell is high and the need for ATP and carbon skeletons small, relative to the suply of substrates[8]. The alternative path may also function as an inefficient mechanism to produce ATP, e.g., upon a sudden increase in demand for metabolic energy[21].

It may be that the operation of an energy overflow has some advantages in a fluctuating environment[10]. In the absence of a comparison of genotypes with and without alternative path capacity there is no conclusive evidence for such a hypothesis. However, such genotypes have recently been described for Pisum sativum[31] and are also available for Triticum aestivum (M.E. Musgrave, pers. comm.). Genotypes without the alternative path capacity gave higher yields in a CO_2-enriched, but not in a normal atmosphere in comparison with genotypes in which the alternative path was present and (partly) engaged[31]. Selection for genotypes lacking alternative path capacity may therefore lead to increased yield, at least under some conditions.

In species where the alternative path functions as an inefficient adjunct to produce ATP[21], selection for genotypes with a greater cytochrome path capacity, rather than reduced alternative path capacity, could be the aim of a plant breeding program.

Mature leaf respiration

Wilson[1,2] has selected genotypes of Lolium perenne, on the basis of their rate of mature leaf respiration. The rate of respiration was negatively correlated with the crop yield and the selected genotypes have subsequently been used to produce a fast growing, slow respiring population, and a slow growing, fast respiring population of Lolium perenne. Contrary to expectation, the faster respiration in the slow growing plants was not due to a greater engagement of the alternative path[3]. In neither of the two populations did the alternative path contribute to respiration in the mature leaves or leaf meristems. Both the difference in yield and that in respiration disappeared when the plants were grown under N-limiting conditions[4].

D. Wilson (pers. comm.) also succeeded in establishing a similar negative correlation between mature leaf respiration and crop growth rate in Festuca arundinacea. R. M. Gifford (pers. comm.) has found a similar negative correlation for some lines of Phalaris aquatica. Obviously we are dealing with one of those rare examples where plant physiologists have provided a clear tool for the plant breeders. It is to be regretted that hardly any plant breeders have picked up this tool.

There is still a further need to investigate the observed negative correlation between respiration and yield to establish if the slow respiration is associated with any agronomically undesirable traits. So far no evidence for such an association has been found[37,38], but an explanation for the high respiration rate is still lacking. It is not due to increased protein turnover in the mature leaves[39], or increased rates of "cycling" of nitrogenous compounds between roots and leaves[39]. Some other metabolic or transport process may require a substantial amount of energy in one and not in the other population, but this remains obscure. We obviously still have very little understanding of the quantitative importance of the sinks for ATP in a cell. It is also possible that the efficiency of ATP production in the fast respiring population is lower than that in the slow respiring line.

It is important to find satisfactory answers to these questions if the results of respiration studies are to be applied by plant breeders. There is clearly a challenge at the biochemical, physiological and agronomic level!

ACKNOWLEDGEMENTS

We would like to thank Loes Spek for her constructive comments on an earlier version of this contribution.

REFERENCES

1. D. Wilson, Variation in leaf respiration in relation to growth and photosynthesis of Lolium, Ann. Appl. Biol. 80: 323-338 (1975).

2. D. Wilson, Response to selection for dark respiration rate of mature leaves in Lolium perenne and its effects on growth of young plants and stimulated swards, Ann. Bot. 49: 303-312 (1982).
3. D.A. Day, O.C. De Vos, D. Wilson, and H. Lambers, The regulation of respiration in the leaves and roots of two Lolium perenne populations with contrasting mature leaf respiration rates and yield, Plant Physiol. 78: 678-683 (1985).
4. C.J. Pilbeam, M.J. Robson, and H. Lambers, Respiration in mature leaves of Lolium perenne as affected by nutrient supply, cutting and competition, Physiol. Plant. 66: 53-57 (1985).
5. J.K.M. Roberts, Study of plant metabolism in vivo using NMR spectroscopy, Annu. Rev. Plant Physiol. 35: 375-386 (1984).
6. J.K.M. Roberts, D. Wemmer, and O. Jardetzky, Measurements of mitochondrial ATP-ase activity in maize root tips by saturation transfer ^{31}P nuclear magnetic resonance, Plant Physiol. 74: 632-639 (1984).
7. M.E. Musgrave, J. Antonovics, and J.N. Siedow, Is male-sterility in plants related to lack of cyanide-resistant respiration in tissues?, Plant Sci. 44: 7-11 (1986).
8. H. Lambers, Cyanide-resistant respiration: A non-phosphorylating electron transport pathway acting as an overflow, Physiol. Plant. 55: 478-485 (1982).
9. H. Lambers, Respiration in intact plants and tissues: Its regulation and dependance on environmental factors, metabolism and invaded organisms, in: Encyclopedia of Plant Physiology, New Series, R. Douce and D.A. Day, eds., Springer-Verlag, Berlin (1985).
10. H. Lambers, T. Blacquiere, and C.E.E. Stuiver, Interactions between osmoregulation and the alternative respiratory pathway in Plantago coronopus as affected by salinity, Physiol. Plant. 51: 63-68 (1981).
11. M.E. Nicolas, H. Lambers, R.J. Simpson, and M.J. Dalling, Effect of drought on metabolism and partitioning of carbon in two varieties of wheat differing in drought tolerance, Ann. Bot. 4: 727-742 (1985).
12. G.R. Schonbaum, W.D. Bonner, Jr., B. Storey, and J.T. Bahr, Specific inhibition of the cyanide-insensitive respiratory pathway in plant mitochondria by hydroxamic acids, Plant Physiol. 47: 124-128 (1971).
13. J.N. Siedow, and J.N. Bickett, Structural features required for inhibition of cyanide-insenstive electron transfer by propyl gallate, Arch. Biochem. Biophys. 207: 32-39 (1981).
14. J. Azcon-Bieto, D.A. Day, and H. Lambers, The regulation of respiration in the dark in wheat leaf slices, Plant Sci. Lett. 32: 313-320 (1983).
15. R. De Visser, and T. Blacquiere, Inhibition and stimulation of root respiration in Pisum and Plantago by hydroxamate. Its consequences for the assessment of alternative path activity, Plant Physiol. 75: 813-817 (1984).
16. A.M. Hemrika-Wagner, E.J. Verschoor, and L.H.W. Van der Plas, Alternative pathway respiration ij vivo of potato tuber callus grown at various temperatures, Physiol. Plant. 59: 369-374 (1983).

17. H. Lambers, D.A. Day, and J. Azcon-Bieto, Cyanide-resistant respiration on roots and leaves. Measurements with intact tissues and isolated mitochondria, Physiol. Plant. 58: 148-154 (1983).
18. A. Theologis, and G.G. Laties, Relative contribution of cytochrome-mediated and cyanide resistant electron transport in fresh and aged potato slices, Plant Physiol. 62: 232-237 (1978).
19. M.K. Upadhyaya, Effects of salicylhydroxamate on . respiration, seed germination and seedling growth in Avena fatua, Physiol. Plant. 67: 43-48 (1986).
20. L.H.W. Van der Plas, and M.J. Wagner, Influence of ethanol on alternative oxidase in mitochondria from callus-forming potato tuber discs, Physiol. Plant. 49: 121-126 (1980).
21. R. De Visser, K. Spreen Brouwer, and F. Posthumus, Alternative path mediated ATP synthesis in roots of Pisum sativum upon nitrogen supply, Plant Physiol. 80: 295-300 (1986).
22. P.R. Rich, N.K. Wiegand, H. Blum, A.L. Moore, and W.D. Bonner, Studies on the mechanism of inhibition of redox enzymes by substituted hydroxamic acids, Biochim. Biophys. Acta 525: 325-337 (1978).
23. K. Spreen Brouwer, T. Van Valen, D.A. Day, and H. Lambers, Hydroxamate-stimulated O_2 uptake in roots of Pisum sativum and Zea mays, mediated by a peroxidase. Its consequences for respiration measurements, Plant Physiol. 82: in press (1986).
24. T. Blacquiere, and R. De Visser, Capacity of cytochrome and alternative path in coupled and uncoupled root respiration of Pisum and Plantago, Physiol. Plant. 62: 427-432 (1984).
25. P.C. Jackson, P.E. Pfeffer, and W.V. Gerasimowicz, Use of ^{31}P NMR to assess effects of DNP on ATP levels in vivo in barley roots, Plant Physiol. 81: 1130-1133 (1986).
26. C.V. Givan, and J.G. Torrey, Respiratory response of Acer pseudoplatanus cells to pyruvate and 2,4-dinitrophenol, Plant Physiol. 43: 635-640 (1968).
27. D.A. Day, and H. Lambers, The regulation of glycolysis and electron transport in roots Physiol. Plant. 58: 155-160 (1983).
28. J. Azcon-Bieto, H. Lambers, and D.A. Day, Respiratory properties of developing bean and pea leaves, Aust. J. Plant Physiol. 10: 237-245 (1983).
29. J. Azcon-Bieto, H. Lambers, and D.A. Day, Effect of photosynthesis and carbohydrate status on respiratory rates and the involvement of the alternative pathway in leaf respiration, Plant Physiol. 72: 598-603 (1983).
30. H. Lambers, The physiological significance of cyanide-resistant respiration in higher plants, Plant Cell Environ. 3: 293-302 (1980).
31. M.E. Musgrave, B.R. Strain, and J.N. Siedow, Response of two pea hybrids to CO_2 enrichment: A test of the energy overflow hypothesis for alternative respiration, Proc. Natl. Acad. Sci. USA submitted.
32. A.M. Rychter, and E. Ciesla, The participation of cyanide resistant pathway in respiration of winter rape plants treated with low temperature, Abstract, 2nd International

Meeting Plant Mitochondria, The University College of Wales, Aberystwyth (1986).

33. L.H.W. Van der Plas, and M.J. Wagner, Regulation of the activity of the alternative oxidase in callus-forming potato tuber tissue discs, Physiol. Plant. 58: 311-317 (1983).
34. H. Lambers, R.K.Szaniawski, and R. De Visser, Respiration for growth, maintenance and ion uptake. An evaluation of concept, methods, values and their significance, Physiol. Plant. 58: 556-563 (1983).
35. B.W. Veen, Energy costs of ion transport, in: Genetic Engeneering of Osmoregulation. Impact on Plant Productivity for Food, Chemicals and Energy, D.W. Rains, R.C. Valentine and A. Hollaender, eds., Plenum, New York (1980).
36. H. Lambers, Growth, respiration, exudation and symbiotic associations: The fate of carbon translocated to the roots, in: Root Development and Function - Effects of the Physical Environment, P.J. Gregory, J.V. Lake and D.A. Rose, eds., Cambridge University Press, Cambridge (1986).
37. M.J. Robson, The growth and carbon economy of selection lines of Lolium perenne with 'fast' and 'slow' rates of mature leaf respiration. 1. Grown as simulated swards during a regrowth period, Ann. Bot. 49: 321-329 (1982).
38. M.J. Robson, The growth and carbon economy of selection lines of Lolium perenne cv. S23 with 'fast' and 'slow' rates of dark respiration. 2. Grown as young plants from seeds, Ann. Bot. 49: 331-339 (1982).
39. A.J. Barneix, H.D. Cooper, I. Stulen, and H. Lambers, Metabolism and translocation of nitrogen in two Lolium perenne populations with contrasting mature leaf respiration and yield, Physiol. Plant. submitted

ON THE INTEGRATION OF PLANT GROWTH AND RESPIRATION

Ries de Visser[1]

Department of Plant Science
University of Alberta
Edmonton, Alberta T6G 2P5 Canada

INTRODUCTION

The aim of this brief review is to discuss some intriguing new developments which should receive more attention, and to indicate some areas where rapid progress could be made if biochemists, physiologists and biophysists would join forces. The subject of plant growth and respiration has been extensively reviewed during the last few years, including quantitative aspects like growth- and maintenance respiration (Lambers 1985, Lambers et al. 1983, Ryle 1984), carbon economy and nitrogen metabolism (Minchin et al. 1981, Ryle 1984, Schubert and Ryle 1984), and the energetics of ion transport (Clarkson 1985, Lin 1985, Veen 1980, 1981) and more specific aspects like respiration and senescence (Rhodes 1980), seed germination and respiration (Perl 1986), influences of biotic and abiotic environmental factors (Lambers 1985, 1986), and the physiological significance of CN-resistant respiration (Day et al. 1980, Lambers 1980,1982,1985, Laties 1982, Siedow and Berthold 1986). In view of the overwhelming amount of published work on the various aspects of the relation between growth and respiration, I will mainly deal with a subjective selection of interesting physiological and biochemical topics that have emerged recently. The title wants to express an ultimate goal, rather than our present knowledge, as noted by Beevers (1985) in his foreword to Vol. 18 of the Encyclopedia of Plant Physiology.

Commonly accepted views on the relation between growth and respiration have changed dramatically over the last half century. The idea that respiration is needed for (plant) growth and life in general, is probably more than a century old, but the supporting evidence and general acceptance came only ca 30 years ago, long after the main respiratory pathways, glycolysis, oxidative pentose phosphate pathway, citric acid cycle and oxydative phosphorylation, had been discovered (cf. Beevers 1961). Until that time, the dominant opinion was that, like growth, respiration was a carbohydrate utilizing process, but without any use. Now, it has been demonstrated that the CN-resistant, non-phosphorylating electron transport path is active in many plant tissues. This pathway has been interpreted by Lambers as an 'energy overflow'. Its main function is proposed to be the 'wasteful' oxidation of an excess of carbohydrates. Therefore, a chimaera type of theory

---

Abbreviations: CCCP, carbonylcyanide m-chlorophenyl hydrazone, NMR, nuclear magnetic resonance; SHAM, salicylhydroxamic acid.

[1]Present address: Centre for Agrobiological Research (C.A.B.O.), P.O. Box 14, 6700 AA Wageningen, The Netherlands.

has to be adopted which combines the 'wasteful' and the 'useful' aspects of respiration (Lambers 1982, 1985).

In the present context, growth will be defined biochemically as follows:

- the complex of processes involved in the synthesis of biomass with the preservation of genetical information,
- the intra- and intercellular transport of solutes, and
- the uptake of molecules from the environment.

Respiration and growth may be distinct entities, macroscopically as well as thermodynamically, but biochemically, these processes are integrated, since glycolysis, oxidative pentose phosphate pathway, and citric acid cycle have major biosynthetic functions, e.g. in the synthesis of amino acids, lipids, nucleic acids and co-enzymes.

Growth processes require a continuous input of free energy, necessary to create (and maintain) order against the spontaneous tendency towards chaos. The free energy in sunlight is trapped by photosynthesis in NTP and reducing equivalents, and stored in the form of carbohydrates. These are oxidized in (dark) respiration to yield CO_2, H_2O and free energy-rich intermediates, i.e. NTP, NAD(P)H/$FADH_2$ and/or carbon skeletons, which are utilized in growth processes and thus form a link between growth and respiration. The metabolic pool of cellular NTP is ca 1 mM, as determined by ^{31}P-NMR (Roberts, this volume). Combining this result with data in the literature on the rates of oxygen uptake and ATP production by intact tissues or plants (e.g. de Visser et al. 1986, van der Plas and Wagner 1984), it follows that the turnover rate of the metabolic ATP pool is once to several times per minute. This shows, that any block of either growth or respiration will affect the rate of the other within a minute. This strict interdependence suggests that both growth may control respiration and vice versa. Which of these two possibilities prevails, depends on the many factors affecting growth and/or respiration.

SOME METHODOLOGICAL ASPECTS OF THE STUDY OF GROWTH AND RESPIRATION

The overall process of growth, i.e. dry matter production, has been widely studied in relation to respiration (for a review, see Lambers 1985). The techniques involved are simple and include measurements of oxygen and/or carbon dioxide exchange and dry matter accumulation. Lambers et al. (1983b) discuss the need to assess the contribution of the alternative, CN-resistant electron transport pathway when the ATP or carbohydrate demands of growth processes are to be measured (see also following section). Root (respiratory) metabolism has to be studied in relation to shoot size and growth rate, because the root exports mineral nutrients and organic nitrogen to the shoot (de Visser and Lambers 1983, de Visser 1985). Disregarding this aspect may hamper the interpretation of data on root metabolism, especially when different species are compared (e.g. Bryce & ap Rees 1985). Little work has been done to determine actual in vivo costs of particular growth processes like ion uptake and other transport processes. Attempts have been made to study the relation between respiration and growth processes like nitrate uptake (Clarkson 1985, Veen 1980), nitrogen metabolism (de Visser 1985), nitrate and ammonium uptake (de Visser et al. 1986), mineral nutrient uptake (Lambers et al. 1981b) and osmoregulation (Lambers et al. 1981a, van der Plas and Wagner 1984). New techniques, for example ^{31}P-NMR (Roberts 1984), and the use of well-characterized mutants may promote further progress in this field. The recent finding by Musgrave et al. (1986) of pea cultivars that lack a CN-resistant respiratory pathway, opens up many promising possibilities. These cultivars will simplify the measurement of in vivo ATP production, and help to find and identify the alternative oxidase (and/or other components involved in CN-resistant respiration; Stegink and Siedow 1986a). Since many growth processes change during ontogeny, it is important that plants are used when in a well-defined growth stage. Different tissues have different functions and respiratory characteristics. Therefore, insight in the energetics of particular growth processes will be more easily gained when the plant material under study performes one major function (e.g. ion uptake, as in roots, or cell division, as in meristems). Instead of specializing on either physiology or biochemistry, in vitro and in vivo studies should be combined.

Determination of the Efficiency of Oxidative Phosphorylation in Vivo

In vitro experiments with isolated mitochondria can only give information on the capacity of the different electron transport pathways, and this only, if the proper combination of substrates is used (see the following section). In order to gain insight in the in vitro activity of the various electron transport pathways, one has to apply specific inhibitors. Alternatively, non-destructive methods (like ^{31}P-NMR; Roberts 1984) may be used in experiments with intact tissues or plants. No significant objections have been made against the use of cyanide, but SHAM and other hydroxamates are still not generally accepted for use with intact tissues (e.g. Møller and Bérczi 1985). However, in the last three years different groups have shown, independently, that high concentrations (10-25 mM) of hydroxamate can be safely used for the assessment of alternative and cytochrome path activities (Lambers et al. 1983a, de Visser and Blacquière 1984). SHAM stimulates the oxygen uptake by intact roots of pea and Plantago as a side effect at low concentrations which block the alternative path. Although confusing, this phenomenon did not interfere with the determination of the activities of the cytochrome and alternative pathways, since high SHAM concentrations blocked the activity of the unknown oxidase without affecting the cytochrome path (de Visser and Blacquière 1984). Recently, these findings have been corroborated by other groups working with intact roots of pea (Spreen Brouwer and Lambers, this Vol.) and barley (Bingham and Farrar, this Vol.), and isolated mitochondria and intact tissue of potato tuber callus (van der Plas and Wagner, this Vol.). The hydroxamate-stimulated oxidase has been shown to be a peroxidase, which requires NAD(P)H, SHAM and oxygen as substrates and can be inhibitied by gentisic acid and high SHAM concentrations (Spreen Brouwer and Lambers, this Vol.). Møller and Bérczi (1985) have reported a similar system in plasmalemma vesicles from wheat roots, but these authors suggest that a cytochrome P420/450 is involved.

Summarizing, the most recent information indicates that SHAM can be used for the assessment of alternative path activity in intact tissues, when used in short-term experiments (<20 min) and at the proper concentration as determined by titration in the absence and presence of a cytochrome path inhibitor (de Visser and Blacquière 1984, Lambers 1985), or when applied in combination with a peroxidase inhibitor (e.g. gentisic acid, Spreen Brouwer and Lambers, this Vol.) or catalase (Møller and Bérczi 1985; van der Plas and Wagner, this Vol.). An alternative method for the determination of the ADP:O ratio of oxydative phosphorylation is ^{31}P-NMR (Roberts 1984, Roberts et al. 1984) which is now also suitable for measurements on intact plants (Roberts, this Vol.). This non-destructive technique will make is possible to detect, for example, in vivo activity of the rotenone-resistant bypass of site I (Palmer and Ward 1985) and an actual increase in ATP synthesis by intact roots after a nitrate gift (de Visser et al. 1986).

The Respiratory Substrates of Mitochondria In Vitro and In Vivo

An interesting recent development is the use of substrate 'cocktails' in studies with isolated mitochondria. Pyruvate is likely to be the main substrate of plant mitochondria in vivo (Wiskich and Dry 1985), although a different view is held by Lance and coworkers (Lance and Rustin 1984, Lance et al. 1985) who suggest that malate is the main substrate. Little is known about the relative importance of possible substrates like pyruvate, malate, succinate, citrate, α-ketoglutarate and NADH in vivo. However, this is an important issue when one wants to relate intact tissue respiration and the properties of isolated mitochondria. For example, not only the rate of

Table 1. The cyanide-resistance of mitochondrial state 3 respiration as dependent on the substrate (combination).

Plant species/ tissues	CN-resistance					
	malate	succinate	NADH	combination	Dimension[a]	Ref.[b]
Petunia hybrida cells	-	6.2	7.7	14.1	(1)	(A)
Potato tuber callus	-	18.7	14.6	33.5	(1)	(A)
Kalanchoe blossfeldiana leaves	78	42	19	-	(2)	(B)
Sweet potato tubers	42	-	-	70[c]	(2)	(C)

[a](1) nmol $O_2.min^{-1}.g^{-1}$ fr wt; (2) nmol $O_2.min^{-1}.mg^{-1}$ prot
[b](A) Hemrika-Wagner et al. 1986; (B) Rustin et al. 1984; (C) Wiskich and Day 1982
[c]pH 7.4; combination = succinate + malate (+glutamate) + citrate + α-ketoglutarate

mitochondrial respiration depends on the substrate (Day et al. 1985; Dry and Wiskich 1985), but also the CN-resistance of mitochondria in vitro is highly dependent on the substrate (combination) used (Table 1; Rustin et al. 1984, Wiskich and Day 1982, Hemrika-Wagner et al. 1986). Even more importantly, Hemrika-Wagner et al. (1986) showed that the CN-resistance of mitochondria isolated from tuber callus and Petunia hybrida cells with succinate + NADH as subtrates is equal to the sum of the CN-resistances in the presence of succinate or NADH separately (Table 1). These authors suggest that different substrates have access to different ubiquinone-pool (for a review on Q-pool behaviour, see Moore and Rich 1985) and/or different alternative pathways. Heterogenity of the inner mitochondrial membrane has indeed been found in mitochondria from potato and Jeruzalem artichoke tubers (Lidén, Sommarin and Møller, this Vol.). This supports the idea of the existence of different 'domains' in the inner mitochondrial membrane, with different oxidase compositions (for a review, see Lance et al. 1985). It is clear that this subject needs further investigation.

THE SIGNIFICANCE OF CN-RESISTANT RESPIRATION FOR PLANT GROWTH

The Regulation of its Activity

Two models for the regulation of electron transport via the cytochrome and alternative paths have been proposed. Bahr and Bonner (1973) showed that the alternative path becomes engaged in isolated mung bean hypocotyl mitochondria when the cytochrome path is saturated. De Troostembergh and Nyns (1978) suggested that the electron flow is partioned between over both pathways according to the relative rate constants of the reactions between ubiquinol and both electron transport pathways. The Bahr and Bonner hypothesis has been most widely accepted. At present, both models are under review again (see Kay and Palmer, this Vol.).

Based on the Bahr and Bonner (1973) model, Lambers and Steingröver (1978) proposed the 'energy overflow' model for the regulation of alternative path activity in higher plants. In this model, the alternative path is engaged when an excess of carbohydrates occurs in the tissue (see also Lambers 1985). A still unexplained aspect of this model is, how glycolysis and tricarboxylic acid cycle can operate at a controlled rate which exceeds the energy demands of the cell. Day and Lambers (1983) have demonstrated that glycolysis is controlled in the roots of several plant species which show alternative path activity. Wiskich and Dry (1985) and Lambers (1985) suggest that the substrate supply of glycolysis is larger than required by the cell's metabolism. This suggestion implies that sucrose and/or glucose alleviate the inhibition of phosphofructokinase and/or pyruvate kinase by ATP. This effect has not been reported, although it may be demonstrated easily in vitro (see also the last section).

The Physiological Functions of Alternative Path Respiration

The fact that the alternative path is non-phosphorylating does not necessarily imply that the 'wasteful' oxidation of carbohydrate is the main function of CN-resistant respiration. Even so, there is strong evidence to support this idea (for a review, Lambers 1982, 1985). Especially the observation by Azcón-Bieto et al (1983ab) that the alternative path is more (only) active in wheat leaves after a period of photosynthesis, lends support to the 'overflow' hypothesis. In slices of wheat leaves harvested after a 6 h light period, the alternative path operated while the cytochrome path activity was limited by adenylates. However, addition of sucrose of glycine to slices of wheat leaves harvested at the end of the night increased both the cytochrome and the alternative path, showing that respiration was limited by the substrate (Azcón-Bieto et al. 1983a). Rychter and Cieśla (this Vol.) also provide evidence for the 'energy overflow' model: in winter rape plants, low temperature induces parallel rises in sugar content and alternative path activity. Two other hypotheses have been proposed for a physiological function of the alternative path: 1. the alternative path permits turnover of the tricarboxylic acid cycle under conditions of low ATP requirements by the cell (Bahr and Bonner 1973, Palmer 1976); 2. the alternative path contributes to ATP-synthesis when the cytochrome path cannot accept more electrons (de Visser et al. 1986, van der Plas and Wagner 1984). Lambers (1985) finds no evidence for the first hypothesis. The evidence provided in support of the latter idea is the following: van der Plas and Wagner (1984) observed that, when osmotic stress impaired the cytochrome pathway, the alternative path activity increased in callus forming potato tuber discs to the extent that total ATP production in oxydative phosphorylation remained constant. However, these results support the 'energy overflow' hypothesis of Lambers (1982, 1985) just as well. In experiments with intact pea plants, de Visser et al. (1986) showed that addition of nitrate to the root medium of N-free grown pea plants increased the cytochrome path activity in the roots when this pathway was not saturated, but such a treatment increased the alternative path when the cytochrome path was saturated (as determined with CCCP + SHAM). For more information on the role of the alternative path in plant carbohydrate metabolism, I refer to the contribution by Siedow and Musgrave (this Vol.). As judged from even the latest reviews, CN-resistant respiration seems to be universal among higher plants. The recent finding of pea cultivars completely lacking alternative path respiration (Musgrave et al. 1986) indicates that plants may survive without this pathway, and even have higher yields (Siedow and Musgrave, this Vol.). Intriguing questions are, why the trait has never been lost in the evolution of higher plants, and what the important function of the pathway may be under natural conditions.

ENERGY REQUIREMENTS OF ION TRANSPORT IN PLANTS

It is important to note that the energy costs of the transport and uptake of mineral ions, for example by roots, are not known. Experimental estimates range from 1 to 4 moles of ATP per mol ions (e.g. nitrate) per membrane passage (Clarkson 1985, Lambers et al. 1983b, Veen 1980). For their theoretical calculations of conversion efficiencies, Penning de Vries et al. (1974) and Lambers and Steingrover (1978) assumed values of 0.3 moles of ATP per mol of ions per membrane passage. Therefore, in theoretical calculations of growth respiration, too low values may have been assumed. If net ion uptake is regulated by ion efflux, as proposed by Deane-Drummond (1984) for nitrate, the high values mentioned above can be easily explained. A major reason for the lack of studies on energetics of ion transport in vivo is probably the unavailability of proper techniques. Possibly, NMR will facilitate further progress in this field. This technique allows simultaneous monitoring of atoms like ^{13}C, ^{15}N and ^{31}P as well as determinations of the rate of ATP synthesis in vivo (Roberts 1984 and this Vol.).

THE REGULATION OF GROWTH AND RESPIRATION

Our knowledge of the regulation of growth and respiration is still scanty, partly because little is known about the organization of respiratory pathways (ap Rees 1985). The subject is nevertheless frequently studied, and thus demands some comments here.

At present it is not clear whether plant growth controls respiration and/or vice versa, at least not for optimal growth conditions. Growth is more sensitive than respiration to a decrease in the supply of an external factor like oxygen (Saglio et al. 1984). Growth and respiration are similarly affected by temperature (Q_{10} = 2; Smakman and Hofstra 1982) and both processes are controlled by substrate (sucrose) supply. Many factors are known to affect growth processes as well as respiratory pathways directly, but the significance of this type of regulation in vivo is not known. For example, kations, especially ammonium and potassium, stimulate the in vitro activity of the two regulatory enzymes of glycolysis, phosphofructokinase and pyruvate kinase (Turner and Turner 1980), but a role of kations in the control of glycolysis in vivo has not been demonstrated.

In animal systems and recently also in plants, fructose 2,6-biphosphate has been shown to be a regulatory metabolite, stimulating glycolysis and inhibiting gluconeogenesis (Huber 1986). There seem to be indications that, in plants, fructose 2,6-biphosphate modulates the sensitivity of starch-degrading enzymes for adenylates, and thus allows glycolysis to respond in a variable manner to changes in energy requirements by the cell (Stitt and Steup 1985). Possibly, fructose 2,6-biphosphate is the missing link in Lambers's 'energy overflow model' (Lambers 1985) which still needs a factor responsible for the 'uncoupling' of glycolysis from adenylate control. In a following contribution, ap Rees will discuss the potential roles of fructose 2,6-biphosphate in plant carbohydrate metabolism. Interestingly, Stitt et al. (1986) report that ethylene enhances fructose 2,6-biphosphate concentrations in carrot roots and potato tubers, after 20 hours. However, high ethylene concentrations were used ($50\ \mu l.l^{-1}$).

When plant physiochemists cannot succeed in finding a suitable metabolic explanation, they seem to resort to hormones (with the 'h' of holy?). And so, we can find many ideas on the role of the different plant hormones in the regulation of not only growth, but also respiration (cf. Gude and van der Plas 1985, Laties 1982, Musgrave and Siedow 1985, Nichols and Laties 1985). As pointed out in the introduction, any effect on growth will affect respiration and vice versa, probably within a minute. Therefore, the direct effects are difficult to identify. Inasmuch as the role of plant hormones has been investigated properly, ethylene and cytokinins have been proposed to play significant roles in the regulation of respiration, although evidence is scarce, if not absent. Cytokinin induces disengagement of the alternative path in several tissues (Musgrave and Siedow 1985), but the concentrations required may not be physiological. Moreover, cytokinin also stimulates ethylene production by pea seedlings (Musgrave and Siedow 1985), suggesting that cytokinins affect many aspects of metabolism. It seems difficult to convince workers on ethylene that this hormone cannot play a direct role in the regulation of respiration. Indeed, ethylene does induce CN-resistance of mitochondrial respiration in pea cotyledons (Duncan and Spencer 1986), and it gives rise to expression of particular genes in carrot roots (Nichols and Laties 1985, Tucker and Laties 1984). Also, its biosynthesis has been reported to be related to the expression of CN-resistant respiration in mitochondria from callus-forming potato tuber discs (Gude and van der Plas 1985) and pea seedlings (Musgrave and Siedow 1985). However, the observed correlations may not necessarily point to a direct relation. Isolated mitochondria can synthesize ethylene from 1-aminocyclopropane-1-carboxylic acid (ACC), but the system does not resemble the physiological mechanism of ethylene formation from ACC as found in intact plants. For example, it lacks the stereospecificity associated with the in vivo conversion of 2-ethyl ACC to 1-butene (Venis 1984), and is similar to other cell-free systems producing ethylene from ACC (Stegink and Siedow 1986b, Venis 1984). Further, in vivo ethylene production rates are at least five orders of magnitude lower than the rate of oxygen uptake by the alternative oxidase (cf. de Visser and Spencer 1986, de Visser et al. 1986). Therefore, a direct link of ethylene biosynthesis with alternative path activity is unlikely. The effects of ethylene (unphysiologically high concentrations in the range of 10 $\mu l.l^{-1}$ and higher) on respiration are not found within ca 10 hours after the start of the treatment (e.g. Day et al. 1978). This is much later than the effects of lower ethylene concentrations (ca 0.1 $\mu l.l^{-1}$) on other processes like root growth (Bucher and Pilet 1982) and de novo synthesis of ethylene-forming enzyme (de Visser and Spencer 1986). This indicates that a direct control of (CN-resistant) respiration by ethylene can be excluded.

ACKNOWLEDGEMENTS

I wish to acknowledge the financial support in the form of a travel grant from the Department of Plant Science, The University of Alberta. I thank Ms. Edna Fleck for the preparation of the camera-ready copy of the manuscript.

REFERENCES

ap Rees T 1985 The organization of glycolysis and the oxidative pentose phosphate pathway in plants. In: Higher Plant Cell Respiration, R. Douce, DA Day, eds. pp. 391-417. Springer-Verlag, Berlin

Azcón-Bieto J, Day DA, and Lambers H 1983a The regulation of respiration in the dark in wheat leaf slices. Plant Sci. Lett. 32:313-320

Azcón-Bieto J, Lambers H, and Day DA 1983b The effect of photosynthesis and carbohydrate status on respiratory rates and the involvement of the alternative path in leaf respiration. Plant Physiol 72:598-603

Bahr JT and Bonner WD Jr 1973 Cyanide-insensitive respiration. I. The steady states of skunk cabbage spadix and bean hypocotyl mitochondria. J Biol Chem 248:3441-3445

Beevers H 1961 Respiratory metabolism in plants. Harper & Row, New York Evanston, London

Beevers H 1985 Foreword In: Higher Plant Cell Respiration, R. Douce, DA Day, eds, pp. V-VI. Springer-Verlag, Berlin

Blacquière T and de Visser R 1984 Capacity of cytochrome and alternative path in coupled and uncoupled root respiration of Pisum and Plantago. Physiol Plant 62:427-432

Bryce JH and ap Rees T 1985 Comparison of the respiratory metabolism of Plantago lanceolata L. and Plantago major L. J Exp Bot 36:1559-1565

Bucher D and Pilet PE 1982 Ethylene effects on growing and gravireacting maize root segments. Physiol Plant 55:1-4

Clarkson DT 1985 Factors affecting mineral nutrient acquisition by plants. Annu Rev Plant Physiol 36:77-115

Day DA, Arron GP and Laties GG 1980 Nature and control of respiratory pathways in plants: The interaction of cyanide-resistant respiration with the cyanide-sensitive pathway. In: The Biochemistry of Plants. A comprehensive Treatise. Vol. 2, Metabolism and Respiration. DD Davies ec., pp. 197-241, Academic Press, New York

Day DA and Lambers H 1983 The regulation of glycolysis and electron transport in roots. Physiol Plant 58:155-160

Day DA, Arron GP, Christoffersen RE and Laties GG 1978 Effect of ethylene and carbon dioxide on potato metabolism. Plant Physiol 62:820-825

Day DA, Neuburger M and Douce R 1985 Interactions between glycine decarboxylase, the tricarboxylic acid cycle and the respiratory chain in pea leaf mitochondria. Aust J Plant Physiol 12:119-130

Deane-Drummond CE 1984 Mechanism of nitrate uptake into Chara corallina cells: lack of evidence for obligatory coupling to proton pump and a new NO_3^-/NO_3^- exchange model. Plant Cell Environ 7:317-323

de Troostembergh JC and Nyns EJ 1978 Kinetics of respiration of cyanide-insensitive mitochondria from the yeast Saccharomycopsis lipolytica. Eur J Biochem 85:423-432

de Visser R 1985 Efficiency of respiration and energy requirements of N assimilation in roots of Pisum sativum. Physiol Plant 65:209-218

de Visser R and Blacquière T 1984 Inhibition and stimuation of root respiration in Pisum and Plantago by hydroxamate. Its consequences for the assessment of alternative path activity. Plant Physiol 75:813-817

de Visser R and Lambers L 1983 Growth and the efficiency of root respiration of Pisum sativum as dependent on the source of nitrogen. Physiol Plant 58:533-543

de Visser R, Spencer MS 1986 Regulation of ethylene-forming enzyme activity in pea roots. Plant Physiol (Suppl) 80:112

de Visser R, Spreen Brouwer K, and Posthumus F 1986 Alternative path mediated ATP synthesis in roots of Pisum sativum upon nitrogen supply. Plant Physiol 80:295-300

Dry IB, Wiskich JT 1985 Characteristics of glycine and malate oxidation by pea leaf mitochondria: evidence of differential access to NAD and respiratory chains. Aust J Plant Physiol 12:329-339

Duncan I, Spencer MS 1987 Changes in respiration of mitochondria isolated from cotyledon of ethylene-treated pea seedlings. Planta (in press)

Gude H and van der Plas LHW 1985 Endogenous ethylene formation and the development of the alternative pathway in potato tuber discs. Physiol Plant 65:57-62

Hemrika-Wagner AM, Kreuk KCM and van der Plas LHW 1982 Influence of growth temperature on respiratory characteristics of mitochondria from callus-forming potato tuber discus. Plant Physiol 70:602-605

Hemrika-Wagner AM, Gude H, Marissen N, van der Plas LHW, and Verleur JD 1986 Comnpartmentation of alternative oxidase in plant mitochondria. Plant Cell Physiol 27:499-503

Huber SC 1986 Fructose 2,6-biphosphate as a regulatory metabolite in plants. Annu Rev Plant Physiol 37:233-246

Lambers H 1980 The physiological significance of cyanide-resistant respiration. Plant Cell Environ 3:292-302

Lambers H 1982 Cyanide-resistant respiration: A non-phosphorylating electron transport pathway acting as an energy overflow. Physiol Plant 55:478-485

Lambers H 1985 Respiration in intact plants and tissues: Its regulation and dependence on environmental factors, metabolism and invaded organisms. In: Higher Plant Cell Respiration, R Douce, DA Day, eds., pp. 418-473. Springer-Verlag, Berlin

Lambers H 1986 Growth, respiration, exudation and symbiotic associations: the fate of carbon translocated to the roots. In: Root development and function. Effects of the physical environment. JV Lake, R Rose, P Gregory, eds., Cambridge Univ Press
Lambers H and Steingröver E 1978 Growth respiration of a flood-tolerant and a flood-intolerant Senecio species as affected by low oxygen tension. Physiol Plant 42:179-184
Lambers H, Blacquière T and Stuiver B (CEE) 1981a Interactions between osmoregulation and the alternative respiratory pathway in Plantago coronopus as affected by salinity. Physiol Plant 51:63-68
Lambers H, Day DA and Azcón-Bieto J 1983a Cyanide-resistant respiration in roots and leaves. Measurements with intact tissues and isolated mitochondria. Physiol Plant 58:148-154
Lambers H, Posthumus F, Stulen I, Lanting L, van de Dijk SJ, and Hofstra R 1981b Energy metabolism of Plantago major as dependent on the nutrient supply. Physiol Plant 51:245-252
Lambers H, Szaniawski RK and de Visser R 1983b Respiration for growth, maintenance and ion uptake. An evaluation of concepts, methods, values and their significance. Physiol Plant 58:556-563
Lance C and Rustin P 1984 The central role of malate in plant metabolism. Physiol Vég 22:625-641
Lance C, Chauveau M, and Dizengremel P 1985 The cyanide-resistant pathway of plant mitochondria. In: Higher Plant Cell Respiration, R Douce, DA Day, eds, pp 202-247. Springer-Verlag, Berlin
Laties GG 1982 The cyanide-resistant, alternative path in higher plant respiration. Annu Rev Plant Physiol 33:519-555
Lin W 1985 Energetics of membrane transport in protoplasts. Physiol Plant 65:102-108
Minchin FR, Summerfield RJ, Hadley P, Roberts EH, and Rawsthorne S 1981 Carbon and nitrogen nutrition of nodulated roots of grain legumes. Plant Cell Environ 4:5-26
Moore AL and Rich PR 1985 Organization of the respiratory chain and oxidative phosphorylation. In: Higher Plant Cell Respiration, R Douce, DA Day, eds. pp. 134-172. Springer-Verlag, Berlin
Møller IM and Bérczi A 1985 Oxygen consumption by purified plasmalemma vesicles from wheat roots. Stimulation by NADH and salicylhydroxamic acid (SHAM). Febs Lett 193:180-184
Musgrave ME and Siedow JN 1985 A relationship between plant responses to cytokinins and cyanide-resistant respiration. Physiol Plant 64:161-166
Musgrave ME, Murfet IC and Siedow JN 1986 Inheritance of cyanide-resistant respiration in two cultivars of pea (Pisum sativum L.). Plant Cell Environ 9:153-156
Nichols SE and Laties GG 1985 Differential control of ethylene-induced gene expression and respiration in carrot roots. Plant Physiol 77:753-757
Palmer JM 1976 The organization and regulation of electron transport in plant mitochondria. Annu Rev Plant Physiol 27:133-157
Palmer JM and Ward JA 1985 The oxidation of NADH by plant mitochondria. In: Higher Plant Cell Respiration, R Douce, DA Day, eds, pp. 173-201. Springer-Verlag, Berlin
Penning de Vries FWT, Brunsting AHM and van Laar HH 1974 Products, requirements and efficiency of biosynthetic processes: A quantitative approach. J Theor Biol 45:339-377
Perl M 1986 ATP synthesis and utilization in the early stage of seed germination in relation to seed dormancy and quality. Physiol Plant 66:177-182
Rhodes MJC 1980 Respiration and senescence of plant organs. In: The Biochemistry of Plants, Vol. 2, PK Stumpf, EE Conn, eds. pp. 419-462. Academic Press, London New York
Roberts JKM 1984 Study of plant metabolism in vivo using NMR spectroscopy. Annu Rev Plant Physiol 35:375-386
Roberts JKM, Wemmer D and Jardetzky O 1984 Measurement of mitochondrial ATPase

activity in maize root tips by saturation transfer ^{31}P nuclear magnetic resonance. Plant Physiol 74:632-639

Rustin P, Dupont J and Lance C 1984 On the involvement of lipid peroxy radicals in the cyanide-resistant pathway. Physiol Vég 22:643-663

Ryle GJA 1984 Respiration and plant growth. In: The Physiology and Biochemistry of Plant Respiration. JM Palmer, ed. Cambridge Univ Press, Cambridge

Saglio PH and Pradet A 1980 Soluble sugars, respiration, and energy charge during aging of excised maize root tips. Plant Physiol 66:516-519

Saglio PH, Rancillac M, Bruzan F and Pradet A 1984 Critical oxygen pressure for growth and respiration of excised and intact roots. Plant Physiol 76:151-154

Schubert KR and Ryle GJA 1984 The energy requirements for nitrogen fixation in nodulated legumes. In: Advances in legume science. RJ Summerfield, AM Bunting, eds., pp. 85-96, Her Majesty's Stat Off, London

Siedow JN and Berthold DA 1986 The alternative oxidase: A cyanide-resistant respiratory pathway in higher plants. Physiol Plant 66:569-573

Smakman G and Hofstra R 1982 Energy metabolism of Plantago lanceolata as affected by change in root temperature. Physiol Plant 56:33-37

Stegink SJ and Siedow JN 1986a Binding of butyl gallate to plant mitochondria. Plant Physiol 80:196-201

Stegink SJ and Siedow JN 1986b Ethylene production from 1-aminocyclopropane-1-carboxylic acid in vitro: A mechanism for explaining ethylene production by a cell-free preparation from pea epicotyls. Physiol Plant 66:625-631

Stitt M and Steup M 1985 Starch and sucrose degradation. In: Higher Plant Cell Respiration, R Douce, DA Day, eds., pp. 347-390. Springer-Verlag, Berlin

Stitt M, Cseke C and Buchanan B 1986 Ethylene-induced increase in fructose-2,6-biphosphate in plant storage tissues. Plant Physiol 80:246-248

Tucker ML and Laties GG 1984 Comparative effects of ethylene and cyanide on respiration, polysome prevalence, and gene expression in carrot roots. Plant Physiol 75:342-348

Turner JF and Turner DH 1980 The regulation of glycolysis and the pentose phosphate pathway. In: The Biochemistry of Plants, Vol. 2. PK Stumpf, EE Conn, eds. pp. 279-316. Academic Press, London New York

van der Plas LHW and Wagner MJ 1984 Influence of osmotic stress on the respiration of potato tuber callus. Physiol Plant 62:398-403

Veen BW 1980 Energy cost of ion transport. In: Genetic engineering of osmoregulation. Impact on plant productivity for food, chemicals and energy. DW Rains, RC Valentine, A Hollaender, eds. pp. 187-195 Plenum Press, New York London

Veen BW 1981 Relation between root respiration and root activity. In: Structure and Function of Plant Roots, R Brouwer, D Gasparikova, J Kolek, BC Loughman, eds., pp 277-280 Martinus Nyhoff/Dr W Junk, The Hague

Venis MA 1984 Cell-free ethylene-forming systems lack stereochemical fidelity. Planta 162:85-88

Wiskich JT and Day DA 1982 Malate oxidation, rotenone-resistance, and alternative path activity in plant mitochondria. Plant Physiol 70:959-964

Wiskich JT and Dry IB 1985 The tricarboxylic acid cycle in plant mitochondria: Its operation and regulation. In: Higher Plant Cell Respiration, R Douce, DA Day, eds, pp 281-313. Springer-Verlag, Berlin

FRUCTOSE-2,6-BISPHOSPHATE AND PLANT RESPIRATION

Tom ap Rees and Jane E. Dancer

Botany School
University of Cambridge
Downing Street
Cambridge, U.K.

One of the most significant recent advances in our understanding of metabolism is the discovery of the importance of the signal compound fructose-2,6-bisphosphate (Fru-2,6-P_2) in the regulation of carbohydrate metabolism. In mammals Fru-2,6-P_2, through its activation of 6-phosphofructokinase [PFK (ATP)] and inhibition of fructose-1,6-bisphosphatase, makes a major contribution to the control of flux between fructose 6-phosphate and fructose-1,6-bisphosphate.[1] In this essay we consider whether Fru-2,6-P_2 regulates higher plant respiration.

We can not rely on comparative biochemistry to reveal the mechanisms that control plant respiration. First, the pathways in plants are differently organized from those in animals. For example, plant respiration begins with sucrose, not hexose; a much higher proportion of the carbon that enters the respiratory pathways is used to provide carbon skeletons for biosynthesis in plants than in animals; in plants the pathways of carbohydrate oxidation are compartmented, in that many of the steps occur in both the cytosol and the plastids.[2] Second, mechanisms of control are likely to differ markedly between animals and plants; plants neither fight nor flee, but have to acquire, retain and persist immobile through long periods of adversity. A good example of differences in control is provided by PFK (ATP); in mammals and plants largely controlled by the ratios ATP:AMP, and phosphoenolpyruvate: P_i, respectively.

The first requirement for any hypothesis that Fru-2,6-P_2 regulates plant respiration is the presence of this compound in plant tissues. Although we still lack the type of proof provided by isolation and rigorous chemical characterization that is available for animal tissues,[3] there is ample evidence that plants contain Fru-2,6-P_2.[2] For example, synthesis and hydrolysis of Fru-2,6-P_2 has been demonstrated with plant preparations;[4] measurements of Fru-2,6-P_2 in leaves show it to play a major role in regulating sucrose synthesis;[5] two plant enzymes, cytosolic fructose-1,6-bisphosphatase and pyrophosphate: fructose 6-phosphate 1-phosphotransferase [PFK(PP_i)] have been shown to respond markedly to Fru-2,6-P_2.[6] The amounts of Fru-2,6-P_2 found in plant tissues range from 0.56 to 0.03 nmol/g fresh wt. in, respectively, the club of the spadix of Arum maculatum[7] and potato tuber.[8] All the available evidence strongly suggests that Fru-2,6-P_2 is confined to the cytosol in plants.[2] The only evidence to the contrary, that plastid PFK(ATP) is sensitive to Fru-2,6-P_2, has not been confirmed.[9] If the cytosol is 10% of the tissue volume then the concentration of Fru-2,6-P_2 is likely to be in the range 0.5 to 0.03 μmolar.

Fru-2,6-P_2 could contribute to the control of plant respiration either by directly affecting regulatory steps in respiration, or by affecting the activity of steps that catalyse a flux in the opposing, gluconeogenic, direction. On present evidence the only candidates for control of plant respiration via a substrate cycle responding to Fru-2,6-P_2 are cytosolic fructose-1,6-bisphosphatase and PFK(PP_i). The role of the bisphosphatase is discussed at this point. Appreciable activities of cytosolic fructose-1,6-bisphosphatase have only been demonstrated for photosynthetic and gluconeogenic tissues. The enzyme from both types of tissue is inhibited by Fru-2,6-P_2. Thus changes in Fru-2,6-P_2 could alter the rate at which fructose-1,6-bisphosphate was converted to fructose 6-phosphate and thus regulate how much of the product of PFK(ATP) was available for glycolysis. We know that Fru-2,6-P_2, acting on cytosolic fructose-1,6-bisphosphatase, plays a crucial role in regulating the conversion of triose phosphate to sucrose in leaves.[5] We know much less about the control of cytosolic glycolysis in leaves. Net flux from fructose 6-phosphate to fructose-1,6-bisphosphate will not occur during the conversion of triose phosphate to sucrose, but probably will in the dark. The extent to which this is due to regulation of PFK(ATP) or to variation in the activity of fructose-1,6-bisphosphatase is not known, but it is difficult to believe that the latter does not make a significant contribution to control.

The situation in gluconeogenic tissues is equally complex. There is little net glycolysis of hexose phosphate during gluconeogenesis. This is likely to be due in part to the low maximum catalytic activity of PFK(ATP)[10], but again control of cytosolic fructose-1,6-bisphosphatase may well contribute to preventing glycolysis.[4]

Perhaps the most obvious way to discover whether Fru-2,6-P_2 regulates plant respiration directly is to determine its effect on cytosolic PFK(ATP). Results of such studies[9,11,12] have proved negative and there is no evidence of effects comparable to those reported for mammalian PFK(ATP). This strongly suggests that Fru-2,6-P_2 does not regulate plant glycolysis via PFK(ATP). We should accept this view cautiously because it is clear that there have been very few definitive studies of plant cytosolic PFK(ATP).[13]

A novel effect of Fru-2,6-P_2 on plant PFK(ATP) was suggested by Balogh et al.,[14] who proposed that the enzyme existed in two kinetically distinct, linear and non-linear, forms and that Fru-2,6-P_2 caused a change from one to another. They suggested that in the presence of Fru-2,6-P_2, cytosolic PFK(ATP) from spinach was converted to a hyperactive form with non-linear kinetics. This was part of a wider proposal that held that plant cytosolic PFK(ATP) and PFK(PP_i) were interconvertible by a mechanism christened metabolite-mediated catalyst conversion. Fru-2,6-P_2 plus ATP or fructose 6-phosphate were held to promote conversion to PFK(ATP), and UDPglucose and PP_i conversion to PFK(PP_i). Such a mechanism could provide a potent means whereby Fru-2,6-P_2 could regulate plant respiration.

Subsequent studies failed to confirm both the above effect of Fru-2,6-P_2 on PFK(ATP), and the evidence for metabolite-mediated interconversion. The initial rise in PFK(ATP) activity in the presence of Fru-2,6-P_2 to give the hyperactive form may be attributed to the presence of PFK(PP_i) in the preparation of PFK(ATP) and of PP_i in the ATP used to assay the PFK(ATP).[9] This torpedos the major evidence for the conversion of PFK(PP_i) to PFK(ATP). The evidence for the reverse change may also be explained by contamination, this time of the coupling enzymes used in the assay of PFK(PP_i) by UDP-glucose pyrophosphorylase. This means that PP_i plus UDPglucose would have generated UTP, which would serve as a substrate for PFK(ATP).[15] Subsequent studies with purified enzymes have provided sound evidence that PFK(ATP) and $PFK(PP_i)$ are quite distinct enzymes[16]. We conclude that there is no reliable evidence that Fru-2,6-P_2 regulates PFK(ATP) from plants.

Our conclusion appears to dispose of the possibility that Fru-2,6-P_2 regulates plant glycolysis directly via one of its component enzymes. The discovery of the widespread distribution of $PFK(PP_i)$ means that this is not necessarily so. This enzyme catalyses:

$$\text{Fru-6-P} + PP_i \rightleftharpoons \text{Fru-1.6-}P_2 + P_i \qquad \Delta G = -\ 2.93\ \text{kJ mol}^{-1}$$

This enzyme is activated by Fru-2,6-P_2 and could catalyse the entry of fructose 6-phosphate into glycolysis and thus provide a means whereby Fru-2,6-P_2 might regulate plant respiration. The view that $PFK(PP_i)$ mediates plant glycolysis, has appreciable support.[17-19]

The properties of plant $PFK(PP_i)$ have been reviewed recently[2,13]; the following are the most pertinent to the present argument. The enzyme is confined to the cytosol and catalyses a readily reversible reaction in vitro. It is specific for PP_i and the common nucleotides do not act as substrates. Activity is largely dependent upon Fru-2,6-P_2 and estimates of Ka are well below the likely concentration of Fru-2,6-P_2 in vivo. Hyperbolic kinetics are found with all four substrates in the presence and absence of Fru-2,6-P_2. The degree of aggregation of the enzyme is affected by Fru-2,6-P_2 and PP_i so that activity in the glycolytic direction is favoured by Fru-2,6-P_2 and in the reverse direction by PP_i.

The main evidence for a glycolytic role for $PFK(PP_i)$ is its widespread distribution at high activity. Carnal and Black[20] found it in leaves of over 100 species of angiosperms that included C3, C4 and CAM plants. The enzyme has also been demonstrated in a range of non-photosynthetic tissues of plants , including some that are gluconeogenic.[6] In general the activities of $PFK(PP_i)$ found in extracts of plant tissues are high relative to the expected rates of respiration. It is likely that there is sufficient $PFK(PP_i)$ in most plant tissues to make a very substantial contribution to the carbohydrate oxidation observed in vivo.

Initially a major bar to accepting a glycolytic role for $PFK(PP_i)$ was the assumption that the concentration of PP_i in plant cells was kept very low by the action of inorganic pyrophosphatase (PPase). Direct proof of the latter assumption was lacking. The specificity of $PFK(PP_i)$ for PP_i prompted the measurement of PP_i in plants and revealed the presence of appreciable amounts in a range of tissues (Table 1). The following arguments suggest that this PP_i may be confined to the cytosol. Acid PPase appears to be located in the vacuole[21], and alkaline PPase in the plastid[22]. Two major enzymes known to be specific for PP_i, UDPglucose pyrophosphorylase[23] and $PFK(PP_i)$[6], are confined to the cytosol. If PP_i is confined to the cytosol, and the latter is 10% of the tissue volume then PP_i is present at concentrations of 90-250 μM. These values exceed the estimates of Km, 10-30 μM, of $PFK(PP_i)$ for PP_i.[6]

A further requirement of the hypothesis that $PFK(PP_i)$ is glycolytic is evidence that such a role is permitted by the equilibrium position of the reaction in the cell. As we do not know the amounts of the relevant reactants in the cytosol we can not determine the mass action ratio for $PFK(PP_i)$ directly. If we make the, admittedly arguable, assumptions that the cytosol is 10% of the tissue, and that most of the PP_i, fructose 6-phosphate and fructose-1,6-bisphosphate is in the cytosol, then from measurements of the tissue content of the latter three compounds we may calculate the concentration of P_i that would be present in the cytosol if $PFK(PP_i)$ catalysed an

Table 1. Estimates of the amount of PP_i in plant tissues

Tissue	PP_i (nmol/g fresh wt.)
Developing embryos of pea[38]	9
Root apices of pea[38]	9
Club of spadix of Arum maculatum[7]	21 - 36
Shoot of pea[39]	16
Maize coleoptile[36]	20
Suspension cultures of sycamore[34]	3

equilibrium reaction in vivo. For the club of the spadix of Arum maculatum[7] and the developing embryo of pea[24] we obtain values of 1.4 and 1.7 mM, respectively. These values are close to the recent estimate of 6mM for non-photosynthetic cells of plants made by NMR.[24] This agreement strongly suggests that in Arum club and pea embryos, at least, the reaction catalysed by PFK(PP_i) is close to equilibrium in vivo. This allows the enzyme to mediate entry into glycolysis, but makes it very difficult to see how Fru-2,6-P_2 could exert any major control of flux as it would have to act via an equilibrium, as distinct from a non-equilibrium step. There is an urgent need for direct and definitive measurements of the concentrations of Fru-2,6-P_2, and the reactants of PFK(PP_i), in plant cytosol.

Even collectively, the above evidence does not amount to proof that PFK(PP_i) catalyses entry of fructose 6-phosphate into glycolysis. The presence of adequate concentrations of substrates and amounts of enzyme is consistent with a glycolytic role but does not prove it, because the data are equally consistent with the view that the enzyme functions in the opposite direction and catalyses the formation of PP_i at the expense of fructose-1,6-bisphosphate. Indeed the evidence that PFK(PP_i) is close to equilibrium in vivo means that the reaction is capable of working in either direction, depending upon the relative concentration of substrates and products. A number of approaches to the question of what does happen in vivo are considered.

The first of the above approaches is comparison of the maximum catalytic activity of PFK(PP_i) with the rate of glycolysis. If the enzyme is responsible for glycolysis then a correlation would be expected. Such correlations have been found, in plant tissues, between glycolysis and the oxidative pentose phosphate pathway and, respectively, the maximum catalytic activities of PFK(ATP) and glucose-6-phosphate dehydrogenase.[25] During the differentiation of the root of pea seedlings the maximum catalytic activities of both PFK(ATP) and PFK(PP_i) vary appreciably (Table 2). The variation of the former, but not the latter, correlates with the rate of CO_2 production. This suggests a glycolytic role for PFK(ATP) but not for PFK(PP_i). This argument is not decisive because the data allow PFK(PP_i) to be specifically associated with the glycolytic flux responsible for provision of intermediates for biosynthesis. Activity is highest in the apex and stele where biosynthesis is likely to be most marked. If this view is correct then a correlation between PFK(PP_i) and CO_2 production might not be expected.

More definitive data are available for the developing club of the spadix of Arum maculatum (Table 3). This tissue passes through a number of arbitrarily designated stages of development that culminate in a rapid and massive rise in respiration that supports a short period of thermogenesis. Development from α stage to pre-thermogenesis takes about 12 days and the onset of thermogenesis takes 2 to 3 hours. During the development of the club there are appreciable increases in the maximum catalytic activities

Table 2. Relationship between maximum catalytic activities of PFK(PP_i) and PFK(ATP), and rate of respiration in different regions of the root of pea[7]

Region of root		Rate (nmol/min per mg of protein) PFK(PP_i)	PFK(ATP)	CO_2 production
Apical 6 mm		297	95	54
6–26 mm from apex:	stele	259	46	30
	cortex	136	52	36
26–46 mm from apex:	stele	200	50	40
	cortex	66	46	36

Table 3. Maximum catalytic activities of enzymes of carbohydrate metabolism during the development of the club of the spadix of Arum maculatum[40]

Enzyme	Activity (μmol/min per g fresh wt.) α Stage	β Stage	Pre-thermogenesis	Thermogenesis
Hexokinase	2.6	5.0	5.6	16.4
PFK(ATP)	1.2	3.8	18.2	30.1
Aldolase	3.7	6.1	13.4	14.6
Glyceraldehyde 3-P dehydrogenase	5.1	12.0	75.2	-
Citrate Synthase	18.6	-	60.8	92.4
PFK(PP_i)	5.6	5.2	4.0	4.3

of respiratory enzymes known to be involved in the rapid thermogenic respiration (Table 3). These include PFK(ATP) and other glycolytic enzymes. No significant change in PFK(PP_i) was detected. In the thermogenic clubs the rate of glycolysis may be estimated from the rate of starch breakdown; rates near 10 μmol hexose/min per g fresh wt. are probable.[7] Thus in the clubs of Arum, not only does PFK(PP_i) not correlate with glycolysis and the other glycolytic enzymes but the activity is insufficient to make any substantial contribution to glycolysis. We suggest that it is most unlikely that PFK (PP_i) is a glycolytic enzyme in the thermogenic clubs of Arum. However it will not do to generalize about other plant tissues on the basis of a study of one rather specialized tissue.

Definitive data on the contribution of PFK(PP_i) to glycolysis have also been obtained for the bundle sheath cells of maize leaves.[26] The enzyme has been shown to be located predominantly, if not exclusively, in the mesophyll cells. On the reasonable assumption that glycolysis is universal, these results show that PFK(PP_i) makes little or no contribution to the glycolysis of the bundle sheath.

Another clearly established instance of a lack of correlation between the maximum catalytic activity of PFK(PP_i) and glycolysis is provided by the gluconeogenic endosperm of germinating castor beans. Enzyme activity is high[10] yet glycolytic activity is minimal[27]. Although it is unlikely that PFK(PP_i) is responsible for a net glycolytic flux in castor bean endosperm, it may still be acting in a glycolytic direction in vivo. Kruger and Beevers[4] have suggested that in such tissue gluconeogenic flux from fructose-1,6-bisphosphate to fructose-6-phosphate may be regulated by a substrate cycle involving cytosolic fructose-1,6-bisphosphatase and PFK(PP_i) with the latter operating in the direction of fructose-1,6-bisphosphate formation. Fru-2,6-P_2 has opposite effects on the two enzymes so the effect of changes in the signal compound might be amplified by such a cycle. In the present context, two important points would arise if this view is correct. First, in the strictest sense the role proposed for PFK(PP_i) is in the regulation of gluconeogenesis, not in mediating glycolysis. Second, the precise role of PFK(PP_i) would have to differ from that in the many tissues where there is no net flux from fructose-1,6-bisphosphate to fructose 6-phosphate. This is because such tissues have been shown to contain appreciable activities of PFK(PP_i).[7] We can not exclude a role for PFK(PP_i) in a substrate cycle in non-gluconeogenic tissues, but if there is one, the enzyme would have to work in the opposite direction to that proposed for castor bean endosperm. This is because significant activities of cytosolic fructose-1,6-bisphosphatase have not been demonstrated in such tissues. On present evidence the only substrate cycle that might involve PFK(PP_i) in non-gluconeogenic tissues would be one in which the enzyme catalysed the conversion of fructose-1,6-bisphosphate to fructose 6-phosphate in opposition to PFK(ATP) in the regulation of glycolysis. At present this seems unlikely as the maximum catalytic activity of PFK(PP_i) in non-gluconeogenic tissues is often higher than that of PFK(ATP). Further, we have no clear proof that the two enzymes respond in opposing directions to any likely regulator.

A second approach to determining the role of PFK(PP_i) in plant cells is to alter the rate of hexose phosphate metabolism and observe whether there are changes in Fru-2,6-P_2 and the substrates and products of PFK(PP_i). It is difficult to interpret the results of such experiments. First, a change in Fru-2,6-P_2 in the same direction as that of the rate of glycolysis does not demonstrate that PFK(PP_i) is acting glycolytically. We know too little about the role of Fru-2,6-P_2 to exclude all other possibilities: as will be shown later Fru-2,6-P_2 could be involved in sucrose breakdown via a mechanism that involves PFK(PP_i) working in the direction of fructose 6-phosphate formation. If this is so then an increase in Fru-2,6-P_2 might accompany increased glycolysis. The second difficulty in interpreting changes in substrates is that it is not clear what changes we should expect if PFK(PP_i) catalyses entry into glycolysis. At first sight falls in PP_i and fructose 6-phosphate might be expected when glycolysis is increased. This assumes that PFK(PP_i) catalyses a non-equilibrium reaction, but the available evidence indicates otherwise. If the reaction is close to equilibrium fructose 6-phosphate and PP_i might be expected to change in the same direction as the glycolytic flux. Finally, changes in fructose 6-phosphate are not diagnostic as this is a substrate for PFK(ATP) and changes in the opposite direction to glycolytic flux would be expected if PFK(ATP) contributes to the regulation of that flux.

Authenticated measurements of Fru-2,6-P_2 and the reactants of PFK(PP_i) when glycolysis varies are rather sparse. In the club of Arum spadix no changes in Fru-2,6-P_2 or PP_i were found when the large increase in glycolysis occurred at thermogenesis.[7] This is further evidence that PFK(PP_i) does not contribute greatly to glycolysis in this tissue.

When thin disks of plant storage tissue are incubated at room temperature there is a marked increase in the rate of respiration and a rapid mobilization of storage carbohydrate. A considerable rise in Fru-2,6-P_2 at the beginning of this increased metabolism has been reported for slices of Jerusalem artichoke.[28] No measurements of respiration were presented but comparison with earlier work[29] does not suggest a close correlation between the increase in Fru-2,6-P_2 and that in respiration. However, a closer correlation is reported as occurring in slices of carrot and potato.[8] In Jerusalem artichoke no fall in hexose 6-phosphate was detected during the rise in Fru-2,6-P_2. When carrot roots and potato tubers were treated with ethylene the content of Fru-2,6-P_2 doubled and there was a modest rise in glucose 6-phosphate.[8] Collectively the above data strongly suggest a role for Fru-2,6-P_2 in the acceleration of the metabolism of storage tissue, but the data do not establish what this role is. In particular, we do not know whether Fru-2,6-P acts on the mobilization of reserve carbohydrates or on glycolysis or both.

Perhaps the best evidence for a glycolytic role for PFK(PP_i) is the observation that illumination of guard cell protoplasts is accompanied by a rise in Fru-2,6-P_2 and a fall in glucose 6-phosphate.[30] Even so, these data do not establish the site of action of Fru-2,6-P_2. The fall in glucose 6-phosphate is equally well explained by proposing PFK(ATP) or PFK(PP_i) as entry points into glycolysis.

Another means of investigating the role of PFK(PP_i) is to measure the degree of cycling between fructose-1,6-bisphosphate and fructose 6-phosphate in the cytosol. Absence of cycling would provide definitive evidence against PFK(PP_i) acting in the direction of fructose 6-phosphate formation. Methods are available for assessing the extent of cycling in mammalian cells[31], but are not readily applicable to plant cells. This is because plants have a well developed oxidative pentose phosphate pathway, the non-oxidative branch of which could lead to movement of label from fructose-1,6-bisphosphate to fructose 6-phosphate and vice versa.

In the club of Arum maculatum at thermogenesis and in maize bundle sheath cells we can be fairly sure that PFK(PP_i) is not a quantitatively significant means of entry of fructose 6-phosphate into glycolysis. For the other tissues studied there is no compelling evidence that PFK(PP_i) is a means of entry into glycolysis, but the available data do not exclude such a role. Thus we suggest that the hypothesis that PFK(PP_i) is a point of entry into glycolysis, whilst remaining a possibility, is not established. This view is strengthened by the fact that alternative proposals for PFK(PP_i) have been put forward and are compatible with available data.

Recently, Dennis and Greyson[13] have shown that the information that we have about PFK(PP_i) is compatible with the view that the enzyme may serve to equilibrate the pools of hexose phosphates and pentose phosphates in the cytosol, particularly during the extensive biosynthesis characteristic of many plant cells. This view is supported by the observation that PFK(PP_i) activity is high in tissues where biosynthesis is most marked.

It has also been suggested that a major role for PFK(PP_i) is the synthesis of PP_i at the expense of glycolytic fructose-1,6-bisphosphate.[32-34] In this hypothesis the formation of one molecule of PP_i would entail the conversion of one molecule of fructose 6-phosphate to fructose-1,6-bisphosphate by PFK(ATP), the use of the fructose-1,6-bisphosphate to make PP_i by PFK(PP_i), followed by the re-entry of the resulting fructose 6-phosphate into glycolysis via PFK(ATP) for conversion to pyruvate. The view that PFK(PP_i) may be a source of PP_i presupposes a significant need for PP_i in the cytosol. Wood et al.[35] have presented the general arguments for the view that PP_i may act as an energy source in metabolism. Although our knowledge of PP_i metabolism in plants is still in its infancy, two possible uses of PP_i as an energy source in plant metabolism are already apparent.

It is now quite clear that plants contain a mechanism for PP_i driven proton transport.[36] Recently, Rea and Poole[37] have reported the separation of a H^+-translocating PPase from the H^+-translocating ATPase in preparations of tonoplast vesicles from Beta vulgaris. Thus plants may use PP_i to generate proton gradients.

The second possible use of PP_i in plant cells is in the breakdown of of sucrose via sucrose synthase.[32-34] The initial metabolism of sucrose delivered to non-photosynthesising cells of plants is mediated by two cytosolic enzymes, alkaline invertase and sucrose synthase. In some tissues, such as developing tubers of potato and embryos of pea, the relative activities of alkaline invertase and sucrose synthase are such that it seems clear that significant amounts of sucrose are converted to hexose 6-phosphate via sucrose synthase. This involves the metabolism of the glucosyl group of UDPglucose, the product of sucrose synthase, to hexose phosphate. Studies with potatoes and peas, and with suspension cultures of sycamore cells are compatible with the view that this is achieved via UDP-glucose pyrophosphorylase working in the direction of glucose 1-phosphate at the expense of PP_i formed by PFK(PP_i). The proposal may be represented as:

$$\text{Sucrose} + \text{UDP} \overset{a}{\rightleftharpoons} \text{UDPglucose} + \text{Fructose}$$
$$\text{Fructose} + \text{UTP} \overset{b}{\longrightarrow} \text{Fru-6-P} + \text{UDP}$$
$$\text{UDPglucose} + PP_i \overset{c}{\rightleftharpoons} \text{Glc-1-P} + \text{UTP}$$
$$\text{Fru-1,6-}P_2 + P_i \overset{d}{\rightleftharpoons} \text{Fru-6-P} + PP_i$$

The enzymes are: a, sucrose synthase; b, fructokinase; c, UDPglucose pyrophosphorylase; d, PFK(PP_i). The need for sucrose breakdown correlates with the widespread distribution of PFK(PP_i). If the above proposal is correct then it can be seen that a rise in Fru-2,6-P_2 when glycolysis is increased could represent an increase in the activity of PFK(PP_i) in the direction of PP_i production in order to produce enough of the latter to permit sufficient sucrose breakdown to provide substrate for glycolysis.

We conclude that, at present, a direct role for Fru-2,6-P_2 in regulating plant respiration has not been established. The changes in Fru-2,6-P_2 seen in some tissues when the rate of respiration is altered suggest some role for this regulator in respiratory metabolism, but there is insufficient information to establish whether this is directly related to the pathways of carbohydrate oxidation or to the provision of substrate for these pathways. The key to further progress in this field seems to be the discovery of the role of PFK(PP_i) in plants.

REFERENCES

1. H.-G. Hers and L. Hue, Gluconeogenesis and related aspects of glycolysis, Annu. Rev. Biochem. 52:617 (1983).
2. T. ap Rees, The organisation of glycolysis and the oxidative pentose phosphate pathway in plants, in "Encyclopedia of Plant Physiology", Vol.18, R. Douce and D.A. Day, eds., Springer-Verlag, Berlin (1985).
3. H.-G. Hers and E. Van Schaftingen, Fructose 2,6-bisphosphate 2 years after its discovery, Biochem. J. 206:1 (1982).
4. N.J. Kruger and H. Beevers, Synthesis and degradation of fructose 2,6-bisphosphate in endosperm of castor bean seedlings, Plant Physiol. 77:358 (1985).
5. M. Stitt and H.W. Heldt, Control of photosynthetic sucrose synthesis by fructose 2,6-bisphosphate, Plant Physiol. 79:599 (1985).
6. E. Kombrink, N.J. Kruger and H. Beevers, Kinetic properties of pyrophosphate: fructose 6-phosphate phosphotransferase from germinating castor bean endosperm, Plant Physiol. 74:395(1984).
7. T. ap Rees, J.H. Green, and P.M. Wilson, Pyrophosphate: fructose 6-phosphate 1-phosphotransferase and glycolysis in non-photosynthetic tissues of higher plants, Biochem. J. 227:299 (1985).
8. M. Stitt, C. Cséke, and B.B. Buchanan, Ethylene-induced increase in fructose 2,6-bisphosphate in plant storage tissues, Plant Physiol. 80:246 (1986).
9. N.J. Kruger, and D.T. Dennis, Reassessment of an apparent hyperactive form of phosphofructokinase from plants, Plant Physiol. 78:645 (1985).

10. N.J. Kruger, E. Kombrink and H. Beevers, Pyrophosphate: fructose 6-phosphate phosphotransferase in germinating castor bean seedlings, FEBS Lett. 153:409 (1983).
11. D.C. Sabularse and R.L. Anderson, D-Fructose 2,6-bisphosphate: a naturally occurring activator for inorganic pyrophosphate: D-fructose 6-phosphate 1-phosphotransferase in plants, Biochem. Biophys. Res. Commun. 103:848 (1981).
12. E. Van Schaftingen, B. Lederer, R. Bartrons, and H.-G. Hers, A kinetic study of pyrophosphate: fructose 6-phosphate phosphotransferase from potato tubers, Eur. J. Biochem. 129:191
13. D.T. Dennis and M.F. Greyson, Fructose 6-phosphate metabolism in plants, Physiol. Plant. (in press).
14. A. Balogh, J.H. Wong, C. Wötzel, J. Soll, C. Cséke, and B.B. Buchanan, Metabolite-mediated catalyst conversion of PFK and PFP: a mechanism of enzyme regulation in green plants, FEBS Lett. 169:287 (1984).
15. N.J. Kruger and D.T. Dennis, A source of apparent pyrophosphate: fructose 6-phosphate phosphotransferase activity in rabbit muscle phosphofructokinase, Biochem. Biophys. Res. Commun. 126:320 (1985).
16. D.T. Dennis, W.E. Hekman, A. Thomson, R.J. Ireland, F.C. Botha and N.J. Kruger, in "Regulation of Carbon Partitioning in Photosynthetic Tissue", R.L. Heath and J. Preiss, eds., Waverly Press, Baltimore (1985).
17. J. Preiss, Starch, sucrose biosynthesis and partitioning of carbon in plants are regulated by orthophosphate and triose-phosphates, Trends Biochem. Sci. 9:24 (1984).
18. D.A. Smyth and C.C. Black, The discovery of a new pathway of glycolysis in plants, What's New in Plant Physiology 15:13 (1984).
19. C. Cséke, A. Balogh, J.H. Wong and B.B. Buchanan, Pyrophosphate fructose 6-phosphate phosphotransferase (PFP): an enzyme relating fructose 2,6-bisphosphate to the control of glycolysis and gluconeogenesis in plants, Physiol. Veg. 23:321 (1985).
20. N.W. Carnal and C.C. Black, Phosphofructokinase activities in photosynthetic organisms, Plant Physiol. 71:150 (1983).
21. R.R. Walker and R.A. Leigh, Mg^{2+}-dependent cation-stimulated inorganic pyrophosphatase associated with vacuoles isolated from storage roots of red beet (Beta vulgaris L.), Planta 153:150 (1981).
22. P. Gross and T. ap Rees, Alkaline inorganic pyrophosphatase and starch synthesis in amyloplasts, Planta 167:140 (1986).
23. F.D. Macdonald and T. ap Rees, Enzymic properties of amyloplasts from suspension cultures of soybean, Biochim. Biophys. Acta, 755:81 (1983).
24. F. Rebeille, R. Bligny, J.-B. Martin, and R. Douce, Relationship between the cytoplasm and the vacuole phosphate pools in Acer pseudoplatanus cells, Arch. Biochem. Biophys. 225:143 (1983).
25. T. ap Rees, Conservation of carbohydrate by the non-photosynthetic cells of higher plants, Symp. Soc. Exp. Biol. 31:7 (1977).
26. N.J. Kruger, S.M. Hemmingsen and D.T. Dennis, Distribution of pyrophosphate: fructose 6-phosphate phosphotransferase in maize leaves, Plant Physiol. 81:12 (1986).
27. M.J. Kobr and H. Beevers, Gluconeogenesis in the castor bean endosperm, Plant Physiol. 47:48 (1971).
28. E. Van Schaftingen and H.-G. Hers, Fructose 2,6-bisphosphate in relation with the resumption of metabolic activity in slices of Jerusalem artichoke tubers, FEBS Lett. 164:195 (1983).
29. W. Stiles and K.W. Dent, Researches on plant respiration, Ann. Bot. 11:1 (1947).
30. R. Hedrich, K. Raschke and M. Stitt, A role for fructose 2,6-bisphosphate in regulating carbohydrate metabolism in guard cells, Plant Physiol. 79, 977 (1985).
31. R.A.J. Challis, J.R.S. Arch and E.A. Newsholme, The rate of substrate cycling between fructose 6-phosphate and fructose 1,6-

bisphosphate in skeletal muscle from cold-exposed hyperthyroid or acutely exercised rats, Biochem. J. 231:217 (1985).

32. S. Morrell and T. ap Rees, Sugar metabolism in developing tubers of Solanum tuberosum, Phytochemistry 25: 1579 (1986).

33. J. Edwards and T. ap Rees, Metabolism of UDPglucose by developing embryos of round and wrinkled varieties of Pisum sativum, Phytochemistry, 25:2033 (1986).

34. S.C. Huber and T. Akazawa, A novel sucrose synthase pathway for sucrose degradation in cultured sycamore cells, Plant Physiol. 81:1008 (1986).

35. H.G. Wood, W.E. O'Brien, and G. Michaels, Properties of carboxytransphosphorylase; pyruvate, phosphate dikinase; pyrophosphate-phosphofructokinase and pyrophosphate-acetate kinase and their roles in the metabolism of inorganic pyrophosphate, Adv. Enzymol. 45:85 (1977).

36. A. Chanson, J. Fichmann, D. Spear, and L. Taiz; Pyrophosphate-driven proton transport by microsomal membranes of corn coleoptiles Plant Physiol. 79:159 (1985).

37. P.A. Rea and R.J. Poole, Chromatographic resolution of H^+-translocating pyrophosphatase from H^+-translocating ATPase of higher plant tonoplast, Plant Physiol. 81:126 (1986).

38. J. Edwards, T. ap Rees, P.M. Wilson and S. Morrell, Measurement of the inorganic pyrophosphate in tissues of Pisum sativum L., Planta 162:188 (1984).

39. D.A. Smyth and C.C. Black, Measurement of the pyrophosphate content of plant tissues, Plant Physiol. 75:862 (1984).

40. T. ap Rees, S. Morrell, J. Edwards, P.M. Wilson and J.H. Green, Pyrophosphate and the glycolysis of sucrose in higher plants, in "Regulation of Carbon Partitioning in Photosynthetic Tissue", R.L. Heath and J. Preiss, eds. Waverly Press, Baltimore (1985).

THE SIGNIFICANCE OF CYANIDE-RESISTANT RESPIRATION TO CARBOHYDRATE METABOLISM IN HIGHER PLANTS

James N. Siedow and Mary E. Musgrave

Department of Botany
Duke University
Durham, NC 27706

INTRODUCTION

Cyanide-resistant respiration was first described in plants over 60 years ago; however its significance to the general physiology of higher plants remains unknown. Biochemical studies carried out in a number of laboratories have established that this respiration is associated with a non-phosphorylating "alternative" electron transfer pathway to oxygen that is present, to some extent, in most plant mitochondria (Siedow and Berthold, 1986). At the level of the whole plant, the temperature increase that accompanies the marked rise in respiration rates during thermogenesis in aroid spadices is well characterized, but it provides the only clear example of a metabolic response directly linked to the operation of the alternative pathway (Laties, 1982). However, most plant tissues show neither the high overall respiration rates nor the extreme levels of cyanide resistance associated with the operation of the pathway during spadix thermogenesis. In an attempt to provide a general explanation for the presence of cyanide-resistant respiration in a wide range of higher plant tissues, Lambers (1980) suggested that the alternative pathway could serve an "energy overflow" function as a means of oxidizing cellular carbohydrate which exists in excess of that needed for growth, storage or to synthesize ATP. Evidence in support of this concept has been provided in several cases (Lambers, 1980; 1985).

A major difficulty affecting attempts to establish a physiological role for the alternative pathway in plants is the fact that the degree of cyanide resistance varies considerably among those plant tissues that show cyanide-resistant respiration. Further, several examples are known where respiration of the intact tissue displays considerable cyanide resistance while the resulting isolated mitochondria are found to be fully cyanide sensitive (Laties, 1982). Recently, we carried out a series of experiments designed to test the effects of cytokinins on the operation of the alternative pathway in a number of cytokinin bioassays (Musgrave and Siedow, 1985). These experiments were derived from Miller's (1982) earlier observation that high concentrations of cytokinins could inhibit the alternative pathway in isolated plant mitochondria. During the course of our investigation, we found that two pea cultivars, which differed with regard to their ability to produce ethylene when treated with added benzyladenine, also differed with respect to the appearance of cyanide-resistant respiration.

Specifically, one cultivar, Progress No. 9, that failed to respond to cytokinins also lacked any SHAM-sensitive, cyanide-resistant respiration in axis tissue while a second cultivar, Alaska, responded to cytokinins and displayed cyanide-resistant respiration.

In the present study, these two pea cultivars were utilized to look at the pattern of inheritance of the alternative pathway and subsequently to generate hybrids differing in the presence or absence of the alternative pathway. The latter material was then used to assess the role of cyanide-resistant respiration in whole plant carbon budgets.

MATERIALS AND METHODS

In the genetic studies, seeds of Pisum sativum L. cvs. Alaska and Progress No. 9 were grown in the Duke University greenhouses and hybrids were produced by reciprocally crossing the two cultivars (Musgrave et al., 1986a). In studies looking at the effects of elevated CO_2 on growth, hybrids of cvs. Alaska and Progress No. 9 were grown in the Duke University Phytotron greenhouses which were maintained at either 350 or 650 ppm CO_2 Musgrave et al., 1986c).

Respiration measurements on either intact tissue (Musgrave and Siedow, 1985) or isolated mitochondria (Siedow and Bickett, 1983) were obtained polarographically using a Clark-type oxygen electrode. For growth analyses, four harvests (9, 16, 23 and 57 days after emergence) were made and total area and dry weight of leaves, stems, roots and any reproductive parts were determined. The change in dry weight (ΔW), relative growth rate (RGR) and net assimilation rate (NAR) were then calculated for the day 9-16 and 16-23 harvest intervals.

RESULTS

Inheritance of the Alternative Pathway

In our initial study (Musgrave and Siedow, 1985), only the respiration of intact axis tissue was measured for the two pea cultivars with Progress No. 9 showing no measurable cyanide-resistant pathway and Alaska having a level of cyanide-resistant respiration equivalent to 14% of the uninhibited rate. This differential expression of the alternative pathway in the two cultivars was maintained in respiration associated with leaf discs and isolated epicotyl mitochondria (Table 1). This indicated that the lack of alternative pathway in Progress No. 9 was not

Table 1. Respiration rates in leaf discs and isolated epicotyl mitochondria from pea cvs. Progress No. 9 and Alaska. Adapted from Musgrave et al. (1986a).

	Progress No. 9		Alaska	
	v_T	v_{alt}(%)	v_T	v_{alt}(%)
Leaf tissue[a]	209	0	174	20
Mitochondria[b]	289	0	292	8

[a]Rates expressed as nmol O_2/g F.W./min.
[b]Rates expressed as nmol O_2/mg protein/min.

a tissue-specific phenomenon but was found across a range of tissue types and was also reflected in the electron transfer chain of isolated mitochondria.

Having ascertained a clear distinction between the two cultivars with respect to the appearance of the alternative pathway, an effort was made to determine the pattern of inheritance of this trait (Musgrave et al., 1986a). Following reciprocal crosses between the two cultivars, the resulting F_1 progeny were grown up and assayed for the presence or absence of the cyanide-resistant pathway by measuring leaf disc respiration. The appearance of the alternative pathway in the F_1 generation showed a distinctly maternal pattern of inheritance (Table 2); only when cv. Alaska provided the maternal parent was the alternative pathway observed in the resulting progeny.

The maternal pattern of inheritance was maintained in F_2 progeny generated following self-pollination of the F_1 plants. This ruled out the possibility that the pattern seen in the F_1 generation was due to a maternal effect as opposed to a true maternally inherited trait. Interestingly, the ability of exogenously applied cytokinin (40 uM benzyladenine) to stimulate ethylene production in pea seedlings also displayed a maternal pattern of inheritance in the F_1 progeny Table 2). As noted above, this combination of traits was also observed in the two parental types which makes for an interesting, if poorly understood, correlation between the appearance of the alternative pathway and the ability of pea seedlings to respond to added cytokinins.

Significance of the Alternative Pathway to Plant Energy Budgets

The F_1 hybrids produced in reciprocal crosses of the two pea cultivars represent a useful experimental material for studying the physiological role of the alternative pathway since they contain comparable nuclear backgrounds and differ mainly in the presence or absence of the alternative pathway. In addition, a wide range of studies have established the general ability of CO_2 enrichment to stimulate increased carbon assimilation; which represents a convenient way of imposing an 'energy overflow' on plants. Therefore, the effects of growth under both normal (350 ppm) and elevated (650 ppm) levels of CO_2 were investigated using the two reciprocal hybrids (Musgrave et al., 1986c). It was felt that these studies would provide a direct measure of the impact of the alternative pathway on the overall carbon budget of the plant as well as a direct test of the energy overflow concept.

Table 2. Pea leaf disc respiration and response to cytokinins of F_1 populations produced by crossing pea cvs. Progress No. 9 and Alaska.

Cross	v_T	v_{alt}(%)	C_2H_4 Formation[a] -BA	C_2H_4 Formation[a] +BA
Progress No. 9 X Alaska[b]	185[c]	0	0	0
Alaska X Progress No. 9	196	13	27.3	43.2

[a]Rates expressed as nl C_2H_4/g F.W./24 hr; assayed in intact seedlings according to Musgrave and Siedow (1985). BA = 40 uM benzyladenine

[b]The maternal parent appears first in each cross.

[c]Rates expressed as nmol O_2/g F.W./min.

Table 3. Specific leaf weight of leaf discs from parental pea cultivars and hybrids grown under normal (350 ppm) and elevated (650 ppm) levels of CO_2. Adapted from Musgrave et al. (1986c).

	Specific Leaf Weight[a]	
	350 ppm	650 ppm
Parental cultivar		
Alaska	2.3	2.4
Progress No. 9	2.6	3.5
Hybrids[b]		
Alaska X Progress No. 9	2.2	2.6
Progress No. 9 X Alaska	2.6	3.7

[a]Expressed as mg D.W./cm^2.
[b]The maternal parent appears first in each cross.

Early morning respiration rates by leaf discs from both parental cultivars and the two hybrids showed the pattern outlined previously. Among those plants which had the alternative pathway, in the parental cultivar Alaska, the pathway was fully engaged ($p = 1.0$) (Laties 1982) at 650 ppm CO_2 but not engaged at all ($p = 0$) at 350 ppm CO_2. A similar result appeared with the F_1 hybrid having Alaska as the maternal parent. However, when leaf tissue in this hybrid had an available sink for carbohydrate (a pod developing in its axil), no engagement of the alternative pathway appeared even at 650 ppm CO_2.

Measurements of specific leaf weight supported the concept that the operation of the alternative pathway could serve to oxidize excess carbohydrate in these leaves (Table 3). With the two parental cultivars, the one lacking the alternative pathway (Progress No. 9) showed a marked increase in specific leaf weight when grown at 650 ppm CO_2 (i.e., relative to that seen at 350 ppm CO_2). No such increase in specific leaf weight appeared in the cultivar having the alternative pathway (Alaska), indicating that some mechanism existed in the latter to remove the excess carbon assimilated when the plants were grown under under elevated levels of CO_2. Identical results were obtained with the two hybrids (Table 3); plants lacking the alternative pathway showed a significant increase in specific leaf weight in response to 650 ppm CO_2 while the hybrid which had the pathway showed only a small increase.

Given the results obtained with specific leaf weights, differences in the patterns of carbon utilization between the two sets of hybrids were further characterized by carrying out a growth analysis. Harvests were made 9, 16 and 23 days after emergence on hybrids grown under both levels of CO_2, and changes in dry weight (ΔW), net assimilation rate (NAR) and relative growth rate (RGR) were calculated for the resulting two harvest intervals. In an effort to compare more directly the effects of elevated CO_2 levels on growth, the ratio of the values obtained at 650 ppm and 350 ppm CO_2 were calculated for each of the growth parameters and expressed as an 'enrichment ratio.' As is seen in Table 4, the resulting enrichment ratio values for the hybrid lacking the alternative pathway (Progress No. 9 as the maternal parent) are consistently much higher than those seen with the hybrid having the pathway; the latter generally showed only a limited response (0-20%) to elevated CO_2. These results substantiated those seen with specific leaf weight (Table 3) in that the hybrid which lacked the alternative pathway showed a more substantial response, in terms of the production of plant dry matter, to elevated

Table 4. Growth analysis data for two harvest invervals (HI) with pea hybrids grown at 350 and 650 ppm CO_2. Adapted from Musgrave et al. (1986c).

Growth Parameter	HI	Enrichment Ratio[a]	
		Progress No. 9[b]	Alaska
ΔW, g	1	1.71	1.15
	2	1.88	1.22
NAR, g/dm^2/day	1	1.57	1.20
	2	1.50	1.17
RGR, g/g/day	1	1.37	1.00
	2	1.21	1.05

[a]Ratio of growth parameter at 650 versus 350 ppm CO_2.
[b]Maternal parent of the cross is indicated.

CO_2 than did the hybrid which was able to oxidize excess carbohydrates via the alternative pathway.

At 57 days after emergence, a final harvest was made and the results appear in Table 5. The hybrid lacking the alternative pathway outperformed the reciprocal cross in total dry weight, leaf area and seed dry weight at both CO_2 levels, but the differences were more pronounced at 650 ppm CO_2. As seen before (Table 3), specific leaf weight showed little response to CO_2 enrichment when Alaska was the maternal parent and the alternative pathway was present but increased markedly at 650 ppm CO_2 when the alternative pathway was absent.

DISCUSSION

The initial observation that respiration in the pea cultivar Progress No. 9 lacked the cyanide-resistant pathway was confirmed by looking at respiration in several different tissues and at the properties of isolated pea mitochondria (Table 1). Once the absence of the alternative pathway was confirmed, the genetics of this trait could be studied by crossing Progress No. 9 with a cultivar known to have the alternative pathway, Alaska. The finding that the alternative pathway is inherited in a maternal fashion in these two pea varieties represents the first observation of any kind regarding the genetics of the alternative pathway in a higher plant. While, in plants, a maternal pattern of inheritance could result from DNA residing in either the chloroplast or the mitochondrion, it seems most reasonable to suspect that, in this case, the respiration-associated alternative pathway would be under the control of the mitochondrial genome. Less clear at this time is whether the mitochondrial genome is coding for a specific protein on the alternative pathway or acts to regulate the expression of some nuclear-encoded component.

Bertrand et al. (1983), using the fungus *Neurospora crassa*, have demonstrated the presence of two, non-allelic nuclear genes which they feel code for structural components of the alternative pathway. Likewise, earlier work by Edwards et al. (1974) indicated the presence of a mitochondrially encoded "repressor" which could act to control the expression of one or more nuclear genes that encode structural components of the pathway. If a similar regulatory system exists is pea, it could

Table 5. Final harvest data for pea hybrids grown under two levels of CO_2. Adapted from Musgrave et al. (1986c).

Parameter	350 ppm	650 ppm
Total dry weight, g		
Alaska[a]	11.1	13.1
Progress No. 9	12.6	17.4
Leaf area, cm^2		
Alaska	449	479
Progress No. 9	516	717
Specific leaf weight, mg/cm^2		
Alaska	2.73	2.70
Progress No. 9	2.94	3.46
Seed dry weight, g		
Alaska	5.2	6.2
Progress No. 9	5.4	7.0

[a]Maternal parent of the hybrid is indicated.

explain the observed maternal pattern of inheritance of the alternative pathway. However, it will remain for future studies to determine the exact mechanism by which the mitochondrial genome is able to regulate the expression of the alternative pathway in pea.

The initial observation of Bahr and Bonner (1973) that the alternative pathway becomes engaged only when the capacity of the main (cyanide-sensitive) pathway is exceeded led Palmer (1976) and, later, Lambers (1980) to suggest that the physiological role of cyanide-resistant respiration is to serve as an energy overflow pathway. This would allow plants to oxidize carbohydrate which accumulates in excess of that needed for growth and maintainence of the plant. Evidence supporting this concept has been provided in a number of cases (Lambers, 1980; 1985). For example, Azcon-Bieto et al. (1983) found a daily cycle associated with engagement of the alternative pathway in several plants. In the morning, before significant levels of carbohydrate had built up in the leaves, the alternative pathway was either not engaged at all or only operating to a limited extent. However, after several hours illumination, in which photosynthetic activity led to an accumulation of leaf carbohydrate, the alternative pathway became engaged to a significant extent.

In the present study, we took advantage of 1) the unique genetic material provided by the Alaska/Progress No. 9 F_1 hybrids and 2) the fact that CO_2 enrichment results in increased carbon assimilation in most C_3 plants to provide a means of imposing conditions of energy overflow on plants to elucidate the effect of alternative respiration on plant carbon budgets. The results, as reflected in a number of different parameters, served to generally support the concept that the alternative pathway is able to oxidize excess carbohydrate.

Respiration measurements using leaf discs showed that the hybrid having the alternative pathway tended to show engagement of the pathway only when leaf carbon reserves were high. Early morning respiration showed an engaged cyanide-resistant component when the plants were grown at 650 ppm CO_2 with no readily available sink for the leaf's carbohydrate reserves. Regardless of whether plants were grown at 350 ppm CO_2 or 650 ppm CO_2 when pods were developing, respiration of subtending leaf material showed no engagement of the alternative pathway. The results obtained measuring specific leaf weight supported this observation (Table 3). At 350 ppm CO_2, both hybrids showed comparable values for specific leaf weight (consistent with only a limited engagement of the alternative pathway in the hybrid having Alaska as the maternal parent) while at 650 ppm CO_2, the hybrid lacking the alternative pathway showed a significant increase in specific leaf weight. The reduced response at 650 ppm CO_2 of the hybrid having the alternative pathway would be associated with the observed engagement of the pathway acting to metabolize much of the excess carbohydrate.

The above pattern was maintained when a growth analysis was carried out on the hybrids grown under each of the two CO_2 levels. When the ratios of the responses at 650 and 350 ppm CO_2 were compared for W, NAR and RGR, the hybrid lacking the alternative pathway showed a significantly greater response to CO_2 enrichment than did the hybrid which had the pathway (Table 4). This suggested that in the presence of the alternative pathway, excess carbon assimilated at 650 ppm CO_2 was being metabolized. The absence of the alternative pathway precluded the 'wasteful' oxidation of this carbohydrate which manifests itself as a greater increase in dry matter production. This pattern also appeared with the final harvest data (Table 5) where the hybrid lacking the alternative pathway showed greater total dry matter production at both CO_2 levels. However, this response was accentuated at 650 ppm CO_2, where a 33% increase was observed (over the hybrid having the pathway), than at 350 ppm CO_2 where only a 14% increase was seen. With respect to seed production (g D.W.), the two hybrids showed no significant difference at 350 ppm CO_2. At 650 ppm CO_2, both hybrids showed increases in total seed yield, but the increase was significantly greater in the hybrid lacking the alternative pathway.

Since the alternative pathway by-passes two energy conservation sites on the cytochrome pathway, it has been hypothesized (Lambers, 1980) that the use of the alternative pathway renders plants less energetically efficient than use of the cytochrome pathway. Kuiper (1983) has provided evidence that indices of vigor might be inversely related to the amount of alternative pathway present in a plant. In *Plantago major*, he found that a line with slow root growth (and few seeds per capsule) had a higher alternative respiration rate than a second line with faster root growth (and many seeds per capsule).

In a recent study of seven male-sterile genetic lines from four plant species (Musgrave et al., 1986b), we found a correlation between a lack of cyanide-resistant respiration and the occurrence of male-sterility. In conjunction with the present data and those of Kuiper (1983), this suggests that the greater vigor often observed in male-sterile genetic lines may be traced to a suppression of cyanide-resistant respiration in the tissues.

In conclusion, the development of genetic lines differing in the presence or absence of the alternative pathway has provided a new probe for studying the effects of this enigmatic pathway on whole plant carbon budgets. Plants lacking the alternative pathway outperform those having the pathway in terms of total dry matter production and overall growth, especially under conditions of CO_2 enrichment. These results support the hypothesis that cyanide-resistant respiration can consume luxury carbohydrates and that respiration via this pathway may be considered energetically wasteful in terms of whole plant carbon budgets, at least under the conditions of the experiments reported here. These results support previous studies which have indicated that the alternative pathway may be able to significantly affect plant productivity and vigor (Kuiper, 1983; Lambers, 1985; Day et al., 1985) and further point to the need for plant breeders to consider this trait as possibly being worth considering in attempts to increase crop productivity. Along this same line, it would be of interest to determine whether other agronomically important crop species contain any naturally occurring varieties which lack the cyanide-resistant respiratory pathway.

ACKNOWLEDGEMENTS

This work was supported by NIH grant GM 26095 to J. N. Siedow and NSF grants BSR82-15533 and BSR83-14925 to the Duke University Phytotron.

REFERENCES

Azcon-Bieto, J., Lambers, H., and Day, D. A., 1983, The effect of photosynthesis and carbohydrate status on respiratory rates and the involvement of the alternative path in leaf respiration, Plant Physiol., 72:598.

Bahr, J. T. and Bonner, W. D., Jr., 1973, Cyanide-insensitive respiration I. Steady states of skunk cabbage spadix and bean hypocotyl mitochondria, J. Biol. Chem., 248:3441.

Bertrand, H., Argan, C. A., and Szakacs, N. A., 1983, Genetic control of the biogenesis of cyanide-insensitive respiration in Neurospora crassa, in: "Mitochondria 1983," R. J. Schweyen, K. Wolf and F. Kaudewitz, eds., Walter de Gruyter & Co., Berlin.

Day, D. A., De Vos, O. C., Wilson, D., and Lambers, H., 1985, The regulation of respiration in the leaves and roots of two Lolium perenne populations with contrasting mature leaf respiration rates and crop yields, Plant Physiol., 78:678.

Edwards, D. L., Rosenberg, E., and Maroney, P. A., 1974, Induction of cyanide-insensitive respiration in Neurospora crassa, J. Biol. Chem., 249:3551.

Kuiper, D., 1983, Genetic differentiation in Plantago major: growth and root respiration and their role in phenotypic adaptation, Physiol. Plant. 57:222.

Lambers, H., 1980, The physiological significance of cyanide-resistant respiration in higher plants, Plant Cell Environ., 3:293.

Lambers, H., 1985, Respiration in intact plants and tissues: its regulation and dependence on environmental factors, metabolism and invaded organisms, in: "Higher Plant Cell Respiration," R. Douce and D. A. Day, eds., Springer-Verlag, Berlin.

Laties, G. G., 1982, The cyanide-resistant, alternative path in higher plant respiration, Annu. Rev. Plant Physiol., 33:519.

Miller, C. O., 1982, Cytokinin modification of mitochondrial function, Plant Physiol., 69:1274.

Musgrave, M. E., and Siedow, J. N., 1985, A relationship between plant responses to cytokinins and cyanide-resistant respiration, Physiol. Plant., 64:161.

Musgrave, M. E., Murfet, I. C., and Siedow, J. N., 1986a, Inheritance of cyanide-resistant respiration in two cultivars of pea (Pisum sativum L.), Plant Cell Environ., 9:153.

Musgrave, M. E., Antonovics, J., and Siedow, J. N., 1986b, Is male-sterility in plants related to lack of cyanide-resistant respiration in tissues?, Plant Sci., 44:7.

Musgrave, M. E., Strain, B. R., and Siedow, J. N., 1986c, Response of two pea hybrids to CO_2 enrichment: a test of the energy overflow hypothesis for alternative respiration, Proc. Natl. Acad. Aci. USA, 83:In Press.

Palmer, J. M., 1976, The organization and regulation of electron transport in plant mitochondria, Annu. Rev. Plant Physiol., 27:133.

Siedow, J. N., and Berthold, D. A., 1986, The alternative oxidase: a cyanide-resistant respiratory pathway in higher plants, Physiol. Plant., 66:569.

Siedow, J. N., and Bickett, D. M., Structural features required for inhibition of cyanide-insensitive electron transfer by propyl gallate, Arch. Biochem. Biophys., 207:32.

NATURE OF THE TERMINAL RESPIRATORY OXIDASES IN SWEET POTATO SLICES

Theophanes Solomos

Department of Horticulture
University of Maryland
College Park, MD 20742

INTRODUCTION

The investigation of the nature of the terminal oxidases in intact plant tissues is complicated by the presence of cytosolic oxidases which alone or in coupled reaction can transfer electrons from reduced pyridine nucleotides to oxygen (James, 1953; Bonner, 1957; Beevers, 1961). In addition the metabolite transport system of the mitochondria may result in the shuttling of reducing equivalents between the mitochondria and cytosol (Wiskich, 1980). The participation, or lack thereof, of the alternative oxidase in the respiration of intact plant tissues has been investigated either by determining the values of ρ according to the model of Bahr and Bonner (1973), or simply by observing the degree of inhibition of tissue respiration by hydroxamic as well as by gallate derivatives (Theologis and Laties, 1978; Siedow and Girvin, 1980; Lambers, 1985; Lance et al. 1985). In this report we have attempted to elucidate the nature of the terminal "oxidases" fresh slices of sweet potatoes, by relying on the differences in their affinities for oxygen.

RESULTS AND DISCUSSION

In fig. 1 the results of a titration experiment with SHAM are shown. It may be seen that ρ is zero, which implies that the participation of the alternative oxidase in intact slice respiration is negligible. In drawing up fig. 1 the fraction of oxygen uptake which is resistant to the combined addition of SHAM and KCN, the "residual" (Theologis and Laties, 1978), has been subtracted. It is implicitly assumed that the "residual" is functioning in intact uninhibited slices. We observed, however, that the magnitude of the "residual" is a function of the external oxygen concentrations and varies from almost zero to 75% of the control slice respiration (table 1). It is obvious that the calculation of the capacities of either the cytochrome or the alternative paths ($V_T = V_{cyt} + \rho V_{al} + Res$) will be unrealistically small in the case where higher-than-air oxygen levels are used. It has been observed that in thick banana segments the magnitude of the "residual" can be as high as that of the control slices, and this would produce zero capacities for both the cytochrome and alternative paths (cf. Solomos, 1983). Such is not the case, of course. These observations

Table 1. Effect of oxygen concentrations on the rate of oxygen uptake of sweet potato slices treated with SHAM and KCN both singly and together.

Experiment No.	Rate of oxygen uptake $\mu\ell\ O_2/g/h$				
	$\mu M\ O_2$	Control	2 mM SHAM	0.2 mM KCN	0.2 mM KCN + 2 mM SHAM
1	260	55.20	55.20	67.80	11.00
	760	54.60	55.20	69.00	41.40
2	249	-	75.90*	-	10.13
	65	-	74.10*	-	~ 2

*Slices were washed with phosphate buffer + 3 mM SHAM. At the indicated concentrations of O_2, 0.2 mM KCN was added.

raise two important questions: a) is the "residual" functioning in uninhibited sweet potato slices? and b) is it possible that the inhibition of the alternative may be masked by the ability of the slices to shuttle reducing equivalents from the mitochondria to the cytosolic "residual oxidases"?

The results in table 1 indicate that the affinity for oxygen of the "residual oxidases" is small since even at 760 µM oxygen the rate of oxygen uptake is dependent on its external concentration (Solomos, 1983).

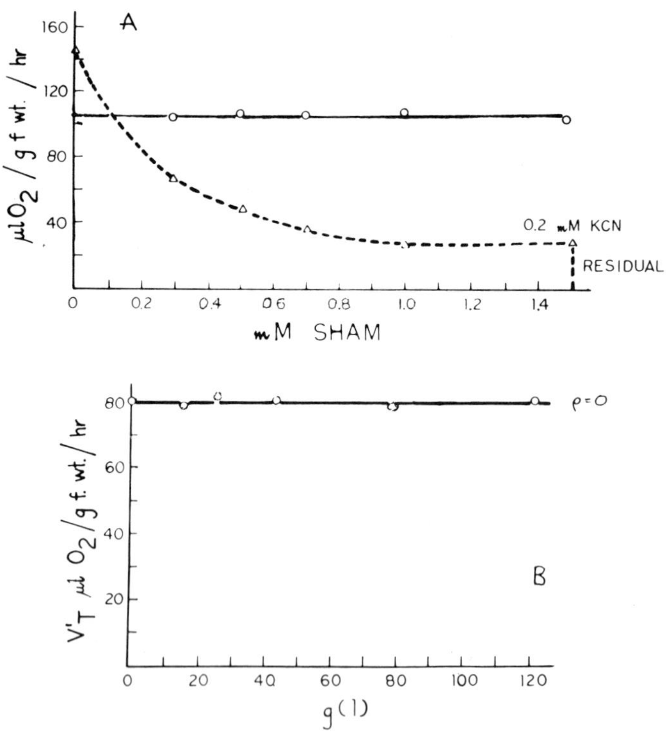

Fig. 1 Rate of oxygen uptake of control or cyanide-treated slices, 0.2 mM, without and with different concentrations of SHAM. Slices were bathed in 1 mM phosphate buffer pH 5.6 at 25°C.

Table 2. Levels of pyruvate and lactate in sweet potato slices kept under either 250 μM O_2 or 10.5 μM O_2.

	Oxygen concentration μM	Time under the respective treatments (mins)	n moles/g F wt	
			Pyruvate	Lactate
Experiment I	250 10.5	60 60	350 ± 48* 522.2 ± 62*	47 ± 12* 356 ± 35*
Experiment II	250 10.5	90 90	380 578	58 642.4

*Average of 2 samples.

Fig. 2 presents the oxygraph trace of 1.5 mm thick control, SHAM and cyanide-treated slices. These results show that the rate of oxygen uptake is zero order until its external concentration decreases to 15.5 μM for both control and SHAM-treated slices. This critical oxygen concentration varied with the experiments from 11 - 39 μM, while that of cyanide-treated slices varied from 60-120 μM. If fresh slices are kept in 10.5 μM oxygen for either 60 or 90 min, lactate accumulates, indicating that the tissue is experiencing hypoxia (table 2). The data thus show that (a) the contribution of the residual respiration to control as well as to both SHAM- and cyanide-treated slices is negligible, since, because of its low affinity for oxygen it would be anticipated that the rate of oxygen uptake would decrease even at the early stages of the oxygraph trace; (b) the affinity for oxygen of the alternative oxidase is smaller than that of cytochrome oxidase. In order to calculate the actual differences between the apparent K_m for oxygen of the two oxidases, the contribution of the rate of oxygen diffusion to the critical levels of O_2 must be calculated. The barriers to oxygen diffusion include the unstirred layers of water at the surface of the slices, the intercellular spaces, the cell walls and cell membranes. Because of the presence of gases in the intercellular spaces, it is not expected that

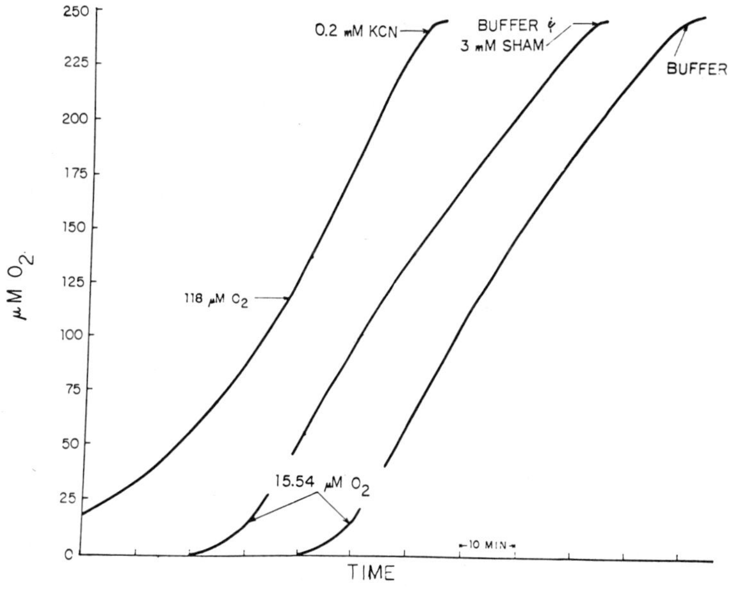

Fig. 2. Oxygraph trace of 1.5 mm thick slices bathed in 1 mM phosphate buffer pH 25°C.

they will be completely flooded with water (Briggs et al. 1961). Thus, oxygen diffuses in part through spaces filled with water and in part through gaseous channels. If it is accepted that (a) the rate of oxygen diffusion can be approximated by assuming that for a short time steady-state conditions prevail, and (b) the diffusion medium is homogenous, then the change in concentration with distance is given by (Hill, 1928):

$$C_i = C_o - \frac{V}{2D} \ell^2 \qquad (1)$$

where C_o and C_i are the concentrations of oxygen at the surface and center of the slices, (μmoles/cm^{-3}), V the rate of oxygen uptake per unit volume of the tissue (μmoles cm^{-3} sec^{-1}), ℓ is half the thickness of the slices in cm, and D is the diffusion coefficient in cm^2 sec^{-1}. If it is further assumed that a) K_m for oxygen of the cytochrome oxidase is 0.05 μM, and b) the departure from linearity of the oxygraph is perceived at 50 x K_m for oxygen, then from its average value of different experiments and eq. 1 a value of 1.4×10^{-4} cm^2 sec^{-1} can be calculated for D. If this value of D and the data of the cyanide-treated slices are used, then an average apparent K_m for oxygen of 58.60 μM can be calculated for the alternative oxidase which may correspond to a real value of 1.17 μM. This is about 20-fold higher than that of cytochrome oxidase.

REFERENCES

Bahr, J. T., and Bonner, W. D., 1973, Cyanide-insensitive respiration. II. Control of the alternative pathway. J. Biol. Chem. 248: 3446-50.

Beevers, H., 1961, "Respiration metabolism in plants," Row, Paterson and Co., White Plains, New York.

Bonner, W. D, 1957, Soluble oxidases and their functions, in "Ann. Rev. Plant Physiol." 8:427-52.

Briggs, G. E, Hope, A. B., and Robertson, R. N., 1961, "Electrolytes and Plant Cells," Blackwell Scientific Publications Co.

Hill, A. V, 1928, Diffusion of oxygen and lactic acid through tissues. Proc. Royal Soc. B104:39-96.

James, W. O, 1953, "Plant Respiration," Oxford at the Clarendon Press.

Lambers, H., 1985, Respiration in intact plant tissues, its regulation and dependence on environmental factors, metabolism and invaded organisms, in: "Higher Plant Cell Respiration," pp. 418-473. "Encyclopedia of Plant Physiology," Vol. 18, Springer-Verlag, Berlin, Heidelberg, New York.

Lance, C, Chauveau, M., and Dizengremel, P., 1985, The cyanide-resistant pathway of plant mitochondria, In: "Higher Plant Respiration," pp. 202-247. "Encyclopedia of Plant Physiology", Berlin-Heidelberg, New York.

Siedow, J., and Girvin, M. E., 1980, Alternative respiratory pathway: Its role in seed respiration and its inhibition by propyl gallate, Plant Physiol. 65: 669-674.

Solomos, T., 1983, Respiration and energy metabolism in senescing plant tissues, in: "Post-Harvest Physiology and Crop Improvement," pp. 61-98, Plenum Publishing Co.

Theologis, A., and Laties, G. G., 1978, Cyanide-resistant respiration in fresh and aged sweet potato slices, Plant Physiol. 62: 243-48.

Wiskich, J. T., 1980, Control of the Krebs Cycle, In: "The Biochemistry of Plants," Vol. 2, p. 243-277. ed. D. D. Davies, Academic Press, New York and London.

THE ALTERNATE OXIDASE IN BARLEY ROOTS OF DIFFERING GROWTH AND RESPIRATION RATES

I.J. Bingham and J.F. Farrar

School of Plant Biology, University College of North Wales, Bangor, Gwynedd LL57 2UW, U.K.

INTRODUCTION

Plant mitochondria possess both a cyanide sensitive and a cyanide resistant, or alternate, electron transfer pathway. The degree of engagement of the alternate pathway is quite variable and, apart from a few specialised cases such as heat generation in aroid spadices, the role of the non-phosphorylating, alternate pathway in intact tissues is unclear (Lambers, 1985). Whatever its physiological role, transfer of electrons to O_2 via the alternate pathway results in a lower ATP yield than equivalent transfer via the cytochrome path. A knowledge of the relative activities of alternate and cytochrome paths *in vivo* is necessary to evaluate the energy available for metabolic processes. Here we describe the response of selectively pruned barley plants to the inhibitors SHAM and KCN in an attempt to assess the involvement of the alternate oxidase. We have shown that the growth rate, respiration rate and carbohydrate status of barley roots is modified by such pruning (Farrar and Jones, 1986).

MATERIALS AND METHODS

Barley (*Hordeum distichum* cv. Maris Mink) was grown hydroponically as described previously (Farrar and Jones, 1986). When 15-16 days old the root system was pruned to a single seminal axis (root-pruned), or the shoot pruned so that all but the 2nd leaf was removed (leaf-pruned). Respiration rate of freshly detached seminal axes was measured polarographically in Clark type O_2 electrodes 16 to 22 hours after pruning.

RESULTS

In roots of control (non-pruned) plants, SHAM alone had no significant influence on respiration at concentrations below 5 mol m^{-3}, but above this substantial inhibition was observed (Fig. 1). In the presence of 0.1 mol m^{-3} KCN, SHAM concentrations below 5 mol m^{-3} were sufficient to inhibit respiration; therefore SHAM was able to penetrate roots at low concentrations. Bahr-Bonner (ρ) plots (Bahr and Bonner, 1973) constructed from these data are non linear.

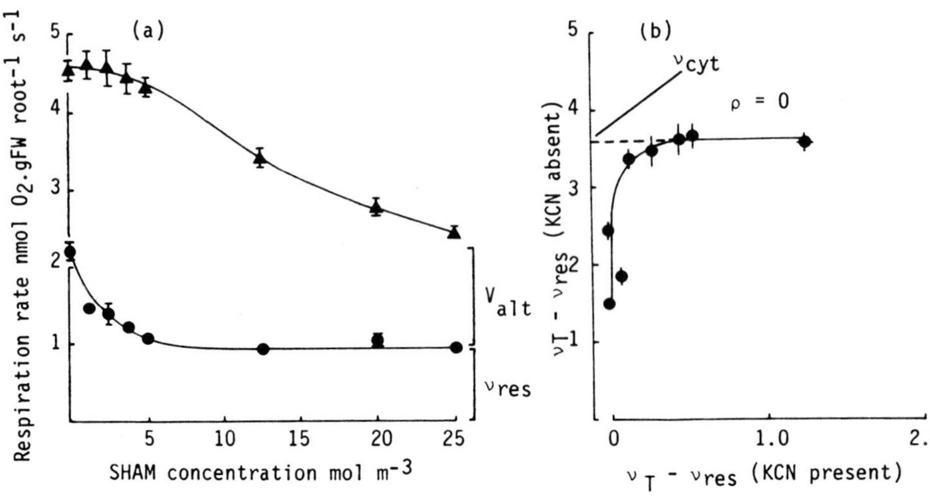

FIG. 1: (a) Effect of SHAM on root respiration of control (non-pruned) plants in the absence (▲) and presence (●) of 0.1 mol m^{-3} KCN. Rates were measured 8 to 10 minutes after placing roots in the electrode; inhibitors were present from the start. Each point represents the mean (±SE) of 3 to 11 independent determinations.

(b) Bahr-Bonner plot of data from (a).

Non-linearity of ρ-plots for roots of pea, plantain and maize has been ascribed to the simultaneous stimulation of a peroxidase (which can use NADH as a substrate) and inhibition of the alternate pathway by low concentrations of SHAM (De Visser and Blacquière, 1984; Spreen Brouwer and Lambers, this volume). Although a similar peroxidase is present in barley roots, there is no SHAM stimulated activity in the absence of exogenous NADH (Bingham and Farrar, unpublished). Non-linearity of ρ-plots here appears to result from non-specific effects of SHAM at high concentrations. Inhibition by 25 mol m^{-3} SHAM in the absence of KCN was nearly double the capacity of the alternate oxidase (Fig. 1). Further, reverse titration of cytochrome path activity in the presence and absence of 25 mol m^{-3} SHAM has shown that cytochrome activity is impaired by such high concentrations of SHAM (Bingham and Farrar, unpublished).

Estimates of the respiratory parameters shown in Fig. 1 are justified on the grounds that only low concentrations of SHAM may be used to assess the activity of the alternate pathway in our system. The roots possessed an alternate path capacity (V_{alt}) of 1.25 nmol O_2 g FW^{-1} s^{-1}, approximately 27% of the total rate, but there was no appreciable engagement (ρ = 0). Residual respiration (that insensitive to both SHAM and KCN) represented some 20% of the total. The major component of respiratory activity (80%) was via the cytochrome pathway (ν_{cyt}).

Pruning roots caused a 25% increase in respiration rate of the sole remaining root and a 30% increase in its soluble sugar content. Leaf-pruning reduced the respiration rate by 35% and soluble sugar content by 39%. With both treatments the capacity of the alternate oxidase was similar to controls, but there was little apparent engagement (Figs. 2 and 3). The increase in respiration rate of root-pruned plants can be accounted for by an increase in cytochrome activity, whilst a decrease in cytochrome activity and a small reduction in residual respiration were associated with leaf-pruning. It should be noted that respiration rates of pruned plants should be compared with controls conducted at the same time (Figs. 2 and 3).

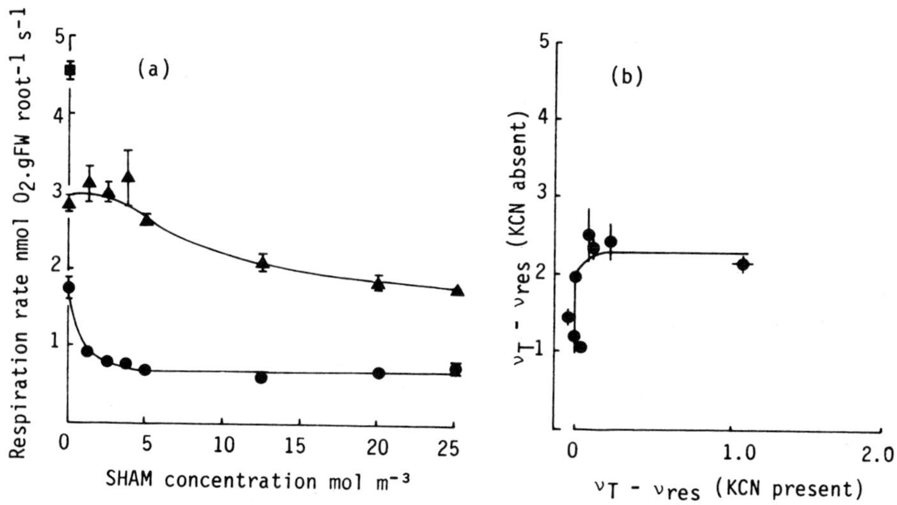

FIG. 2: (a) Effect of SHAM on root respiration of leaf-pruned plants in the absence (▲) and presence (●) of 0.1 mol m^{-3} KCN. Each point represents the mean (±SE) of 3 to 11 independent determinations. (■) rate of control plants in the absence of KCN.

(b) Bahr-Bonner plot of data from (a).

DISCUSSION

The alternate pathway is non-phosphorylating. Since it is only engaged when the cytochrome pathway is fully saturated (Bahr and Bonner, 1973; Theologis and Laties, 1978) it has been suggested that this energetically inefficient pathway operates as an energy overflow in situations where the supply of carbohydrates exceeds the demands for growth, maintenance or storage (Lambers, 1985). The hypothesis is supported by correlations between sugar content, respiration rate and engagement of the alternate path in a number of tissues (Lambers, 1985).

In pruned barley plants, rates of root respiration also correlate with changes in soluble sugar content. However, there was no appreciable engagement of the alternate pathway with any of the treatments; adjustments in respiration rate were almost entirely due to changes in cytochrome path activity. Lack of alternate path activity cannot be ascribed to substrate limitation since sucrose (25 mol m^{-3}) supplied to roots of control and leaf-pruned plants failed to stimulate respiration. Furthermore, respiration could be uncoupled in the absence of exogenous sucrose (Bingham and Farrar, unpublished) implying that the rate was under some degree of adenylate control.

The data are difficult to reconcile with the hypothesis that the alternate path of barley roots acts as an energy overflow. Possibly, in spite of the clear changes in soluble sugar content, the capacity for storage was never exceeded and that much higher tissue contents might result in carbohydrates being oxidised wastefully. However, plants will rarely experience conditions as severe as those imposed here, and so energy overflow is probably not an important function of the alternate path <u>in vivo</u>. A more important role for the significant alternate capacity present in these roots may be in allowing either, a flux of carbon skeletons through the respiratory network when the energy demand is low, or a rapid response to an increase in energy demand when the cytochrome path is operating at maximum capacity (Lambers, 1985). It is now necessary to investigate the control of the capacity of the cytochrome path.

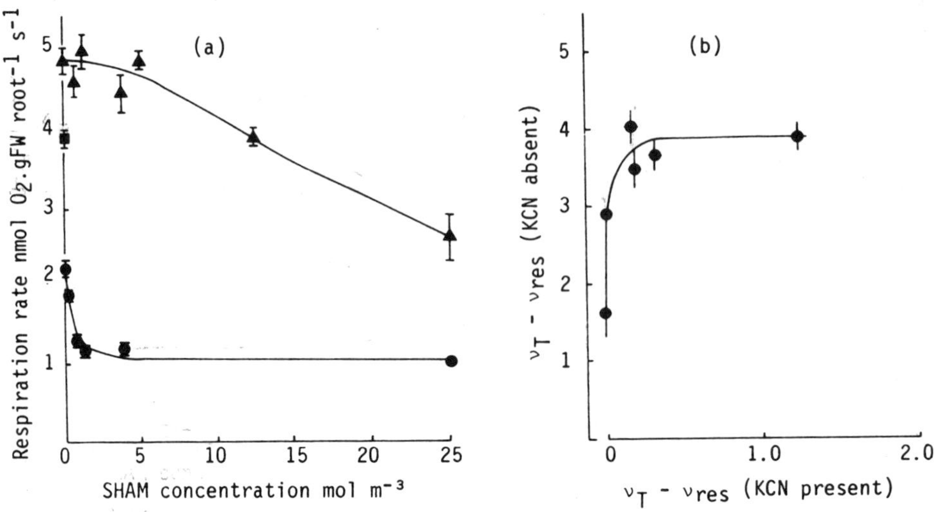

FIG. 3: (a) Effect of SHAM on root respiration of root-pruned plants in the absence (▲) and presence (●) of 0.1 mol m^{-3} KCN. Each point represents the mean (±SE) of 3 to 11 independent determinations. (■) rate of control (non-pruned) plants in the absence of KCN.

(b) Bahr-Bonner plot of data from (a).

REFERENCES

Bahr, J. T., and Bonner, W. D., 1973, Cyanide-insensitive respiration. I. The steady states of skunk cabbage spadix and bean hypocotyl mitochondria, J. Biol. Chem., 248:3441.

De Visser, R., and Blacquière, T., 1984, Inhibition and stimulation of root respiration in Pisum and Plantago by hydroxamate. Its consequences for the assessment of alternative path activity, Plant Physiol., 75:813.

Farrar, J. F., and Jones, C. L., 1986, Modification of respiration and carbohydrate status of barley roots by selective pruning, New Phytol., 102:513.

Lambers, H., 1985, Respiration of intact plants and tissues: its regulation and dependence on environmental factors, metabolism and invaded organisms, in: "Higher Plant Cell Respiration", R. Douce and D. A. Day, eds., Encyclopedia of Plant Physiology, New Series, Vol. 18, Springer-Verlag, Berlin.

Theologis, A., and Laties, G. G., 1978, Relative contribution of cytochrome mediated and cyanide resistant electron transport in fresh and aged potato slices, Plant Physiol., 62:232.

THE PARTICIPATION OF CYANIDE RESISTANT PATHWAY IN RESPIRATION OF WINTER RAPE PLANTS TREATED WITH LOW TEMPERATURE

Anna M. Rychter and Elżbieta Cieśla

Institute of Botany, University of Warsaw
Warsaw, Poland

INTRODUCTION

Low temperature treatment in so-called chilling resistant species brings about two types of effects concerning the respiration - a direct response in respiration rate and indirect one resulting from cold induced modifications in physical and metabolic properties of the cell[1].

Winter rape plants subjected to cold pretreatment (2°C) showed increased respiration rate[1], also the accumulation of soluble sugars was observed[2]. It was noted that leaves with high levels of free sugars have higher rates of O_2 uptake and an active alternative pathway[3,4]. The purpose of this study was to check the possibility of involvement of the alternative pathway in respiration of rape plants during cold hardening in relation to sugar content in the leaves.

MATERIAL AND METHODS

Winter rape plants were grown in sand supplied with aqueous Hoagland solution, as described by Sikorska and Kacperska-Pałacz[5], in a growth cabinet at a day/night temperature 20°/16°C with light/dark periods of 16/8 hr. After 5 weeks of growth half of the plants were subjected to continuous low temperature, 5°C. The second and third leaves were removed from plants at the end of dark period and kept in dark at room temperature or in refrigerator for respiratory measurements. One cm segments from upper part of the leaf were cut into 1 mm thin slices by razor blade in solution as described by Azcón-Bieto at al,[3]. The slices were then rinsed 3 times for 15 min in the same solution changed each time, the last one containing 5% albumin. Fifty slices were taken for respiratory measurements (in solution used for last rinsing) by Clark oxygen electrode at 25°C. Respiratory inhibitors, 1mM KCN and 2.5 mM SHAM were used and the activity of cytochrome and alternative pathways determined. The actual participation of the alternative pathway in total oxygen uptake was estimated by method of Bahr and Bonner[6].

RESULTS

It was found that in cold pretreated plants the respiration rate was higher than in the control plants. The increase in respiration was particularly high after 4 days of cold treatment, 60% higher than in control plants grown at 20°C in which respiration rate did not change during the culture, Fig.1.

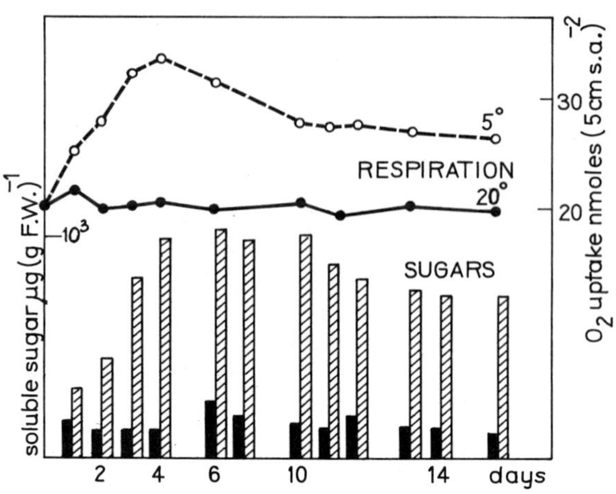

Fig.1. Respiration of leaf slices cut from rape plants cultured for 16 days at 5°C (open circles and dashed line) or 20°C (closed circles) Soluble sugar content of leaves of plants grown at 5°C (hatched bars) or 20°C (black bars) for 16 days.

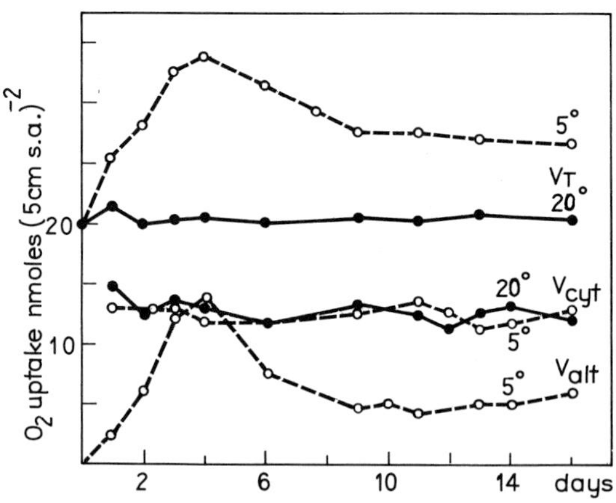

Fig. 2. Changes in total oxygen uptake (V_T), cytochrome path activity (V_{cyt}) and alternative path activity (V_{alt}) of leaf slices cut from rape plants cultured for 16 days at 5°C (open circles and dashed line) or 20°C (closed circles and continuous line).

During the culture of plants at 5°C an increase in soluble sugar was observed in the leaves. After 4 days of cold treatment soluble sugar increased from 260 µg to 1000 µg per g of fresh weight (300% higher than in plants kept at 20°C), Fig.1.

Using the inhibitors of cytochrome (KCN) and alternative (SHAM) pathways it was estimated that the activity of cytochrome path does not change during cold treatment, similarly as the activity of cytochrome path of plants grown at 20°C, Fig.2. On the other hand, activity of the alternative path is present in cold grown plants only; it alters during the cold treatment, being the highest at the 4 day, when the contribution of the alternative path is 56% of the total respiration, Fig.2.

SHAM titration of the alternative path gave ρ value of 0.56, Fig.3, indicating that the observed increase in respiration of plants grown at 20°C is caused by activity of the alternative pathway.

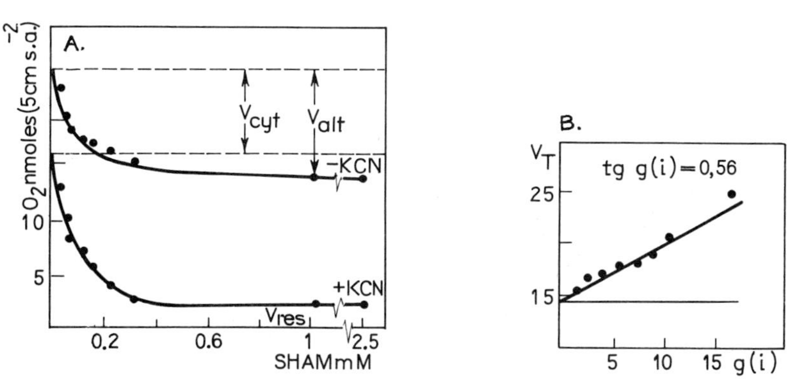

Fig.3. A. Effect of SHAM concentration on the rate of oxygen uptake of leaf slices in the presence or absence of 1 mM KCN.
B. A plot of rates of oxygen uptake in the presence of SHAM alone against those obtained in the presence of both KCN and SHAM.
A slope of this line represents the extent to which alternative path is engaged in the absence of inhibitors (ρ). Mean values ± S.E. of three experiments are shown.

In order to show the influence of sugar level in the tissue on respiratory activity, sucrose was added to leaf slices cut from plants grown at 20°C, Fig.4. Sucrose caused about 44% increase in respiration; the slices after sucrose addition were less sensitive to KCN and more sensitive to SHAM, indicating the increase of participation of alternative path in total respiration.

CONCLUSIONS

The activity of the alternative pathway is solely responsible for the observed increase in respiration rate in plants grown at 5°C. Stimulation

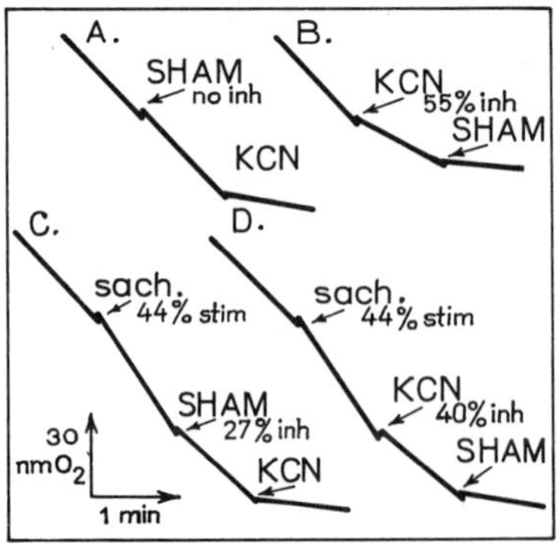

Fig.4. Typical polarographic traces of O_2 uptake of leaf slices from plants grown at 20°C. Rates shown are expressed as nmol O_2 min^{-1}
A. Control with 2.5 mM SHAM added before 1mM KCN. B. Control with 1 mM KCN added before 2.5 mM SHAM. C. 60 mM sucrose added first followed by 2.5 mM SHAM and 1mM KCN. D. 60 mM sucrose followed by 1 mM KCN and 2.5 mM SHAM

of the alternative path activity is caused by high sugar level which is the result of metabolic changes caused by cold treatment. Our results support the hypothesis that sugar level controls the participation of the alternative pathway in leaf respiration[7].

REFERENCES

1. U.Maciejewska, J.Tomczyk and A.Kacperska, Effect of cold on CO_2 exchange in winter rape leaves. Physiol.Plant. 62: 15 (1984).
2. A. Kacperska-Palacz and B.Wciślinska, The effect of CCC on the nitrogen compounds content in rape plants and their frost hardiness. Biol.Plant. 14 : 9 (1972)
3. J.Azcón-Bieto, D.A.Day and H.Lambers, The regulation of respiration in the dark in wheat leaf slices. Plant Sci.Lett. 32: 313 (1983)
4. J.Azcón-Bieto, H.Lambers and D.A.Day, Effect of phothosynthesis and carbohydrate status on respiratory rates and the involvement of the alternative pathway in leaf respiration. Plant Physiol.72: 598 (1983)
5. E.Sikorska and A.Kacperska-Palacz, Phospholipid involvement in frost tolerance. Physiol.Plant. 47: 144 (1979)
6. J.T.Bahr and W.D.Bonner Jr. Cyanide-insensitive respiration.II. Control of the alternative pathway. J.Biol.Chem. 248: 446 (1973)
7. H.Lambers, Respiration in intact plants and tissues: its regulation and dependence on environmental factors, metabolism and invaded organisms in Encyclopedia of Plant Physiol. R.Duce and D.A.Day ed. Springer-Verlag Berlin Heidelberg (1985)

NADPH OXIDATION IN POTATO TUBER CALLUS MITOCHONDRIA AND ITS PHYSIOLOGICAL SIGNIFICANCE DURING IN VIVO RESPIRATION

Anneke M. Wagner, Dirk-Jan Leek and Linus H.W. van der Plas

Department of Plant Physiology, Free University,
De Boelelaan 1087, 1081 HV Amsterdam, The Netherlands

INTRODUCTION

In callus-forming potato tuber discs glucose can be oxidized via the glycolysis/citric acid cycle or via the pentose phosphate pathway (PPP). The nature of the reduction equivalents formed during glucose oxidation is pathway dependent. Glycolysis/citric acid cycle supplies mainly NADH, while the PPP supplies NADPH. It has been considered that NADPH is primarily used in biosynthetic reactions in the cytoplasm. However, recently evidence has accumulated indicating that NADPH can also be oxidized through the mitochondrial electron transport chain via a specific dehydrogenase located on the outer surface of the mitochondrial inner membrane (Palmer and Ward, 1985). The aim of the present study was to investigate the mitochondrial respiration with respect to the possible physiological role of this NADPH oxidation. Therefore the affinity of the mitochondrial respiration for NADPH was determined and the effect of NADPH as a respiratory substrate in combination with substrates derived from glycolysis (NADH) or citric acid cycle (succinate and malate) was studied.

MATERIAL AND METHODS

Material and Treatment of the Tissue

Callus-forming potato tuber discs (Solanum tuberosum L. cv. Bintje) were cultured at 28°C as described before (Hemrika-Wagner et al., 1982).

Isolation of Mitochondria and Measurements of Mitochondrial Respiratory Characteristics

Mitochondria were isolated from callus-forming potato tuber discs as described before (Hemrika-Wagner et al., 1982). Oxygen uptake of the mitochondria was measured in a reaction medium containing 0.7 M mannitol, 10 mM potassium phosphate, 10 mM Tris-HCl and 0.1% BSA, pH 6.8. Dependent of the type of experiment, respiration was measured after addition of different combinations of the following substances: 20 µmol succinate, 20 µmol malate (+ 1.8 µmol NAD), 1.8 µmol NADH, 1.8 µmol NADPH, 0.1-0.15 µmol ADP, 0.1 µmol KCN, 3.6 nmol AA, 0.7 µmol 8-OHQ or 6.3 µmol BHAM.

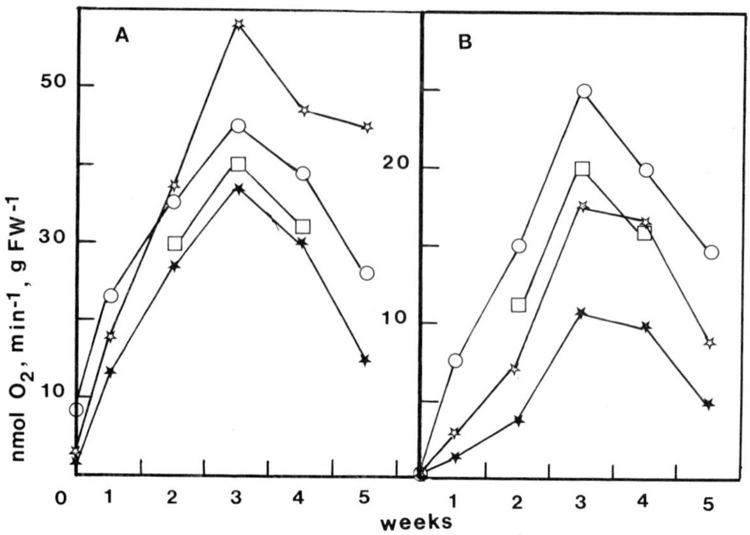

Fig. 1. The course of uninhibited state 3 (A) and CN-resistant (B) respiration with NADH (☆), NADPH (★), succinate (○) or malate (□) as the respiratory substrate of mitochondria isolated from callus-forming potato tuber discs.

RESULTS AND DISCUSSION

In mitochondria, isolated from callus-forming potato tuber discs, two electron transport pathways are present: the CN-sensitive cytochrome pathway and the CN-resistant alternative pathway. A considerable increase in uninhibited respiration and CN-resistant respiration can be observed in mitochondria from potato tuber discs after incubation on a callus-forming medium. Uninhibited state 3 respiration with succinate, malate, NADH or NADPH is depicted in Fig. 1A. State 3 rates during callus-formation were highest with NADH and lowest with NADPH. Capacities of the CN-resistant alternative pathway were also substrate dependent, but in a different way than the uninhibited respiration (Fig. 1B): the highest alternative pathway capacities were observed with succinate, the lowest with NADPH. The difference in CN-resistant respiration with NADH or NADPH strongly suggests the existence of a separate dehydrogenase for NADPH (cf. Møller and Palmer, 1981).

It appears that NADPH can be respired by potato tuber callus mitochondria and therefore consequently the question arises:
- Does NADPH oxidation take place in vivo?

Two factors are important:
1. The affinity of the mitochondrial dehydrogenases for NADPH.
2. The interaction with respiration caused by other, glycolytic or citric acid cycle derived substrates.

1. The Apparent K_m of the Mitochondrial Dehydrogenase for NADPH

In Table 1. it is shown that the apparent K_m for NADPH of the respiration of mitochondria isolated from callus-forming potato tuber discs is very high compared to that of NADH (exogenous). In several plant species the K_m for NADPH has been reported to be about twice as high as that for NADH. Only in Arum mitochondria a comparable result has been obtained (Møller and Palmer, 1981).

Table 1. Apparent K_m of State 3 Respiration in Mitochondria Isolated from Callus-forming Potato Tuber Discs

Substrate	Apparent K_m of State 3 Respiration
NADPH	223 μM
NADH	14 μM

2. Interactions with Other Substrates

Table 2. gives the results of experiments in which the respiratory characteristics of mitochondria isolated from 14-28 days old callus have been determined. When succinate or malate were combined with NADPH, the state 3 rates were higher, but less than the sum of the values obtained with the substrates added separately. However, when succinate and malate were given together as respiratory substrates, respiration rates were about equal to those obtained with succinate alone. When NADPH was combined with NADH, values were obtained about equal to those with NADH alone. A combination of NADH with succinate or malate resulted in state 3 rates higher, but again less than the sum of the values obtained with the substrates given separately.

In conclusion it can be said that when an exogenous substrate is combined with another exogenous substrate (NADH with NADPH) or when alternatively an endogenous substrate is combined with another endogenous substrate (succinate and malate), state 3 rates do not exceed the highest rate of the two when given separately. However, when an exogenous substrate is combined with an endogenous substrate (succinate or malate with NADH, succinate or malate with NADPH) an increase of the state 3 rate can be observed. In the exogenous combination, the rate-limiting step lies in a part of the respiratory chain(s) shared by both substrates. While addition of NADPH to NADH oxidizing mitochondria does not increase the respiratory rate, addition of NADPH to succinate or malate oxidizing mitochondria increases the respiration. Apparently the rate limiting step with an endogenous/exogenous combination

Table 2. State 3 Respiration (nmol O_2, min^{-1}, g FW^{-1}) with several Substrates, of Potato Tuber Callus Mitochondria

Substrate	State 3
Succinate	38
Malate	30
NADH	56
NADPH	23
Succinate + NADH	81
Succinate + NADPH	50
Malate + NADH	73
Malate + NADPH	46
Succinate + Malate	39
NADH + NADPH	55

lies on another part of the respiratory chain than with the exogenous/exogenous or with the endogenous/endogenous combinations.
A further step in this study will be the determination of the contribution of each substrate after combined addition to the cytochrome pathway and the alternative pathway. Preliminary results suggest that (at least part of) the NADPH is oxidized via the cytochrome pathway when it is cooxidized with succinate.

GENERAL CONCLUSION

The high K_m for NADPH oxidation in potato tuber callus mitochondria makes it unlikely that NADPH oxidation plays an important role in vivo under normal conditions, but under conditions of high NADPH concentrations in the cell, NADPH can be respired, even when high concentrations of citric acid cycle derived substrates are present at the same time.

REFERENCES

Hemrika-Wagner, A. M., Kreuk, K. C. M., and Van der Plas, L. H. W., 1982, Influence of growth temperature on respiratory characteristics of mitochondria from callus-forming potato tuber discs, Plant Physiol., 64:754.

Møller, I. M., and Palmer, J. M., 1981, Properties of the oxidation of exogenous NADH and NADPH by plant mitochondria. Evidence against a phosphatase or a nicotinamide nucleotide transhydrogenase being responsible for NADPH oxidation, Biochim. Biophys. Acta, 638:225.

Palmer, J. M., and Ward, J. A., The oxidation of NADH by plant mitochondria, in: "Encyclopedia of Plant Physiology," R. Douce and D. A. Day, eds. Springer-Verlag, Berlin (1985).

THE ROLE OF SOME PHYTOHORMONES IN THE REGULATION OF RESPIRATORY PATHWAYS IN POTATO TUBER TISSUE

Henk Gude, Paul van 't Padje
Ellen Rietveld and Linus H.W.
van der Plas

Dept. of Plant Physiol., Vrije Univ.
De Boelelaan 1087, 1081 HV
Amsterdam, The Netherlands

INTRODUCTION

Incubating potato tuber tissue discs on nutrient media with various phytohormones makes it possible to manipulate the development of the tissue and is thus a useful tool for assessing the role of phytohormones in development and concomittant physiological changes. In the present study tissue discs were forced into two major directions of development: callus-forming discs and non callus-forming discs. In these two types of tissue the role of the phytohormones ethylene and abscisic acid (ABA) with respect to their regulation of the induction and functioning of respiratory pathways (in particular alternative oxidase) was studied.

MATERIAL AND METHODS

Potato tuber discs (S.tuberosum var.Bintje), were sterilized and incubated on Petri-dishes with a modified B_5-medium as described by Gude and Van der Plas (1985). Callus-forming discs were obtained by adding 54 µM Naphthylacetic acid (NAA, a synthetic auxin) and 0.5 µM kinetin (a cytokinin) to the nutrient medium. Non-callus-forming discs were the result of incubation on the same medium without these phytohormones. Dependent on the experiment 1-aminocyclopropane-1-carboxylic acid (ACC, the precursor of ethylene; final concentration 10 µM) or abscisic acid(ABA, 40 µM) were added to the nutrient medium. Inhibition of ethylene action was achieved by injecting 2,5-norbornadien (NBD, a competitive inhibitor of ethylene action) into an exsiccator, in which the discs were kept(concentration : 4000 ppm). Measurements of in vivo respiration were carried out as decribed by Van der Plas and Wagner (1983). Mitochondria were isolated and mitochondrial respiratory characteristics were measured according to Gude and Van der Plas (1985).

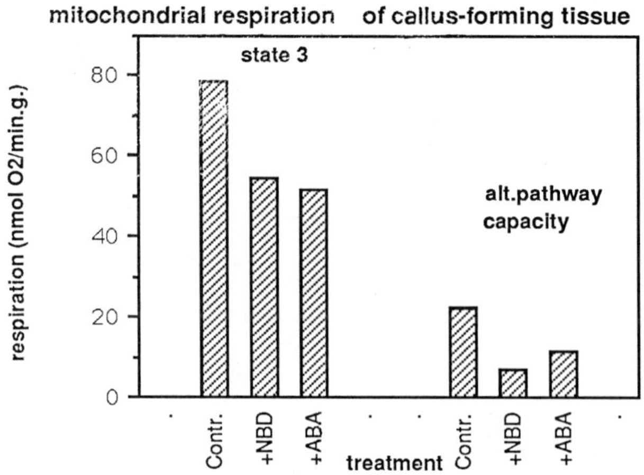

Fig.1. Oxygen uptake of mitochondria isolated from callusforming tissue after 6-20 days of incubation Respiratory substrate: Succinate + NADH.

RESULTS AND DISCUSSION

The induction of alternative oxidase was specifically affected by ethylene. Mitochondria from callus-forming tissue showed a high alternative pathway capacity as compared to the non callus-forming tissue (Fig's 1 and 2). The presence of auxin and cytokinin in the nutrient medium of these discs causes an ethylene formation that is much higher than the amount produced by non callus-forming discs (1600 vs. 60 pl/h.gram). Incubation of callus-forming discs in an atmosphere containing 4000 ppm NBD (2,5-Norbornadien, a competitive inhibitor of ethylene action) caused an inhibition of both alternative pathway capacity and state 3 respiration. Alternative path capacity, however, was inhibited to a

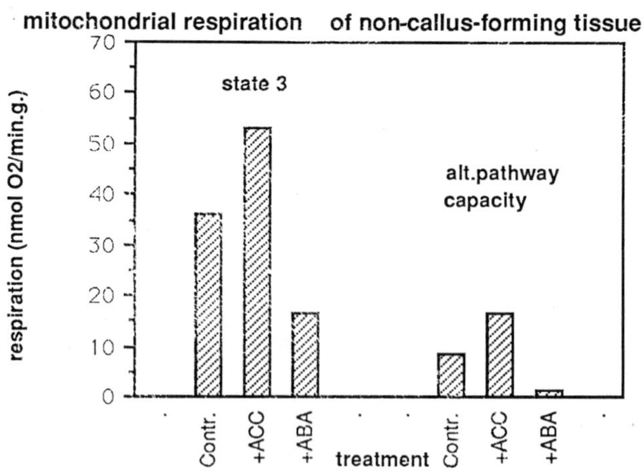

Fig.2. Oxygen uptake of mitochondria isolated from non-callusforming tissue, after 6-20 days of incubation. Respiratory substrate: Succinate + NADH.

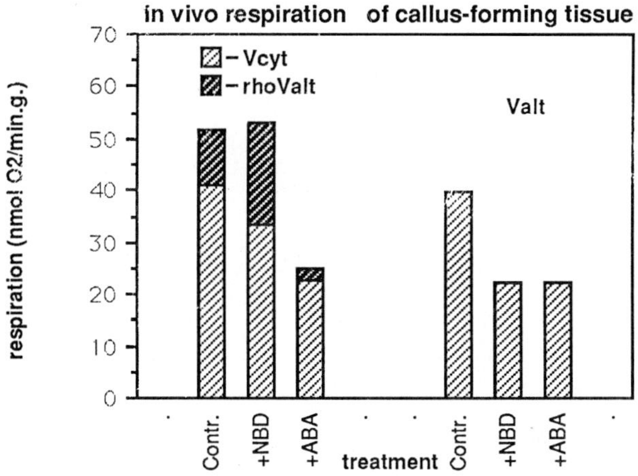

Fig.3. In vivo respiration of callus-forming tissue, measured after 6-20 days of incubation. Vcyt: respiration via the cytochrome path, which is preferentially used. rhoValt: activity of the alternative path. Vcyt+rhoValt = total respiration. Valt: capacity of the alternative pathway

larger extent (Fig.1). In mitochondria from the non callus-forming discs, with their relatively low ethylene formation, the level of alternative oxidase could be increased by adding ACC (the direct precursor of ethylene) to the nutrient medium, which enhances ethylene formation from 60 to 150 pl/h.gram (Fig.2). These results suggest that the induction of the alternative oxidase is (at least partly) regulated by the endogenous ethylene formation of the tissues. Abscisic acid, the phytohormone functioning in dormancy induction in potato tubers, had inhibitory effects on the formation of both respiratory pathways (in both tissue types), but in non-calllus-forming tissue a relatively specific inhibition of alternative oxidase induction was found (Fig's 1 and

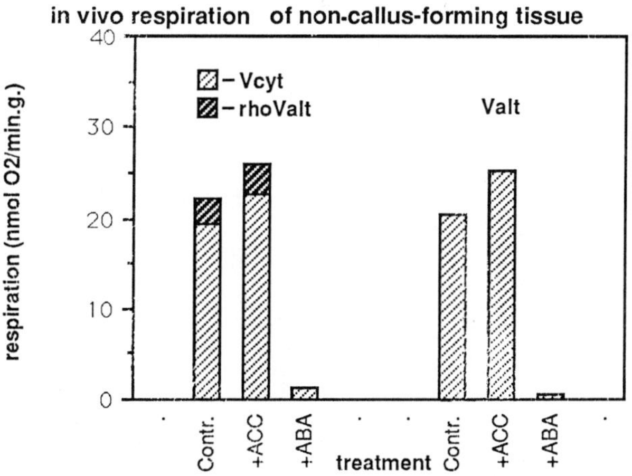

Fig.4. In vivo respiration of non-callus-formimg tissue, measured after 6-20 days of Incubation. Legend as in Fig.3

2). Another striking effect of ABA in non callus-forming tissue was a dramatical reduction of in vivo respiration (Fig.4). Total respiration of ABA-treated tissue decreased from 22 to 1 nmol O_2/min.gram, a respiratory activity far below that of mitochondria isolated from this tissue (Fig.2). Apparently ABA blocks the supply of respiratory substrates to the mitochondria. By this treatment a situation was achieved, which closely resembles the situation in resting potato tubers (with very low respiratory activity and hardly any alternative oxidase present).
The effects of ABA, ACC and NBD on the alternative pathway capacity, measured in vivo, were roughly the same as the effects observed in isolated mitochondria. The participation of the alternative pathway in uninhibited in vivo respiration depends on the overflow from the cytochrome pathway, which is preferentially used (Bahr and Bonner, 1973). In both types of control tissue a small part of alternative pathway capacity is used (Fig's 3 and 4). In NBD-treated callus-forming tissue, where ethylene action was inhibited, a large increase in alternative path activity could be observed, although alternative path capacity was markedly decreased (Fig.3). This was apparently caused by a slight decrease in cytochrome path activity and a non affected uninhibited respiration (supply of respiratory substrates). In ABA-treated tissues the alternative pathway activity was (almost) completely inhibited. In that case total respiration was inhibited to such an extent (by an inhibited glycolysis?) that the cytochrome pathway could account for the resulting supply of respiratory substrates.
Conclusion: The induction of alternative oxidase in potato tuber tissues is regulated by the phytohormones ethylene and abscisic acid. Effects of these phytohormones on the participation of alternative respiration are only secondary: they are caused by effects on the cytochrome pathway activity and the total supply of respiratory substrates.

Acknowledgement

The investigation was supported (in part) by the Foundation for Fundamental Biological Research (BION), which is subsidized by the Netherlands Organization for the Advancement of Pure Research (ZWO).

REFERENCES

Bahr, J.T. and Bonner, W.D., 1973. Cyanide-insensitive respiration. II. Control of the alternative pathway, J. Biol. Chem., 248:3446.
Gude, H. and Van der Plas, L.H.W., 1985. Endogenous ethylene formation and the development of the alternative pathway in potato tuber discs. Physiol. Plant., 65:57.
Lau, O. and Yang, S.F., 1973. Mechanism of a synergistic effect of kinetin on auxin-induced ethylene production. Suppression of auxin conjugation. Plant Physiol., 51:1011.
Van der Plas, L.H.W. and Wagner, M.J., 1983. Regulation of the activity of the alternative oxidase in callus-forming discs from potato tubers. Physiol. Plant., 58:311.

AGEING OF CARROT SLICES; RESPIRATION OF DISCS AND MITOCHONDRIA OBTAINED FROM THEM AS COMPARED WITH FRESH

Vera Vaz-Pinto

Department of Plant Biology
Faculty of Science
1294 Lisboa, Portugal

INTRODUCTION

The process of ageing in reserve tissues has been the subject of intense study and generally found to involve the appearance of a cyanide-resistant respiration. See Laties (1982) for a review of this problem. Most of the studies concerning both ageing and cyanide-resistant respiration, have been done in potato which thereby become a suitable frame of reference for this purpose (Day et al., 1980). Ethylene is known to trigger a cyanide-resistant respiration in different plant tissues (Rychter et al., 1978).

It has been known for quite a time that in carrot cyanide-resistance can be found in leaves (Marsh and Goddard, 1939). Besides during germination, the alternative pathway decreases with time as sugar accumulates in the taproot (Steingrover, 1981).

Our aim was to establish whether or not in carrot root ageing is a process similar to other reserve tissues, namely potato.

MATERIALS AND METHODS

Plant material: Carrot roots (*Daucus carota* L.), 25-35mm in diameter, were obtained from a local market. Within a few hours of purchase, 1Kg of roots, after peeling, were cut transversally in slices of 1mm thickness, rinsed briefly with distilled water and either used immediately as fresh slices or let to age for 24 hours.

Ageing treatment: Slices were immersed in two 4-liter jars, each one containing 2 liters of 0.1mM $CaSO_4$ and chloramphenicol (50 μg/ml). The medium was changed six times, four of them in the first seven hours of ageing. The system was shakened and aerated with an aquarium pump.

Ethylene treatment: Intact roots were immersed five minutes into, or sprayed with a solution 4,8 g/1 of ethephon (2-chlorethylphosphonic acid), an ethylene releasing substance.

CO_2 evolution: CO_2 exchange were monitored by an infrared gas analyzer (ADG Type 225 MARK II) in a closed system at 300ppm CO_2.

Oxygen uptake of discs: Discs 7mm diameter were prepared from slides, previously cut. Respiratory rates were measured polarographically in a 7.5ml thermostated cell at 25ºC in distilled water (YSI Model 53). Inhibitors of electron transport were 1mM KCN and 1.5mM SHAM.

Mitochondrial isolation: Washed mitochondria were obtained by differential centrifugation, as previously described (Vaz-Pinto et al., 1983). Washed mitochondria were purified in a Percoll 18% self-generated gradient. Outer membrane integrity was estimated in a spectrophotometer by succinate-cytochrome c oxidoreductase, as 95% intactness.

Oxygen uptake of mitochondria: Oxygen was measured in a 1ml thermostated cell (Hansatech electrode) at 25ºC. Respiration rates were measured in standard reaction medium containing 0.3 mannitol, 10mM KCl, 5mM $MgCl_2$, 10mM phosphate buffer pH 7.2 and BSA (1mg/ml). The O_2 concentration of the air saturated medium was taken as 240µM at 25ºC. Substrates used were 10mM succinate, 30mM malate, 0.3mM TPP, 10mM pyruvate and 1mM NADH.

Protein determination: Mitochondrial protein was determined by the Lowry method after solubilizing in 5% deoxicholate (Lowry et al., 1951).

RESULTS AND DISCUSSION

Fig. 1 where the respiration of discs was plotted against time, shows that the percentage of cyanide inhibition was the same all along the experimental period. However, it should be noted that the application of cyanide plus SHAM almost completely inhibited oxygen uptake also at all measurements times.

Similar experiments were carried out with potatos in order to see how the results compared with those obtained with carrot. It is clear from Fig. 2 that there is a considerable difference of the two kinds of discs. Although in both carrot and potato discs is an increase in the respiratory rates due to the ageing; nevertheless such an increase in the potato discs can be attributed mainly to the appearance of a cyanide-resistant respiration, which is not the case in carrot discs.

It should be stressed that although the percentage of cyanide inhibition varies with the storage temperature, in each storage condition used that percentage kept the same throughout time (Table 1).

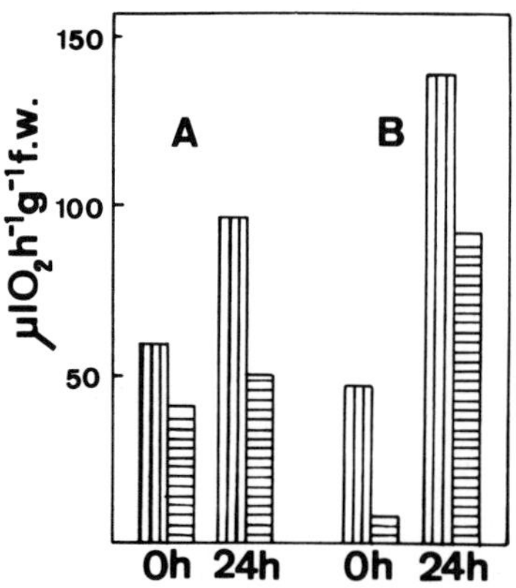

Fig. 1. Carrot roots discs respiration during ageing.
○ Control
◐ + 1 mM KCN
● + 1 mM KCN+ 1mM SHAM

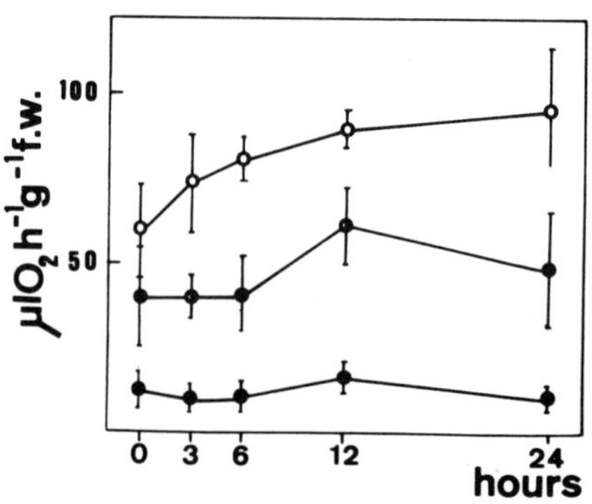

Fig. 2. Fresh and aged discs respiration in carrot roots.
Control
+ 1mM KCN

Table 1. Respiration of control and cyanide-treated carrot discs under two storage conditions

Storage Conditions	Time Hours	$\mu l\ O_2\ h^{-1}.\ g.\ f.\ w.$ Control	+1mM KCN	Inhibition (%)
	0	30.7	6.8	77.8
25°C	6	99.0	38.8	60.8
	24	111.2	41.8	62.4
	0	58.6	40.3	31.1
0°C	6	81.1	40.6	49.9
	24	96.1	49.2	48.8

With a view to induce an increase of cyanide-resistance, ethylene was applied. Fig. 3 shows that fresh slices from ethylene-treated roots leads to an increase in CO_2 output. However, this increase is not accompanied by a similar increase in O_2 uptake and that the percentage of cyanide inhibition remains the same during the experiment.

An improved technique for the isolation and purification of carrot mitochondria (Fig. 4 and Fig. 5) was used in order to see whether ageing and alternative pathway would go together. That was disproved by the results shown in Fig. 6 which are self-explanatory.

On the whole the data reported in this paper indicated that the process of ageing in carrot does not involve a significant development of a mitochondrial cyanide-resistant respiration which is in contrast with what happens in potato.

Acknowledgment: I am very grateful to Dr. C. Tuquet (Laboratoire de Biologie Végétale IV, Université Pierre-et-Marie-Curie, Paris VI) for kindly providing the electron micrograph of carrot root mitochondria.

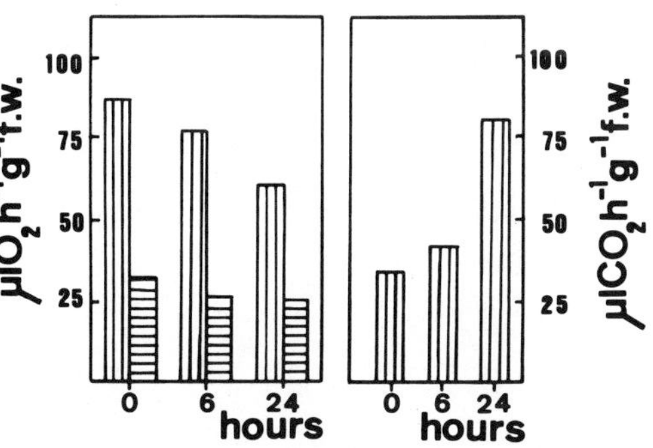

Fig. 3. Carrot roots respiration under ethylene
control
+ 1mM KCN

Fig. 4. Electron micrograph of Percoll-purified carrot mitochondria

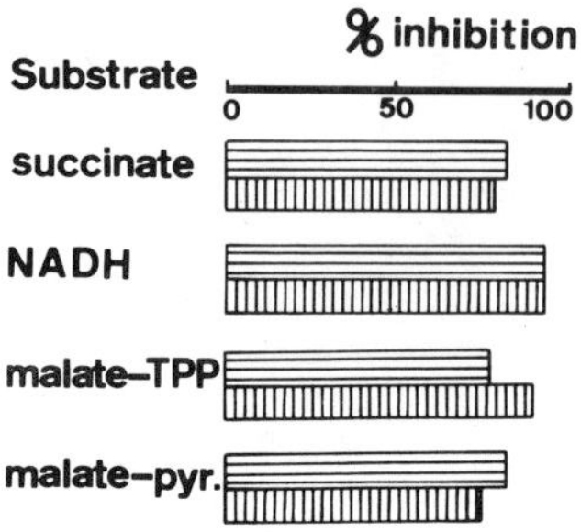

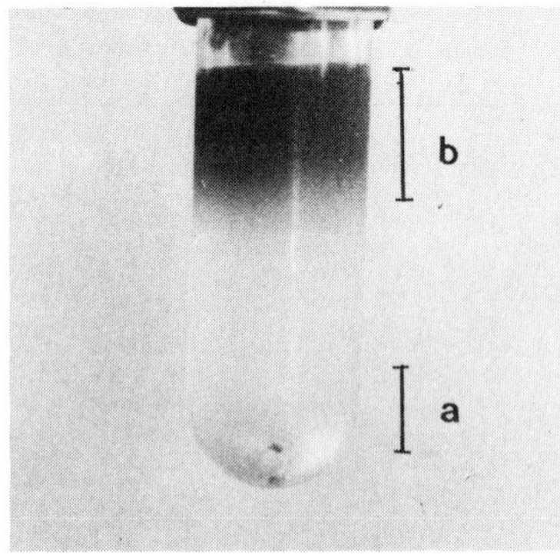

Fig. 5. Carrot mitochondria purification in an 18% Percoll self.regenerated gradient
a - mitochondria
b - carotenoids

Fig. 6. Mitochondrial state 3 inhibition by cyanide (500uM)

▤ Fresh slices mitochondria
▥ Aged slices mitochondria

REFERENCES

Day, D. A., Arron, G. P., and Laties, G. G., 1980, Nature and control of respiratory pathways in plants: The interaction of cyanide-resistant pathway, in: The Biochemistry of Plants, 2, D. D. Davies, ed., Academic Press, New York.

Laties, G. G., 1982, The cyanide-resistant, alternative path in higher plant respiration, Ann. Rev. Plant Physiol. 33:519.

Lowry, O. H., Rosebrough, N. J., Farr, A. L. and Randall, R. J., 1951, J. Biol. Chem., 193:265.

Marsh, P. B. and Goddard, D. R., 1939, Respiration and fermentation in the carrot, Daucus carota, Am. J. Bot., 26:724.

Rychter, A., Janes, H. W. and Frenkel, C., 1978, Cyanide-resistant respiration in freshly cut potato slices, Plant Physiol. 61:667.

Steingrover, E., 1981, The relationship between cyanide-resistant root respiration and the storage of sugars in the taproot in Daucus carota L., J. Exp. Bot., 32:911.

Vaz-Pinto, V., Gonçalves, M. P., Arrabaça, J. D., 1983, Communication submitted at 15th FEBS Meeting, Brussels.

RELATIONSHIP BETWEEN STRUCTURAL MITOCHONDRIAL PARAMETERS AND RESPIRATION RATES IN VARIEGATED LEAVES OF *Dieffenbachia* 'CAMILLA'

J.L. Araus, R. Calafell, G. Burgos, J.F. Aguila and J. Azcón-Bieto

Departament de Fisiologia Vegetal (J.L.A., R.C., G.B., J.F.A.)
Càtedra de Fisiologia General (J.A.B.), Facultat de Biologia
Universitat de Barcelona, Diagonal 645, Barcelona-08028, Spain

INTRODUCTION

Little is know about the relationship between structure and function of mitochondria during cell differentiation, although these organelles are usually considered to be relatively conservative compared to other cell organelles, like chloroplasts[1]. During greening, changes in mitochondrial activity seem to be restricted mainly to the early stages[2,3], and these changes seem to reflect transient shifts in the adenylate metabolism rather than a photomorphogenetic event[1]. A variegated plant like *Dieffenbachia* 'Camilla' show special features in the phenomenon of greening. For instance, greening develops slowly, and there are different degrees of greening in the same leaf. These characteristics make *Dieffenbachia* a very adequate work material for greening studies. The objective of this paper has been to investigate the relationship between mitochondrial number and volume and respiration rate in different zones of variegated adult leaves of *Dieffenbachia* grown under two different light regimes.

MATERIAL AND METHODS

Plant Material. *Dieffenbachia* 'Camilla' plants were grown in a glasshouse. Maximum photosynthetic photon flux density (PPFD) was 650 $\mu molm^{-2}s^{-1}$ (high light type plants, HL). Some of these plants were transferred to a growth chamber during 4 weeks. The illumination in the growth chamber was supplied by flourescent tubes (OSRAM-L-FLORA), giving a PPFD of 130 $\mu molm^{-2}s^{-1}$ at the plant level (low light type plants, LL). In both cases, the plants were adequately fertilized and irrigated. The fifth fully expanded leaf from the apex was used for measurements. Four different pigmented zones from the central nerve to the leaf margin were considered: "white", "yellow", "pale green", and "green".

Ultrastructural studies and chlorophyll determination. Small segments from the four zones considered were fixed, embedded in plastic, thin sectioned, and examined with transmission electron microscopy to document organelle size, and the quantity of mitochondria surrounding every mesophyll chloroplast, using techniques described previously[4]. Additionally, the number of chloroplast sections per mm of leaf transverse section were determined using light microscopy photomicrographs. Mitochondrial volume per leaf area was calculated by considering the mean size of mitochondrial sections, the number of mitochondrial sections per chloroplast and the number of chloroplast sections per unit of abaxial epidermis length in leaf cross sections. Leaf chlorophyll content was measured according to Lichtenthaler and Wellburn[5].

Respiratory oxygen uptake measurement. The rate of dark oxygen uptake of slices from the different zones considered was measured with a Rank O_2 electrode as described by Burgos et al.[6] in this book. Measurement temperature was 24ºC.

RESULTS AND DISCUSSION

Chlorophyll content and chloroplast sectional area in the four zones were normally higher in LL leaves than in HL leaves (Fig. 1a,b). Similar results were obtained by other authors in non-variegated leaves[7]. However, the unexpected higher number of chloroplasts per mm of leaf section in the LL leaf (Fig. 1c) could be related to the variegated condition of this species, that is, low light promotes greening in *Dieffenbachia* 'Camilla'. In all cases, these three parameters increased sharply from the "white" region to the "green" region, which is consistent with the existence of a differential greening process in the leaves. The number of mitochondria per chloroplast was also higher in the LL leaf (Fig. 1d). This ratio remained fairly constant in all zones of the leaf. Obviously, the number of mitochondria per leaf section increases from the "white" to the "green" zones.

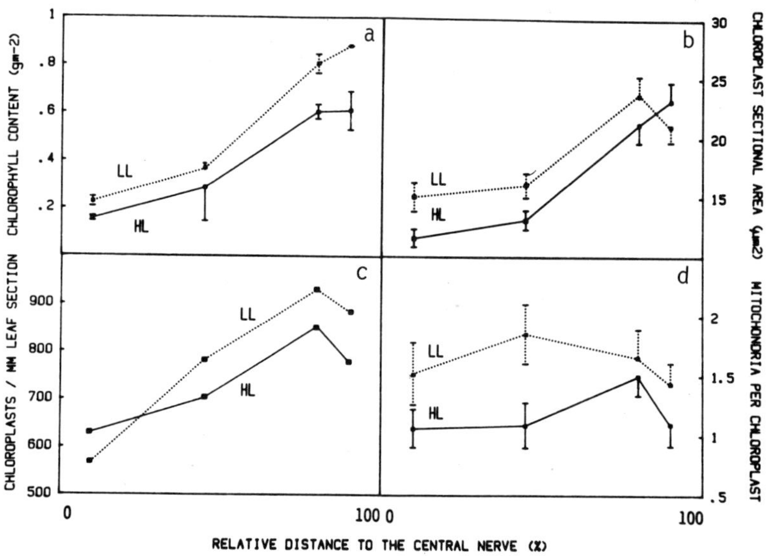

Fig. 1. Effect of light growth regime on several chloroplast and mitochondrial parameters of different parts of variegated leaves of *Dieffenbachia* 'Camilla'. Abcisse axis referres to the relative distance (%) to the central nerve of the zones considered: 10 "white", 45 "yellow", 80 "pale green", and 90 "green". *a*: Chlorophyll content per leaf area. Values are means ±SE of two replicates. *b*: Area of individual chloroplast sections measured from projections of electron microscopy negatives. Values are means ±SE of at least 25 replicates. *c*: Number of chloroplast sections per mm abaxial epidermis length in leaf cross sections. *d*: Mitochondrial sections per chloroplast section. It has been measured from electron microscopy photographies. Values are means ±SE of at least 25 replicates.

The HL leaf mitochondria showed an electron-dense matrix (Fig. 2) with a large number of randomly arranged cristae. Cristae were mostly elongated. Some cristae had almost no matrix, specially in the "white" and "yellow" regions. In the LL leaf, the cristae decreased in number and size, and fragmented into small vesicles; the matrix was less dense. Similarly to our results, Selga et al.[8] showed that mitochondria with similar ultrastructural features than those from the HL leaf exhibited active multiplication: the amount of mitochondria per chloroplast increased concomitantly with the number of chloroplasts. A similar mitochondrial ultrastructure to that of the LL leaf was observed by Vartapetian and Andreeva[9] under degradative anaerobic conditions. The mitochondrion sectional area was similar in both HL and LL leaves.

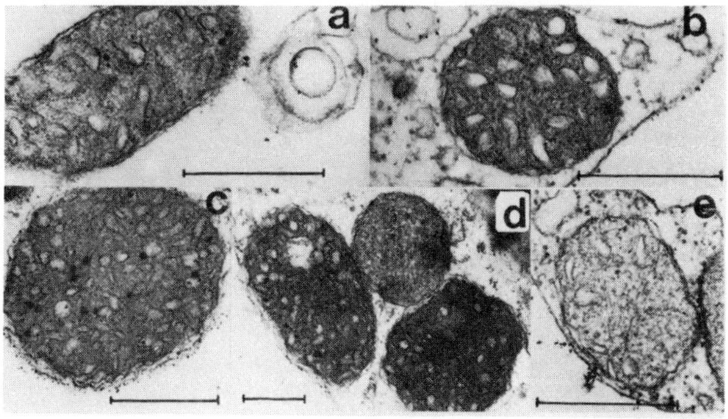

Fig. 2. Ultrastructure of mitochondria in mesophyll cells of leaves of *Dieffenbachia* 'Camilla'. *a,b*: Mitochondria from the "pale green" (*a*) and "white" (*b*) zones of a HL leaf. *c,d,e*: Mitochondria from the "green" (*c*), "pale green" (*d*) and "white" (*e*) zones of a LL leaf. Bars = 4 μm.

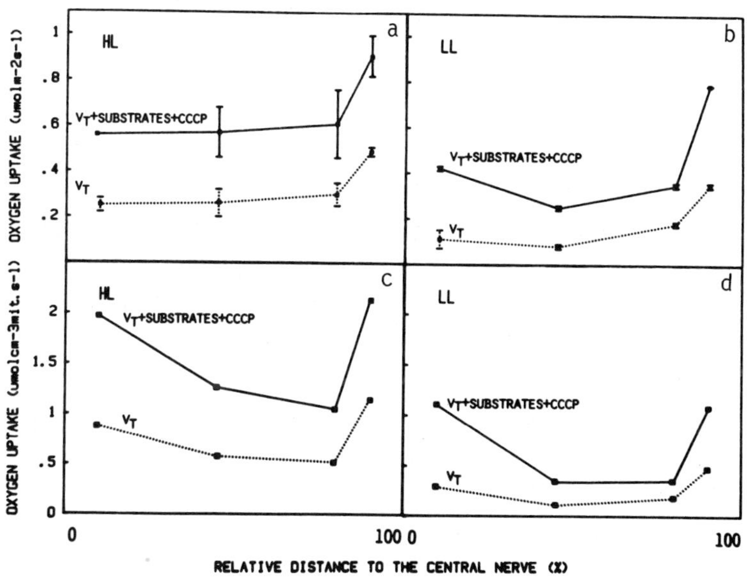

Fig. 3. Effect of light growth regime on O_2 uptake of leaf slices of *Dieffenbachia* 'Camilla' from different zones. The concentrations used were: sucrose 40 mM, glycine 20 mM and malate 20 mM as substrates; and carbonyl-cyanide m-chlorophennyl hydrazone (CCCP), 2 μM as uncoupler. *a,b*: O_2 uptake per leaf area. Values are means ±SE of 2-3 replicates. *c,d*: O_2 uptake per mitochondrial volume.

The rate of O_2 uptake of HL and LL leaves per area and per mitochondrial volume was similar in the "white", "yellow" and "pale green" regions, and increased significantly in the "green" region (Fig. 3). When respiration was measured in the presence of substrates (sucrose, malate plus glycine) and the uncoupler CCCP, with the object of avoiding substrate and adenylate limitations, the rate of O_2 uptake increased largely in all parts of the leaf (Fig. 3). However, the relative increase of respiration by substrates plus CCCP was higher in the less coloured zones, specially in the "white" zone. This suggests that the capacity of respiration is more expressed in the "green" zone. High light leaves presented higher respiration rates in the absence and in the presence of substrates plus CCCP (per leaf area and per mitochondrial volume), than LL leaves (Fig. 3). This suggests that the potential capacity of respiration is significantly lower in the LL leaves, in all zones studied. These results strongly contrast with some results obtained in the structural studies, which indicate that the LL leaves had a higher number of mitochondria per leaf section and per chloroplast than the HL leaves. It is evident that the biochemical activity of LL mitochondria should be lower than that of HL mitochondria and it seems not to be correlated with the amount of respiratory machinery. We have not isolated mitochondria for testing this point yet, but the mitochondrial ultrastructural features (e.g. high cristae number and dense matrix) of HL mitochondria are apparently consistent with their higher biochemical activity[10].

In conclusion, the low light induced greening in variegated leaves of *Dieffenbachia* 'Camilla' was associated with a decrease in respiration rates in all parts of the leaf, which did not correspond to the changes in the number of chloroplasts during the process of low light induced greening.

Acknowledgements. We are grateful to the "Servei de Microscopia Electronica de la U.B." for technical assistance, and to the "CIRIT, Ge. Cat." and "CAICYT" for financial support.

REFERENCES

1. P. Schopfer and K. Apel, Intracellular photomorphogenesis, in: "Photomorphogenesis. Encyclopedia of Plant Physiology," Vol. 16A, W. Shropshire, Jr. and H. Mohr, eds., Springer-Verlag, Berlin, Heidelberg (1983).
2. A. R. Wellburn and R. Hampp, Appearance of photochemical function in prothylakoids during plastid development, Biochim. Biophys. Acta, 547:380 (1979).
3. R. Hampp, Kinetics of mitochondrial phosphate transport and rates of respiration and phosphorylation during greening of etiolated Avena leaves, Planta, 144:325 (1979).
4. J. L. Araus, L. Alegre, L. Tapia, R. Calafell and M. D. Serret, Relationships between photosynthetic capacity and leaf structure in several shade plants, Am. J. Bot., in press, (1986).
5. H. K. Lichtenthaler and A. R. Wellburn, Determinations of total carotenoids and chlorophyll a and b of leaf extracts in different solvents, Bioch. Soc. Trans., 603:591 (1983).
6. G. Burgos, J. L. Araus and J. Azcón-Bieto, The effect of temperature on respiratory pathways of Fatsia japonica leaves, in: "Proceedings 2nd International Meeting on Plant Mitochondria Structural, Functional and Physiological Aspects," R. B. Beechey and A. L. Moore, eds., Plenum Publishing Corporation, New York (1986).
7. O. Björkman, Response to different quantum flux densities, in: "Physiological Plant Ecology I. Encyclopedia of Plant Physiology," Vol. 12A, O. L. Lange, P. S. Nobel, C. B. Osmond and H. Ziegler, eds., Springer-Verlag, Berlin, Heidelberg (1981).
8. M. Selga, A. Vitola, S. Kristkalne, G. Arente and L. Kattai, Ultrastructural features of cellular organelles in photosynthesizing tissues of rye, cucumber and pea plants treated with Camposan M, Photosynthetica, 19:19 (1985).
9. B. B. Vartapetian and I. N. Andreeva, Mitochondrial ultrastructure of three hygrophyte species at anoxia and in anoxic glucose-supplemented medium, J. Exp. Bot., 37:685 (1986).
10. J. L. Hall, T. J. Flowers and R. M. Roberts, The mitochondrion, in: "Plant Cell Structure and Metabolism," Longman, London, New York (1982)

THE EFFECT OF TEMPERATURE ON RESPIRATORY PATHWAYS OF *Fatsia japonica* LEAVES

G. Burgos, J.L. Araus and J. Azcón-Bieto

Departament de Fisiologia Vegetal (G.B. and J.L.A.) and
Catedra de Fisiologia General (J.A.B.), Facultat de Biologia
Universitat de Barcelona, Diagonal 645, Barcelona-08028, Spain

INTRODUCTION

The effect of temperature on plant respiration is a subject that has attracted much attention. There are many studies on the effect of temperature on mitochondrial membrane activities, specially in relation to the "chilling" stress[1,2]. The contribution of the cytochrome and alternative pathways to respiration at different temperatures has been studied at the isolated mitochondria level[3,4]. In general, the results show that electron flow through the alternative oxidase is more active at lower temperatures than at higher temperatures. Similar results were obtained when intact tissues were used, such as callus[5], roots[6] and hypocotyls[7]. However, there is no information in the case of intact leaves, and we have investigated in this paper the effect of temperature upon the activities of the respiratory pathways of leaf slices of *Fatsia japonica*.

MATERIALS AND METHODS

Plant Material. *Fatsia japonica*, a commercial C_3 plant, was grown outdoors under a shadehouse during spring time, in the Experimental Fields of the Faculty of Biology, University of Barcelona. The plants were adequately fertilized and irrigated. Almost fully mature leaves were used in the experiments.

Leaf slice preparation and oxygen consumption measurements. Methods were adapted from those described by Azcón-Bieto, Lambers and Day[8]. Leaves were transversely cut into 1-mm thick slices with a sharp razor blade under a 0.2 mM $CaCl_2$ solution. The slices were washed for 30-60 min in the same solution, which was renewed several times. The rate of O_2 uptake was measured with a Rank oxygen electrode in 5 ml of an air saturated solution (20 mM MES*buffer, pH 6.7, 0.2 mM $CaCl_2$). A nylon net separated the slices from the stirrer and the electrode. The cuvette of the O_2 electrode was surrounded by a water-jacket, and temperature changes were made by manipulating the temperature of a water bath connected with the jacket. SHAM was added from a stock solution of 1 M in 2-methoxyethanol. CCCP was added from a stock solution of 2 mM in ethanol.

---

*Abbreviations: CCCP, carbonyl-cyanide m-chlorophenyl hydrazone; SHAM, salicyl-hydroxamic acid; MES, 2-(N-morpholino)ethanesulfonic acid.

RESULTS AND DISCUSSION

The rate of O_2 uptake in the absence of inhibitors (V_t) of F.japonica leaves increased with temperature until about 25 ºC, and then significantly decreased at higher temperatures (Tables 1 and 2). This relatively low temperature optimum (25 ºC) for respiration is unusual, and generally it occurs at higher temperatures (above 35 ºC) in other plant species[9]. Interestingly, the degree of stimulation of respiration by the uncoupler CCCP was very low at temperatures higher than 30 ºC (Table 1, Expt. 3). It appears that leaves of F. japonica have an apparent difficulty to respire at high temperatures ($>$25 ºC).

The rates of O_2 uptake in the presence of KCN at different temperatures were very high compared to the uninhibited rates, and even some of them were stimulated by KCN (Table 1, Expt. 1). The capacity of the alternative pathway (V_{alt}) at different temperatures was estimated from these data, and it seems that V_{alt} was very high at temperatures between 6 and 30 ºC, and significantly declined at higher temperatures (Table 2). These results are similar to those obtained by McCaig and Hill[3] using isolated wheat mitochondria, and are in line with the results of other authors (see Introduction). Note that the estimates of V_{alt} presented in the Table 2 corresponding to temperatures between 6 and 30 ºC (except 25 ºC) do not probably reflect the maximum capacity of the alternative pathway, because KCN stimulated respiration at these temperatures (see above).

Similarly to the alternative pathway, the capacity of the cytochrome pathway (V_{cyt}), which was estimated by the rate of O_2 uptake in the presence

TABLE 1. The effect of temperature on O_2 uptake of leaf slices of F. japonica. The concentrations used were: sucrose, 40 mM; CCCP, 2 µM; SHAM, 8 mM; KCN, 1 mM. For other details see Materials and Methods. The values shown are means ± S.E. of 3-8 measurements.

Sequential additions	Respiration at different temperatures (ºC)						
	6	12	18	25	30	36	40
	$\mu mol\ O_2.m^{-2}.s^{-1}$						
Expt. 1							
None	0.64±0.11	0.83±0.06	1.03±0.15	1.12±0.06	0.81±0.09	0.53±0.08	0.68±0.02
+ KCN	0.99±0.02	1.05±0.08	1.04±0.08	1.05±0.09	0.83±0.08	0.46±0.08	0.66±0.04
+ SHAM	0.12±0.02	0.11±0.03	0.07±0.03	0.14±0.05	0.02±0.02	0.06±0.04	0.06±0.03
Expt. 2							
None	0.72±0.08	0.82±0.10	0.92±0.07	1.14±0.11	0.58±0.04	0.49±0.01	0.56±0.07
+ SHAM	0.66±0.06	0.82±0.10	0.90±0.03	0.99±0.03	0.55±0.03	0.45±0.01	0.56±0.07
+ KCN	0.03±0.03	0.07±0.03	0.06±0.02	0.10±0.04	0.02±0.02	0	0.06±0.04
Expt. 3							
None	0.65±0.18	0.92±0.01	0.88±0.08	1.05±0.13	0.85±0.02	0.71±0.03	0.66±0.07
+CCCP+SUCROSE	1.04±0.15	1.17±0.04	1.15±0.14	1.50±0.16	0.85±0.02	0.75±0.05	0.66±0.07
+ SHAM	0.92±0.06	0.96±0.13	0.96±0.11	1.02±0.13	0.69±0.02	0.71±0.05	0.66±0.03
+ KCN	0.10±0.03	0.12±0.07	0.09±0.02	0.12±0.03	0.11±0.01	0.06±0.03	0.13±0.02

TABLE 2. Estimation of respiratory pathways of F. japonica leaves at different temperatures in the absence of CCCP and sucrose. The values shown were calculated from the experiments described in Table 1. V_t is the measured rate of O_2 uptake in the absence of inhibitors. V_{res} is the rate of residual respiration, and was estimated by the rate of O_2 uptake resistant to both KCN and SHAM. v_{cyt} is the activity of the cytochrome pathway, estimated by measuring O_2 uptake in the presence of SHAM minus V_{res}. v_{alt} is the rate of O_2 uptake sensitive to SHAM alone and represents the activity of the alternative pathway. V_{alt} is the capacity of the alternative pathway, estimated by measuring O_2 uptake in the presence of KCN minus V_{res}. The asterisks indicate that the capacity of the alternative pathway may have been underestimated at the temperatures shown, because KCN stimulated respiration (see Table 1). The parameter ρ is the fraction of the alternative pathway operating, and is calculated by the ratio v_{alt}/V_{alt}.

TEMPERATURE	V_t	v_{cyt}	v_{alt}	V_{alt}	V_{res}	ρ
ºC			$\mu mol\ O_2.m^{-2}.s^{-1}$			ratio
6	0.67 ±0.07	0.62 ±0.07	0.06 ±0.02	*0.88 ±0.01	0.08 ±0.02	0.07
12	0.83 ±0.05	0.75 ±0.11	0	*0.94 ±0.06	0.09 ±0.02	0
18	0.97 ±0.08	0.84 ±0.01	0.02 ±0.01	*0.98 ±0.08	0.06 ±0.02	0.02
25	1.14 ±0.05	0.87 ±0.05	0.22 ±0.05	0.89 ±0.09	0.10 ±0.03	0.25
30	0.69 ±0.06	0.53 ±0.03	0.03 ±0.01	*0.81 ±0.09	0.02 ±0.01	0.04
36	0.51 ±0.04	0.45 ±0.01	0.04 ±0.02	0.32 ±0.05	0.03 ±0.02	0.13
40	0.62 ±0.04	0.50 ±0.05	0	0.60 ±0.06	0.06 ±0.02	0

TABLE 3. Estimation of the activities of respiratory pathways of F. japonica leaves at different temperatures in the presence of CCCP plus sucrose. The values shown were calculated from the experiments described in Table 1. Symbols and definitions are the same as in Table 2, except that V_{cyt} is the capacity of the cytochrome pathway, estimated by measuring O_2 uptake in the presence of CCCP (2 μM), sucrose and SHAM. The values of V_{alt} used for calculations were the same as in Table 2.

TEMPERATURE	V_t	V_{cyt}	v_{alt}	V_{res}	ρ	v_{cyt}/V_{cyt}
ºC		$\mu mol\ O_2.m^{-2}.s^{-1}$			ratio	ratio
6	1.04 ±0.15	0.83 ±0.06	0.23 ±0.09	0.10 ±0.03	0.26	0.75
12	1.17 ±0.04	0.84 ±0.06	0.21 ±0.09	0.12 ±0.07	0.22	0.89
18	1.15 ±0.14	0.87 ±0.10	0.19 ±0.08	0.09 ±0.02	0.19	0.97
25	1.50 ±0.16	0.88 ±0.05	0.48 ±0.05	0.11 ±0.03	0.54	0.99
30	0.85 ±0.02	0.58 ±0.04	0.16 ±0.02	0.11 ±0.01	0.20	0.91
36	0.75 ±0.05	0.64 ±0.05	0.05 ±0.03	0.06 ±0.03	0.16	0.70
40	0.66 ±0.05	0.54 ±0.04	0	0.13 ±0.02	0	0.93

an uncoupler (CCCP), sucrose and SHAM, was also higher at lower temperatures than at higher temperatures (Table 3). The cytochrome pathway was used at full capacity (that is, $v_{cyt}/V_{cyt}=1$; v_{cyt} is the activity of the cytochrome path, estimated by the rate of O_2 uptake in the presence of SHAM -but not CCCP) only at 25 ºC (Table 3). At any other temperature, the cytochrome pathway was somewhat restricted by the adenylate system ($v_{cyt}/V_{cyt} < 1$).

The alternative pathway was practically not expressed in the absence of KCN and uncoupler, except at 25 ºC, where it was expressed about 25 % of its estimated capacity (Table 2). However, the expression of the alternative pathway increased in the presence of CCCP (plus sucrose), reaching the maximum engagement (54 % of its capacity) again at 25 ºC (Table 3). Interestingly, the largest measured activity of the alternative pathway was obtained when the cytochrome pathway was fully expressed (that is, at 25 ºC). These results suggest that the expression of the alternative pathway in *F. japonica* leaves at different temperatures was not correlated with its capacity (see Table 2), and they are consistent with the model of Bahr and Bonner[10]. The results also suggest that the expression of the alternative pathway seems to be dependent to some extent on the energy conditions, because the uncoupler stimulated the activity of this pathway on top of a fully saturated cytochrome pathway. Maybe the phosphorylation site I, which is rotenone-sensitive, is involved in the control of the alternative pathway by adenylates.

Acknowledgements. We are grateful to the "Comisión Asesora de Investigación Científica y Técnica" (CAICYT) of the Spanish Government for financial support.

REFERENCES

1. J.K.Raison, Effect of low temperature on respiration, *in*:"The Biochemistry of Plants. A Comprehensive Treatise," Vol. 2, D.D. Davies, ed. Academic Press, New York (1980).
2. D. Graham and B.D. Patterson, Response of plants to low, non-freezing temperatures: proteins, metabolism and acclimation, *Annu. Rev. Plant Physiol*., 33:347 (1982).
3. T.N. McCaig and R.D. Hill, Cyanide-insensitive respiration in wheat: cultivar differences and effects of temperature, carbon dioxide, and oxygen, *Can. J. Bot*., 55:549 (1977).
4. N.D. Cook and R. Cammack, Effects of temperature on electron transport in *Arum maculatum* mitochondria, *Plant Physiol*., 79:332 (1985).
5. S. Yoshida and F. Tagawa, Alteration of the respiratory function in chill sensitive callus due to low temperature stress. I. Involvement of the alternative pathway, *Plant Cell Physiol*., 20:1243 (1979).
6. G. Smakman and J.J. Hofstra, Energy metabolism of *Plantago lanceolata*, as affected by change in root temperature, *Physiol. Plant.*, 56:33 (1982).
7. C.M. Kiener and W.J. Bramlage, Temperature effects on the activity of the alternative respiratory pathway in chill-sensitive *Cucumis sativus*, *Plant Physiol.*, 68:1474 (1981).
8. J. Azcón-Bieto, H. Lambers and D.A. Day, The effect of photosynthesis and carbohydrate status on respiratory rates and the involvement of the alternative path in leaf respiration, *Plant Physiol.*,72:598 (1983).
9. D.F. Forward, Effects of temperature on respiration, *in* "Encyclopedia of Plant Physiol." Vol. XII (2), W. Ruhland, ed., Springer, Berlin (1960).
10. J.T. Bahr and W.D. Bonner, Cyanide-insensitive respiration. II. Control of the alternate pathway, *J.Biol. Chem.*, 248:3446 (1973).

SALT TOLERANCE - DOES LEAF RESPIRATION HAVE A CONTRIBUTION TO MAKE?

Simon Brown, David A. Day and Christa Critchley

Botany Department, Australian National University
GPO Box 4, Canberra, ACT, 2601, Australia

INTRODUCTION

A saline environment imposes considerable energetic demands upon a plant (Yeo 1983). Paradoxically, however, in spite of the increased costs associated with maintenance respiration (Schwarz and Gale 1981), ion uptake (Veen 1980) and osmotic adjustment (McCree 1986), photosynthesis is depressed in salt-stressed plants (Papp et al. 1983). On the other hand, respiration may rise (Livne and Levin 1967) or fall (Morozovskii and Kabanov 1970) perhaps depending upon leaf age (Azcon-Bieto et al. 1983). Generally, however, little is known about the effects of salinity upon leaf respiration.

Here we compare the regulation of respiration in the leaves of several salt tolerant species with that of peas grown in the presence or absence of 80mM NaCl.

MATERIALS AND METHODS

The salt tolerant species *Avicennia marina*, *Aegiceras corniculatum*, *Limonium australe*, *Hydrocotyle bonariensis* and *Suaeda australis* were either grown under glasshouse conditions and watered with nutrient solution supplemented with 250mM NaCl or collected from a mangrove swamp on the southern coast of New South Wales. The different environments did not significantly alter the respiratory properties of leaves from any given species. Pea plants (*Pisum sativum* cv "Green Feast") were similary treated but the nutrient solution of the salinised plants was supplemented with 80mM NaCl. This concentration of NaCl significantly reduced plant growth.

Leaf slice respiration was assayed polarographically in 4ml of buffer (50mM HEPES, 10mM MES pH 6.6, 0.2mM $CaCl_2$) (Azcon-Bieto et al. 1983). SHAM was added from a 1M solution in 2-methoxyethanol. The pattern of regulation was analysed as described by Azcon-Bieto et al. (1983).

Leaf mitochondria were isolated from peas as described by Day and Wiskich (1980) and similarly from *Limonium* leaves except that 1M osmoticum was used.

RESULTS

It can be seen from Table 1 that *in vivo* salinisation (80mM NaCl applied during growth) of peas reduced the total rate of oxygen uptake (Vt) and the capacity of the cytochrome pathway (V_{cyt}) was much reduced. *In vitro* salinisation (1M NaCl included in the assay medium) of pea leaf slices stimulated Vt and reduced V_{cyt}. Such treatment had no effect on *Avicennia* or *in vivo* salinised pea leaf slices. Salinisation generally increased alternative pathway engagement (α). In *Avicennia* the alternative oxidase was almost completely engaged in the absence of salt *in vitro*. Furthermore, *Avicennia* respiration could not be stimulated by the addition of uncouplers and/or substrates.

In order to see whether the respiratory pattern exhibited by *Avicennia* was typical we examined other halophytes. From Table 2, however, it can be seen that there is considerable variation in the pattern of respiratory regulation between various halophytes. It is of particular note that the two mangrove species, *Avicennia marina* and *Aegiceras corniculatum*, had very different α values.

Finally it can been seen from Figure 1 that there is little difference in the salt response of NADH oxidation between isolated pea and *Limonium* mitochondria

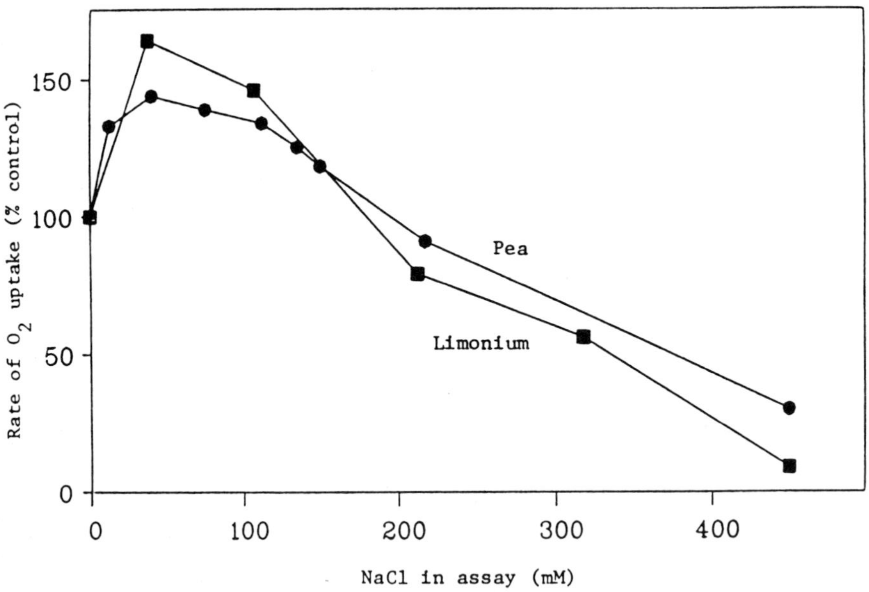

Figure 1. Effect of the concentration of NaCl on the rate of oxygen uptake of isolated mitochondria from pea and *Limonium* leaves. Osmotic strength was maintained at 1M for *Limonium* and at 0.3M for peas, except for the final point where it was 1M. Assay medium was osmoticum (NaCl, sorbitol), 1mM NADH, 20mM HEPES (pH 7.4), 5mM KH_2PO_4, 2mM $MgCl_2$, 1% BSA.

Table 1. Respiratory characteristics of the leaves of peas grown in the presence or absence of 80mM NaCl, and *Avicennia marina* and the effect of assaying respiration in the presence of 1M NaCl (values are means with less than 10% error except where marked (*) where the error is less than 15%).

	Pisum sativum				*Avicennia marina*	
	Control		Salinised			
[NaCl] *in vitro* (M)	0	1	0	1	0	1
	Rate of O_2 uptake (moles O_2 h^{-1} g^{-1} DW)					
Total	73.4	112.2	49.2	56.4	44.1	42.3
Cytochrome pathway						
activity	58.2	73.4	28.8*	33.6*	35.2	31.7
capacity	145.5	88.7	49.2*	44.4*	35.2	31.7
Alternative pathway						
activity	15.2*	38.8*	20.4*	22.8*	8.9*	10.6*
engagement	0.29	0.67	0.59	0.95	0.96	1.00

Table 2. Respiratory characteristics of the leaves of several halophytes (Values are means with less than 10% error).

Species	Total Rate of O_2 uptake	Alternative Pathway Activity (mole O_2 h^{-1} g^{-1} DW)	Cytochrome Pathway Activity	Engagement of Alternative Pathway
Hydrocotyle bonariensis	82.3	17.8	64.5	0.30
Aegiceras corniculatum	33.0	2.0	31.7	0.06
Limonium australe	147.1	15.7	131.3	0.20
Suaeda australis	76.3	22.1	35.8	0.62
Avicennia marina	44.1	8.9	35.2	0.96

DISCUSSION

It is known that the growth of peas is considerably retarded by as little as 100mM NaCl (Nieman 1962) but *Avicennia* grows successfully in the presence of 500mM NaCl (Burchett et al. 1984) and can withstand at least 850mM NaCl (Connor 1969). Consequently halophytes must have specific properties which enable them to tolerate such huge salt concentrations. The fact that mitochondria from peas and *Limonium* were not obviously different in their response to salt suggests that they themselves are not inherently salt tolerant in halophytes.

The question remains, however, as to whether the regulation of respiration *in vivo* is at all different due to salt tolerance. It is generally supposed that ions are sequestered in the vacuole and that compatible solutes accumulate in the cytoplasm to balance the osmotic potential of the vacuole. Lambers et al. (1983) observed a correlation between the synthesis of compatible solutes and a reduction in alternative pathway activity. This is an example of respiration not contributing to tolerance. It is, however, conceivable that an increase in respiration might accompany solute synthesis thus implying an active role in salt tolerance.

Our observation that there is no discernible pattern to the regulation of respiration in a number of halophytes from a single ecosystem suggest that respiration does not directly contribute to the differences between salt tolerant and salt sensitive species. It must be concluded, therefore, that there is no obvious respiratory strategy which is peculiar to halophytes. Furthermore it is unlikely that salt tolerance is a subcellular adaptation. Rather, it is probable that the salt is excluded from the cytoplasm.

REFERENCES

Azcon-Bieto, J., Lambers, H. and Day, D.A. (1983) Aust. J. Plant Physiol. 10, 237-245.

Burchett, M.D., Field, C.D. and Pulkownik, A. (1984) Physiol. Plant. 60, 113-118.

Connor, D.J. (1969) Biotrop. 1, 36-40.

Day, D.A. and Wiskich, J.T. (1980) FEBS Letts. 112, 191-194.

Lambers, H., Blacquiere, T. and Stuiver, B. (1981) Physiol. Plant. 51, 63-68.

Livne, A. and Levin, N. (1967) Plant Physiol. 42, 407-414.

McCree, K.J. (1986) Aust. J. Plant. Physiol. 13, 33-43.

Morozovskii, V.V. and Kabanov V.V. (1970) Sov. Plant Physiol. 17, 482-486.

Nieman, R.H. (1962) Bot. Gaz. 123, 279-285.

Papp, J.C., Ball, M.C. and Terry, N. (1983) Plant Cell Environ. 6, 675-677.

Schwarz, M. and Gale, J. (1981) J. Exp. Bot. 32, 933-941.

Yeo, A.R (1983) Physiol. Plant. 58, 214-222.

Veen, B.W. (1980) In: Rains, D.W., Valentine, R.C. and Hollaender, A. (eds) Genetic Engineering of Osmoregulation, New York: Plenum Press. pp. 187-195.

EFFECT OF SALT STRESS ON PROPERTIES OF MITOCHONDRIA ISOLATED FROM ETIOLATED AND GREENING MAIZE SEEDLINGS

Yves Jolivet and Pierre Dizengremel

Laboratoire de Physiologie Cellulaire Végétale
Université de Nancy I, B.P. 239
54506 Vandoeuvre-les-Nancy Cedex, France

INTRODUCTION

The enhancement of respiration in the leaves of plants submitted to a salt stress, particularly salt sensitive species, is well known[1]. However, such a result is often found to be dependent on the age of plants and the severity of the applied stress. Otherwise, in some plants, drastic environmental conditions result in an increase in proline content[2]. The biochemical mechanisms, contributing to proline accumulation are still controverted . The loss of oxidation of this amino acid was usually thought as one of the main causes of its accumulation.

In the experiments reported here, a severe NaCl stress (150 mM) was applied and changes in respiratory metabolism during the growth of etiolated and greening maize seedlings (Zea mays L.) were followed. The oxidation of substrates was investigated in mitochondria isolated from the shoots of control and treated seedlings. The results were discussed taking account the levels of sodium, potassium and proline in the different organs.

MATERIAL AND METHODS

Maize seeds were germinated in the dark at 28°C. After 87 hours, the etiolated seedlings were transferred to continuous light (30 Wm^{-2}) for further 48 hours. The salt treatment (150 mM) was applied at the onset of seed imbibition. Na^+ and K^+ contents were determined by flame photometry. Free proline was determined according to the method described by Troll and Lindsley[3].

Mitochondria were isolated from mesocotyls, coleoptiles and leaves using classical techniques[4]. Mitochondrial proteins were estimated by the method of Lowry and the protein content was corrected for contamination by thylakoids proteins. O_2 uptake was measured at 25°C using Hansatech O_2 electrode. Twenty mM succinate, 30 mM malate (+2mM glutamate and 0.4 mM NAD) and 20 mM proline were used as substrates. Cyanide and salicylhydroxamic acid (SHAM) were used as inhibitors of respectively the cytochrome and alternative pathways. Succinate dehydrogenase was assayed by the PMS-DPIP method (600 nm). Malate dehydrogenase (EC 1.1.1.37) was measured by following NADH oxidation (340 nm). To assess outer membrane integrity, the activity of cytochrome oxidase ($\pm$0,02 % Triton X-100) was measured at 550 nm.

Table 1. Sodium, potassium and proline contents of shoots from control and NaCl-treated maize plants

	Etiolated seedlings				Greening seedlings					
	Foliar organs		Mesocotyls		Leaves		Coleoptiles		Mesocotyls	
	Co	Tr	Co	Tr	Co	Tr	Co	Tr	Co	Tr
K^+ [a]	808	941	648	818	1253	967	754	719	962	778
Na^+ [a]	15	209	16	404	9	206	13	339	23	795
Proline [b]	12	9	11	35	15	26	18	10	12	31

Co : Control plants ; Tr : 150 mM NaCl-treated plants.

[a] : µeq/g dry matter ; [b] : µmole/g dry matter

RESULTS

Effect of NaCl Treatment on Growth and Metabolite Levels

Salt treatment affected drastically the growth of etiolated seedlings. The dry matter or the mesocotyls and foliar organs was reduced to 50 % compared to the control seedlings. The transition to light did not make worse this salt effect. The water deficit induced in the different tissues of etiolated seedlings was more pronounced in the foliar organs (87 % on the basis of fresh matter weight).

Upon salt treatment the K^+ level was not considerably modified (Table 1). By contrast, Na^+ content increased. However, Na^+ was preferably accumulated in mesocotyls, allowing the leaves to maintain a K^+/Na^+ ratio higher than 1. Proline content chiefly increased in mesocotyls of etiolated and greening seedlings growing in the presence of NaCl (Table 1). In leaves, such an increase was only observed after the 48 h light-grown period.

Effect of NaCl Treatment on Mitochondrial Properties

Mitochondria isolated from mesocotyls. In control plants, succinate was actively oxidized by mitochondria isolated from mesocotyls of etiolated seedlings (Table 2). On the contrary, malate and proline were slowly oxidized. The 48 h light grown period resulted in a decrease of the oxidation rates, particularly with succinate and proline as substrates. Moreover, in mitochondria isolated from greening seedlings, the phosphotylative efficiency slightly increased, whatever the engaged substrates. In all cases, higher ADP/O ratios occurred with malate in agreement with an occasional additional site for energy conservation.

In etiolated seedlings, NaCl treatment induced nearly a two fold rate of succinate and malate oxidation, whereas it was reduced by half with proline as substrate (Table 2). The phosphorylative efficiency was poorly modified with succinate and malate as substrates, while it significantly diminished with proline as substrate. After light exposure, the rates of oxidation of the different substrates were not modified. By contrast, the rate of proline oxidation diminished upon salt treatment. The phosphorylative efficiency decreased with all the substrates in mitochondria isolated from mesocotyls of greening seedlings submitted to 150 mM NaCl. The slight cyanide resistance commonly observed in control tissues (20 - 30 %) slightly increased upon salt treatment only with malate as substrate.

Table 2. Changes in properties of mitochondria isolated from mesocotyls of maize seedlings treated with 150 mM NaCl.

Seedlings		Substrate							
		Succinate			Malate			Proline	
		State 3[a]	ADP/O	KCN[b]	State 3[a]	ADP/O	KCN[b]	State 3[a]	ADP/O
Etiolated	Co	156	1.25	27	56	1.63	21	56	1.28
	Tr	248	1.31	24	82	1.52	27	23	0.92
Greening	Co	51	1.34	33	31	1.82	27	14	1.53
	Tr	236	1.15	24	65	1.68	41	18	1.30

[a]The rates are expressed in nmol $O_2.min^{-1}.mg^{-1}$ protein.
[b]KCN refers to cyanide resistance in percent of the state 3 rate.

Mitochondria isolated from leaves and coleoptiles. Succinate, malate and proline were oxidized in foliar organs of control plants to the same extent than in mesocotyls. By contrast, after salt treatment, the rates of oxidation of succinate and malate did not increase in mitochondria isolated from leaves and coleoptiles of etiolated seedlings while slightly higher values than in the control were observed in mitochondria from greening seedlings (Fig. 1). Whatever the substrate, the phosphorylative efficiency was significantly depressed in mitochondria isolated from etiolated and greening plants submitted to 150 mM NaCl. Moreover, an engagement of the alternative pathway did not seem to occur in response to salt treatment.

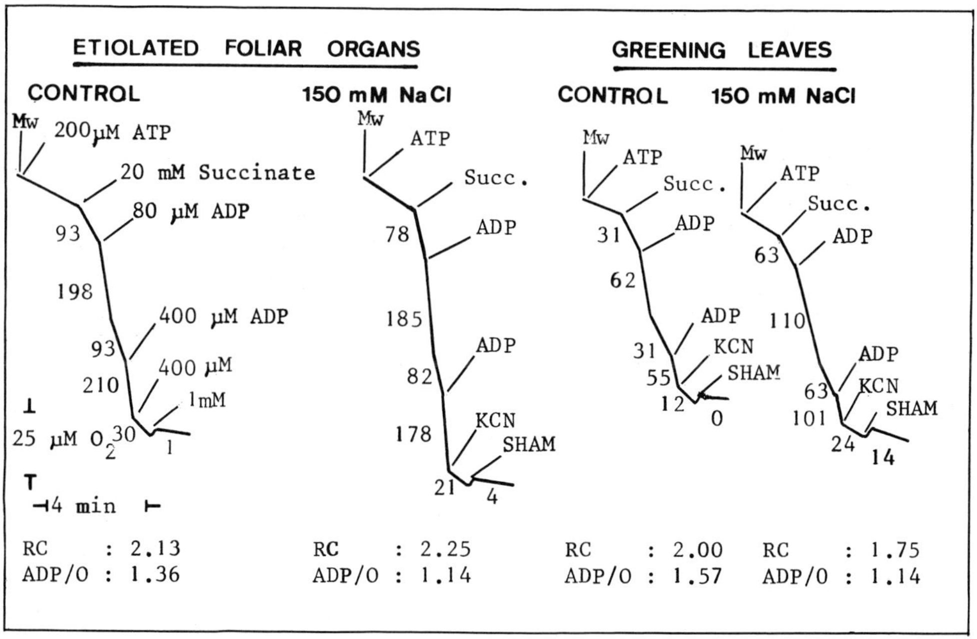

Fig. 1. Succinate oxidation by mitochondria isolated from foliar organs of control and treated maize plants.

Attempt to Identify the Target of Salt Stress on Mitochondria

Assays were made to appreciate the effect of the salt stress on mitochondrial membrane integrity as well as on enzyme activities taking part in the oxidations of succinate and malate.Cytochrome oxidase was measured on mitochondria isolated from green leaves. The comparison of cytochrome oxidase activity in presence and absence of Triton X-100 shows that the percentage of intact mitochondria slightly decreased from 92 to 88% upon salt treatment. Succinate DPIP reductase activity was 185 nmol. min^{-1}. mg^{-1} protein in mitochondria from control green leaves and malate dehydrogenase activity was 1.94 μmol. min^{-1}. mg^{-1} protein. These activities were not markedly modified in mitochondria from leaves of treated plants.

DISCUSSION

The growth of mesocotyls and foliar organs of both etiolated and greening maize seedlings was inhibited in the same way by a treatment with 150 mM NaCl. Higher rates of succinate and malate oxidations were observed in mitochondria isolated from mesocotyls of etiolated and greening seedlings treated with NaCl without an higher involvement of the alternative pathway. The Na^+ content in mesocotyls of etiolated and greening plants increased upon NaCl treatment . In this case, the regulation of ion transport required an higher need for energy which could be met with the increased activity of the cytochrome path. Moreover, this increased rate is necessary to counteract the low efficiency of the phosphorylating process. In foliar organs, slight higher rates with malate and succinate as substrates occurred after 48h of greening. It must be noted that, in these organs, Na^+ accumulation was avoided while a slight water deficit was observed. A lower efficiency of the phosphorylation in mitochondria isolated from leaves could be related to the water deficit although it is not a general agreement[5,6].Whatever the organ, proline oxidation in mitochondria from etiolated seedlings decreased upon salt treatment. This was associated with lower ADP/O ratios compared to control. This decrease in the rate of proline oxidation was also shown when etiolated maize shoots are submitted to a water stress[6]. Thus, salt treatment, as water stress, could directly affect mitochondria responsible for proline oxidation. Furthermore, there seems to be no correlation between the increase in proline content and the inhibition of the oxidation of this aminoacid.

REFERENCES

1. A. Livne and N. Levin, Tissue respiration and mitochondrial oxidative phosphorylation of NaCl-treated pea seedlings. Plant Physiol. 42: 407 (1967).
2. D. Aspinall and L.G. Paleg, Proline accumulation: physiological aspects in: "The Physiology and Biochemistry of Drought Resistance in Plants," L.G. Paleg and D. Aspinall, eds, Academic Press, Sydney (1981).
3. W. Troll and R. Lindsley, A photometric method for the determination of proline. J. Biol. Chem. 215: 655 (1955).
4. C. Lance, Les mitochondries de l'inflorescence de Chou-fleur.I. Propriétés oxydatives. Physiol. Vég. 9: 259 (1971).
5. V. N. Zolkevitch and A.Y. Rogacheva, P/O ratio in mitochondria isolated from wilting plant tissues. Sov. Plant Physiol. 15: 450 (1968).
6. G. D. Sells and D.E. Koeppe, Oxidation of proline by mitochondria isolated from water stressed maize shoots. Plant Physiol. 68: 1058 (1981).

OPTICAL PROPERTIES AND RESPIRATORY ACTIVITY IN THE MARINE DINOFLAGELLATE *Crypthecodinium cohnii*

Michel Denis, Ted Packard* and Martine Rodier

Centre d'Océanologie de Marseille, CNRS UA 41, Faculté des Sciences de Luminy, Case 901, 13288 Marseille Cedex 9 - France
* Permanent address : Bigelow Laboratory for Ocean Sciences, McKown Point, West Boothbay Harbor, Maine 04575

INTRODUCTION

Biological processes in the ocean are receiving more and more attention because they interact with oceanic circulation to regulate the distribution of organic and inorganic chemicals. Being common to all organisms respiration emerges as one of the most important biological processes operating as oceanic regulators.

Whereas a great deal of work has been focused on photosynthesis and nitrogen metabolism, two major biological processes, far less attention has been paid to respiration, especially in phytoplankton. In this poorly investigated field, the level of knowledge on dinoflagellate respiration is even lower.

The present study is the initial part of a broader effort dedicated to the investigation of the structure and the function of respiratory systems in the marine phytoplankton. For this attempt, we choose the dinoflagellate *Crypthecodinium cohnii* as working material. This colorless species presents the advantage to comply with parallel investigations of the energy metabolism when using an oxygen electrode as well as spectrometric techniques.

MATERIAL AND TECHNIQUES

The axenic strain was purchased from the Culture Collection of Marine Phytoplankton of the Bigelow Laboratory for Ocean Sciences (Maine, USA). Cells were grown on medium h/2 according to Guillard[1] plus 1 g/l peptone. Experiments were conducted with whole cells as well as with cell-wall-free fractions. Oxygen uptake was monitored with a Clark-type electrode connected to a chart recorder (Sefram PED). Spectra were run in the visible range with a Johnson-Foundation-built dual wavelength spectrometer. Cell counts were done by using a Thoma cell.

RESULTS AND DISCUSSION

Optical properties

Absolute spectra related to oxidized and reduced states exhibit absorption bands easily attributable to c- and b-type cytochromes. The contribution of cytochrome c oxidase is not clearly resolved on absolute spectra but is made evident on difference spectra. The peak position of the Soret absorption bands in the absolute spectra of the oxidized and reduced states are respectively 416 and 422 nm. The α band of the absolute reduced spectrum has the same maximum location as in the reduced minus oxidized spectrum : 555 nm. A significant absorption band at 470 nm is present in all absolute spectra (not shown). It has been tentatively assigned to some flavoprotein. The fully reduced (dithionite) minus fully oxidized (oxygenated) spectrum is displayed in figure 1. It presents a peak in the Soret region at 427 nm with a trough at 404 nm. The presence of cytochrome c oxidase is evidenced by a significant shoulder at 440 nm in the γ band and a characteristic α band centered at 604 nm. From such a spectrum, it can be estimated that, in *Crypthecodinium cohnii*, the cytochrome c oxidase to cytochrome c ratio is about 30% the one in higher plant mitochondria[2]. This could be interpreted as an equivalent lower content of cytochrome c oxidase in this dinoflagellate under the above conditions.

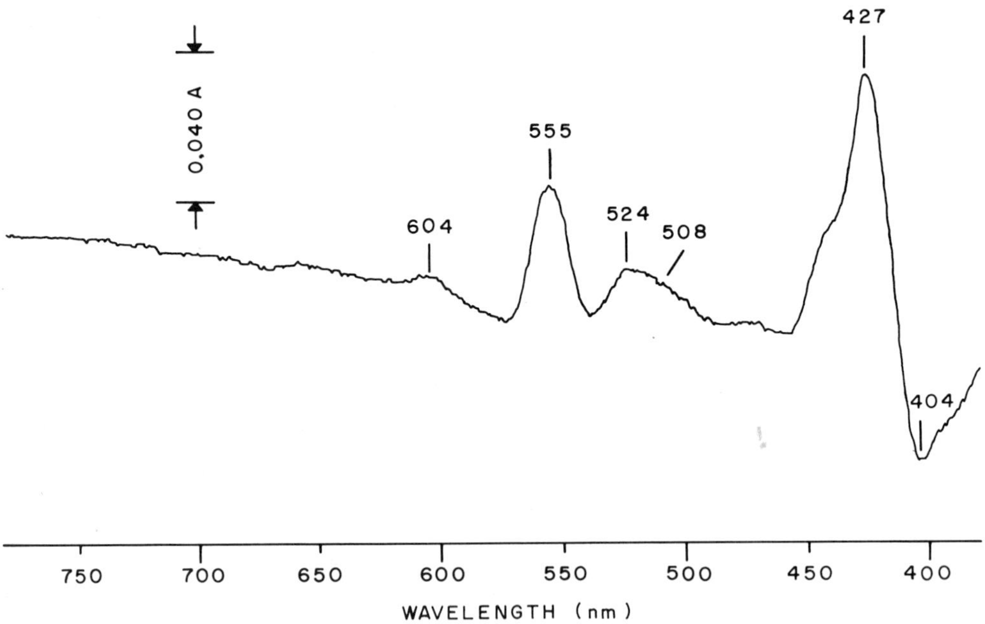

Fig. 1. Difference spectrum at 24 °C of a cell suspension of *Crypthecodinium cohnii* between the dithionite reduced state and the oxidized ($+O_2$) one. Cell concentration : 65 10^6 cells ml^{-1}. Culture medium, pH 8. Optical path : 1 cm. Reference wavelength : 540 nm, wavelength bandwidth : 4 nm.

Cell respiration

Oxygen uptake by cells harvested in the exponential growth phase has been determined from several distinct batches within the range 5 to 20 pl_{O2} hr^{-1} $cell^{-1}$. This rate is about 1/100 the one reported for the yeast Saccharomycopsis lipolytica[3] and it falls within the range of diatom respiration, showing a good correlation with cell volume (see Table 1). Cell respiration was partially insensitive to cyanide as might be expected from a plant system[4]. With a suspension of 0.863 10^6 cells ml^{-1}, addition of 1.25 mM NaCN left the oxygen uptake at 50% of its original level. In contrast, cell respiration was magnified by 3.3 to 4.3 fold upon addition of antimycin A (2.9 μg AA ml^{-1}) to the cell suspension. This uncommon behavior which is illustrated in figure 2 was not conserved with cell-wall-free fractions. To the opposite, addition of 29 μg AA ml^{-1} in these fractions inhibited 60% of the initial oxygen uptake sustained by 6.25 mM succinate and 12.5 mM NADH oxidation (not shown).

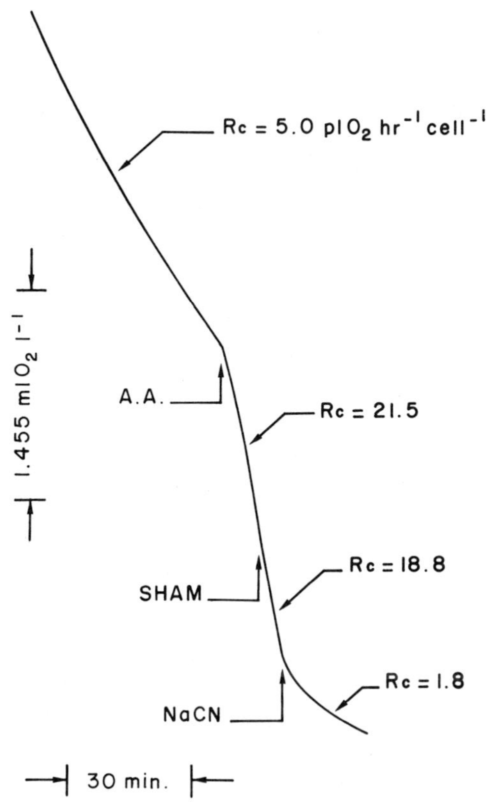

Fig. 2. Oxygen uptake by Crypthecodinium cohnii. Cell concentration : 0.465 10^6 cells ml^{-1}. Culture medium, pH 8. Additions as pointed by arrows : 2.9 μg ml^{-1} Antimycin A, 1.25 mM salicylhydroxamic acid (SHAM), 1.25 mM NaCN.

Table 1. Respiration per cell and cell volume of *Crypthecodinium cohnii* and 7 species of marine diatoms (from ref. 5).

Species	Respiration $pl_{O2}\ h^{-1}$	Biovolume um^3
Crypthecodinium cohnii	20.00	2 043
Phaedactylum tricornutum	0.19	55
Thalassosiria pseudonana	0.17	77
Skeletonema costatum	0.12	127
Thalassosiria weissflogii	3.19	529
Ditylum brightwellii	13.00	118 000
Coscinodiscus sp. C388	81.40	275 000
Coscinodiscus sp. CoA	987.10	6 200 000

CONCLUSIONS

The above findings, though preliminary, demonstrate if necessary that dark respiration in dinoflagellate deserves more attention and that *Crypthecodinium cohnii* is a suitable species for such an investigation and for characterizing the peculiarities of the electron transport system in these marine organisms. The study of oxygen uptake by purified mitochondria will constitute the necessary next step and should be the source of valuable information.

ACKNOWLEDGEMENTS

This work was partially supported by the Centre National de la Recherche Scientifique, the Ministère de l'Education Nationale and a grant to M. D. from the Ministère de l'Industrie et de la Recherche, n° 83.V.0638.

REFERENCES

1. R. R. L. Guillard, Culture of Marine Invertebrate Animals, W. L. Smith and M. H. Chanley, eds, Plenum, N. Y. (1975).
2. R. Douce, E. L. Christensen and D. W. Jr. Bonner, "Preparation of intact plant mitochondria", *Biochim. Biophys. Acta* 275:148 (1972).
3. J. C. de Troostenbergh, M. J. Ledrut-Damanet and E. J. Nyns, "Acetate, a trigger for the induction of the alternative cyanide insensitive respiration in the yeast *Saccharomycopsis lipolytica* ", *in* : "Plant Mitochondria", G. Ducet and C. Lance, eds, Elsevier/North-Holland Biomedical Press (1978).
4. C. Lance, M. Chauveau, and P. Dizengremel, "The cyanide resistant pathway in plant mitochondria", *in* : "Higher Plant Cell Respiration", R. Douce, and D. A. Day, eds, Springer-Verlag, Berlin, Heidelberg, New York, Tokyo (1985).
5. D. Blasco, T. T. Packard and P. Garfield, "Size dependence of growth rate, respiratory electron transport system activity and chemical composition in marine diatoms in the laboratory", *J. Phycol.* 18:58 (1982).

NITRATE REDUCTASE ACTIVITY IN SEEDS AND SEEDLINGS OF TROPICAL SPECIES

[1]Agnes M. Udoh, Alyson K. Tobin, Michael O. Proudlove and Anthony L.Moore

[1]Department of Biological Sciences, University of Lagos Lagos, Nigeria and Department of Biochemistry, University of Sussex, Falmer, Brighton BN1 9QG, U.K..

INTRODUCTION

Apart from those plants which have a symbiotic association with nitrogen-fixing bacteria, most plants depend solely on nitrate ions taken in from the soil for their nitrogen requirements. During nitrogen assimilation in plants, this nitrate is reduced to nitrite by the enzyme nitrate reductase (NR). In higher plants, nitrate reductase is extremely unstable and is subject to rapid turn-over, especially at room temperature. The adaptive nature of this enzyme, however, means that it is easily induced, by its substrate nitrate, in roots, stem and leaves of plants[1-3].

Much work has been done on NR activity in roots and shoots of plants but the mechanism of NR activity in seeds of most tropical species has not yet been fully studied. In this report, the induction and distribution of NR activity in some economically very important tropical seeds has been monitored. Special emphasis has been made of the effect of ageing on NR in upland ecotypes of rice seeds.

MATERIALS AND METHODS

Materials

Old and new, upland ecotypes of rice seeds (harvested in 1982, 1985 respectively) were collected from the Federal Department of Agriculture,

Lagos and The International Institute for Tropical Agriculture, Ibadan, Nigeria. Peanuts and okra seeds were obtained locally, from farmlands in the eastern states of Nigeria.

Methods

Seeds were surface sterilized in 1% sodium hypochlorite for 15 min. Rice seeds (Oryza sativa) were planted in washed sawdust while peanuts (Arachis hypogaea) and okra seeds (Abelmoschus esculentus) were planted in washed perlite. Activity of NR was determined according to Stewart et al.[4]. Seeds or germinating seedlings (1g f wt) were sectioned (2mm) and incubated in side-armed test tubes containing 10ml of 0.15M KNO_3, 0.1M phosphate, pH 7.5, and 1.5% (v/v) propanol. After vacuum infiltration, samples were incubated at 30°C in the dark. Aliquots (1ml) were taken at 1h intervals, incubated for 30min with 1ml of 1% (w/v) sulphanilic acid, in concentrated HCl, and 1ml 0.02% (w/v) α-naphthylethylediamine dihydrochloride and the absorbance measured at 540nm.

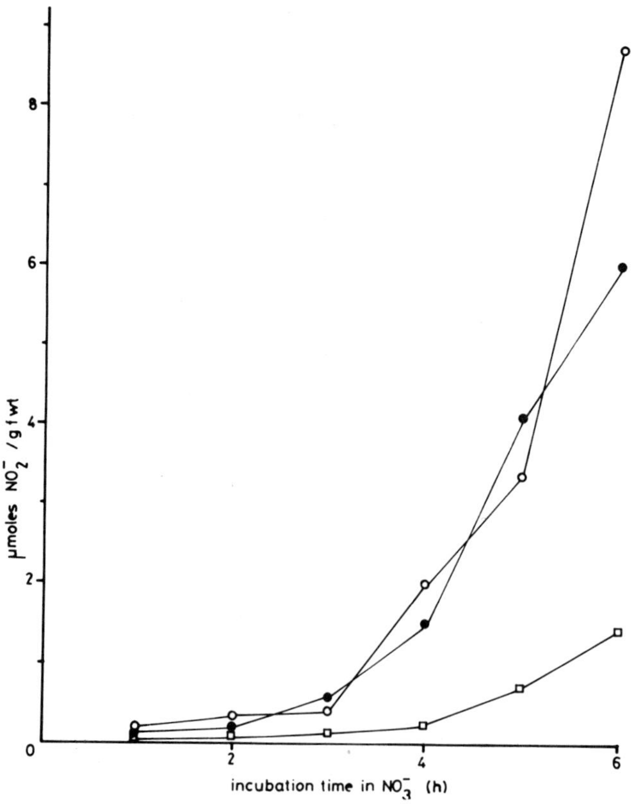

Fig 1. Induction of NR activity in rice (Oryza sativa) □—□ , okra (Abelmoschus esculentus) ○—○ and peanut (Arachis hypogaea) ●—● seeds after 6 days germination, as described in Materials and Methods.

RESULTS AND DISCUSSION

After six days germination (denoted by the emergence of the radicle) in distilled water, the addition of 0.15M nitrate solution caused a rapid increase in NR activity in the seeds tested (Fig 1), as found for other species[2,3]. Enzyme activity was distributed in all parts of the germinating seeds, although in rice seedlings the activity in the endosperms was up to 50% greater than that found in the embryo. In okra and peanut seedlings the cotyledon was also found to have a higher activity than the embryo (results not shown). This increase in NR activity could be explained by -

a) nitrate activating NR already present in the seed or

b) nitrate inducing de novo synthesis of NR.

Treatment with the RNA synthesis inhibitor, 6-methylpurine, resulted in a reduction in the measured activity of NR in freshly harvested rice seeds (Fig 2B). In contrast, there was no effect of this compound on NR activity in seeds harvested and stored for 3 years (Fig 2A). This suggests that the stimulation of nitrite formation in the newer seeds was due to a *de novo* protein synthesis, as previously shown for barley aleurone layers[5].

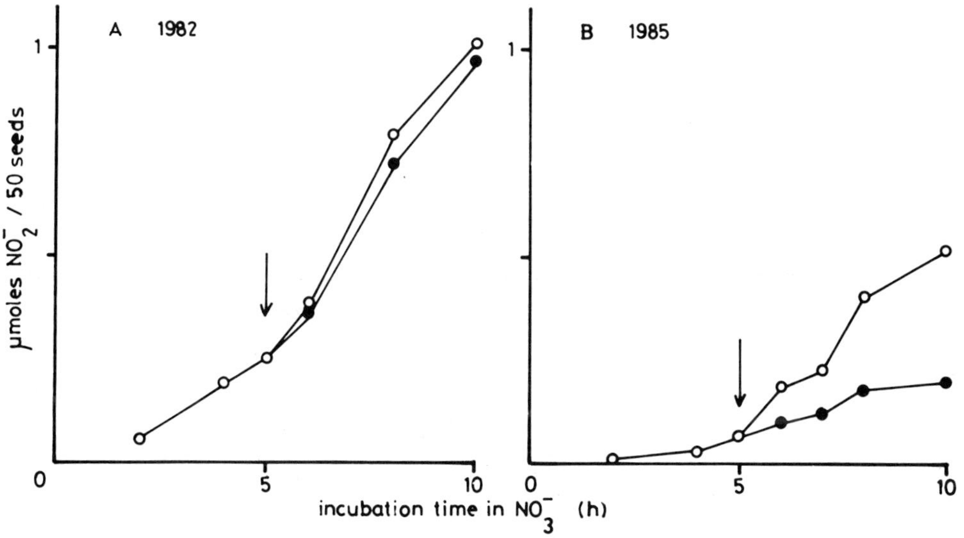

Fig 2. Effect of 6-methyl purine (RNA synthesis inhibitor) on induction of NR in 24h imbibed seeds of rice, harvested in the year shown.

●—● 6-methyl purine (50µg/ml) added after 5h incubation in 0.15MKNO_3.

○—○ without inhibitor.

For the older seeds, the lack of inhibition indicates that they had lost the capacity to synthesize NR and that nitrite formation was due to the activity of pre-existing enzyme. This possibility is further supported by the observation (Fig.2) that the older seeds exhibited higher initial NR activities than the newer seeds.

CONCLUSIONS

1. Treatment with nitrate induced NR activity in seeds/germinating seedlings of rice, okra and peanut.
2. Two different mechanisms of regulation of NR levels exist in old and new rice seeds. In new seeds, induction of NR is a result of protein biosynthesis whereas in older seeds, with reduced viability, nitrite production is due to the activation of pre-formed enzyme.

ACKNOWLEDGEMENTS

This work is partially funded by a Sir Adam Thomson scholarship to AMU.

REFERENCES

1. T. O. Orebamjo and G. R. Stewart, Some unusual characteristics of nitrate reductase in Erythrina senegalensis D.C., New Phytol. 83:311 (1979).
2. P. Mehta and H. S. Srivastava, Comparative stability of ammonium- and nitrate-induced nitrate reductase activity in maize leaves, Phytochem. 19:2527 (1980).
3. W. Wallace, Distribution of nitrate assimilation between the root and shoot of legumes and a comparison with wheat, Physiol. Plant. 66:630 (1986).
4. G. R. Stewart, J. A. Lee and T. O. Orebamjo, Nitrogen metabolism of halophytes. II. Nitrate availability and utilization, New Phytol. 72:539 (1973).
5. T. E. Ferrari and J. E. Varner, Substrate induction of nitrate reductase in barley aleurone layers, Plant Physiol. 44:85 (1969).

CONTRIBUTORS

INDEX